Theory of Thermoluminescence and Related Phenomena

Theory of Thermoluminescence and Related Phenomena

Reuven Chen
School of Physics and Astronomy, Tel-Aviv University

Stephen W.S. McKeever
Department of Physics, Oklahoma State University

World Scientific
Singapore • New Jersey • London • Hong Kong

Published by

World Scientific Publishing Co. Pte. Ltd.

P O Box 128, Farrer Road, Singapore 912805

USA office: Suite 1B, 1060 Main Street, River Edge, NJ 07661

UK office: 57 Shelton Street, Covent Garden, London WC2H 9HE

Library of Congress Cataloging-in-Publication Data

Chen, R. (Reuven)

Theory of thermoluminescence and related phenomena / Reuven Chen, Stephen W. S. McKeever.

p. cm.

Includes bibliographical references and indexes.

ISBN 9810222955

1. Thermoluminescence. I. McKeever, S. W. S., 1950– II. Title.

QC479.C46 1997

535'.356--dc21 97-18873 CIP

British Library Cataloguing-in-Publication Data

A catalogue record for this book is available from the British Library.

This book is printed on acid-free paper.

Printed in Singapore by Uto-Print

Preface

The study of *thermally stimulated phenomena* (TSPs) is, to a large degree, a stable science, the field having grown to maturity over several decades of development and use. Its understanding and theoretical base have been established, and its mathematical description is well developed. The simplicity of its experimental measurement has led to its enormous popularity over the years - and this, perhaps, is the technique's greatest enemy. Despite the in-depth analysis, the detailed understanding, the wider appreciation of what can and cannot happen under given sets of circumstances during thermal stimulation, and an appreciation of its limitations, it is an unfortunate fact that one still sees - more often than is desirable - inappropriate, and almost careless analysis leading to erroneous, even pointless conclusions. The result has, in certain instances, been to undermine the confidence in all measurements and applications of the technique, no matter how well these may have been carried out.

The most popular TSP is that of thermoluminescence (TL) - which may be considered to be the flagship method of all TSPs. A enormous amount of energy and intellectual effort has been expended since the pioneering work of the group at the University of Birmingham (UK) in the late nineteen forties (Randall, Garlick, and others) to extract model parameters from TL glow peaks. Despite major advances in our understanding of the process of TL, this goal still remains one of the most difficult to attain. The basic difficulty lies in the inherent "degeneracy" of the analytical results in that more than one set of parameters and, often, more than one model, can yield theoretical curves which fit the experimentally obtained curves with acceptable accuracy (within experimental limits). However, it would be wrong of us to suggest that such an analysis is impossible. It is one of the themes of the present book that in order to fully understand a particular TL system, and in order to attain an acceptable level of confidence in the model and the parameters deduced, one has to employ a wide arsenal of experiments in which the TL signal is monitored and probed as a function of parameters such as excitation dose, heating rate, emission wavelength, radiation (excitation) quality, etc. Furthermore, it is preferable, where possible, to compare the TL data with other, related TSPs, such as thermally stimulated conductivity (TSC), or thermally stimulated exoelectron emission (TSEE). Only in this way can a complete picture be developed of the system under study.

It would also be wrong of us to imply that extracting model parameters from a TL curve is the only worthwhile goal of TL studies. TL provides much, much more than that. Using TL, experimenters have been able to study, for example, the radiation absorption processes in insulating materials as a function of radiation quality and linear energy transfer; or the age of ancient ceramic artifacts, or geological deposits; or the dose administered to people during intentional or accidental radiation exposure. It is our intention that this book will provide the reader with all the fundamental tools necessary to embark upon an application of TL in any of these arenas armed with the full complement of needed theoretical weapons. We fully adhere to the concept that to apply reliably a physical phenomenon, one has to understand it at a fundamental level, the more detailed the better. We hope this books assists in achieving this goal.

The book is arranged in the following way. After an *Introduction* (Chapter 1) in which the different types of TSP are introduced, along with their relationship to TL, there follows the first chapter of substance, dealing with the detailed theory of TL (Chapter 2 - *Thermoluminescence*). The methods available for the analysis of TSPs are described in the next chapter (Chapter 3 - *Analysis of TL Glow Curves*) in which we describe, compare and contrast the analytical tools available for extracting model parameters from experimental TL peaks. Chapter 4 (*Dose Dependence of TL; Nonlinearities*) deals with the important topic of how the intensity of a given TL peak can vary with dose, highlighting and contrasting the various processes that may occur during the two necessary phases of TL production - namely the excitation phase and the heating phase.

Although this book is concerned with *thermally* stimulated phenomena, it would be incomplete without a discussion of the related processes of *optical* stimulation of luminescence. Thus, Chapter 5 (*Optical Phenomena*) deals with various optical phenomena and discusses their relationship with TL. Some of the mathematical tools required for a thorough study of TL are described next (Chapter 6 - *Mathematical Considerations*) and a consideration of applications follow in Chapter 7 (*Applications of Thermoluminescence*). Although many applications have been suggested over the years - not all of which are in the scientific "mainstream" - we make no apologies for discussing here only the main two - namely, dosimetry and dating. Whole books have been written individually about these uses, and we can here only mention the outlines of these two very important applications. The reader is referred to specialist texts by Horowitz, McKeever et al., Aitken and others for more informative detail in this regard.

Since TL is only one of several TSPs, *Related Phenomena* are discussed next, in Chapter 8. Here, the details of TSC, TSEE, TSDC/TSPC, TG, DTG, DTA, and many others are introduced and contrasted, where appropriate, with TL. Chapter 9 deals with *Simultaneous Measurements of TL and Other TSPs* in which we describe the advantages to be gleaned from a study of more than one TSP simultaneously. Several *Miscellaneous Effects* (thermal and athermal fading, unusual excitation processes, etc.) are discussed in Chapter 10. Finally, some *Concluding Remarks* (Chapter 11) round off the work.

We take this opportunity to thank those who have aided in the completion of this book. First and foremost, we thank our wives and families, who have patiently listened to our complaints and watched us burn many gallons of midnight oil in our struggles to complete the work. Their ceaseless encouragement cannot be underestimated, or forgotten. We also thank all of the publishers who graciously allowed the reproduction of all of the previously published figures used in this book. Finally we thank World Scientific Publishing for the opportunity to publish this labor of love.

Contents

Theory of Thermoluminescence and Related Phenomena

Chapter 1
INTRODUCTION

Thermally Stimulated Processes (TSPs) are a group of physical phenomena in which a certain property of the material under study is measured as a function of temperature. Usually, the property is initially perturbed from a stable configuration, and is thermally stimulated back to equilibrium by raising the temperature. Typically, the property in question is measured during the heating of the sample, although isothermal measurements also qualify as TSPs. In many instances, the initial perturbation occurs when the specimen under investigation undergoes a process of excitation at the starting temperature, which may be ambient temperature (typically, room temperature, RT) but, in principle, it can be any other temperature. Typically, the process of excitation may be an irradiation of the sample by ionizing radiation or ion bombardment, but other possibilities exist such as applying pressure on the specimen. It is important to realize that, in these circumstances, the temperature-dependent property is a *functional*, not simply a *function*. That is the value of the property during heating is not repeated during cooling. Re-excitation of the specimen is necessary in order to repeat the measurement.

In most cases, the heating of the sample is performed at a linear rate although other heating schedules can be utilized. As will be explained later, there is a theoretical reason why a hyperbolic heating rate has some advantages over the linear as well as other heating functions for the analysis of the results.

The resulting thermal curve, sometimes termed the *thermogram* - namely the measured property as a function of temperature - is, typically, a curve consisting of one or more peaks. The shape of the whole curve depends on the initial treatment. If this is, for example, an irradiation of the sample by a certain source (UV, γ, β etc.), the intensity of the measured peaks will depend in one way or another on the excitation dose. When this dose gradually increases, different peaks may depend in different ways on the dose, which would cause the shape of the thermal curve to be dose dependent. In particular, new peaks may appear only at or above a certain dose. Some parts of the curve may reach saturation when others keep growing. In some rare cases, certain peaks may reach a maximum as a function of the dose and decrease from that point on. In this respect, the possible interaction between different parts of the curve is to be considered; this interaction may be rather complex.

It is obvious that a most desirable feature is reproducibility; one hopes that repeating a measurement several times would yield the same results. It sometimes happens, however, that a cycle of excitation and heating changes the sample in such a way that the sensitivity to the next excitation changes. Such a change may be detrimental to the basic analysis of the measured effect; however, following such a change of sensitivity in detail may add information about the studied sample. The same can be said about the applications in which one tries to evaluate the dose imparted on the sample; changes in sensitivity may disrupt this goal. As will be shown, in thermoluminescence (TL) dating, an independent method has been developed based on such a sensitivity change.

An investigation of the shape of the thermal curve (sometimes called "glow curve" although, originally, this name was associated with thermoluminescence only) may yield information on the physical parameters of the material in question. These parameters include always a typical "activation energy", usually given in eV, and a "pre-exponential factor" in units of s^{-1} - termed a "frequency factor". Other parameters are involved in the various thermally stimulated phenomena, and some of them can be evaluated from the specific process under investigation.

Some of this family of phenomena can be studied on the same material. For example, *thermoluminescence* (TL), *thermally stimulated conductivity* (TSC) and *thermally stimulated electron emission* (TSEE) can be measured in the same material, and the combined information can yield a better picture of the studied sample. In particular, simultaneous measurements of two or more thermally stimulated phenomena in the same material may yield information beyond the mere sum of what is learned from each phenomenon separately.

The above-mentioned study of the dose dependence of the whole glow curve as well as different individual peaks, is an important tool in the investigation of the processes involved. At the same time, this study is of great importance in the main application of TL, and to a smaller extent some other TSPs, namely dosimetry and dating of archaeological findings and geological samples. Although in this book we focus more on the theory and the processes, these applications are discussed in some detail in Ch. 7.

1.1 Thermoluminescence (Thermally Stimulated Luminescence)

The most broadly investigated and utilized of all thermally stimulated phenomena is the emission of light during the heating of a solid sample, previously excited. This is usually termed thermoluminescence (TL) and is the main subject of the present book. The term thermoluminescence is somewhat unfortunate since it sounds as if the heating is the main energy source of the emitted light, which is not the case. The initial excitation (typically by irradiation) is the source of energy, whereas the heating serves only as a trigger which helps in releasing this accumulated energy. The term Thermally Stimulated Luminescence (TSL) is more descriptive; however, "Thermoluminescence" is traditionally more often utilized and popularly accepted.

The phenomenon of thermoluminescence (TL) can be described as follows. A solid sample, usually an insulating material, is excited, characteristically by ionizing radiation at a certain "low" temperature. This irradiation may take place in the laboratory or, alternatively, in some of the applications, when the sample is inserted in a radiative environment (as, for example, in personal or environmental dosimetry). Yet another version is when it is a naturally occurring material irradiated by the radiation field in its natural surrounding. At the end of this stage, the sample is placed in an appropriate oven, unless the irradiation itself is performed when it is in the oven, and the sample is heated. In cases where the initial "low" temperature is below the freezing point of water, say, liquid nitrogen temperature, the sample is placed in a cryostat which is cooled and the irradiation (say x-rays) is applied when the sample is under vacuum, and at the low temperature.

The next period of heating is also called the "readout" stage. At this stage, the temperature of the oven is raised gradually, typically at a constant heating rate, and the light emission as a function of temperature (or time) is recorded using a light sensitive detector such as a photomultiplier. This emitted light as a function of temperature yields the so-called thermoluminescence "glow curve". Normally, the shape of the glow curve will be of one or more peaks of emitted light, some of which may be overlapping. As will be shown, the shape of the glow curve depends on the light sensitive device, and in particular, on its spectral response. Moreover, even with a given device, the glow curve may look different if different filters are inserted between the sample and the detector. Also, as will also be shown, the shape of the TL glow curve depends on the heating rate used. As noted, except for very unusual cases, the occurrence of the TL curve following a given irradiation is a "one shot" effect. Cooling the sample and re-heating it normally does not result in a second TL emission.

All the glow curve characteristics mentioned, such as the number of peaks, their shapes and their emission spectral composition, may depend in a given sample on the nature of the initial excitation. For example, excitation with UV light may result in a different glow curve than that recorded following, say, α, β, or γ irradiations. Moreover, even when the same kind of excitation agent, say UV light is used, different excitation energies (different wavelengths in UV light), may result in different TL properties. If the property in question is the emission intensity, one can talk about the "excitation spectrum" while monitoring the emission intensity, preferably in a limited wavelength range of the emission light, when different wavelengths of the excitation light are used, keeping the excitation light intensity constant.

Another important feature of glow curves is their dependence on the excitation dose. In the applications in dosimetry and archaeological dating of ceramic samples, a linear dependence on the dose in broad ranges is highly desirable, but this is by no means always found to be the case. Typically, at high doses, the dose dependence curve approaches some saturation value. At low doses, linear dose dependence is quite prevalent, however superlinear (or supralinear) dependence as of the lowest doses sometimes occurs. In some cases (though quite rarely), one has to consider also the dependence upon dose rate as well.

Thermoluminescence is usually observed in ordered crystals, although the phenomenon also occurs in disordered materials such as glasses. A relatively small number of researchers reported the results of TL from biological materials such as tomato seeds and spinach, as well as teeth and bones. The physical models explaining the occurrence of TL glow peaks are based on the assumption that the thermoluminescent material is an ordered crystal lattice; the effect occurring in amorphous and biological materials can be explained only in very general terms.

In a given crystal, the study of TL is actually the study of imperfections in the lattice. In principle, a perfect crystal with no impurities or defects is not expected to exhibit any TL. Of course, such perfect crystals do not exist at all. The association of specific imperfections with, say, a certain TL peak is often rather tricky. It may very well happen that a certain impurity or defect is abundant in a given sample (say, ~1%), but does not contribute to the measured TL. On the other hand, other imperfections, sometimes undetectable by other means due to their low concentrations, are found to be responsible for the measured TL signal. One can thus say that TL is a very sensitive tool to detect imperfections in small quantities, but often with no simple way to identify them. For the applications, mainly dosimetry, this inability to identify the relevant imperfections is usually unimportant; if the effect is reproducible (and, of course, one tries to employ only reproducible materials), the only important thing to have is a reliable calibration. From the physical point of view, however, understanding of the role of defects and impurities is of great importance, and thus, indirectly it may influence the considerations of choosing the appropriate dosimetric materials.

In order to explain the occurrence of a single TL peak, one has to assume that at least two kinds of imperfections exist in the crystal. During the initial excitation, one of these captures electrons and the other captures holes; the electrons and holes are produced in pairs by the applied radiation. The energies of these captured carriers are related to trapping states associated with the imperfections, and the energies are within the forbidden gap of the given crystal. These states are metastable, and normally, at the "low" temperature of excitation, their lifetime is very long. For the sake of discussion, the electron trapping state is often considered to be rather close to the conduction band and the hole trapping state is far from the valence band. In this model, while heating the sample, the electrons are thermally released, usually to the conduction band, and then they may recombine with trapped holes, thus producing emitted photons. In this situation, the hole trapping state is called the *luminescence* or *recombination center* and the electron trapping state is simply termed the *trap*. It may very well happen, however, that the hole trapping state is close to the valence band whereas the electron trapping state is far from the conduction band. In this case, during the heating, holes are "raised" (energetically) into the valence band and then recombine with trapped electrons. In this case, one would talk about hole traps and electron recombination centers. Note that in the older literature, the term *activator* is sometimes used for the luminescence center and *co-activator* for the trap. In the literature, there are a few works mentioning the possibility that electrons and holes are released into the conduction and valence bands, respectively, in the same temperature range, in which case a certain trapping state can serve as both a trap and a recombination center.

Another possibility that sometimes occurs is that the trapped electron and hole are situated in close proximity. In this situation, if recombination of trapped electrons and holes does not take place directly, in which case no TL is expected, it may happen that a direct transition is forbidden, and the electron (or hole) has to be thermally elevated into an excited state from which it may recombine with the trapped hole (electron), thus emitting TL. The consequences concerning the kinetics of the process, as well as related phenomena such as thermally stimulated conductivity (TSC) and thermally stimulated electron emission (TSEE) will be discussed in the following chapters.

The detailed theory of thermoluminescence will be given in Ch. 2. However, for the sake of initial illustration, the basic simplified model by Randall and Wilkins[1] will be described here.

The two levels shown in Fig. 1.1 are a trap n and a center m. In this simplified model, we assume that at the end of the excitation, n_0 (in m^{-3}) electrons are accumulated in the trap, and at least the same concentration of holes in the center. At a temperature T (in K), the rate of release of trapped electrons is proportional to the Boltzmann function $\exp\{-E_t/kT\}$ where E_t is the "activation energy" (in eV) of the trap and k is Boltzmann's constant (in eV/K). Of course, when the temperature increases, this function grows quite rapidly. The number of electrons released per unit time is also proportional to

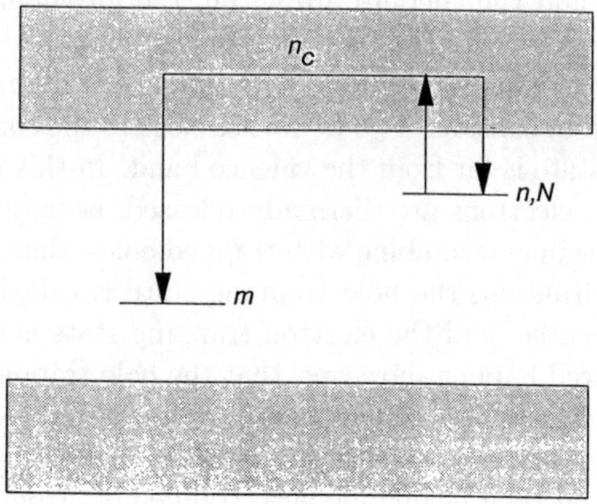

Figure 1.1 Energy Levels in a simple model. n and N are the occupancy and total concentration of trapping states, respectively. In the simple Randall-Wilkins model, the retrapping shown is negligible, but it is of importance in other cases.

the total concentration of trapped electrons. For the sake of this preliminary discussion (and following Randall and Wilkins), we assume that once an electron is in the conduction band, it immediately recombines with a hole in the center, which yields an emitted photon. The possible retrapping of excited electrons is also shown here although at this stage it is assumed to be negligible. The rate of production of emitted photons I in the Randall-Wilkins model is, therefore,

$$I(t) = -dn/dt = sn \exp\{-E_t/kT\} \tag{1.1}$$

where s is a constant, characteristic of the trap in question, and having the dimensions of s^{-1}. This rather simple case is called "first-order" kinetics. $I(t)$ is given as the number of photons per unit volume per unit time ($m^{-3}s^{-1}$). If the use of other units (say, energy) is desirable, $I(t)$ should be multiplied by an appropriate transformation constant. It is to be noted here that the time t (s) and temperature T are connected by the heating function $T = T(t)$. Quite often, a linear heating function $T = T_0 + \beta t$ is utilized in which β (K/s) is the (constant) heating rate.

The solution of the simple differential equation in 1.1 is

$$n(t) = n_0 \exp\left[-s\int_{t_o}^{t} \exp\{-E_t/kT\}\, dt\right] \tag{1.2}$$

which is a decreasing function of time. Using the linear heating function, we get

$$n(T) = n_0 \exp\left[-\frac{s}{\beta}\int_{T_0}^{T} \exp\{-E_t/k\theta\}d\theta\right] . \tag{1.3}$$

where θ represents the varying temperature (dummy variable). Inserting this expression in Eq. 1.1 yields

$$I(T) = sn_0 \exp\{-E_t/kT\} \exp\left[-\frac{s}{\beta}\int_{T_0}^{T} \exp\{-E_t/k\theta\}d\theta\right] . \tag{1.4}$$

This model is usually an oversimplification of a more complex situation since it assumes the existence of only one trapping state and one kind of recombination center, and since it does not allow for any retrapping of free electrons or even their spending some time in the conduction band before recombining with holes. The main features of a glow peak do exist, however, in expression 1.4. Whereas the first exponent $\exp\{-E_t/kT\}$ grows with T, the second exponent starts at unity, varying very slowly as long as $T \approx T_0$, and decreases substantially for larger values of T. As a result, the initial range of the glow peak behaves like $\exp\{-E_t/kT\}$ which, in fact, is always true, even in more complicated non-first-order situations. This is the basis for the simple "initial rise" method for evaluating the activation energy. At higher temperatures, the decrease of the second exponent becomes faster than the increase of the first, and therefore, the curve reaches a maximum and decreases to negligible values at higher temperatures. Stated differently, the probability of release of electrons from the trap increases with temperature while the number of electrons left to be released decreases with temperature. As a result, the *rate* of electron release (and, therefore, $I(t)$) reaches a maximum at some intermediate temperature. As will be discussed in the next chapter, the shape of the glow peak given in Eq. 1.4 is asymmetric, where the low temperature "half" of the peak is broader than the high temperature one. A typical first-order peak is shown in curve I of Fig. 1.2, whereas curve II is a second-order kinetics curve, discussed in detail in Ch. 2.

In fact, the general features noted above can be seen from Eq. 1.1. Since n starts at a given value n_0 and goes down to zero while the sample is heated, and since $\exp\{-E_t/kT\}$ grows with T, their product should yield the peak shape. The temperature of the maximum in terms of the trap parameters E_t and s and the heating rate β can be found by equating the derivative of Eq. 1.4 to zero which yields

$$\beta E_t/(kT_m^2) = s \exp\{-E_t/kT_m\} \tag{1.5}$$

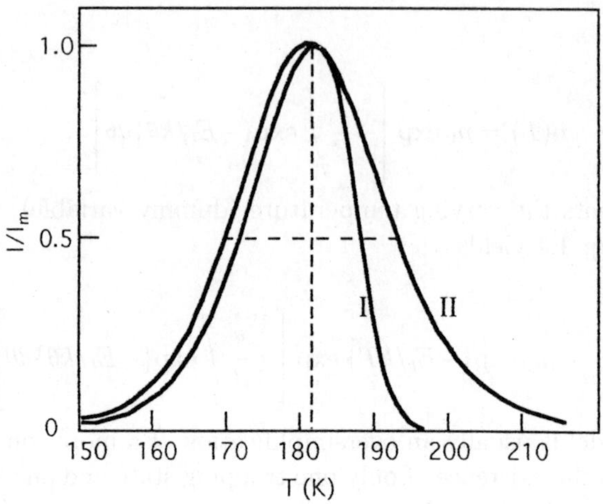

Figure 1.2 A typical first-order peak calculated by the Randall-Wilkins equation (curve I) and a second-order peak (curve II). The latter will be discussed in Ch. 2. Both curves are simulated numerically with $E_t = 0.4$ eV. In curve I, the frequency factor is $s = 10^{10}$ s^{-1}. In curve II, the pre-exponential factor is $s' = 10^{-11}$ m^{3}s^{-1}.

where T_m is the temperature at the maximum. Since β is determined in the experiment, and varied at will, and since T_m is very easily found, a method can be devised to evaluate E_t and s by making some measurements at different heating rates. This also will be described in a subsequent chapter.

The generalizations to this simple model will be discussed in Ch. 2. These include allowing for the thermally stimulated charge carriers to spend a finite length of time in the delocalized bands (conduction band for electrons, or valence band for holes) before they either recombine with a trapped carrier of the opposite sign, or be retrapped into the same kind of trap they came from, or be trapped in a different kind of trap which also captures carriers of the same sign. Alternatively, local transitions can be considered where the charge carriers are stimulated into an excited state which is not the conduction band prior to the light-emitting recombination. A further generalization takes into consideration a series of TL glow peaks, some of which may be overlapping to some extent. Even more complicated situations discussed in the literature are those of a continuum of trapping states with a given distribution. Also discussed in the literature are situations in which, in the same temperature range, electrons and holes can be raised into the delocalized bands so that the two entities, electron and hole trapping states, have a dual function of being both traps and centers. Finally, the spectral decomposition of the glow emission is also of great importance. All these will be discussed in the following chapters.

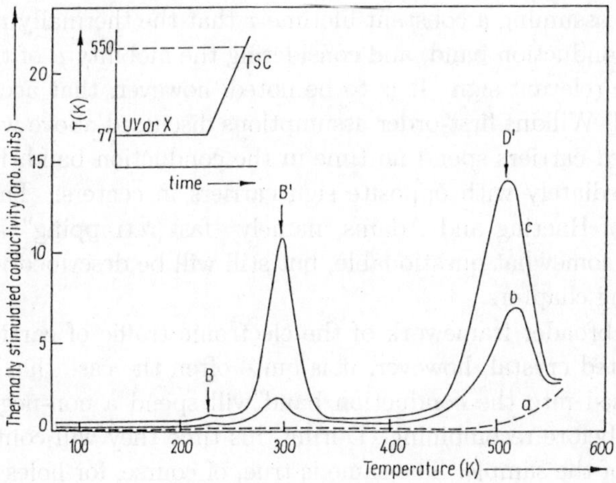

Figure 1.3 Typical TSC curve, measured in α alumina. (a) Before any irradiation, (b) after UV excitation (206 nm, 4 min.) at 77 K, (c) after x-irradiation (40 kV, 86 Gy) at 77 K.

(After Lapraz et al.[2].)

1.2 Thermally Stimulated Conductivity

The most closely associated phenomenon to thermoluminescence is Thermally Stimulated Conductivity (TSC). A basic description of this effect is as follows. Electric contacts are formed on a solid sample, typically an insulating crystal. The sample is excited at a given "low" temperature, usually by applying radiation. When the excitation stage is terminated, an electric field is applied across the contacts on the sample, and the sample is heated, usually at a constant rate. During this heating, transient electric currents in the sample may be measured, usually in the form of peaks, basically similar in shape to the TL peaks. In fact, quite often these peaks can be associated with parallel TL peaks, although the theory does not predict exact correspondence. In other cases, however, TL peaks may occur where no TSC counterparts are observed, and vice-versa. Figure 1.3 shows a typical TSC curve.

Let us exclude, for the moment, phenomena which, from the experimental point of view, look nearly the same as TSC- such as Thermally Stimulated Depolarization Currents (TSDC) which will be discussed later. The understanding of the TSC curve has to do with transitions of electronic charge carriers (electrons and holes) between trapping states, the conduction (or valence) band and recombination centers. Some early researchers (e.g. Haering and Adams[3]) ascribed the first-order kinetics of Randall and Wilkins[1] to TSC curves by just assuming a constant lifetime τ that the thermally raised carriers spend in the conduction band, and considering the mobility μ of the sample for carriers of the relevant sign. It is to be noted, however, that according to the simple Randall-Wilkins first-order assumptions discussed above with respect to TL, the excited carriers spend no time in the conduction band, but rather recombine immediately with opposite sign carriers in centers. The two sets of assumptions of Haering and Adams, namely "fast retrapping" and "slow retrapping" are somewhat questionable, but still will be described in some detail in the following chapters.

In the broader framework of the electronic traffic of carriers in a thermally stimulated crystal, however, it is quite often the case that the electrons thermally raised into the conduction band will spend a non-negligible length of time there before recombining. During this time they will contribute to the conductivity of the sample. The same is true, of course, for holes thermally released from a hole trapping state into the valence band. An important point to note, however, is that the assumption that the lifetime τ is constant is usually questionable since the lifetime of a released carrier is strongly dependent on the instantaneous occupancies of traps and centers. These, in turn, depend on the initial filling of these entities, and vary during the heating when the occupancies change.

The simple expression connecting the instantaneous conductivity of the sample and the concentration of free carriers in the band is

$$\sigma = e\mu n_c \tag{1.6}$$

where n_c is the concentration of free carriers, e is the electronic charge and μ is the mobility of the specimen to the relevant carriers. Although μ may be slightly temperature dependent, it is usually assumed that the conductivity is proportional to the concentration of free carriers.

A rather serious problem is how to measure the conductivity σ as a function of temperature. The apparently obvious solution is to use a simple geometric shaped sample, e.g. a rectangular box where two contacts are formed on two opposite faces. A voltage is applied across the sample and the current is monitored as the sample is heated. If the area of each of the contacts is A then, obviously, the relation between the conductivity σ and the measured current is

$$I = \sigma A. \tag{1.7}$$

Of course, the relation between the current and conductivity is more complicated in different kinds of geometry of the sample. However, the main difficulty in this concern has to do with the contacts. It is usually stated that it is desirable that the contacts are ohmic. In most general terms this means that the contacts do not form any barrier, or, in other words, that the simple Ohm's Law of linear dependence of the current on the voltage holds and that the measured resistivity is that of the sample only. Actually it is not strictly necessary that the current-voltage relationship be linear since most TSC experiments are conducted at a constant applied voltage. What is required is that the contact and barrier resistance be small compared to the sample resistance so that most of the applied voltage is dropped across the sample. Another important feature is that the properties of the contacts do not change with the temperature. Although as shown by Henisch[5], the occurrence of the "good" qualities of contacts cannot be guaranteed, some measurements can be made which would indicate that the properties are desirable in the sense that Eq. 1.7 is really fulfilled - including using the same contacts with different thicknesses of the sample. A "good" result is when the resistance depends linearly on the thickness. Also, if different contact materials are used with the same sample and the results are the same, this is an indication that the contacts are ohmic. Finally, if similar TSC curves are measured with a given sample following a given initial excitation but with different applied voltages, differing only in the magnitude of the current, the above-mentioned complications are probably insignificant. Some contactless measurements of the conductivity of semiconductors have also been suggested in the literature. We have no knowledge, however, of such measurements performed while studying TSC curves. One should mention finally that deliberately blocking contacts can be used with semiconductors to provide injection of a non-equilibrium charge population, which then becomes trapped. In this way, a trapped charge population can be induced without the need for external irradiation.

1.3 Thermally Stimulated Electron Emission

Whereas in the phenomena of TL and TSC it has been stated that the mobile charge carriers can be either electrons or holes, there is another thermally stimulated phenomenon in which, normally, only one type of mobile carrier - namely, electrons - take part. This is thermally stimulated electron emission (TSEE). This effect can be observed when a sample, usually an insulator, is placed in a vacuum cryostat and is connected to an electrode. Another electrode is also present in the tube at a certain distance from the sample. A given potential difference is maintained between the two electrodes, the latter electrode being positive with respect to the sample. The sample is now excited by one of the means mentioned while discussing TL and TSC, namely, x-rays, γ rays, etc. During this time electrons as well as holes may be trapped in trapping levels and luminescence centers. When the sample is heated, electrons and/or holes may be released into the conduction band and valence band, respectively.

These two entities, electrons and holes, could behave in exact symmetry with one another as far as TL and TSC were concerned. At this point, where TSEE is the phenomenon under study, this symmetry breaks. Under appropriate conditions, the electrons can be released from the sample, move in the electric field in the vacuum, and be captured by the positive electrode whereas nothing of the sort can happen with the thermally released holes even if the polarity of the electrodes is inverted. This is an obvious inherent difference between electrons and holes, the latter, after all, being missing electrons in the crystal lattice. As already mentioned, this substantial difference between the two kinds of charge carriers cannot be detected directly when only TL and TSC measurements are concerned, but can be easily determined here, in the study of TSEE. One possible difficulty, however, arises when holes are indeed the released charges but transfer their energy to electrons via Auger recombination. The electrons may then be released as exoelectrons. Thus, whereas normally, if one associates a certain TL or TSC peak with a specific TSEE peak, measured in the same sample and the same temperature range, one may use this association to identify the TL and TSC peaks as being related to the traffic of electrons, rather than holes, this conclusion may not always be true.

The condition for freeing an electron from a sample once it is in the conduction band, has to do with the work function ϕ which can be considered as an additional energy barrier which the electron has to overcome in order to be able to be released into the vacuum. This has been discussed by numerous authors, and can be given by the expression

$$J = n_c(T)\,[kT/(2\pi m^*)]^{1/2}\exp\{-\phi/kT\} \tag{1.8}$$

where J is the current density out of the sample, m^* is the effective mass of the conduction electron and $n_c(T)$ the momentary concentration of electrons in the conduction band resulting from all the traps and centers involved, and including retrapping. With respect to Eq. 1.8, however, it should be noted that if, in a certain temperature range, the function $n_c(T)$ (representing, more or less, the TSC peak) has a peak shape, this function should be multiplied by another function, increasing with temperature - namely $\exp\{-\phi/kT\}$ - which has the effect of shifting the TSEE peak to a higher temperature. The relation between the TL peak and its associated TSEE peak is complicated and will be discussed in some detail in Chapters 8 and 9. We would like to mention here that whereas the TL properties and, in most cases, the TSC properties, of a specimen are related to the bulk of the sample, it is not unusual that the TSEE peaks are related to surface states. This should be carefully considered when TSEE results are compared to TL and TSC; this point will also be elaborated upon in Ch. 8.

1.4 Thermally Stimulated Depolarization Currents

The measurement of thermally stimulated conductivity, related to the occurrence of free electrons or holes in the conduction/valence band respectively, has been described above and will be further elaborated upon in Ch. 8. In many cases, however, thermally stimulated currents have been observed in which it is quite obvious that free charge carriers have very little or no effect at all. This group of phenomena includes polarization and depolarization currents measured in dielectrics during their heating, and are termed *thermally stimulated polarization currents* (TSPC) and *thermally stimulated depolarization currents* (TSDC), also called TSD, respectively. At this stage, we shall only briefly describe the TSDC phenomenon.

A sample previously polarized, typically by cooling it down while in an electric field, is heated from a certain initial "low" temperature T_0. Two electrodes on opposite faces of the sample are then short-circuited through a sensitive current detector. The current thus appearing in the external circuit due to depolarization processes is recorded as a function of temperature or time. As opposed to the case of TSC, the measurement is performed without any external field. The similarity between the two phenomena, however, is obvious, namely, that in both cases one measures current as a function of temperature during heating. Also, in most cases, both phenomena behave like a "one shot" effect; the peak or series of peaks are seen only after the first heating following the initial excitation (in TSC) or polarization (in TSDC). The similarity between TL, TSC and TSDC goes much further. Here too, in ideal cases, single current peaks can be observed for each separate relaxation process, whereas different amounts of overlapping occur in many situations. An important similarity is that, here too, the effect is associated with two main physical parameters - the activation energy E and the frequency factor s. This is so despite the fact that the physical processes are quite different; in most cases the current is not due to the transport of electrons and holes, and the activation energy here has a different physical meaning than in TL and TSC. In fact, the situation with TSDC is often simpler than with TL and TSC since only first-order kinetics apply.

The effect of TSDC in its different versions will be discussed in Ch. 8, so we here will briefly make only two important remarks. One is that since the experimental setup in TSC and TSDC is practically the same, sometimes the effect of free carriers characteristic to TSC can make some contribution to what is supposed to be a TSDC measurement. This is so in spite of the fact that no external field is applied in the latter. The internal polarization field may assist the motion of thermally released carriers, thus contributing to the measured current.

The other remark is that, although the physical parameters mentioned

- namely, the activation energy and frequency factor - are inherently different in TL and TSDC, the same methods for their evaluation can be utilized. We will come back to this point when discussing in more detail the polarization and depolarization processes, and after having developed the methods for evaluating TL parameters.

1.5 Deep Level Transient Spectroscopy

A method has been developed by Lang[5] for the evaluation of trapping parameters in *semiconductors* from the measurement of the decay of pn-junction capacitance. This method, termed *Deep Level Transient Spectroscopy (DLTS)* is an improved version of a previous technique, *capacitance transient*, which is used to obtain information about an impurity level in the depletion region of a Schottky barrier or pn junction. In both methods, one observes the capacitance transient associated with the return to thermal equilibrium of the occupation of a given energy level, following a nonequilibrium condition. In DLTS, one can measure the time constant of this transient (assuming an exponential decay) as a function of temperature, which is the main novelty of DLTS; thus, one can find the activation energy. This method can distinguish between majority and minority carrier traps. Also, the technique has been extended to enable the calculation of the concentration and cross-section of the trap in question.

An important drawback of DLTS has been pointed out by Lang[5] himself, which has some similarity to problems in drawing conclusions from phosphorescence decay. The point is that one *assumes* an exponential decay, which is basically similar to assuming first-order kinetics.

1.6 Other Thermally Stimulated Processes

As pointed out, some other chemical and physical processes behave, at least as far as their general appearance and basic mathematical analysis are concerned, in a way similar to TL and the other thermally stimulated processes described above. We will briefly discuss these processes here, and elaborate upon them in Ch. 8.

The two most important and extensively utilized effects in this group are *Thermogravimetry* (TG) and *Differential Thermal Analysis* (DTA). These chemical TSPs are methods devised to monitor phase transitions, and are important tools in the characterization of materials; in particular, this kind of measurement is of importance in the classification of newly prepared materials or comparison between different batches of nominally identical material.

In TG, one usually measures the mass (through a measurement of the weight) of the sample in an environment heated in a controlled manner; this is recorded as a function of temperature or time. The TG curve is typically a step-wise curve in which each step indicates a process of either a decrease or an increase of the mass of the sample. Processes causing a decrease of the mass (which are more prevalent during heating) result from sublimation of the sample, its dehydration or chemical decompositions which release gaseous products. An increase of mass occurs by adsorption or absorption of gases and vapors, as well as chemical reactions between the sample and gaseous reactants.

A DTG curve may be evaluated from a TG curve; it is usually recorded along with (or instead of) the TG curve in course of the measurement by an electronic system which differentiates the TG signal with respect to time. The TG signal can be considered as a thermally stimulated process in that the "thermogram" of DTG is analogous to any of the above-mentioned effects (TL, TSC, TSEE, etc.), in particular in the more common cases where TG is a decreasing function. The DTG curve usually has a shape of one or more peaks, and its analysis can be performed with mathematical tools similar to those utilized for TL and TSC.

Another important member of this group of chemical thermally stimulated processes is differential thermal analysis (DTA). In this technique the difference between the temperature of the sample (T_s) and that of a reference (T_r) is recorded as a function of time, or of T_r (or T_s), as both the sample and reference are kept in a furnace and heated or cooled, usually at a constant rate. The reference is an inert material, the thermal properties of which (heat conduction and heat capacity) do not undergo substantial changes in the temperature range of the measurement. Thus, the plot of $\Delta T = T_s - T_r$ vs. t or T_s is a nearly horizontal line, unless abrupt enthalpic changes occur in the sample. These changes can result from endothermic or exothermic chemical reactions, phase transitions, etc. Each such enthalpic change appears as an upward or downward peak in the DTA curve. The analysis of these peaks can sometimes be done using methods developed for TL and DTG peaks. This must be done with care since, in some cases, the basic equations governing a DTA peak are not identical with those of TL and DTG.

A somewhat different and more modern related method is *differential scanning calorimetry* (DSC). In this technique, the sample and reference are both heated at a certain rate in such a way that their temperatures are equal at all the times. The energy flow necessary to establish a zero temperature difference between the two is plotted against either time or temperature. The resulting curve is similar to the DTA curve, since enthalpic processes in the sample appear as peaks in the DSC plot.

Other thermal analytical phenomena related to these are the following: The term *thermal annealing* is used in the literature for two different, though related, phenomena. One is the annealing of irradiation-induced defects during the heating of a sample as monitored by calorimetry. The recorded signal is the rate at which energy, previously stored in the sample is released upon heating. This, of course, is very similar to the DTA measurement except that in the present case, the measured signal is specifically associated with the previous irradiation. The other effect, termed by the same name by other researchers, is the measurement of optical absorption of a sample during its heating. In this case, the recorded quantity is the first derivative of the optical density with respect to temperature, which has a peak shape generally resembling that of a TL peak.

Another technique, different in nature but similar in appearance is *thermal desorption.* Here, a known gas adsorbs on the surface of a sample at some "low temperature". The temperature of the sample is then raised in a controlled rate, and the gas is desorbed. The desorption rate is usually measured by its effect on the pressure in a vacuum system, and recorded versus temperature. The result is normally a thermogram which, similarly to a TL curve, may include one or more peaks. From such a peak, the energy of adsorption can be evaluated using methods very similar to those employed in the study of other thermally stimulated processes. Another related method is *evolved gas analysis* (EGA) in which gaseous products emitted by a sample during heating are analyzed. Here the evolving gas is not a result of a previous adsorption at a low temperature.

Some more thermally stimulated processes will be described in brief in Ch. 8. These include *thermally stimulated creep, partial thermoremanent magnetization* and *dilatometry.*

1.7 References

1. J.T. Randall and M.H.F. Wilkins, *Proc. Roy. Soc. A* **184** (1945) 366.
2. D. Lapraz, P. Boutayer, P. Iacconi, R. Bindi and P. Rostaing, *Phys. Stat. Sol. (a)* **136** (1993) 497.
3. R.R. Haering and E.N. Adams, *Phys. Rev.* **117** (1960) 451.
4. H.K. Henisch, *J. Electrostatics* **3** (1977) 233.
5. D.V. Lang, *J. Appl. Phys.* **45** (1974) 3023.

Chapter 2
THERMOLUMINESCENCE (TL)

2.1 Basic Concepts of Luminescence in Solids

Luminescence is the emission of light from a material following the initial absorption of energy from an external source - e.g. ultra-violet or high energy radiation. The emission can be categorized as either fluorescence or phosphorescence, depending upon the characteristic lifetime τ between absorption of the excitation energy and emission of the luminescence. For luminescence emission with $\tau \leq 10^{-8}$s, one can easily classify the process as fluorescence, whereas for $\tau \geq$ a few seconds the process is certainly phosphorescence. In between, however, the distinction between the two processes is not always clear and is most easily made using the temperature dependence of τ. Consider a ground state energy level g and an excited state level e, illustrated in Fig. 2.1(a). Fluorescence is the emission of light which follows the excitation of an electron from g to e and its subsequent return to level g. Lifetimes in the excited state may be very short or, if the transition is either spin or parity forbidden, may be as long as a few milliseconds. (For example, the d-electron in Mn^{2+} ions in the thermoluminescent phosphor CaF_2:Mn can undertake a parity and spin forbidden transition from the $^6A_{1g}(^6S)$ ground state to the $^4T_{1g}(^4G)$ first excited state due to absorption of light. The excited state electron has a lifetime of $\tau \approx 5$ ms before it undergoes a transition back to the ground state with the emission of a photon (of wavelength $\lambda = 500$ nm).)

On the other hand, if the return to the ground state is delayed by a transition into and out of a metastable level m, then much longer delays between excitation and emission can result (Fig. 2.1(b)) and in this case the process is known as phosphorescence. If the transition into level m occurs at a temperature T, where the energy E of separation between m and e is such that $E \geq$ several kT (k is Boltzmann's constant) then the electron is likely to reside in m for a considerable period. In this case, assuming a Maxwellian distribution of energies, the probability p per unit time for thermal excitation from the trap is exponentially dependent upon temperature according to

$$p = s \exp\{-E/kT\} \tag{2.1}$$

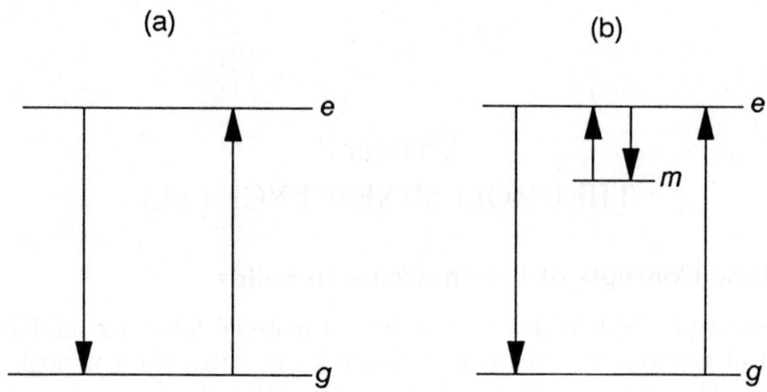

Figure 2.1 (a) Excited e and ground state g energy levels showing absorption and emission for the process of fluorescence. (b) Metastable level m giving rise to a delay between excitation and emission during the process of phosphorescence.

where s is a constant, with dimensions of reciprocal time. Thus, the time between excitation and final relaxation back to the ground state is delayed by the residence of the electron in the metastable state. It is clear that such a process would be sensitively dependent upon temperature because of the exponential function in Eq. 2.1, and this provides a means for distinguishing between the weakly temperature dependent process of fluorescence, and the strongly temperature dependent process of phosphorescence. The key feature is the intervention of the metastable level m.

The form of the excitation energy used to stimulate the emission gives rise to a wide variety of names for luminescence phenomena. Thus we have *photoluminescence* excited by the absorption of light (usually visible or ultra violet), *cathodoluminescence* excited by the absorption of energy from an electron beam, and *radioluminescence* due to the absorption of nuclear energy (from x-rays, gamma rays, etc.). We also find luminescence excited by chemical energy (*chemiluminescence*), mechanical energy (*triboluminescence*), electrical energy (*electroluminescence*), biochemical energy *(bioluminescence*), and even by energy from sound waves *(sonoluminescence)*. In all of these cases the emission may occur directly from the excited state, or may be delayed by the participation of a metastable state.

We should note in this regard that *thermo*luminescence is something of a misnomer since, unlike the examples above, heat is not the primary excitation source. Because of this, one often finds the more technically accurate term *thermally stimulated luminescence* (TSL) rather than TL, signifying that heat merely stimulates the emission, not the excitation. However, the word thermoluminescence is now thoroughly accepted in all its fields of application and, as a consequence, we prefer in this book to keep this accepted, traditional usage, originally coined as long ago as 1895 by Wiedemann and Schmidt[1].

2.2 Phosphorescence

In the processes to be described and analyzed in this book, the involvement of the metastable energy level (or levels) is critical in understanding the phenomena. The lifetime in this level, the temperature dependence of which is given by Eq. 2.1, can be anything - up to the age of the material - as long as the temperature T is low enough and/or the activation energy E is high enough to slow down the thermal emission rate. Consider some arbitrary time after the cessation of the excitation at which point there is a concentration of electrons n in level m. Following Randall and Wilkins[2] we may write that the rate of thermal excitation of electrons from level m back to e is

$$-\frac{dn}{dt} = np = ns\exp\left\{\frac{-E}{kT}\right\} \tag{2.2}$$

where the negative sign indicates a loss of electrons. Under conditions of no retrapping back into this level, the intensity I of the luminescence which results from the decay of the electron from e to the ground state g is governed by the rate of $m \to e$ transitions and thus:

$$I\left(t\right) = -\eta\frac{dn}{dt} = \eta ns\exp\left\{\frac{-E}{kT}\right\} \tag{2.3}$$

where η is a constant. Integration yields

$$I\left(t\right) = I_0\exp\{-tp\} \tag{2.4}$$

where I_0 is the initial intensity at $t = 0$. The decay of the phosphorescence under these circumstances is thus a simple exponential function of time, at a constant temperature. If $E \gg kT$ then the residence time in the metastable state becomes extremely long and essentially this means that the electrons in m can remain there indefinitely. For example, if $E = 1.5$ eV and $s = 10^{12}$ s^{-1}, then, at 298 K, $p^{-1} = 7.3 \times 10^5$ years.

In practice one often finds more complex decays than a simple exponential. Randall and Wilkins[3] (and later Garlick and Gibson[4]) also discuss the potential of having an equal probability that the excited electron will be retrapped into level m or will recombine in level g. Under these circumstances we have, instead of Eq. 2.3,

$$I\left(t\right) = -\eta\frac{dn}{dt} = \alpha n^2 \tag{2.5}$$

where α is a constant at constant T. Integration now yields

$$I(t) = \frac{I_0}{(n_0\alpha t + 1)^2}. \tag{2.6}$$

Since, in Eqs. 2.2 and 2.3, the decay rate is proportional to n we have a *first-order* process. In Eq. 2.5, however, we have a decay rate proportional to n^2, which is a *second-order* process. The constant α is related to the mean time the electron spends in m, and to a term describing the relative probabilities of retrapping at level m, versus recombination at level g.

The constant s requires some further discussion. Generally s is called the "pre-exponential factor", but for the present application it is more descriptively called the "attempt-to-escape frequency"[5],[6]. Using simple models for the localization of an electron in a potential well, s is interpreted as the number of times per second ν that an electron interacts with the lattice, multiplied by a transition probability κ, multiplied by a term which accounts for the change in entropy ΔS associated with the transition from m to e[6]-[9]. Thus, s may be written

$$s = \nu\kappa \exp\left\{\frac{\Delta S}{k}\right\}. \tag{2.7}$$

One can expect, therefore, that s will take on a value similar to the lattice vibrational frequency, i.e. $10^{12} - 10^{14}\ \mathrm{s}^{-1}$. In the processes to be described in this book the excited state e is usually a delocalized band (i.e. the conduction band for electrons, or the valence band for holes). In these cases, from detailed balance arguments, the frequency factor can also be equated to the capture cross-section σ of the metastable state[5], thus:

$$s = N_s v \sigma \tag{2.8}$$

where N_s is the effective density of states in the delocalized band, and v is the free carrier thermal velocity. Since N_s, v and σ all have a temperature dependence, so s also exhibits a temperature dependence. This particular property of s will be examined in more detail in a later chapter.

The above general description of phosphorescence has been extended to the band theory of solids by Randall and Wilkins[2],[3], Garlick and Gibson[4] and others[10]. In this picture phosphorescence from insulators and semiconductors is described in terms of electronic transitions between the delocalized bands and localized states (the latter being created by deviations in stoichiometry within the host lattice due to the presence of impurities and other lattice imperfections). This type of transition is shown schematically in Fig. 2.2. It is this description that will act as our vehicle for the analysis of thermoluminescence and related phenomena in the chapters and sections to follow.

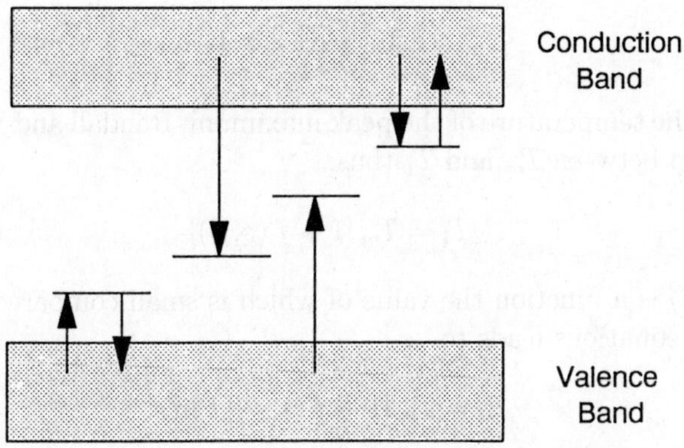

Figure 2.2 Transitions of electrons and holes to and from the conduction and valence bands in an insulator or semiconductor. Localized states act as trapping states (metastable levels) and recombination sites (ground states).

2.3 Elementary Theories of Thermoluminescence

2.3.1 TL as a Thermally Stimulated Process

Consider the case of electrons trapped at an energy E_t below the conduction band in which case the activation energy for phosphorescence $E = E_t$. Keeping the numerical values used above, namely that of trap with $E_t = 1.5$ eV and $s = 10^{12}$ s^{-1}, the calculation of p at 298 K indicates that, for all practical situations, the trap would never thermally release its trapped electron and, as a result, phosphorescence would never be observed at this temperature. However, if the temperature were to be raised, a temperature may be reached at which p is high enough to ensure that electron release occurs and luminescence is observed. Thus, consider raising the temperature of the system at some arbitrary rate $\beta = dT/dt$. As T rises, so p increases according to Eq. 2.1, producing enhanced luminescence emission. Unlike phosphorescence, which is normally assumed to occur at a fixed temperature, this emission is stimulated in a non-isothermal situation and is thus termed thermoluminescence (TL or TSL). As the temperature continues to rise, so the TL intensity increases until such time as the population of trapped electrons in the metastable state is sufficiently depleted at which point the TL intensity decreases with further increase in temperature. The result is the characteristic TL peak, for which the temperature at the peak maximum is related to the trap depth E_t, the frequency factor s, and the rate of warming β. Randall and Wilkins[2] define a temperature, which we will term T_1, at which the probability for charge release $p = 1$, i.e.

$$s \exp\left\{\frac{-E_t}{kT_1}\right\} = 1. \tag{2.9}$$

With T_m as the temperature of the peak maximum, Randall and Wilkins defined a relationship between T_m and T_1 thus:

$$T_1 = T_m \left[1 + f(s, \beta)\right] \tag{2.10}$$

where $f(s, \beta)$ is a function the value of which is small compared to one. Combining these equations leads to

$$E_t = T_m \left[1 + f(s, \beta)\right] k \ln s. \tag{2.11}$$

Unraveling the relationship between E_t and T_m is one of the challenges addressed in this book and expanded upon in later sections.

The above description of TL can be formalized in the following statements - statements which describe most thermally stimulated relaxation phenomena: *TL requires the perturbation of the system from a state of thermodynamic equilibrium, via the absorption of external energy, into a metastable state. This is then followed by the thermally stimulated relaxation of the system back to its equilibrium condition.* As a description of a thermally stimulated process (TSP), this statement is quite general. In the specific case of TL, the perturbation is usually the absorption of energy from ionizing radiation, and the thermally stimulated relaxation back to equilibrium is followed by monitoring the emission of luminescence from the system during the transitions of the freed charges back to the ground state. As will become clear in the sections to follow, the intensity of the emitted luminescence is related to the *rate* at which the system returns to equilibrium. The result is a characteristic "glow curve" consisting of a luminescence peak (or peaks) in a plot of luminescence against sample temperature, as was illustrated schematically in Fig. 1.2. In other TSPs the form of the perturbation may differ along with the property being monitored during heating. For example, in Thermally Stimulated Conductivity (TSC), ionizing radiation is still used as the excitation source, but one detects the thermal relaxation back to equilibrium by monitoring the freed charges during their passage through the delocalized excited state (i.e. conduction band for electrons, or valence band for holes). For Thermally Stimulated Exoelectron Emission (TSEE) one monitors the exoemission of electrons, usually from surface traps, during the relaxation process. Alternatively, for either Deep Level Transient Spectroscopy (DLTS) or Thermally Stimulated Capacitance (TSCap), the excitation can either be ionizing radiation or electrical energy, and one monitors the change in capacitance across a pn semiconductor

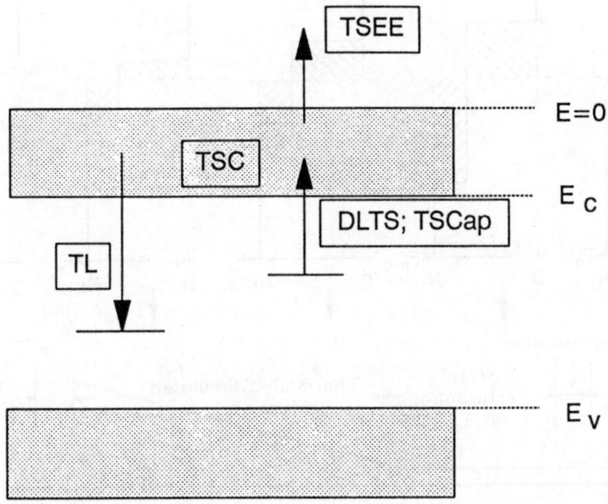

Figure 2.3 Schematic illustration of the distinction between TL, TSC, TSEE, TSCap and DLTS.

junction, or a metal-semiconductor junction, during the thermally stimulated transition of the trapped charge from the traps into the delocalized bands. The distinction between these processes is illustrated schematically in Fig. 2.3.

Each of the above processes, and all related processes, can be described in terms of the perturbation of the equilibrium Fermi Level in the material under study. At 0 K and in thermodynamic equilibrium one can expect from Fermi-Dirac statistics that all states above the Fermi Level E_F are empty while all states below E_F are filled. The situation is illustrated in Fig. 2.4 in which we use "filling diagrams", as described by Bräunlich[11], to represent the extent to which the energy states are filled $F_n(E)$ for a uniform trap distribution during a thermally stimulated relaxation experiment. Beginning with the left-hand figure we see the Fermi-Dirac filling function at 0 K. After perturbation we find a new filling function showing a distribution of electrons trapped at localized states within the band gap above E_F and an equal concentration of holes trapped below E_F. One can define two Quasi-Fermi Levels, one each for electrons E_{Fn} and for holes E_{Fp}. These are useful devices for describing the nonequilibrium state which follows the perturbation in terms of equilibrium statistics by making the assumption that the trapped electron and hole populations are in thermal equilibrium over their available energy levels. During thermally stimulated relaxation, i.e. during heating, the filling function $F_n(E)$

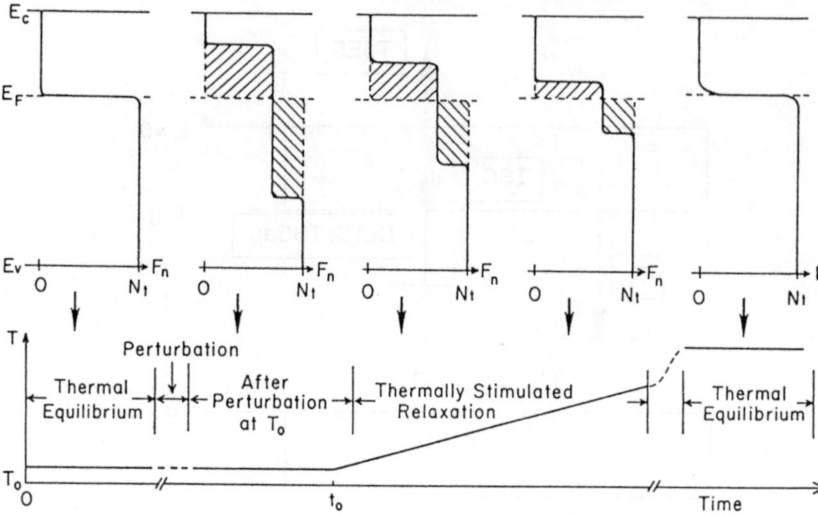

Figure 2.4 Filling functions for a uniform trap distribution before and after perturbation by an external stimulus, and during thermally stimulated relaxation. (After Bräunlich[11].)

gradually returns to its pre-perturbation state, albeit now at a higher temperature. During this process the Quasi-Fermi Levels gradually move back toward the equilibrium Fermi Level as the trapped charge concentrations decay back to their equilibrium values. In Fig. 2.4 the heating is performed according to the function $T(t) = T_0 + \beta t$, where T_0 is the initial temperature, i.e. 0 K in the above example, and β is the (constant) rate of heating. In principle, however, any other heating function could be used.

To develop this picture into a simple description of TL we note several additional definitions and assumptions:

(a) All transitions into or out of the localized states involve passage of the charge (electrons or holes) through the delocalized bands - i.e. for the present we do not include direct transitions between localized states.

(b) We adopt the usual definition of "trapping" states to be those at which the probability of thermal excitation from the level into the respective delocalized band is greater than the probability of recombination of the trapped charge with a free charge carrier of opposite sign. Conversely, a "recombination center" is defined as one in which the probability of recombination with an opposite sign charge carrier is greater than that of thermal excitation of the trapped carrier. One can therefore define a *demarcation level* to be that energy level at which these two probabilities are equal. Thus, we have a demarcation level for electrons E_{Dn}, and one for holes E_{Dp}.

(c) Transitions of electrons from the conduction band ($E \geq E_c$) into electron traps of energy E, for which $E_c > E > E_{Dn}$, are nonradiative, emitting phonons. Similarly, hole transitions from $E \leq E_v$ to hole traps at $E_{Dp} > E > E_v$ are also nonradiative. Here, E_c and E_v are the bottom of the conduction band and the top of the valence band, respectively.

(d) Transitions of free electrons from $E \geq E_c$ to trapped hole recombination sites at $E_F > E > E_{Dp}$, or free holes from $E \leq E_v$ to trapped electron recombination centers at $E_{Dn} > E > E_F$ are radiative, emitting photons.

(e) Once a trap is emptied, the freed carrier can no longer distinguish between it and all other traps of the same type.

In Fig. 2.4 the filling diagrams were illustrated for a uniform trap distribution. For any arbitrary density of states function $N(E)$, however, we may generalize the equations of Adirovitch[10], Haering and Adams[12], and Halperin and Braner[13] and write a series of rate equations describing the flow of charge into and out of the delocalized bands during thermal stimulation *en route* to equilibrium, thus:

$$dn_c/dt = \int_{E_{Dn}}^{E_c} p_n(E)\,N(E)\,f(E)\,dE - n_c v_n \int_{E_{Dn}}^{E_c} \sigma_n(E)\,N(E)\,(1-f(E))\,dE - n_c v_n \int_{E_{Dp}}^{E_F} \sigma_{mn}(E)\,N(E)\,(1-f(E))\,dE \tag{2.12}$$

and

$$dn_v/dt = \int_{E_v}^{E_{Dp}} p_p(E)\,N(E)\,(1-f(E))\,dE - n_v v_p \int_{E_v}^{E_{Dp}} \sigma_p(E)\,N(E)\,(f(E))\,dE - n_v v_p \int_{E_F}^{E_{Dn}} \sigma_{np}(E)\,N(E)\,f(E)\,d(E) \tag{2.13}$$

where $p_n(E)$ and $p_p(E)$ are the probabilities for thermal excitation from the traps, as given by Eq. 2.1; $N(E)$ is the density of states function; $f(E)$ is the occupation, or filling function; n_c is the concentration of free electrons in the conduction band, and n_v is the concentration of free holes in the valence band; v_n and v_p are free electron and free hole thermal velocities; $\sigma_n(E)$ and $\sigma_p(E)$ are the capture cross-sections for the retrapping of free carriers; and $\sigma_{mn}(E)$ and $\sigma_{np}(E)$ are the recombination cross-sections for the free carriers. Several authors (e.g.[10]-[13]) simplify the above equations by assuming only two types of localized state - namely a discrete electron trap at an energy E_t, where $E_c > E_t > E_{Dn}$, and a discrete recombination (trapped hole) center at an energy E, where $E_{Dp} > E > E_v$.

With the above conditions the concentration of trapped electrons simplifies thus:

$$\int_{E_{Dn}}^{E_c} N(E) f(E)\, dE \to n, \tag{2.14}$$

the concentration of empty traps becomes

$$\int_{E_{Dn}}^{E_c} N(E)(1 - f(E)) dE \to N - n, \tag{2.15}$$

and the concentration of available hole states (for recombination) becomes

$$\int_{E_{Dp}}^{E_F} N(E)(1 - f(E)) dE \to m. \tag{2.16}$$

Furthermore, since the trap energy and the recombination center energy are both single valued, then σ_n and σ_{mn} are also single valued. Finally, we assume that only trapped electrons are thermally freed during thermal stimulation (i.e. $p_p = 0$); this in turn means that $n_v = 0$. (Note that in the treatment which follows we will always assume that the TL signal is stimulated by the release of trapped electrons. Alternatively, one could just as easily assume that the holes were the thermally unstable charge carrier type. However, in this case the form of the equations would be identical to those which follow, and only the definitions of the terms would need to be modified.) Using Eq. 2.1, along with these simplifications, Eqs. 2.12 and 2.13 become

$$\frac{dn_c}{dt} = ns \exp\left\{-\frac{E_t}{kT}\right\} - n_c (N - n) A_n - n_c m A_{mn} \tag{2.17}$$

and

$$\frac{dn_v}{dt} = 0, \tag{2.18}$$

where $A_n = v_n \sigma_n$ is the retrapping probability and $A_{mn} = v_n \sigma_{mn}$ is the recombination probability (both expressed in units of volume per unit time). With the same notation we may also write

$$\frac{dn}{dt} = n_c (N - n) A_n - ns \exp\left\{-\frac{E_t}{kT}\right\} \tag{2.19}$$

and

$$\frac{dm}{dt} = -n_c m A_{mn}, \tag{2.20}$$

from which we see that $dn_c/dt = dm/dt - dn/dt$. Furthermore, charge neutrality dictates that

$$n_c + n = m. \tag{2.21}$$

Finally we note that the intensity of TL emitted during the return of the system to equilibrium is given by

$$I_{TL} = -\eta \frac{dm}{dt}. \tag{2.22}$$

where η is the radiative efficiency and is the same constant as used in Eqs. 2.3 and 2.5. If all recombination events produce photons and all photons are detected then $\eta = 1$. Unless specifically stated we shall assume $\eta = 1$ in the discussions to follow. Equations 2.17-2.20 represent the system of rate equations describing the traffic of electrons during heating (at rate β) of a system initially perturbed from equilibrium. They are coupled, first-order, non-linear, differential equations and, in general, are analytically insoluble. Note that the concentrations n_c, n and m are all time and temperature dependent, with the t and T dependencies linked through the heating rate function $T = T(t)$. (In fact, n_c, n and m should each be considered as functionals $n_c\{T(t)\}$, $n\{T(t)\}$, etc., since they depend on the history of the system. They are solutions of the differential equations and as such they depend upon t, T, and the initial values n_{c0}, n_0 and m_0.) In writing the above equations we have placed severe restrictions on the transitions which are allowed during the return of the system to equilibrium and upon the energy levels which are available within the system. This simple scheme - known as the one-trap/one-center model - is depicted in Fig. 2.5 and forms the basis of many, if not most, analyses of TL phenomena. Despite these obvious limitations the challenge now is to use these equations to develop an analytical expression which gives explicitly the relationship between the TL intensity and the temperature - i.e. to determine the form of $I_{TL}(T)$. To do so, several simplifying assumptions concerning the relative sizes of the various transition rates must be introduced into the equations. A variety of different assumptions have been suggested and used in the literature, and these are now described.

2.3.2 Quasiequilibrium

Perhaps the most important of all the assumptions introduced into the rate equations is that of quasiequilibrium (QE). The QE assumption is

$$\left|\frac{dn_c}{dt}\right| << \left|\frac{dn}{dt}\right|, \left|\frac{dm}{dt}\right|. \tag{2.23}$$

This assumption is of profound importance and, stated simply, it requires that the free electron concentration in the conduction band is quasistationary. It is this inequality which allows a simplification of the rate equations 2.17-2.20. If

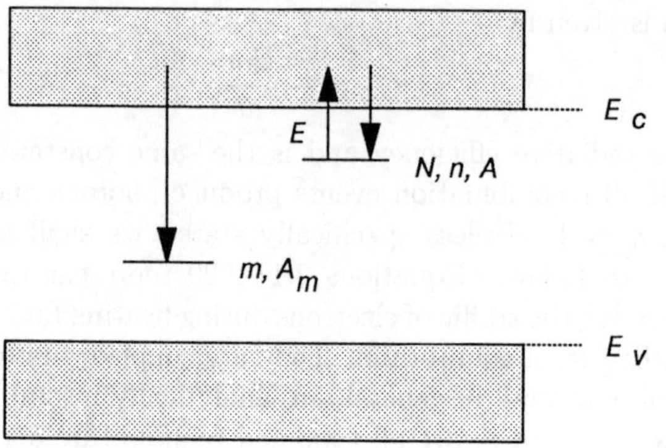

Figure 2.5 The allowed transitions (thermal excitation, retrapping and recombination) for the simple one-trap/one-center model for TL.

combined with the additional dictate that the initial free carrier concentration is small (i.e. $n_{c0} \simeq 0$), it means that free charge never accumulates in the conduction band during thermal stimulation. Applying the above inequalities leads to

$$-\frac{dn}{dt} \simeq -\frac{dm}{dt} = I_{TL} \tag{2.24}$$

and thus:

$$I_{TL} = \frac{ns\exp\{-E_t/kT\}m\sigma_{mn}}{[(N-n)\,\sigma_n + m\sigma_{mn}]}\,, \tag{2.25}$$

or

$$I_{TL} = ns\exp\left\{-\frac{E_t}{kT}\right\}\left[1 - \frac{(N-n)\,\sigma_n}{(N-n)\,\sigma_n + m\sigma_{mn}}\right]. \tag{2.26}$$

Equations 2.25 and 2.26 have been termed the "General One Trap" (GOT) expression for TL emission[14],[15]. In equation 2.26 the term in square brackets is the probability that the thermally released electrons will *not* be retrapped, and the ratio $(N-n)\sigma_n/m\sigma_{mn}$ is the ratio of the retrapping probability to the recombination probability. Further approximations are normally introduced at this stage concerning the relative sizes of these terms.

2.3.3 First-Order Kinetics (Slow Retrapping)

Randall and Wilkins [2],[3] assumed negligible retrapping during the thermal excitation period, i.e. they assumed that $m\sigma_{mn} >> (N-n)\sigma_n$. Note that this is a condition between functions $n(t)$ and $m(t)$ and thus while it may be true at the beginning of a peak, the possibility exists that it may become progressively untrue at later stages. Under these "slow-retrapping" conditions the GOT equation becomes

$$I_{TL} = ns\exp\left\{-\frac{E_t}{kT}\right\}. \tag{2.27}$$

We note also from Eq. 2.24 that $-dn/dt \propto n$, and thus we have a first-order reaction. Integrating from $t=0$ to t, using a constant heating rate, $\beta = dT/dt$, yields the well-known Randall-Wilkins first-order expression for the function $I_{TL}(T)$, namely

$$I_{TL} = n_0 s\exp\left\{-\frac{E_t}{kT}\right\}\exp\left\{-\left(\frac{s}{\beta}\right)\int_{T_0}^{T}\exp\left\{-\frac{E_t}{k\theta}\right\}d\theta\right\} \tag{2.28}$$

where n_0 is the initial values of n at $t=0$, and θ is a dummy variable representing temperature. The properties of this function are illustrated in Fig. 2.6 for (a) variations with n_0, for constant E_t and β, (b) variations with E_t for constant n_0 and β; and (c) variations with β for constant E_t and n_0. In Fig. 2.6(a) the heating rate β is set at 1.0 K/s, s is 10^{12} s^{-1}, and E is 1.0 eV. The peak height varies with n_0 as this parameter changes from 2 to 8 arbitrary units. (In all of the plots in this figure the peak height has been normalized to give a height of 1 intensity unit for $n_0 = 1$ concentration unit.) It is to be noted that the peak position stays fixed, while the height of the peak (directly proportional to the area) scales with n_0. This is an important characteristic of all first-order TL curves. The other characteristic property is the asymmetry of the peak, it being wider on the low temperature side than on the high temperature side. In Fig. 2.6(b) n_0 is fixed at 1.0 and E_t varies from 0.8 eV to 1.2 eV. To be noted here is the shift of the peak to higher temperatures, along with a decrease in the height and an increase in the width (keeping the area constant) as E_t increases. In Fig. 2.6(c), we see the variation in the peak shape and position as a function of heating rate β. As β increases (from 0.25 to 4.0 K/s) the peak shifts to higher temperatures and the size of the peak increases. Note that the peak area (on a temperature axis) scales with β, but if plotted as a function of time the area would remain the same as the heating rate changes.

2.3.4 Second-Order Kinetics (Fast Retrapping)

Garlick and Gibson[4] considered the alternative possibility that retrapping dominates, vis: $m\sigma_{mn} \ll (N-n)\sigma_n$. Applying this inequality to the GOT

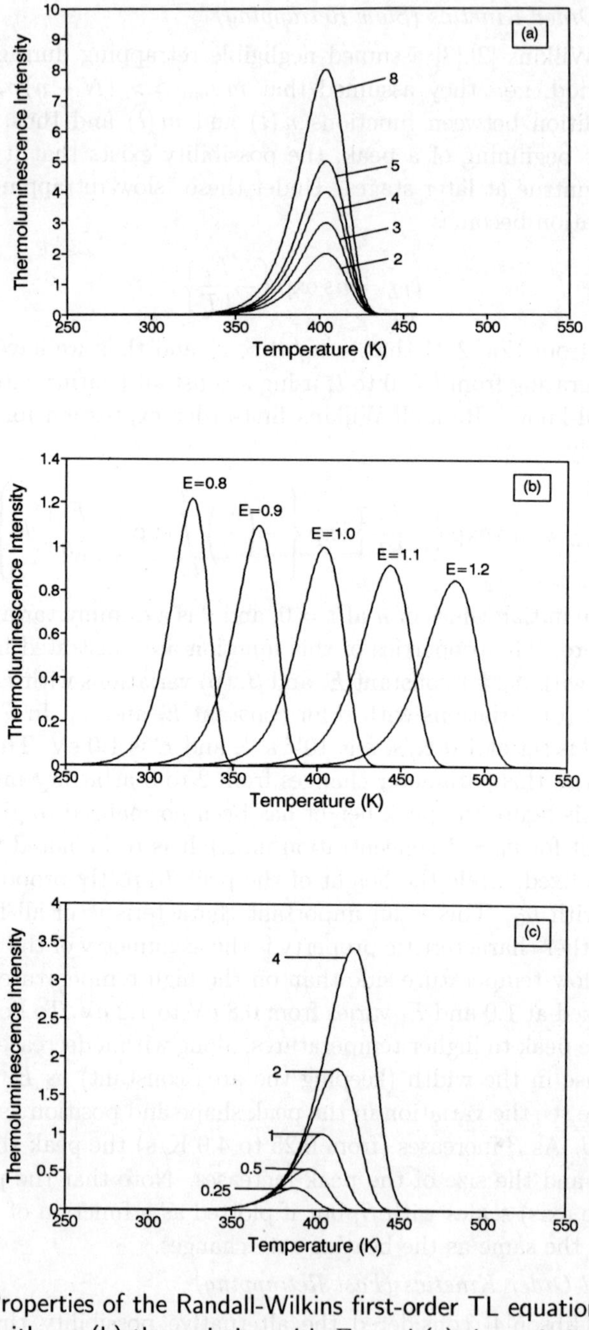

Figure 2.6 Properties of the Randall-Wilkins first-order TL equation, showing (a) the variation with n_o; (b) the variation with E_t; and (c) the variation with β.

expression, along with $N \gg n$ and $n = m$, yields

$$I_{TL} = -\frac{dn}{dt} = s\left(\frac{\sigma_n}{N\sigma_{mn}}\right)n^2 \exp\left\{-\frac{E_t}{kT}\right\}. \tag{2.29}$$

In contrast to the first-order case the inequality $m\sigma_{mn} \ll (N-n)\,\sigma_n$ will remain true throughout the TL peak, if it is true at the beginning. Note now that $dn/dt \propto n^2$ and, thus, this represents a second-order reaction. With the additional assumption that $\sigma_n = \sigma_{mn}$, integration of Eq. 2.29 gives

$$I_{TL} = \left(\frac{n_0^2}{N}\right) s \exp\left\{-\frac{E_t}{kT}\right\}\left[1 + \left(\frac{n_0 s}{\beta N}\right)\int_{T_0}^{T} \exp\left\{-\frac{E_t}{k\theta}\right\} d\theta\right]^{-2}. \tag{2.30}$$

where all the terms retain their previous meanings. This equation is the Garlick-Gibson TL equation for second-order kinetics. Note that one can also arrive at this equation directly from Eq. 2.25, using *only* the assumption $\sigma_n = \sigma_{mn}$, as long as $n = m$.

The properties of Eq. 2.30 are illustrated in Fig. 2.7. In Fig. 2.7(a) we see the change in the peak as a function of n_0, for constant E_t and β. The obvious difference between the behavior of the Garlick-Gibson second-order TL peak, compared to a first-order Randall-Wilkins peak, is that the peak grows - nearly proportionally to n_0 - and shifts to lower temperature as n_0 increases. This can be understood from an examination of Eq. 2.30 which may be recast as

$$I_{TL} = n_0^2 s' \exp\left\{-\frac{E_t}{kT}\right\}\left[1 + \left(\frac{n_0 s'}{\beta}\right)\int_{T_0}^{T} \exp\left\{-\frac{E_t}{k\theta}\right\} d\theta\right]^{-2} \tag{2.31}$$

in which we have defined $s' = s/N$ (or $s' = s\sigma_n/N\sigma_{mn}$, if $\sigma_n \neq \sigma_{mn}$,) as an "effective" pre-exponential factor (in units of $s^{-1}m^3$). Eq. 2.31 can also be written with s'' replacing $n_0 s'$, in which s'' (in units of s^{-1}) is equivalent (in units) to s in the first-order case. Because the position of a TL peak depends upon both E_t and s (or s'') one can see that, for a second-order peak, as n_0 increases, so too does s'', which, in turn, causes the TL peak to shift to lower temperatures. Additional characteristics of a second-order TL peak are that it is wider and more symmetric than a first-order peak. This can be understood from a consideration of the fact that in a second-order reaction, significant concentrations of released electrons are retrapped before they recombine, in this way giving rise to a delay in the luminescence emission and a spreading out of the emission over a wider temperature range. Note that both first-order and second-order peaks are characterized on the low temperature side by a function of the form $a\exp\{-E_t/kT\}$, whereas on the high temperature side the decay of

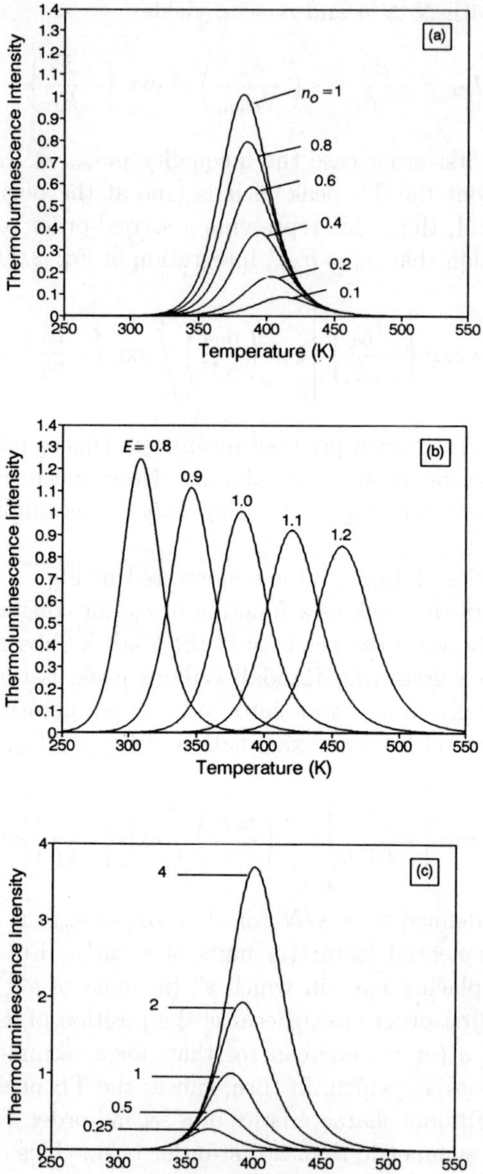

Figure 2.7 Properties of the Garlick-Gibson second-order TL equation, showing (a) the variation with n_o; (b) the variation with E_t; and (c) the variation with β. The figures have been normalized to give 1 intensity unit for $n_o = N = 1$ concentration unit at $b = 1$ K/s and $E_t = 1$ eV.

the TL with temperature is of the form $\exp\{-(a'/\beta)\int_{T_0}^{T}\exp\{-E_t/k\theta\}d\theta\}$ for a first-order peak, and $\left[c+(d/\beta)\int_{T_0}^{T}\exp\{-E_t/k\theta\}d\theta\right]^{-2}$ for a second-order peak (where a, a', c and d are constants).

Figures 2.7(b) and (c) show the variation in the size and position of a second-order peak as function of E_t and β, respectively. Note that the peak height is no longer directly proportional to the peak area, although the deviation is small.

2.3.5 General-Order Kinetics

The Randall-Wilkins and Garlick-Gibson forms of the TL equation have been derived with the use of specific assumptions concerning the relative sizes of the retrapping and recombination probabilities. Considering the form of the expressions produced from these assumptions, specifically Eqs. 2.27 and 2.29, May and Partridge[16] wrote an empirical expression for general-order TL kinetics, namely

$$I_{TL} = n^b s' \exp\left\{-\frac{E_t}{kT}\right\} \tag{2.32}$$

where s' has the dimensions of $m^{3(b-1)}s^{-1}$ and b is defined as the general-order parameter and is not necessarily 1 or 2. Integration of 2.32 for $b \neq 1$ produces

$$I_{TL} = s'' n_0 \exp\left\{-\frac{E_t}{kT}\right\}\left[1+\left((b-1)\frac{s''}{\beta}\right)\int_{T_0}^{T}\exp\left\{-\frac{E_t}{k\theta}\right\}d\theta\right]^{-\frac{b}{b-1}} \tag{2.33}$$

where now $s'' = s'n_0^{(b-1)}$. May and Partridge[16] developed this expression in order to apply it to those cases where the inequalities used above do not apply. Indeed, one can easily envision many combinations of values for N, n, m, σ_{mn} and σ_n for which the first- or second-order assumptions may be invalid, especially when it is remembered that each of these parameters is temperature dependent and in order for the Randall-Wilkins or Garlick-Gibson expressions to be valid one needs the restrictions concerning the relative sizes of these terms to apply across the whole temperature range of interest.

The difficulty with this empirical development of the general-order TL equation lies in the meaning of the parameter s', the dimensions of which change with the kinetic-order. To overcome this problem Rasheedy[17] proposed a recasting of Eq. 2.32 in the following form, thus:

$$I_{TL} = -\frac{dn}{dt} = \left(\frac{n^b}{N^{b-1}}\right) s \exp\left\{-\frac{E_t}{kT}\right\} \tag{2.34}$$

which clearly reduces to the first- and second-order forms with $b = 1$ and $b = 2$,

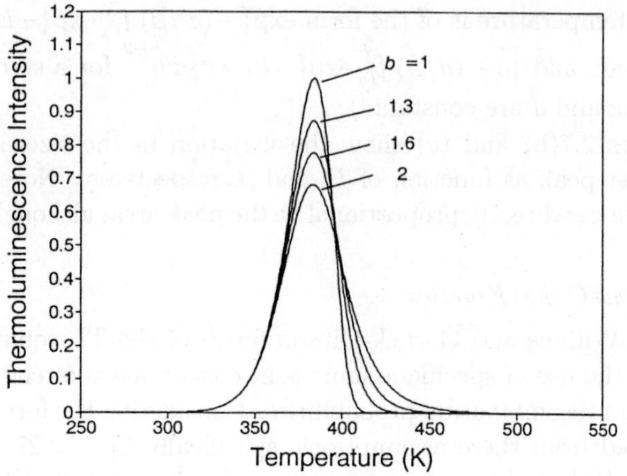

Figure 2.8 Comparison of first-order ($b = 1$), second-order ($b = 2$), and intermediate-order ($b = 1.3$ and $b = 1.6$) TL peaks, using $E_t = 1$ eV, $n_o = N = 1$ concentration unit, and $\beta = 1$ K/s. The figures have been normalized to 1 intensity unit for the first-order peak.

respectively. Integration now gives

$$I_{TL} = n_0^b s \exp\left\{-\frac{E_t}{kT}\right\} N^{(1-b)} \left[1 + \frac{s\,(b-1)\,(n_0/N)^{(b-1)}}{\beta} \int_{T_0}^{T} \exp\left\{-\frac{E_t}{k\theta}\right\} d\theta\right]^{-\frac{b}{b-1}}. \tag{2.35}$$

It should be stressed that this too is an empirical expression but its recasting in this manner removes the difficulty of interpreting the meaning of s'. Note that neither Eq. 2.33 nor Eq. 2.35 are valid for $b = 1$, but both reduce to the Randall-Wilkins equation (Eq. 2.28) as $b \to 1$[18].

TL peaks generated using Eq. 2.35, for $b = 1.3$ and $b = 1.6$, are compared with a first- and a second-order TL peak in Fig. 2.8. One can see that the intermediate-order peaks retain some of the character of each of the two extremes.

Finally we note that the question of interpreting the meaning of s' and the kinetic-order parameter b in the May and Partridge equation was dealt with by Christodoulides[19] who considered the case of a single activation energy, *first*-order kinetics, and a distribution in the frequency factor s (i.e. $n(s)$). If one assumes a Γ-distribution function in s (actually in $\ln(s)$) and applies this to the case of first-order kinetics, one can obtain the "general-order" equation - i.e. Eq. 2.35. With this interpretation s' and b are characteristic parameters of the distribution function for s. This is dealt with in greater detail in Section 2.4.3

2.4 Advanced Theories of TL

2.4.1 Multiple Trapping and Recombination States

The treatment presented so far relies on the simple one-trap/one-center model of Fig. 2.5 from which has been developed expressions describing the primary TL features observed for real materials - namely the shape of the observed TL peaks and their dependence upon trapped charge concentration, heating rate and trap depth. However, despite the utility of this model and its subsequent treatment, the fact remains that there are no actual materials for which one can accurately claim that there is only one trap and only one recombination site. An important consequence of the model discussed so far is that $n = m$ - i.e. the number of trapped electrons equals the number of trapped holes. This is a consequence of the charge neutrality condition (Eq. 2.21) with the added assumptions of quasiequilibrium and $n_{c0} \simeq 0$ (which together lead to $n_c \simeq 0$). In most real materials, however, there inevitably exist additional trapping sites for which the thermal stability of the trapped charge is greater than that for the TL signal being monitored. In other words, at most temperatures, there are likely to exist deep traps which retain their trapped electron population during emptying of the shallower traps and readout of the TL signal. The result of this is that $n \neq m$. Indeed, retaining the condition $n_{c0} \simeq 0$, the charge neutrality expression now becomes

$$n + h = m \tag{2.36}$$

where h is the concentration of electrons trapped in these deeper traps. The deep traps are referred to as being "thermally disconnected" and several authors have preferred to include these in the analysis of TL processes[20]-[22]. The rate equations now include an additional term

$$\frac{dh}{dt} = (H - h)\, v\sigma_h \tag{2.37}$$

and, as a result,

$$\frac{dn_c}{dt} = \frac{dm}{dt} - \frac{dn}{dt} - \frac{dh}{dt} \tag{2.38}$$

where H is the total number of available, deep, thermally disconnected traps and σ_h is the capture cross-section of the deep traps. The expressions for dn/dt and dm/dt remain as in Eqs. 2.19 and 2.20, respectively. Writing the equations in this way takes account of the fact that the deep, thermally disconnected traps may capture electrons released from the shallow trap at energy E_t. In this way the traps are said to be "interactive", with the deep trap competing with the recombination sites for the capture of electrons freed from the shallow trap. Under the special circumstances that $h \simeq H$, then $dh/dt \simeq 0$ and we may write, following Kelly and Bräunlich[22], the equivalent of the "General One Trap" equation (i.e. the equivalent of Eq. 2.26) for this "two trap" case, thus:

$$I_{TL} = \frac{ns\exp\{-E_t/kT\}\,(n+h)\,\sigma_{mn}}{[(N-n)\,\sigma_n + (n+h)\,\sigma_{mn}]}. \tag{2.39}$$

With this restriction, application of the slow-retrapping assumption (that is, $(N-n)\,\sigma_n << (n+h)\,\sigma_{mn}$) to Eq. 2.39 leads straightforwardly to the Randall-Wilkins, first-order TL equation (i.e. Eq. 2.28)[23]. Alternatively, the fast-retrapping case (i.e. $(N-n)\,\sigma_n >> (n+h)\,\sigma_{mn}$) along with $n << N$, leads to

$$I_{TL} = \frac{ns\exp\{-E_t/kT\}\,(n+h)\,\sigma_{mn}}{N\sigma_n}. \tag{2.40}$$

Or, for $\sigma_n = \sigma_{mn}$, we see that Eq. 2.39 becomes

$$I_{TL} = \frac{ns\exp\{-E_t/kT\}\,(n+h)}{(N+h)}. \tag{2.41}$$

Equations 2.40 and 2.41 can both be presented in the form

$$I_{TL} = s'n\,(n+h)\exp\left\{-\frac{E_t}{kT}\right\} \tag{2.42}$$

with $s' = s\sigma_{mn}/N\sigma_n$, or $s' = s/\,(N+h)$, respectively[24]. Expanding Eq. 2.42 produces

$$I_{TL} = s'nh\exp\left\{-\frac{E_t}{kT}\right\} + s'n^2\exp\left\{-\frac{E_t}{kT}\right\}, \tag{2.43}$$

and, as pointed out by Chen et al.[24], this looks like a mixture of first- and second-order kinetics. Specifically, if $h << n$, the equation reduces to the second-order form, while for $h >> n$ the equation reduces to the first-order form. For this reason this has been termed "mixed-order" kinetics by Chen et al.[24]. The solution of Eq. 2.43 is

$$I_{TL} = \frac{s'h^2\alpha\exp\{(hs'/\beta)\int_{T_0}^{T}\exp\{-E_t/k\theta\}d\theta\}\exp\{-E_t/kT\}}{\left[\exp\{(hs'/\beta)\int_{T_0}^{T}\exp\{-E_t/k\theta\}d\theta\} - \alpha\right]^2} \tag{2.44}$$

where $\alpha = n_0/\,(n_0 + h)$. The shape of Eq. 2.44 becomes identical to that of the Randall-Wilkins equation as $\alpha \to 0$, and is the same as that of the Garlick-Gibson equation when $\alpha = 1$.

From the foregoing treatments it is clear that the size, shape, position (in temperature) and behavior (as a function of trapped charge concentration and heating rate) can be encapsulated in one of several basic equations, depending upon the initial assumptions used. In each case, the TL peak is described by just four basic parameters - i.e. n_0, E, s and b (or α) - and the complex rate equations have been reduced to either first-order, second-order, or intermediate-order (or mixed-order) kinetics by the application of several assumptions. Perhaps the two most fundamental of these assumptions are $dn_c/dt \simeq 0$ (i.e. quasiequilibrium (QE)) and $h \simeq H$ (i.e. non-interactive kinetics). The validity of the QE assumption will be dealt with in a later section, but the non-interactive kinetics assumption requires more discussion at this point.

Opanowicz[25] took Eq. 2.39 and compared it with the General-Order expression, Eq. 2.32 to give

$$n^b s' = ns\gamma \tag{2.45}$$

where

$$\gamma = \frac{(n+h)\,\sigma_{mn}}{[(N-n)\,\sigma_n + (n+h)\,\sigma_{mn}]} \tag{2.46}$$

for the case where $h = H$. With the additional conditions $n_0 = N$ and $s' = sn_0^{1-b}$, Opanowicz developed the temperature-dependence of the kinetic-order parameter, $b(T)$, thus

$$b = \frac{\ln(\gamma n/N)}{\ln(n/N)}. \tag{2.47}$$

Since γ and n/N are strongly temperature dependent, so too is b, in contrast to the usual assumption of this parameter being a constant. The implication is that b is a shape parameter, with no simple physical meaning, and this, in turn, questions the usefulness of the General-Order formalisms of Eqs. 2.32 and 2.33. In this regard, Eq. 2.44 is more meaningful.

2.4.2 *Interactive Kinetics*

The most general way of writing the rate equations to describe the flow of electrons into and out of the conduction band for a system of multiple traps and multiple recombination centers was given in Eq. 2.12. For the case of a set of discrete electron traps of index $i = 1$ to u and a set of hole traps (recombination centers) of index $j = 1$ to v we may rewrite the complete set of rate equations as follows.

For $i = 1$ to u

$$\frac{dn_i}{dt} = -n_i s_i \exp\left\{-\frac{E_{ti}}{kT}\right\} + n_c\,(N_i - n_i)\,A_{ni}. \tag{2.48}$$

For $j = 1$ to v

$$\frac{dm_j}{dt} = -n_c m_j A_{mnj}. \tag{2.49}$$

where $A_{ni} = v_n \sigma_{ni}$ and $A_{mnj} = v_n \sigma_{mnj}$.

The time rate of change of the free electron concentrations may be written

$$\frac{dn_c}{dt} = \sum_{i=1}^{u} n_i s_i \exp\left\{-\frac{E_{ti}}{kT}\right\} - n_c \left(\sum_{j=1}^{v} m_j A_{mnj} + \sum_{i=1}^{u} (N_i - n_i) A_{ni}\right) \tag{2.50}$$

and since only the thermal release of trapped electrons is considered we still have $dn_v/dt = 0$.

To analyze this set of equations one can proceed in several ways. One way was developed by Levy[14],[15] who retained the quasiequilibrium approximation and developed the equivalent of the "General One Trap" equation for this more complex case, namely

$$I_{TL} = \sum_{j=1}^{v} E \frac{m_j A_{mnj}}{R + U} \tag{2.51}$$

where

$$E = \sum_{i=1}^{u} n_i s_i \exp\left\{-\frac{E_{ti}}{kT}\right\}, \tag{2.52}$$

$$R = \sum_{j=1}^{v} m_j A_{mnj} \tag{2.53}$$

and

$$U = \sum_{i=1}^{u} (N_i - n_i) A_{ni}. \tag{2.54}$$

In writing Eq. 2.51 we have assumed that all electron-hole recombination processes are radiative and contribute to the TL signal. If this is not so, then only a subset of the recombination terms would be used. Furthermore, the equation also assumes that all emitted photons are detected with the same efficiency. Otherwise each recombination term would have to be multiplied by an efficiency factor $\zeta_j \leq 1$.

An example of a set of TL glow curves, as computed by Levy[14] using Eq. 2.51 with $u = 3$ and $v = 1$, is shown in Fig. 2.9. The figure illustrates the potential curve shapes that are possible with an interactive system and the clear deviations that can be obtained with respect to the standard first- or second-order curves. For the data illustrated we have $\sigma_{ni}/\sigma_{mnj} = 0.1$ and $N_i = 10^{22}$ m^{-3}. For this data set one can see that the "interactive" TL curves are second-order-like for low initial values of the trapped charge concentration, but become more first-order-like as the traps become filled.

The alternative approach is not to assume quasiequilibrium. Although we will not be discussing the detailed ramifications of the QE approximation until a later section, it is useful at this stage to explore the properties of interactive

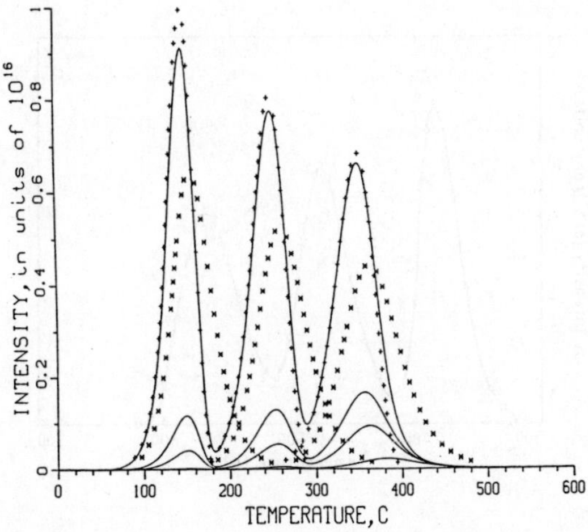

Figure 2.9 Computed TL curves using interactive kinetics for 3 electron traps and one recombination center. The curves (full lines) are obtained with initial trapped charge concentrations in each trap varying from 1×10^{20} to 5×10^{21} m^{-3}, and for $\sigma_{ni}/\sigma_{mnj} = 0.1$. Also shown are first-order (+) and second-order (*) TL curves for an initial trapped charge population of 5×10^{21} m^{-3}.
(After Levy[14].)

kinetics with this restriction removed. Without the QE approximation the rate equations - Eqs. 2.48-2.50 - cannot be decoupled and hence cannot be solved analytically, as was done above. Instead one has to examine the various glow curve shapes that are possible using numerical analysis of the rate equations. Such an exercise was undertaken by Bull et al.[26] who numerically solved the rate equations for a series of electron traps and recombination centers, again using $u = 3$ and $v = 1$. Some example data are shown in Fig. 2.10. Here we see that when recombination dominates (curve (a)) the 'interactive" glow curve can be well represented as the sum of three first-order curves. This is to be expected since, if retrapping is slow, little interaction (i.e. trapping into other traps) will occur compared to recombination. For the case where recombination and retrapping are equally strong, however, the "interactive" glow curve is significantly different from that obtained from a sum of second-order components (Fig. 2.10(b)). In Fig. 2.10(a), the $n_{oi} = N_i = 10^{15}$ m^{-3}, and $\sigma_{ni}/\sigma_{mnj} = 10^8$. For Fig. 2.10(b) $\sigma_{ni}/\sigma_{mnj} = 1$, $N_i = 10^{15}$ m^{-3} and the n_{oi}/N_i ratios are given in the caption.

The conclusion to be reached from the above analyses is that in a system in which the recombination processes dominate over retrapping processes, the TL glow curve can be represented accurately by the superposition of individual

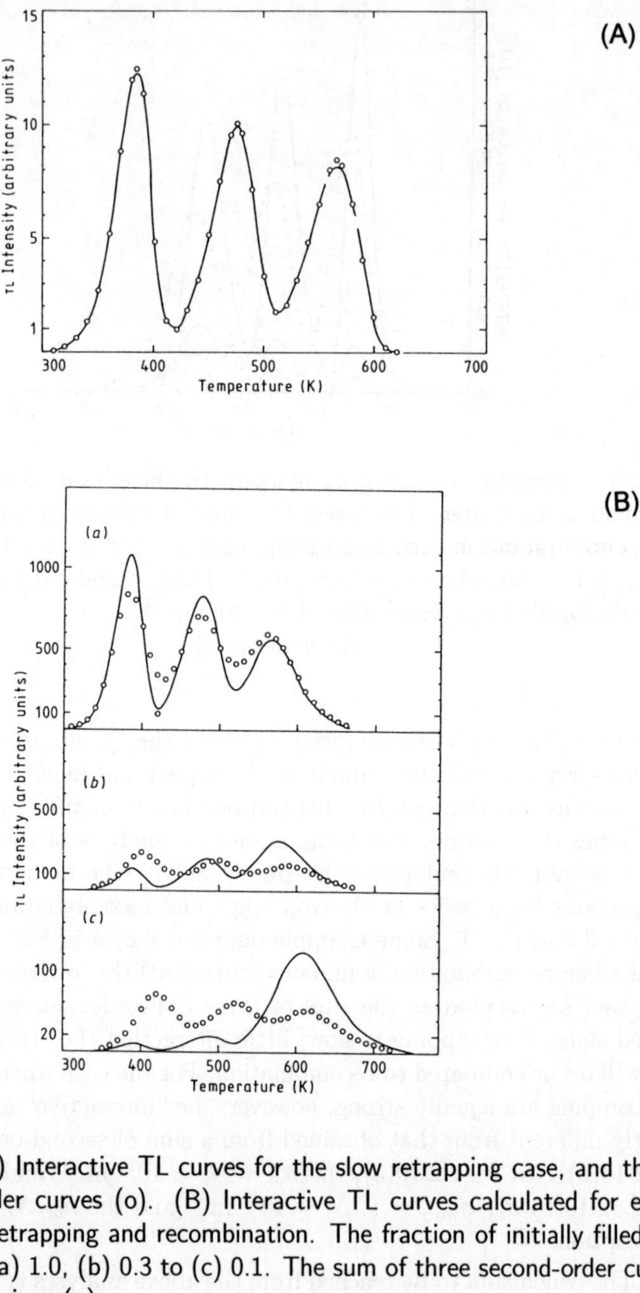

Figure 2.10 (A) Interactive TL curves for the slow retrapping case, and the sum of three first-order curves (o). (B) Interactive TL curves calculated for equal probabilities of retrapping and recombination. The fraction of initially filled traps decreases from (a) 1.0, (b) 0.3 to (c) 0.1. The sum of three second-order curves is shown for comparison (o).

(After Bull et al.[26].)

first-order peaks of the Randall-Wilkins type (vis: Eq. 2.28). However, for systems in which appreciable retrapping is likely, the actual glow curve shape will show significant departures from the superposition of second-order processes. Under these circumstances analysis of the data based on the simple equations, such as the first-, second-, intermediate- or mixed-order, may prove to be unreliable.

2.4.3 *Continua; Trap Distributions*

The discussions of the foregoing sections have assumed that the trap depths associated with the localized states are single-valued in energy. While this may be expected to be true for high quality single crystal materials, one can expect that in highly defective or, especially, amorphous or vitreous materials the trap depths associated with particular defects will be spread over a range of values rather than being uniquely defined. In these latter materials the structure of the lattice surrounding the defect giving rise to the TL signal may exhibit random variations in the nearest neighbor bond angles and bond lengths. The result is that the activation energies (trap depths) tend to be distributed, rather than discrete. For amorphous materials it is no longer possible to talk about the forbidden energy gap that one normally associates with crystalline materials. Instead one refers to the "mobility gap" within which there can exist a finite density of states localized near the Fermi level[27].

Given the variety of lattice fluctuations that one might expect in highly defective or amorphous materials, several kinds of energy distribution have been discussed to describe these localized states[28]-[33]. Included among these are:

(1) Uniform distributions such that $N(E < E_A) = N(E > E_B) = 0$ and $N(E_A < E < E_B) = N_t$ = uniform trap density (in $\mathrm{m^3eV^{-1}}$) between E_A and E_B.

(2) Linear distributions of the type

$$N(E) = N_m \left[\frac{E - E_A}{E_B - E_A}\right], \tag{2.55}$$

where $N(E) = 0$ at E_A and N_m is the maximum trapped charge density at energy E_B.

(3) Exponential distributions, e.g.

$$N(E) = N_e \exp\left\{\frac{-E}{kT_c}\right\}, \tag{2.56}$$

where N_e is a constant ($\mathrm{m^3eV^{-1}}$) and T_c is a characteristic temperature for the distribution.

(4) Gaussian distributions:

$$N(E) = N_o \exp\{-a(E - E_o)^2\}, \tag{2.57}$$

where N_o is the maximum density at energy E_o and a is a constant.

Thermoluminescence from a distribution of traps depths has been observed in many materials - both amorphous and crystalline - e.g. $ZnIn_2S_4$[34], polystyrene[35],[36], $ZnSiO_4$:Mn[37], ZnS[36], silica[38], and quartz[39]. Recombination between distributed states has been dealt with in general by Fowler[40], and Randall and Wilkins[3] discussed an analysis of phosphorescence and thermoluminescence. Hornyak and colleagues have discussed the detailed features of phosphorescence decay curves and thermoluminescence glow curves based on assumptions of Gaussian[42] or uniform[43] trap distributions. Following these latter authors we will now examine the rate equations and TL glow curves that result from an adoption of these energy distributions.

The general expression for the rate of change of free electrons in the conduction band has already been given in Eq. 2.12. We consider here only a distribution of trapping energies and so replace $N(E)$ with $N(E_t)$. Furthermore, we consider only the thermal release of trapped electrons recombining with trapped holes, and we rewrite the integrals as $\int_{E_{D_p}}^{E_c} dE \rightarrow \int_0^\infty dE_t$ (recalling that the trapped electrons, by definition, only exist between E_{D_p} and E_c). We may now rewrite Eq. 2.12 to be

$$\frac{dn_c}{dt} = \int_0^\infty p_n(E_t) N(E_t) f(E_t) dE_t - n_c A_n \int_0^\infty N(E_t)(1 - f(E_t)) dE_t - n_c m A_{mn}. \tag{2.58}$$

In writing Eq. 2.57 we have further assumed that the capture cross-section σ_n is energy independent.

Consider the population of trapped electrons

$$n(E_t) = f(E_t)N(E_t). \tag{2.59}$$

Consider further a segment dE_t of the distribution $N(E_t)$; this infinitesimal region contains $n(E_t)dE_t$ electrons at time t. The total number of trapped electrons at time t is

$$n = \int_0^\infty f(E_t)N(E_t)dE_t \tag{2.60}$$

and the total number of available traps is

$$N = \int_0^\infty N(E_t)dE_t. \tag{2.61}$$

At time $t = 0$

$$n_0 = \int_0^\infty f_0(E_t) N(E_t)dE_t. \tag{2.62}$$

It follows that at $t = 0$, $n_0 = m_0$.

Thus, we may also write

$$\frac{dn\,(E_t)}{dt}dE_t = N(E_t)\frac{df\,(E_t)}{dt}dE_t$$

$$= -p_n(E_t)f\,(E_t)\,N\,(E_t)\,dE_t + n_cA_n\,[1 - f\,(E_t)]\,N(E_t)dE_t \tag{2.63}$$

from which we see that

$$\frac{df\,(E_t)}{dt} = -p_n(E_t)f\,(E_t) + n_cA_n\,[1 - f\,(E_t)]\,. \tag{2.64}$$

Integration of Eq. 2.64 immediately gives the time dependence of the filling function

$$f(E_t) = f_0\,(E_t)\exp\{-p_n\,(E_t)\} + A_n\exp\{-p_n\,(E_t)\}\int_0^t n_c\left(t'\right)\exp\{p_n\,(E_t)\}\,dt'. \tag{2.65}$$

By incorporating the assumption of quasiequilibrium it is acceptable to assume that $n_c\left(t'\right)$ is a slowly varying function and may be considered constant. In this case Eq. 2.65 becomes

$$f(E_t) \simeq f_0\,(E_t)\exp\{-p_n\,(E_t)\} + \frac{n_cA_n}{p_n\,(E_t)}\,[1 - \exp\{-p_n\,(E_t)\}] \tag{2.66}$$

which may be further reduced to

$$f\,(E_t) \simeq f_0\,(E_t)\exp\{-p_n\,(E_t)\} \tag{2.67}$$

by assuming that $n_cA_n << p_n\,(E_t)$.

By proceeding with the assumption of quasiequilibrium one may derive the equivalent of the GOT equation for this system, namely

$$I_{TL} = \int_0^\infty p_n\,(E_t)\,N(E_t)f(E_t)dE_t\left[\frac{m\sigma_{mn}}{(N-n)\,\sigma_n + m\sigma_{mn}}\right] \tag{2.68}$$

where N and n are defined in Eqs. 2.60 and 2.61. The introduction of the slow-retrapping (first-order) approximation leads to the equivalent of the Randall-Wilkins equation, thus

$$I_{TL} = \int_0^\infty N\,(E_t)\,f_0\,(E_t)\,s\exp\left\{-\frac{E_t}{kT}\right\}\exp\left\{-\left(\frac{s}{\beta}\right)\int_{T_0}^T\exp\left\{-\frac{E_t}{k\theta}\right\}d\theta\right\}dE_t. \tag{2.69}$$

where β retains its usual definition of heating rate. To proceed further we must introduce specific functions for the distribution $N(E_t)$.

Hornyak and Chen[43] considered a uniform trap distribution in which case Eq. 2.69 transforms easily into

$$I_{TL} = \frac{n_0 s}{\Delta E} \int_{E_A}^{E_B} \exp\left\{-\frac{E_t}{kT}\right\} \exp\left\{-\frac{s}{\beta} \int_{T_0}^{T} \exp\left\{-\frac{E_t}{k\theta}\right\} d\theta\right\} dE_t \tag{2.70}$$

where n_0 is the total concentration of trapped electrons at $t = 0$, and $\Delta E = E_B - E_A$ is the distribution width. Using an asymptotic series for evaluating the temperature integral[41],[42]

$$\int^{T} \exp\left\{-\frac{E_t}{k\theta}\right\} d\theta \simeq T \exp\left\{-\frac{E_t}{kT}\right\} \sum_{\alpha=1} \left(\frac{kT}{E_t}\right)^{\alpha} (-1)^{\alpha-1} \alpha! \tag{2.71}$$

Eq. 2.70 can be approximated

$$I_{TL} \approx \frac{n_0 s}{\Delta E} \int_{E_A}^{E_B} \exp\left\{-\frac{E_t}{kT} - \frac{skT^2}{\beta E_t} \exp\left\{-\frac{E_t}{kT}\right\} \left[1 - \frac{2kT}{E_t} + \frac{6k^2T^2}{E_t^2} - \ldots\right]\right\} dE_t. \tag{2.72}$$

After making the further approximation

$$\frac{skT^2}{\beta E_t} \left[1 - \frac{2kT}{E_t} + \frac{6k^2T^2}{E_t^2} - \ldots\right] \simeq \frac{skT^2}{\beta E_o} \left[1 - \frac{2kT}{E_o} + \frac{6k^2T^2}{E_o^2} - \ldots\right] = \gamma, \tag{2.73}$$

Eq. 2.72 can be integrated to give

$$I_{TL} \approx \frac{n_0 s}{\Delta E} \frac{kT}{\gamma} \left[\exp\left\{-\gamma \exp\left(-\frac{E_B}{kT}\right)\right\} - \exp\left\{-\gamma \exp\left(-\frac{E_A}{kT}\right)\right\}\right] \tag{2.74}$$

where $E_o = (E_A + E_B)/2$ is the center energy of the distribution. This approximation is good provided ΔE is small (≤ 0.1 eV)[43]. Examples of glow curves evaluated using Eq. 2.74 are shown in Fig. 2.11. One can observe that the wider the distribution, the wider the TL peak. For small ΔE the peak shape is essentially that of a first-order peak for a discrete energy $E = 0.7$ eV. However, as the distribution width increases, the peak becomes wider and smaller, taking on a shape which is reminiscent of a second-order peak and care has to be taken to distinguish between these two. The way to distinguish between a distribution and a second-order peak is to vary the initial trapped electron concentration n_0 since the peak described by Eq. 2.17 scales with n_0 but does not shift, unlike the second-order peak behavior, as already shown in Fig. 2.7(a)).

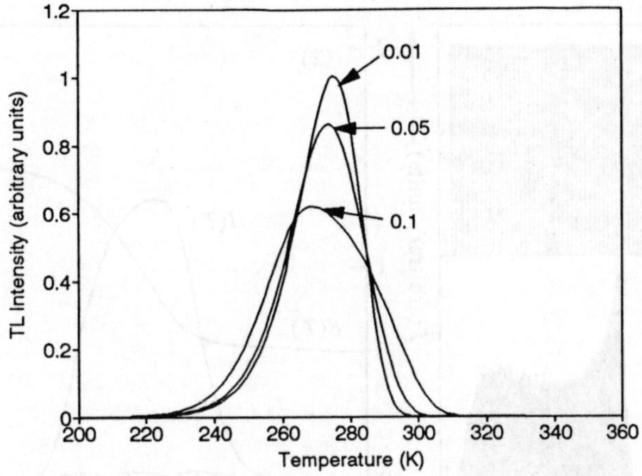

Figure 2.11 Examples of TL glow peaks obtained assuming a uniform distribution of traps for $E_o = 0.7$ eV, $s = 1 \times 10^{12}$ s^{-1}, $\beta = 10$ K/s, and $\Delta E = 0.1, 0.5$ and 0.01 eV. The peaks have been normalized to give 1 intensity unit for the $\Delta E = 0.01$ eV peak.

Other authors treated the same type of distribution by dividing the distribution into multiple (50 to 100) small intervals in order to calculate the glow curve for each interval using the standard Randall-Wilkins equation (i.e. Eq. 2.28) and then adding the resulting curves to obtain the glow curve[43],[44]. This approach is clearly more time-consuming than using the approximate solution given in Eq. 2.74, but is, nevertheless, valid over a wider range of ΔE values.

Other distribution functions $N(E_t)$ are less tractable and not solvable analytically. With these functions one again uses the same method as above - namely summing multiple glow curves to produce the glow curve required. Examples of the curve shapes which are possible using this method for uniform and Gaussian distributions are shown in Fig. 2.12.

At the start of the above analysis it was stated that we were considering only the case of a distribution in trapping energies and that the recombination center energies were assumed to be discrete, and, further, that the capture cross-sections were also considered to be single valued. - that is, σ_n (and σ_{mn}) are both independent of energy. A distribution in recombination center energies does not lead to a necessity to reformulate the equations since all it does is lead to a wider distribution of emission energies - for example, a broad TL emission band may result. In this sense we can treat the parameter m in the above equations to be the total number of available recombination centers in

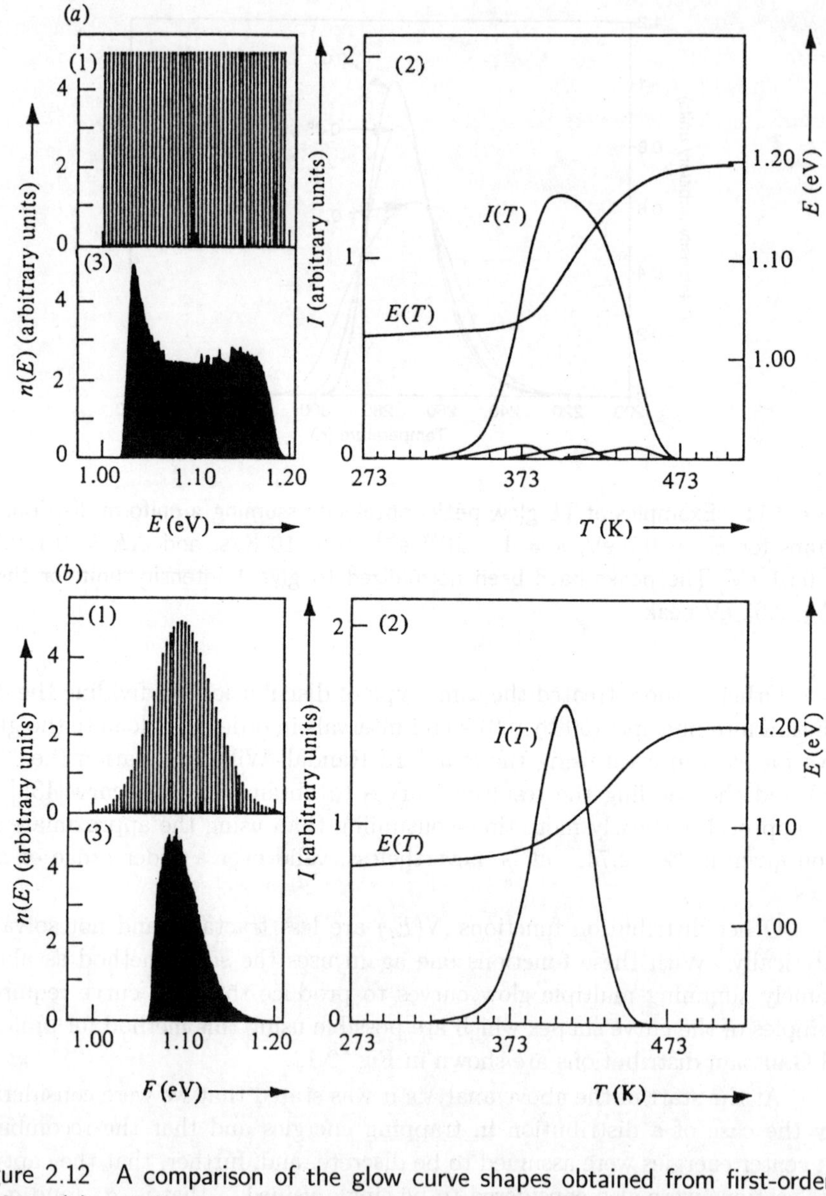

Figure 2.12 A comparison of the glow curve shapes obtained from first-order, uniform (a) and Gaussian (b) trap depth distributions. The distributions used are shown in (1); the change in trap depth $E_t(T)$ with temperature across the glow peak. In (3) are shown the distributions obtained after the Fractional Glow Analysis (discussed in Chapter 3).

(After Rudlof et al.[44].)

which case the equations themselves remain the same. However, a distribution of recombination cross-sections $m(\sigma_{mn})$ (or $m(A_{mn})$) may lead to distortions in the expected curve shape since those centers with the largest recombination probabilities will recombine first and TL emissions with the corresponding emission spectrum will appear at lower temperatures than those emissions corresponding to the smaller cross-section. Hence, if the TL were to be monitored through selective emission filters, the TL peak would appear at slightly different temperatures, according to the filter used. This is discussed further in section 2.5.1.

A distribution in capture-cross sections for the trap (i.e. $n(\sigma_n)$ or $n(A_n)$), however, leads in turn to a distribution in the frequency factor $n(s)$, through Eq. 2.8. A distribution in s has been dealt with by few authors. Rudlof et al.[44] simulate the TL curve shapes using the same methods as used above, but for an assumed Gaussian distribution in s. In Section 2.3.5 it was stated that if one assumes a Γ-distribution in s and applies this to the case of first-order kinetics and a single value for E_t one can obtain the intermediate-order kinetic equation, Eq. 2.33[19]. We now examine this case in more detail.

Consider a frequency factor distribution function $g(\ln s)$. This function is normalized (i.e. $\int_{-\infty}^{+\infty} g(\ln s)\, d(\ln s) = 1$) and is a function of time. At $t = 0$ we denote its value by $g_0(\ln s)$. Assuming a constant heating rate β, then at time t we may write, assuming first-order kinetics,

$$n = n_0 \int_0^\infty g_0(\ln s) \exp\left\{ -\frac{s}{\beta} \int_{T_0}^{T} \exp\left\{ -\frac{E_t}{k\theta} \right\} d\theta \right\} \frac{ds}{s} \tag{2.75}$$

from which we have

$$I_{TL} = n_0 \exp\left\{ -\frac{E_t}{kT} \right\} \int_0^\infty g(\ln s) \exp\left\{ -\frac{s}{\beta} \int_{T_0}^{T} \exp\left\{ -\frac{E_t}{k\theta} \right\} d\theta \right\} \frac{ds}{s}. \tag{2.76}$$

We saw in Section 2.3.5 that for intermediate-order kinetics we may write

$$I_{TL} = s'' n_0 \exp\left\{ -\frac{E_t}{kT} \right\} \left[1 + \left((b-1) \frac{s''}{\beta} \right) \int_{T_0}^{T} \exp\left\{ -\frac{E_t}{k\theta} \right\} d\theta \right]^{-\frac{b}{b-1}}$$

where the parameters s'' and b retain their meaning from Section 2.3.5. Equating Eqs. 2.76 and 2.33 then leads to that distribution function for $g_0(\ln s)$ which would make a first-order kinetics peak look like an intermediate-order kinetics peak, viz.

$$g_0(\ln s) = \frac{(a_0 s)^{1/(b-1)}}{\Gamma(1/(b-1))} \exp\{-a_o s\} \tag{2.77}$$

where

$$a_0 = \frac{1}{(b-1)\, s' n_0^{b-1}}. \tag{2.78}$$

Eq. 2.77 is a Γ-distribution function in $\ln s$ and becomes a delta function (i.e. a single value for s) when $b = 1$. The function 2.77 is, of course, modified by heating during the TL readout, but it retains its form as a Γ-function, but with a different parameter a replacing a_0. In fact

$$a = a_0 + \frac{1}{\beta} \int_{T_0}^{T} \exp\left\{-\frac{E_t}{k\theta}\right\} d\theta. \tag{2.79}$$

Note that the distribution is a function of $\ln s$ rather than s. However, as pointed out by Christodoulides[19], a Γ-function in $\ln s$ is also a Γ-function in s. This distribution in s must arise from a similar distribution in σ_n. However, the physical process that can give rise to a Γ-distribution in capture cross-sections is unknown.

Although it was mentioned in the foregoing discussion that, during the course of the TL readout, the distribution function in s remains a Γ-function, the same cannot be argued for a distribution in E_t. That is, if the occupancy of the available states is (say) uniform at the start of the heating cycle, it will not necessarily remain uniform during the course of the heating. Retrapping of electrons released from low-E_t traps into high-E_t traps may take place, with a result that a dynamic distortion of the distribution function will take place during heating. In other words, interactive kinetics may become important - but this time the interaction is among the states within the distribution. However, if retrapping is weak - as is the case for first-order kinetics - such effects are minor and distortions of the distribution during heating need not be considered.

2.4.4 Schön-Klasens Kinetics

The entire discussion so far has been based on the consideration of electron release from traps only. In this context we wrote in Section 2.3.1 (Eq. 2.8) that $dn_v/dt = 0$, and since that point we have ignored hole transitions, stating only that the equations developed for the release of trapped electrons - if the trapped holes are stable - are the identical to those for the release of trapped holes - if the trapped electrons are stable. However, what we have not considered is that *both* electrons and holes may be released from their traps at the same time, in the same temperature interval. Even this is not necessarily a difficulty; we can still treat the TL signals caused by hole and electron traps in the same way as we would treat multiple electron traps, or multiple hole traps. The difficulty presents itself when (say) the holes are being thermally released from the same centers as are acting as recombination sites for the thermally released electrons, and/or vice-versa. These situations are summarized in Fig. 2.13.

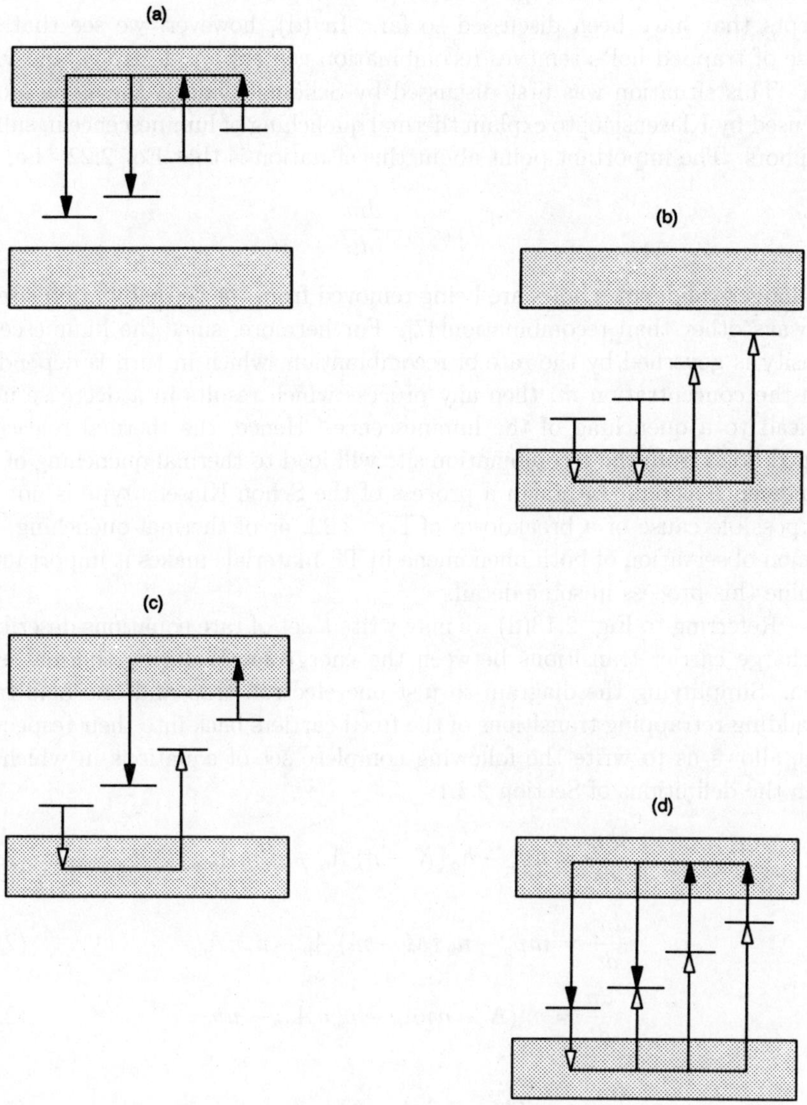

Figure 2.13 (a) Multiple electron traps; (b) multiple hole traps; (c) multiple electron and hole traps; (d) Schön-Klasens kinetics.

The situations depicted in Figs. 2.13(a)-(c) can each be treated using the same concepts that have been discussed so far. In (d), however, we see that the release of trapped holes removes recombination centers for electrons, and vice-versa. This situation was first discussed by Schön[45] and is the same model later used by Klasens[46] to explain thermal quenching of luminescence in sulfide phosphors. The important point about this situation is that Eq. 2.22 - i.e.

$$I_{TL} = -\frac{dm}{dt}$$

is no longer valid, since holes are being removed from the recombination site by pathways other than recombination[47]. Furthermore, since the luminescence intensity is governed by the rate of recombination, which in turn is dependent upon the concentration m, then any process which results in a decrease in m will lead to a quenching of the luminescence. Hence, the thermal release of trapped holes from the recombination site will lead to thermal quenching of the luminescence signal. Although a process of the Schön-Klasens type is not the only possible cause of a breakdown of Eq. 2.22, or of thermal quenching, the common observation of both phenomena in TL materials makes it important to examine this process in some detail.

Referring to Fig. 2.13(d) we may write a set of rate equations describing the charge carrier transitions between the energy levels during thermal excitation. Simplifying the diagram to just one electron trap and one hole trap, and adding retrapping transitions of the freed carriers back into their respective traps, allows us to write the following complete set of equations in which we retain the definitions of Section 2.3.1:

$$\frac{dn_c}{dt} = np_n - n_c\left(N-n\right)A_n - n_c m A_{mn} \tag{2.80}$$

$$\frac{dn_v}{dt} = mp_p - n_v\left(M-m\right)A_p - n_v n A_{np} \tag{2.81}$$

$$\frac{dn}{dt} = n_c\left(N-n\right)A_n - n_v n A_{np} - np_n \tag{2.82}$$

and

$$\frac{dm}{dt} = n_v\left(M-m\right)A_p - n_c m A_{mn} - mp_p \tag{2.83}$$

where $A_p = v_p\sigma_p$ and $A_{np} = v_p\sigma_{np}$. Here v_p is the free hole thermal velocity, and σ_p and σ_{np} are the hole trapping cross-section and the hole-electron recombination cross-section, respectively. M is the total concentration of available hole traps.

Eqs. 2.80-2.83 are the rate equations describing the traffic of charge between the two energy levels. In order to define the TL intensity we must stipulate which of the two recombination transitions is radiative. In the most general case, we may consider both to be radiative, and, with the assumption that all the photons are detected with equal efficiency, the total TL intensity is then given by

$$I_{TL} = I_{TLn} + I_{TLp} = n_c m A_{mn} + n_v n A_{np}. \tag{2.84}$$

Of course, it is possible that only one of these transitions may be radiative, or, if both are radiative, the emission wavelengths may be different such that only one is detected. The important point, however, is that, from Eqs. 2.82 and 2.83, the recombination center concentrations (m and/or n) are changing by mechanisms other than the process of recombination.

Bräunlich and Scharmann[48] considered these equations and introduced several simplifying assumptions in order to solve them. Following these authors let us define the following two parameters

$$R = A_n / A_{mn} \tag{2.85}$$

and

$$R^* = A_p / A_{np} \tag{2.86}$$

each of which expresses the ratio of the probabilities of retrapping, compared to recombination. Four cases can thus be defined: (i) $R << 1$ and $R^* << 1$; (ii) $R >> 1$ and $R^* << 1$; (iii) $R << 1$ and $R^* >> 1$, and (iv) $R >> 1$ and $R^* >> 1$. For each of these cases the assumptions of $n << N$ and $m << M$ are also made, while the neutrality condition

$$n_c + n = n_v + m \tag{2.87}$$

holds throughout. Finally, the QE assumption is made for both electrons and holes (i.e. dn_c/dt, $dn_v/dt << 1$). Using these assumptions and conditions, for case (i) we may write $n\, p_n \approx n_c m A_{mn}$ and $n_v \approx p_p / A_{np}$. Thus

$$\frac{dn}{dt} \approx -np_n - n_v n A_{np} \approx -(p_n + p_p)n. \tag{2.88}$$

Writing

$$p_n = s_n \exp\left\{-\frac{E_{tn}}{kT}\right\} \tag{2.89}$$

and

$$p_p = s_p \exp\left\{-\frac{E_{tp}}{kT}\right\} \tag{2.90}$$

we have

$$n \approx n_0 \exp\left[-\frac{1}{\beta}\int_{T_0}^{T}\left(s_n \exp\left\{-\frac{E_{tn}}{k\theta}\right\} + s_p \exp\left\{-\frac{E_{tp}}{k\theta}\right\}\right) d\theta\right]. \tag{2.91}$$

Thus we have, from Eq. 2.84

$$I_{TL} \approx n_0 \left(s_n \exp\{-E_{tn}/kT\} + s_p \exp\{-E_{tp}/kT\}\right) \times \exp\left[-(1/\beta) \int_{T_0}^{T} \left(s_n \exp\{-E_{tn}/k\theta\} + s_p \exp\{-E_{tp}/k\theta\}\right) d\theta\right] \tag{2.92}$$

which is the same form as the Randall-Wilkins equation for slow retrapping, namely Eq. 2.28 and becomes identical to the Randall-Wilkins equation if we ignore either the transitions to the conduction band (i.e. $p_n \approx 0$) or to the valence band (i.e. $p_p \approx 0$). Furthermore, if we allow only the free electron → trapped hole transition to be radiative, we should ignore the second term in the first bracket. Conversely, if we allow only the free hole → trapped electron transition to be radiative, we should ignore the first term in the first bracket.

For case (ii) we may write $n_c \approx np_n/A_nN$ and $n_v \approx p_p/A_{np}$. Thus

$$n \approx m \approx n_0 \exp\left[-\frac{1}{\beta} \int_{T_0}^{T} s_p \exp\left\{-\frac{E_{tp}}{k\theta}\right\} d\theta\right]. \tag{2.93}$$

The TL intensity is thus

$$I_{TL} = (n_0^2 R/N)\, s_n \exp\{-E_{tn}/kT\} \exp\left[-(2/\beta) \int_{T_0}^{T} s_p \exp\{-E_{tp}/k\theta\} d\theta\right] + n_0 s_p \exp\{-E_{tp}/kT\} \exp\left[-(1/\beta) \int_{T_0}^{T} s_p \exp\{-E_{tp}/k\theta\} d\theta\right] \tag{2.94}$$

which is reminiscent of the Randall-Wilkins form. It becomes the Randall-Wilkins equation exactly if we only consider free hole → trapped electron transitions, or if we ignore the thermal release of trapped electrons (i.e. $p_n = 0$).

Case (iii) is symmetrical with case (ii), and yields the equivalent equation

$$I_{TL} = m_0 s_n \exp\{-E_{tn}/kT\} \exp\left[-(1/\beta) \int_{T_0}^{T} s_n \exp\{-E_{tn}/k\theta\} d\theta\right] + (m_0^2 R^*/M)\, s_p \exp\{-E_{tp}/kT\} \exp\left[-(2/\beta) \int_{T_0}^{T} s_n \exp\{-E_{tn}/k\theta\} d\theta\right] \tag{2.95}$$

which also reduces exactly to the Randall-Wilkins equation if we consider only free electron → trapped hole transitions to be radiative, or if we ignore the thermal release of holes (i.e. $p_p = 0$).

For the remaining case (case (iv)) we have $n_c \approx np_n/A_nN$ and

$$\frac{dn}{dt} \approx -\frac{p_n R n^2}{N} - \frac{p_p R^* m^2}{M} \tag{2.96}$$

which gives

$$I_{TL} = \frac{(R/N)\, s_n \exp\{-E_{tn}/kT\} + (R^*/M)\, s_p \exp\{-E_{tp}/kT\}}{\left[\left(\frac{1}{n_0}\right) + \left(\frac{1}{\beta}\right) \int\limits_{T_0}^{T} \left(\left(\frac{R}{N}\right) s_n \exp\left\{-\frac{E_{tn}}{k\theta}\right\} + \left(\frac{R^*}{M}\right) s_p \exp\left\{-\frac{E_{tp}}{k\theta}\right\}\right) d\theta\right]^2}. \quad (2.97)$$

This is similar to the Garlick-Gibson second-order equation (Eq. 2.30) and becomes identical to it if we neglect all transitions to the valence band (i.e. $p_p = 0$). Note, however, that the Garlick-Gibson equation was obtained with $R = 1$.

Later analyses of the Schön-Klasens model[49] solved Eqs. 2.80 to 2.83 numerically without any of the assumptions of Bräunlich and Scharmann[48] and led to the same conclusions - namely that the TL glow curve for this more complicated kinetic model retains the simple Randall-Wilkins or Garlick-Gibson shape, depending upon the chosen values of the parameters; in particular, good fits between numerically generated data and the Randall-Wilkins and Garlick-Gibson equations (i.e. Eqs. 2.28 and 2.30) were obtained. The difficulty arises in the interpretation of the E_t and s values used in the Randall-Wilkins or Garlick-Gibson equations in order to obtain the fit. McKeever et al.[49] conclude that the E_t value used in the simple equations is equal to the smaller of E_{tn} and E_{tp} when both electron-hole (e-h) and hole-electron (h-e) recombination transitions are radiative. If e-h transitions only are radiative, then $E_t \simeq E_{tn}$, if $E_{tn} < E_{tp}$, or is somewhere between E_{tn} and E_{tp}, if $E_{tn} > E_{tp}$. The opposite situation is true for h-e radiative transitions. Example data are illustrated in Fig. 2.14.

2.4.5 Q- and P-Function Analysis

All of the preceding analyses have been centered on the use of the quasiequilibrium (QE) approximation, namely that the free carrier density remains quasistationary during the process of TL production. This assumption allows the decoupling of the rate equations and the development of all of the analytical solutions determined thus far. Maxia[50]-[52] examined the concept of QE by considering the TL emission to be the result of two non-equilibrium phase transformations - namely, the release of an electron into the conduction band, and the recombination of the freed electron with a trapped hole. By assuming a linear relationship between the rate of entropy production and the transformation velocities for each of these reactions, Maxia applied Gibbs' principle of minimum entropy production to arrive at the conclusion that the free carrier concentrations do indeed remain approximately steady throughout the TL process. However, the treatment is unsatisfactory in the sense that the minimum entropy production principle is in fact just another way of imposing the limitation of quasi-steady-state on the free carrier concentrations and begs the question of whether the principle is in fact valid during TL processes. Furthermore, the assumption of a linear relationship between the entropy production

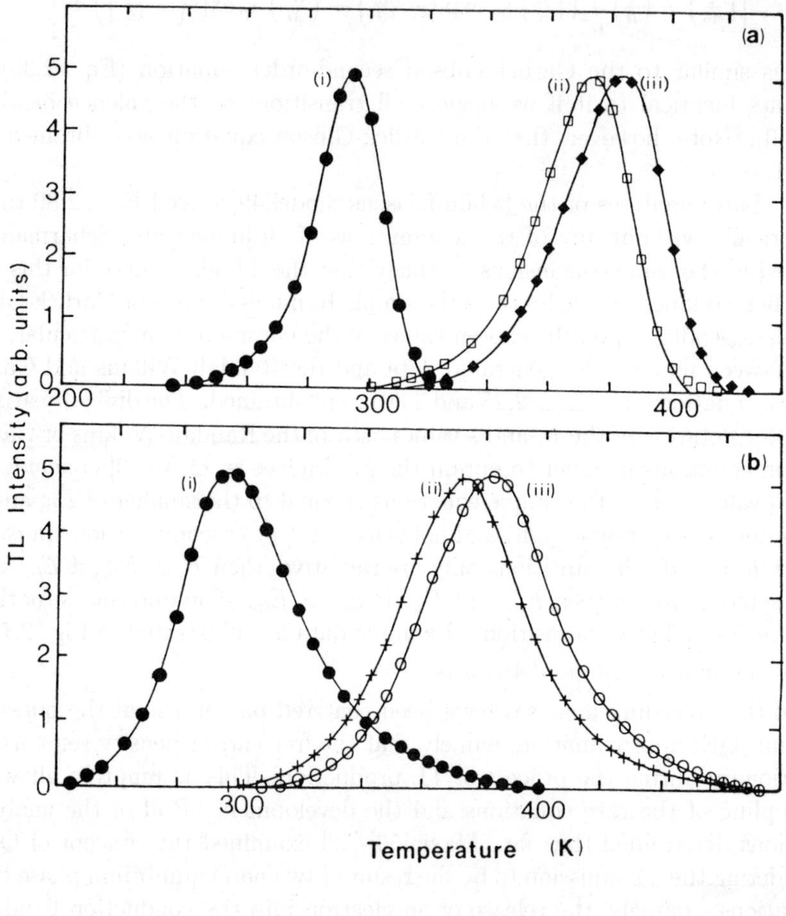

Figure 2.14 (a) First-order TL curves (data points) generated from Eqs. 2.80-2.83 with $E_{tn} = 0.9$ eV and E_{tp} equal to (i) 0.6 eV, (ii) 0.8 eV, and (iii) 1.0 eV. The other parameter values are $n_o = m_o = 10^{15}$ m^{-3}, $M = N = 10^{16}$ m^{-3}, $n_{c0} = n_{v0} = 0$, $s_n = s_p = 10^{10}$ s^{-1}, $R = R^* = 0.01$, and $\beta = 5$ Ks^{-1}. The lines are fits to the data using Eq. 2.28. (b) Same as (a), except $R = R^* = 1$. The kinetics are approximately second-order.

(After McKeever et al.[49].)

rate and the transformation velocities is also questionable, and is in fact invalid at the start of the reaction, only becoming reliable towards the end of the process (i.e. when the concentrations have been restored to their equilibrium values anyway).

Other efforts to test the validity of the quasiequilibrium approximation have centered on the use of numerical solutions to the rate equations for the various models. Shenker and Chen[53] conclude that the simplified expressions for TL arrived at using the QE approximation are valid solutions only if $|dn_c/dt| < 10^{-3}\,|dn/dt|$. They further show that, for the parameter sets examined by them, the ratio of $|dn_c/dt|$ to $|dn/dt|$ varies from $5{\times}10^{-5}$ to 1.0, with the higher values of the ratio appearing toward the end of the peak - in opposition to the conclusions of the minimum entropy production principle. An extensive study by Kelly et al.[54] supports the conclusions of Shenker and Chen, and concludes that the QE approximation is valid for only a part of the range of physically plausible parameters. Exact solutions to the rate equations lead to a wider variety of peaks shapes, intensities and positions than can be predicted from use of the approximate solutions. Thus, application of the simplified solutions is potentially undermined by uncertainty surrounding the QE approximation. Similar pessimism surrounds the application of the kinetic-order assumptions also since these too rely upon the assumption of inequalities between the various rate processes. Furthermore, a degree of "degeneracy" was found in that different sets of parameters inserted into the model were found to yield similar TL (and TSC) curves. In fact Kelly et al.[54] arrived at the altogether pessimistic conclusion that application of simplified phenomenological solutions to real systems is of limited value and is almost valueless at quantifying the trapping parameters (E_t, s, etc.; see also Bräunlich[11]).

An entirely different approach was adopted by Lewandowski and colleagues. Here the QE and kinetic order (KO) assumptions were abandoned and they were replaced by two physically meaningful functions - the Q and P functions, respectively[55]-[58]. Instead of the QE assumption (Eq. 2.23), we have the equality

$$-\frac{dn_c}{dt} = q\frac{dm}{dt} \tag{2.98}$$

or

$$Q\frac{dm}{dt} = \frac{dn}{dt} \tag{2.99}$$

where $Q = q + 1$. Q and q are actually functionals $Q\{T(t)\}$ and $q\{T(t)\}$ since they depend on all the other functionals in the system. As such they depend on the history of the system and are defined so that Eqs. 2.98 and 2.99 are exactly true at all temperatures. The value of Q is a measure of the degree to which QE is maintained, as a function of T. QE means that $q \approx 0$, or $Q \approx 1$.

We may also define a KO function, $P(T)$. In Sections 2.3.3 and 2.3.4 the distinction between first- and second-order kinetics was determined by the ratio $(N-n)\,\sigma_n/m\sigma_{mn}$. Thus we define $P(T)$

$$P(T)=\frac{(N-n)\,\sigma_n}{m\sigma_{mn}} \tag{2.100}$$

where we also emphasize the fact that the value of this ratio is a function of temperature. Slow-retrapping is then given by $P << 1$, while fast retrapping is given by $P >> 1$. The main advantage of defining the Q- and P-functions is to separate the concepts of QE and KO and to allow the degree of adherence to these restriction to vary with temperature. All analyses so far assume QE before the KO approximation is introduced, and require that the approximations be fixed for all temperatures.

If we define

$$\begin{aligned} R_{recom} &= \text{ rate of recombination} \\ &= -\frac{dm}{dt} = n_c v_n \sigma_{mn} m, \end{aligned} \tag{2.101}$$

$$\begin{aligned} R_{ex} &= \text{ rate of (thermal) excitation} \\ &= n s_n \exp\{-\frac{E_t}{kT}\}, \end{aligned} \tag{2.102}$$

and

$$\begin{aligned} R_{recap} &= \text{ rate of recapture} \\ &= n_c\,(N-n)\,v_n\sigma_n, \end{aligned} \tag{2.103}$$

then we may write Q and P in terms of these reaction rates, thus

$$Q=\frac{1}{R_{recom}}\,(R_{ex}-R_{recap}) \tag{2.104}$$

and

$$P=\frac{R_{recap}}{R_{recom}}. \tag{2.105}$$

We also note that

$$Q+P=\frac{R_{ex}}{R_{recom}} \tag{2.106}$$

and

$$\frac{Q}{P}=\frac{R_{ex}}{R_{recap}}-1. \tag{2.107}$$

The Q- and P-functions are illustrated graphically in Fig. 2.15. Note that Q varies from $>> 1$ at low T to ~ 0 at high T. Note also that $Q \approx 1$ at the peak maximum. (In fact, $Q = 1$ exactly at the maximum of the TSC peak; we

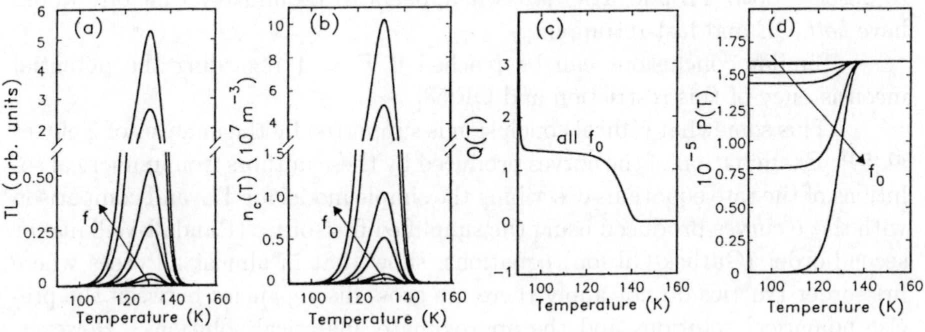

Figure 2.15 Curve shapes for (a) TL, (b) TSC (actually, n_c), (c) Q, and (d) P. In the figure f_o is the degree of trap filling, and values for f_0 range from 0.005 to 1.0. The curves have been obtained from numerical solution of the rate equations, with parameters chosen such that $P << 1$. Note that even though P varies dramatically with both T and f_0, the absolute value of P is always $<< 1$.
(After Lewandowski and McKeever[55].)

discuss in Ch. 9 how the TSC and TL maxima do not always coincide.) Finally we note that Q is independent of the level of trap filling, but depends only upon temperature. P on the other hand, increases as we progress through the peak, owing to the increased probability of retrapping as the trap empties. For the parameters chosen, however, $P << 1$ at all T.

Using these functions it is now possible to make some general statements regarding the various assumptions which are used when arriving at simplified solutions, each of which assume $Q \approx 1$.

(a) *Slow-Retrapping*: This requires that $Q \approx 1$ and $P << 1$. This in turn requires that $R_{recom} >> R_{recap}$ and $R_{ex} >> R_{recap}$. This constitutes a legitimate, self-consistent set of inequalities. Such relationships may exist over wide temperature ranges. It is not obvious from this consideration alone if the pessimism of Kelly et al.[54] is valid for this set of assumptions.

(b) *Fast-Retrapping:* This requires that $Q \approx 1$ and $P >> 1$. In other words, we require $R_{ex} \approx R_{recap}$ over all T. If this were strictly true (i.e. $R_{ex} = R_{recap}$) then the trap would not empty. Furthermore, from Eq. 2.104, in order to have $Q \approx 1$, we must have a very small value for R_{recom} suggesting a poor likelihood of observing a TL signal in the first place (and furthermore, the TSC signal which accompanies the TL peak would not be in the form of a peak at all, but would form a step with a long, slowly decaying tail[58]). Thus, in order to observe both TL and TSC *peaks* it appears to be unlikely that one would have *both* QE and fast-retrapping.

Similar conclusions can be reached if $P = 1$ regarding the potential inconsistency of this restriction and QE[58].

This somewhat radical conclusion is supported by the analysis of Kelly et al.[54]. Examination of the curves produced by these authors from numerical solutions of the rate equations describing the simple model for TL, and comparison with those curves produced using the simplified first-order (Randall-Wilkins) or second-order (Garlick-Gibson) equations, show that in almost all cases where first-order kinetics do not apply there are gross discrepancies between the precise numerical solutions and the approximate analytical solutions. However, much improved agreement is observed when slow-retrapping conditions prevail.

A question arises concerning the generality of the above results, however. Numerical solutions can only be obtained after selection of particular parameter sets (i.e. specific values for E_t, s, σ_n, σ_{mn}, N, n_0, m_0, etc.) and since there is an infinite number of such sets to choose from one inevitably questions the practice of arriving at general statements from a limited number of calculations. Lewandowski et al.[57],[58] attempted to generalize the discussion by solving the rate equations *without* any assumption regarding QE and KO. They started with a model of one trap, one recombination center and one deep, thermally disconnected trap. From the rate equations 2.17-2.20, with the addition of Eq. 2.37, one arrives at

$$n = n_0 \exp\left[-\frac{1}{\beta}\int_{T_0}^{T}\left(\frac{Q}{Q+P}\right) s \exp\left\{-\frac{E_t}{k\theta}\right\} d\theta\right] \tag{2.108}$$

from which one has for the TL intensity

$$I_{TL} = n_0\left(\frac{s}{Q+P}\right)\exp\left\{-\frac{E_t}{kT}\right\}\exp\left[-\frac{1}{\beta}\int_{T_0}^{T}\left(\frac{Q}{Q+P}\right) s \exp\left\{-\frac{E_t}{k\theta}\right\} d\theta\right]. \tag{2.109}$$

This is the *perfectly general* equation for TL - within the confines of the chosen model - since it assumes neither QE nor a particular KO. It is easy to see how it reverts to the Randall-Wilkins equation when we impose $Q \approx 1$ and $P << 1$. However, it is not obvious how the Garlick-Gibson form may be obtained if $Q \approx 1$ and $P >> 1$. Despite its generality, however, Eq. 2.109 is of little practical use since analytical expressions for Q and P, in terms of macroscopic, measurable parameters, do not exist. These functions are defined only in terms of microscopic parameters, the values and temperature dependencies of which are unknown. What is required is a truly analytical solution to the rate equations in which values for Q and P are not predetermined, and which can be expressed purely in terms of measurable parameters. Unfortunately, at this time, this has not been achieved for the perfectly general case - but it has been performed for the special case of slow-retrapping, i.e. $P << 1$[57],[58]. To see how this is done, we first of all set $P << 1$ in Eq. 2.109. Since for most temperatures $Q >> P$, we may rewrite Eq. 2.109

$$I_{TL} = \frac{n_0 s}{Q} \exp\left\{-\frac{E_t}{kT}\right\} \exp\left[-\frac{s}{\beta}\int_{T_0}^{T} \exp\left\{-\frac{E_t}{k\theta}\right\} d\theta\right]. \tag{2.110}$$

If we represent Eq. 2.28 as I_{TL}^{QE}, then we see that

$$I_{TL}^{QE} = QI_{TL}. \tag{2.111}$$

To develop the analysis we make use of the free carrier recombination lifetime τ_n defined as the lifetime of a free electron before it undergoes recombination. τ_n is given by

$$\tau_n^{-1} = mA_{mn} = (n+h)\,A_{mn} \tag{2.112}$$

where we have made use of Eq. 2.36 which is appropriate for the chosen model - i.e. one which includes a deep, thermally disconnected trap. If we further simplify the analysis by stipulating that $h >> n$, then τ_n can be considered as a constant during the thermal emptying of the TL trap. Thus, from the definition of Q in Eq. 2.99 we may write

$$Q = \frac{\beta\tau_n}{I_{TL}}\frac{dI_{TL}}{dT} + 1. \tag{2.113}$$

It is clear from Eq. 2.113 that QE is only exact at the TL peak maximum (when $dI_{TL}/dT = 0$) but also this expression points to a potential experimental method to discover the shape of the Q-function. However, the addition of unity poses difficulties since the absolute values of τ_n and I_{TL} are both unknown. In principle, one could determine Q by recalling that $Q = 1$ at the peak maximum and $Q = 0$ at high temperatures (i.e. at temperatures beyond the peak - see Fig. 2.15). However, an alternative, analytical approach may be devised by combining Eqs. 2.111 and 2.113 to give

$$I_{TL}^{QE} = \beta\tau_n\frac{dI_{TL}}{dT} + I_{TL} \tag{2.114}$$

and by recognizing that this is a first-order, linear, differential equation with the solution

$$I_{TL} = \frac{1}{\beta\tau_n}\exp\left\{-\frac{T}{\beta\tau_n}\right\}\int_{T_0}^{T}\exp\left\{\frac{\theta}{\beta\tau_n}\right\} I_{TL}^{QE}\,d\theta. \tag{2.115}$$

With I_{TL}^{QE} being given by the Randall-Wilkins equation (Eq. 2.28) we have finally

$$\begin{aligned} I_{TL} = (n_0/\beta\tau_n) \exp\{-(T/\beta\tau_n)\} \int_{T_0}^{T} \exp\{(\theta/\beta\tau_n)\} s \exp\{-E_t/k\theta\} \\ \times \exp\left[-(1/\beta)\int_{T_0}^{\theta} s \exp\{-E_t/k\Omega\}\, d\Omega\right] d\theta. \end{aligned} \tag{2.116}$$

where Ω is a dummy variable representing temperature. The approximations used in the development of this equation are (i) the model (one trap, one center, one deep trap), (ii) $P << Q$, and (iii) τ_n = constant (i.e. $h >> n$). The last restriction can be removed by placing all the τ_n terms inside the integrals, but the form of the temperature function would need to be known. Lewandowski et al.[58] suggest that this can be overcome by parameterizing the lifetime, thus $\tau_n = AT^{\alpha}$, where A and α are constants. $P << Q$ is an entirely reasonable assumption for slow-retrapping, and the model will always be an approximation, no matter which model is adopted.

Although the above equation is complex, it represents more accurately the true shape of a slow-retrapping TL peak than does the simple Randall-Wilkins form. Lewandowski et al.[58] tested the validity of Eq. 2.116 by comparing TL curve shapes produced using it, with those produced using the Randall-Wilkins equation, and with those produced using exact numerical solutions of the rate equations. Some example results are shown in Fig. 2.16. As observed in the figure, for certain parameter sets, all three solutions agree, while for other sets some slight discrepancies exist between the exact solutions and the Randall-Wilkins solutions. In all cases, however, Eq. 2.116 gives the same solution as the numerical analysis. However, it is interesting to note that fitting the Randall-Wilkins equation to the numerical data always results in an excellent fit[58], with associated errors in E_t only of the order of 1 or 2% in most cases. Thus, the reservations of Kelly et al.[54] for the case of slow-retrapping do not appear to be serious.

The above analysis is for slow-retrapping processes only (i.e. $P << Q$). Equivalent expressions to Eq. 2.116 for other situations (e.g. $P \approx Q$ or $P >> Q$) have not been developed, which is unfortunate since it is here that one suspects the arguments expressed by Kelly et al. may have greater validity. Furthermore, generalization of the Q-function to the case of multiple traps and centers, with interactive kinetics and multiple, overlapping TL peaks has not yet been forthcoming.

2.4.6 Tunneling

In the models discussed so far, the recombination of electrons from traps with holes in centers has occurred via the conduction band. In this way substantial diffusion of the electron through the crystal can occur with the recombination

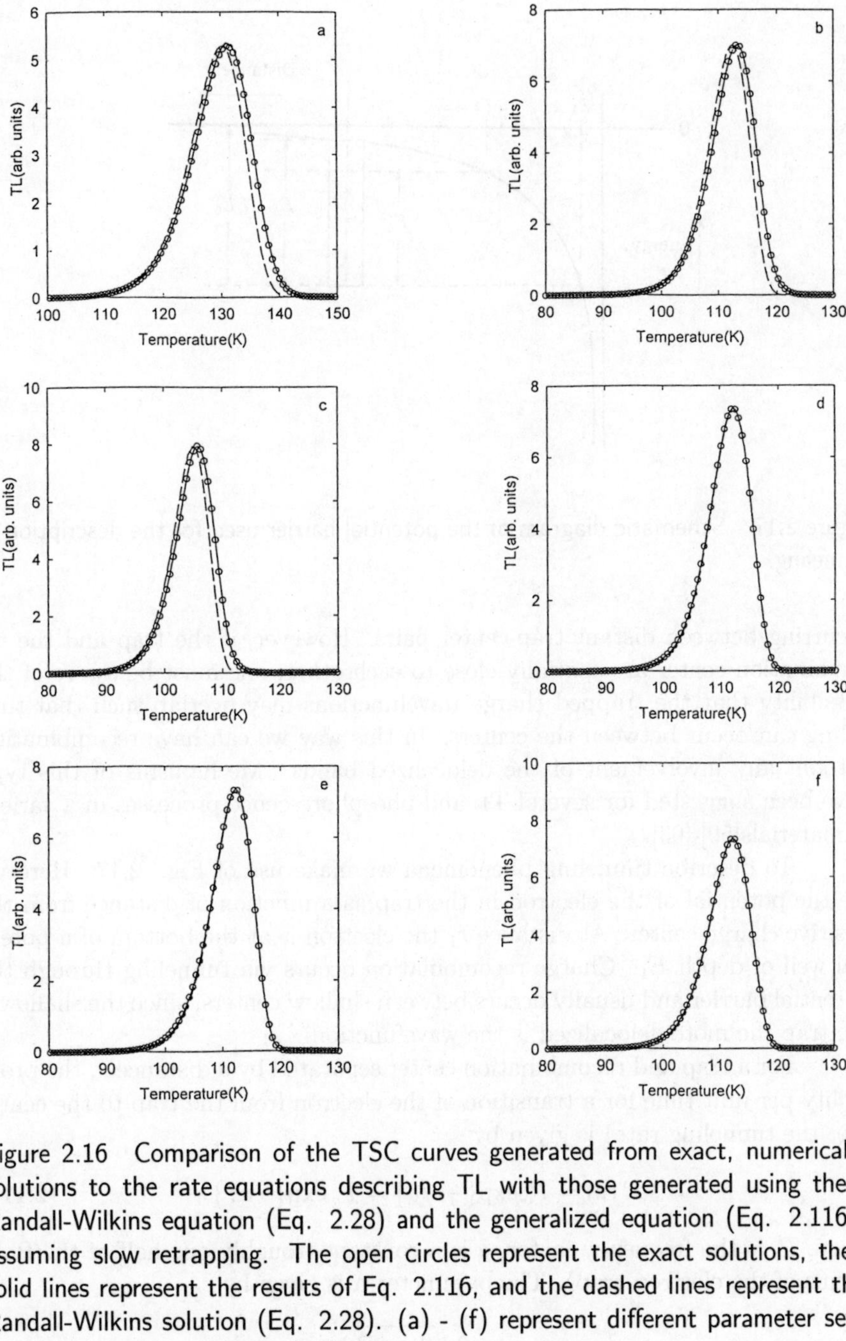

Figure 2.16 Comparison of the TSC curves generated from exact, numerical solutions to the rate equations describing TL with those generated using the Randall-Wilkins equation (Eq. 2.28) and the generalized equation (Eq. 2.116) assuming slow retrapping. The open circles represent the exact solutions, the solid lines represent the results of Eq. 2.116, and the dashed lines represent the Randall-Wilkins solution (Eq. 2.28). (a) - (f) represent different parameter sets used to obtain the numerical solutions.

(After Lewandowski et al.[58].)

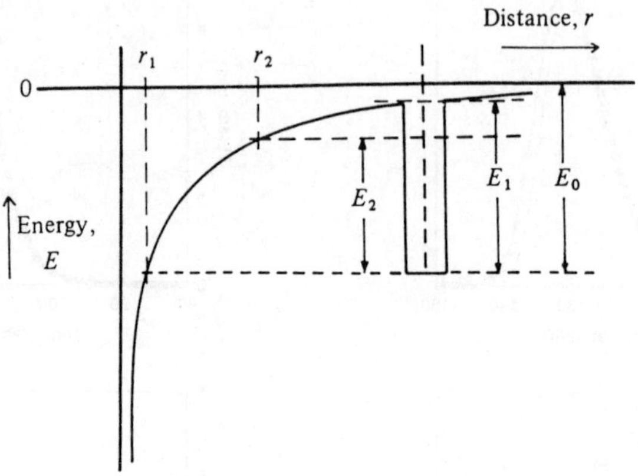

Figure 2.17 Schematic diagram of the potential barrier used for the description of tunneling.

occurring between distant trap-center pairs. However, if the trap and the recombination center are spatially close to each other, one must be aware of the possibility that the trapped charge wavefunctions may overlap such that tunneling can occur between the centers. In this way we can have recombination without any involvement of the delocalized bands. Mechanisms of this type have been suggested for several TL and phosphorescence processes in a variety of materials[59]-[63].

To describe tunneling phenomena we make use of Fig. 2.17. Here we see the potential of the electron in the trap as a function of distance from the positive charge center. At distance r_1 the electron is at the bottom of a potential well of depth E_1. Charge recombination occurs via tunneling through the potential barrier and usually occurs between shallow centers, since the shallower the trap the more delocalized is the wavefunction.

For a trap and recombination center separated by a distance r, the probability per unit time for a transition of the electron from the trap to the center (i.e. the tunneling rate) is given by

$$P(r) = P_o \exp\{-r/a\} = P_o \exp\{-\alpha r\} \tag{2.117}$$

where P_o is the frequency factor, a is a constant (roughly one half of the Bohr radius of the electron trap). The parameter α is given by

$$\alpha = \frac{2\sqrt{(2m^* E_1)}}{\hbar}. \tag{2.118}$$

Here m^* is the electron effective mass, $\hbar = h/2\pi$ where h is Planck's constant.

For randomly distributed traps and recombination sites the probability that the separation between them is equal to r is[66]

$$S(r) = 4\pi n_0 r^2 \exp\left\{-4\pi n_0 r^3/3\right\} \tag{2.119}$$

where $n_0 = m_0$ is the number of electrons in traps or holes in recombination centers at time $t = 0$. If $n_0 r^3 << 1$, we can ignore separations that are not nearest neighbor and write that the concentration of nearest-neighbor pairs having a separation between r and $r + \Delta r$ is

$$dn(r) = n_0^2 4\pi r^2 dr. \tag{2.120}$$

Thus, this concentration of trap-center pairs will decay with rate $P(r)$ given in Eq. 2.117. From this we have the number of pairs surviving after time t

$$dn(r,t) = n_0^2 \exp\{-P(r)t\} 4\pi r^2 dr. \tag{2.121}$$

This describes a fairly sharp step after long times at a distance r_c, defined by $P(r_c) = 1$.

Among the consequences of tunneling between traps and centers is that the number of trapped electrons diminishes with time after the irradiation (perturbation) of the samples has ceased. This gives rise to a "fading" of the subsequent TL signal - i.e. the longer the time interval between irradiation and heating to record the TL, the weaker the remaining TL signal. One should note that this type of fading is athermal - i.e. it is independent of temperature. It depends primarily upon the concentrations of trapped electrons and holes (i.e. n_0 and m_0) which in turn depends upon the amount of radiation absorbed by the sample. However, thermally-assisted tunneling can also occur if there exists a state within the potential well at an energy E_2 (Fig. 2.17). Here the barrier width is reduced to r_2 and so the probability of tunneling recombination is enhanced. We discuss more on fading in Ch. 10.

As discussed by Templer[64], a distribution of r values with the same E_1 values will give rise to a distribution of $E_o = E_t$ values (see Fig. 2.17). The effect of a distribution $n(E)$ upon the TL glow curve has been discussed in Section 2.4.3. Suffice it to remark here that the above-mentioned fading of the TL signal will occur from the *highest* E_t ($= E_o$) first, since these correspond to the lowest values of r.

2.4.7 Localized Transitions

Another process by which recombination can take place without a transition of the electron into the conduction band is depicted in Fig. 2.18. Here the electron is thermally stimulated into an excited state from which a transition into the

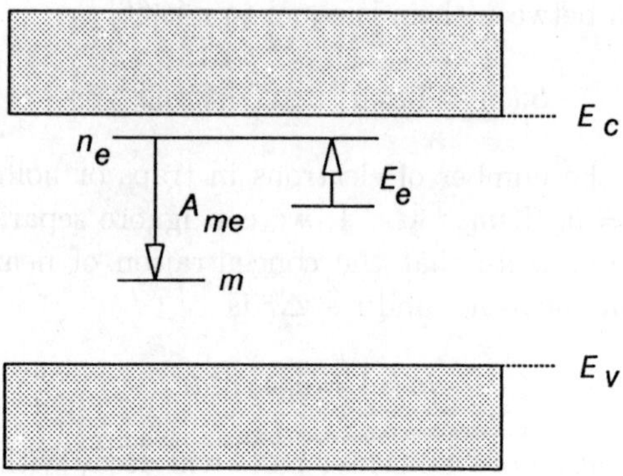

Figure 2.18 Localized transition scheme for recombination, involving n trapped electrons (per unit volume), m trapped holes and n_e electrons in an excited state.

recombination center is allowed. Halperin and Braner[13] gave the following rate equations to describe this process

$$I_{TL} = -\frac{dm}{dt} = n_e m A_{me} \tag{2.122}$$

$$\frac{dn}{dt} = sn_e - s\exp\left\{-\frac{E_e}{kT}\right\} \tag{2.123}$$

and

$$\frac{dn_e}{dt} = sn\exp\left\{-\frac{E_e}{kT}\right\} - n_e\left(mA_{me} - s\right). \tag{2.124}$$

Here, n_e is the concentration of electrons in the excited state, A_{me} is the transition probability for the recombination of an electron from the excited state into the recombination center, and E_e is the energy difference between the trap and the excited state.

In the above equations sn_e appears rather than the usual term of the form $n_e(N-n)A_e$ which might have been expected (where here we define A_e to be the probability of retrapping from the excited state into the ground state). However, since the excited state is not delocalized, and the excited state electrons do not move through the lattice, the total number of available empty ground states - i.e. $(N-n)$ - is irrelevant since retrapping can only occur into the ground state from which the electron was excited. Detailed balance arguments[13] reveal that the rate of recapture is, in fact, given by s.

By similar argument we see that, since the excited state is localized, a transition into the recombination site can only occur if the trap and recombination center are spatially localized as a nearest-neighbor pair - analogous to the above discussion regarding tunneling. To account for this Chen[65] replaced the term mA_{me} in equations 2.122 and 2.124 with a *constant* probability ν (in units of s^{-1}). With the additional assumption that the population of the excited states is always considerably less than that of the ground state - i.e. $m = n + n_e \approx n$, we have

$$I_{TL} = -\frac{dn}{dt} = -\frac{dm}{dt} = n_e \nu \tag{2.125}$$

where quasiequilibrium ($dn_e/dt \approx 0$) is implied. The above set of equations produces

$$I_{TL} = \left[\frac{\nu s}{\nu + s}\right] m \exp\left\{-\frac{E_e}{kT}\right\} = \overline{s} m \exp\left\{-\frac{E_e}{kT}\right\}. \tag{2.126}$$

Upon integration, with the assumption that $\nu >> s$, Eq. 2.126 produces an expression of the same form as the Randall-Wilkins first-order TL equation (Eq. 2.28) with a frequency factor $\overline{s} \approx s$. However, if the recombination rate is very low ($\nu << s$) then we have

$$I_{TL} \approx m\nu \exp\left\{-\frac{E_e}{kT}\right\} \tag{2.127}$$

and integration of this equation produces an expression which is also of the same form as Eq. 2.28, but with ν replacing s. Thus we arrive at the conclusion that localized transitions of the type described should always be governed by first-order kinetics. Note that the value of ν can be small, such that the typical TL peak seen in this case is very broad. Analysis of the peak using any of the methods to be described in Ch. 3 will produce a small value for the frequency factor (perhaps as low as (say) 10^5 s^{-1}, compared to the more usual 10^{12}-10^{14} s^{-1}).

It should be recognized that the assumption of only nearest-neighbor transitions during recombination is itself an approximation which may not be entirely correct. Overlap of wavefunctions between, for example, next-nearest-neighbors may also be possible, with a correspondingly reduced probability, ν. A curious prediction arising from this consideration is that under such circumstances we may expect to have a second (and possibly, third, etc.) TL peak at higher temperatures.

We should also not forget that the above description uses quasiequilibrium ($dn_e/dt \approx 0$) and thus the questions raised in Section 2.4.5 regarding the compatibility of this approximation with the assumption of fast retrapping ($\nu \ll s$ in this case) must be raised again.

During heating in a TL experiment, for the fast retrapping case, the combination of a small frequency factor and a value of $E_e < E_t$ (see Fig. 2.18) means that the lifetime of the electrons in the trap can be much reduced over what is expected from a trap with a depth below the conduction band of E_t. For example, consider a trap with E_t = 1.0 eV and $s = 10^{12}$ s^{-1}, but with E_e = 0.5 eV and $\nu = 10^5$ s^{-1}. Using the standard TL equations and assuming recombination via the conduction band - i.e. using the above values for E_t and s - one can predict that a TL peak should appear at approximately 120 °C (for a heating rate of 1.0 °Cs^{-1}). However, if one considers recombination via the excited state - i.e. using E_e and ν - one would predict a TL peak at $\sim$ 160 °C. Thus, recombination via the conduction band will occur first when the sample is heated at this rate. However, the lifetime of the electrons in the trap at room temperature, using $\tau = p^{-1} = s^{-1}\exp\{E/kT\}$ is approximately 17 hours if we use $E = E_t$ = 1.0 eV and $s = 10^{12}$ s^{-1}, but only $\sim$ 0.5 hours if we use $E = E_e$ = 0.5 eV and $s = 10^5$ s^{-1}. Thus the electrons leak away from the trap via the excited state and, as a result, the TL peak at 120 °C will be smaller the longer one waits between irradiation and TL readout. The key issue here is the heating rate. The fact that recombination via the excited state can be more probable than via the conduction band if the sample is held at a constant temperature, but less probable if heated at a few degrees per second is related to the extreme dependence of the excited state recombination route on heating rate. This in turn is a result of the very low value for the frequency factor under conditions of fast retrapping. Templer[64],[67] used this to discuss the "anomalous fading" characteristics of several materials. In Ch. 10 we discuss this type of fading, and lifetimes and fading in general, in greater detail, while in Ch. 3 we discuss more fully the behavior of TL peaks as a function of heating rate.

2.5 Emission Spectra

2.5.1 TL-versus-Temperature-versus-Wavelength

Conventional experimental measurements of TL glow curves record the luminescence emission using a broad band detector such as a photomultiplier tube, with the TL emission either unfiltered or after it has passed through a broad band filter. The resulting emission intensity is then displayed in a glow curve as a function of temperature. In this way the different traps active in the material are monitored in a temperature-dispersive manner, with each trap giving rise to its own TL peak. However, what is not apparent in such a measurement is the wavelength of the emitted light. Whereas temperature resolution can separate the different detrapping process occurring in the material, wavelength measurement can resolve the different recombination processes which are occurring. The equations developed in the preceding sections simply talk about recombination, without discussing the possibility of several recombination sites existing within the material. One can argue, however, that the important point in these mathematical analyses is the *net* recombination rate (compared with the retrapping rate) at any given temperature; whether this recombination is occurring at one site, or at multiple sites, is immaterial. This can be understood from the rate equations for the case of a single electron trap, with a trapped electron concentration of n, and two recombination centers, with concentrations m_1 and m_2 and recombination probabilities A_{mn1} and A_{mn2}, respectively. Of course, $n = m_1 + m_2$. Here we simply replace the recombination rate mA_{mn} in equations 2.17 and 2.20 with $(m_1 A_{mn1} + m_2 A_{mn2})$ and this leads to the equivalent GOT equation

$$I_{TL} = \frac{ns\exp\{-E_t/kT\}\,(m_1\sigma_{mn1} + m_2\sigma_{mn2})}{[(N-n)\,\sigma_n + (m_1\sigma_{mn1} + m_2\sigma_{mn2})]} \tag{2.128}$$

where σ_{mn1} and σ_{mn2} are the respective capture cross-sections. One then simply assumes either of the inequalities $(m_1\sigma_{mn1} + m_2\sigma_{mn2}) >> (N-n)\sigma_n$, or $(m_1\sigma_{mn1} + m_2\sigma_{mn2}) << (N-n)\sigma_n$ and proceeds as before to develop expressions identical to the first- or second-order equations. These latter equations (Eqs. 2.28 and 2.30) do not contain the terms m or A_{mn} and thus the peak position and shape should be independent of the number of recombination processes occurring.

Nevertheless, one can certainly imagine that if, say, $m_1\sigma_{mn1} > m_2\sigma_{mn2}$, then recombination center #1 (emitting light at wavelength λ_1) should capture electrons more efficiently to begin with than center #2 (emitting at wavelength λ_2). The latter should only become involved as m_1 is depleted. In this way, one might expect that the wavelength of the TL emission will change as the trap empties. In the hypothesized case, TL of wavelength λ_1 would be expected to appear at a lower temperature than the TL emitting at λ_2. In principle, if one had a detector that was equally efficient at detecting photons of wavelength λ_1 as it was of detecting photons of λ_2 (which in fact is implicit in all of the equations), then it would not matter what the wavelength was and a perfectly normal TL peak will result. However, in reality, photomultiplier and other detectors have a wavelength dependent efficiency such that distortions will be introduced into the recorded signal as the emission wavelength changes during the course of the peak. On the other hand, if, instead of detecting the panchromatic emission from the sample we only detect monochromatic emission - either at wavelength λ_1 or at λ_2 - then the recorded glow peak will still be distorted. For example, if the glow curve is measured at wavelength λ_1, then the detection efficiency will decrease as the emission changes as the peak is readout. The distortion will appear on the high temperature side of the peak as this recombination pathway becomes less efficient. Conversely, distortions will appear on the low temperature side of the TL peak if it is recorded at λ_2 as this process becomes more favored. In either case, distortions will occur resulting in a narrowed and shifted peak.

Experimental observations of this type are infrequent. An early example is given by Bonfiglioli et al.[68], but a particularly clear example was described by Hornyak et al.[39]. These authors observed "green" and "UV" emissions from a TL peak in natural quartz. They interpreted the results on the basis of a model with "one trap" and two recombination centers, although in their case they consider a distribution of trapping states. If recorded monochromatically using filters centered in either the green or the UV the TL curves were distorted with *apparent* peaks at different temperatures, $\sim 20\,^{\circ}$C apart. Hornyak et al.[39] note that in a model of this type the two TL emissions I_{TL1} and I_{TL2} at, say, λ_1 and λ_2, respectively, are related to the rates of recombination by

$$I_{TL1} = -\frac{dm_1}{dt} = n_c m_1 A_{mn1} \tag{2.129}$$

and

$$I_{TL2} = -\frac{dm_2}{dt} = n_c m_2 A_{mn2} \tag{2.130}$$

from which it follows that

$$\frac{I_{TL1}}{I_{TL2}} = \frac{dm_1/dt}{dm_2/dt} = \frac{A_{mn1} m_{10}}{A_{mn2} m_{20}} \exp\left\{ (A_{mn1} - A_{mn2}) \int_0^t n_c dt \right\} \tag{2.131}$$

where m_{10} and m_{20} are the initial trapped hole concentrations at the two recombination centers at $t = 0$. Thus, unless $A_{mn1} = A_{mn2}$ the ratio of the TL emission from the two centers will vary with time and temperature in a complex manner. We should note that this behavior is considerably different from (and more complicated than) what we would get if we chose to modify the original TL equations with a simple branching ratio for the two transitions. A simple analysis of this type is not valid.

We can also see from Eqs. 2.129 and 2.130 that

$$\frac{A_{mn1}}{A_{mn2}} = \frac{\sigma_{mn1}}{\sigma_{mn2}} = \frac{1}{m_1}\frac{dm_1}{dt} / \frac{1}{m_2}\frac{dm_2}{dt}. \tag{2.132}$$

This can be related to the integrated TL emissions at the two wavelengths since, in general, $m = \int_0^\infty I_{TL} dt$ (the "light sum"). Applying Eq. 2.132 to the TL data from quartz, Hornyak et al.[39] produced the data of Fig. 2.19. In this figure curve (a) was obtained from Eq. 2.132 assuming that the light sums at any instant were equal to m_1 and m_2. However, as can be seen, this assumption

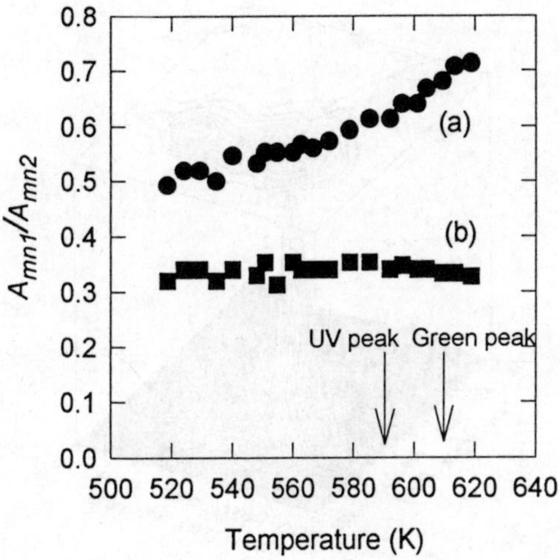

Figure 2.19 The ratio of A_{mn1} to A_{mn2} versus temperature for the green and UV TL emission from quartz. For explanations of curves (a) and (b) see text. (Replotted from Hornyak et al.[39].)

produces a temperature-dependent ratio of A_{mn1}/A_{mn2} - contrary to the initial assumptions of constant recombination cross-sections. However, by introducing a deep, thermally disconnected electron trap into the model, one can now assume that a fraction of the initial trapped hole concentration remains constant during the emission. In fact, by assuming that 60% of m_{10} remains constant, Hornyak et al. produced curve (b) in the figure. This reinforces the view that analysis of real materials cannot ignore the possibility of concentrations of deep, thermally disconnected traps.

This description of multiple emission wavelengths during the production of TL should not be confused with another mechanism by which more than one wavelength may be emitted in a single TL peak. If the recombination process leads to the excitation of an impurity ion (for example a rare earth ion) one can get multiple emission wavelengths due to the de-excitation of the impurity from its various excited states. For example, in, say, CaF_2:Dy, during TL one can get the excitation of Dy^{3+} ions leading to emission at 480 nm ($^4F_{9/2} - ^6H_{15/2}$ transitions), 575 nm emission ($^4F_{9/2} - ^6H_{13/2}$ transitions), 660 nm emission ($^4F_{9/2} - ^6H_{11/2}$ transitions), and 750 nm ($^4F_{9/2} - ^6H_{9/2}$ transitions) - all occurring at the same temperature. Nevertheless, each of these emissions has been excited by the *same* recombination process, and not different processes as described in the above example. Thus, only one mA_{mn} term applies. Similarly, one might get partial re-absorption of the primary emission by another impurity leading to excitation of the second impurity and multiple emission wavelengths

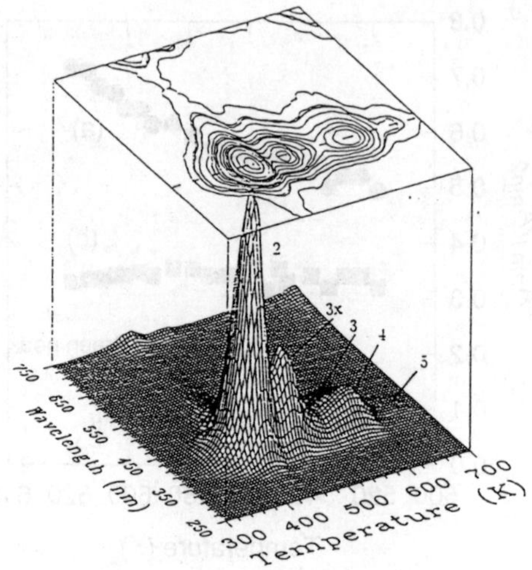

Figure 2.20 TL intensity-versus-wavelength-versus-temperature for x-irradiated Lu_2SiO_5:Ce. The heating rate was 1 Ks^{-1}.
(After Dorenbos et al.[75].)

(e.g. CaF_2:Mn,Ce[69]). Again, only one nA_{mn} term applies.

The only way to discern if one is dealing with situations such as those described is to record a TL emission spectrum as a function of temperature. The need for this type of data for kinetic studies was recognized early on by Levy and colleagues[70], but advances with instrumentation were required before the potential for routine emission spectral analysis became a real possibility[71]. Sensitive spectrometers capable of recording spectral data corrected for the response of the detection system are now proving to be popular additions to many TL laboratories[72]-[74]. An example data set is shown in Fig. 2.20. Data of this type immediately show if the emission spectrum is varying across the TL peak, or if it is constant. In the contour map illustrated in the figure one can see that, in this example, each TL peak has exactly the same emission band and that there appears to be no variation in the emission wavelength with temperature.

To obtain data for kinetic analysis one would sum each wavelength "slice" to produce the total emission as a function of temperature. For data such as that in Fig. 2.20 each "slice" would yield an identically shaped glow curve (differing only in the height of the peaks). For other materials, however, the wavelength does change with temperature and each slice will produce a slightly different glow peak. This is demonstrated using computer simulated data in Fig. 2.21. These data have been produced using a first-order TL peak (Eq. 2.28), with $E_t = 1.0$ eV and $s = 1 \times 10^{12}$ s^{-1}, and convoluting it with two

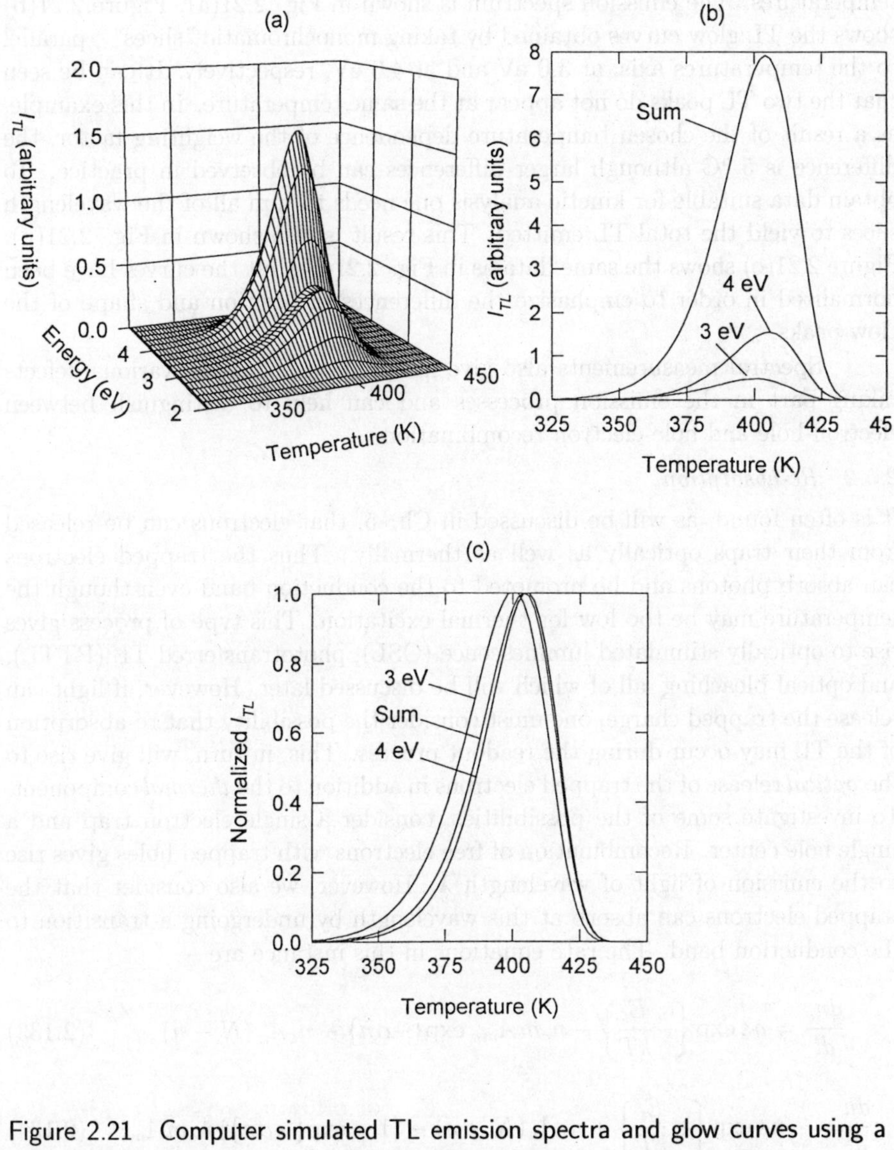

Figure 2.21 Computer simulated TL emission spectra and glow curves using a linear heating rate of 1 Ks^{-1}. For an explanation of the figures see text.

Gaussian emission bands - one centered at 3.0 eV and the other at 4.0 eV. The emission bands have been weighted so that the 3.0 eV band is more favored at low temperatures whereas the 4.0 eV band becomes more favored at higher temperatures. The emission spectrum is shown in Fig. 2.21(a). Figure 2.21(b) shows the TL glow curves obtained by taking monochromatic "slices" , parallel to the temperatures axis, at 3.0 eV and at 4.0 eV, respectively. It can be seen that the two TL peaks do not appear at the same temperature. In this example, as a result of the chosen temperature dependence of the weighting factor, the difference is 5 °C although larger differences can be observed in practice. To obtain data suitable for kinetic analysis one needs to sum all of the wavelength slices to yield the total TL emitted. This result is also shown in Fig. 2.21(b). Figure 2.21(c) shows the same data as in Fig. 2.21(b), but the curves have been normalized in order to emphasize the differences in position and shape of the glow peaks.

Spectral measurements also give information about the various defects taking part in the emission processes and can help to distinguish between electron-hole and hole-electron recombination.

2.5.2 Re-absorption

It is often found, as will be discussed in Ch. 5, that electrons can be released from their traps optically as well as thermally. Thus the trapped electrons can absorb photons and be promoted to the conduction band even though the temperature may be too low for thermal excitation. This type of process gives rise to optically stimulated luminescence (OSL), phototransferred TL (PTTL), and optical bleaching, all of which will be discussed later. However, if light can release the trapped charge, one must consider the possibility that re-absorption of the TL may occur during the readout process. This, in turn, will give rise to the *optical* release of the trapped electrons in addition to the *thermal* component. To investigate some of the possibilities, consider a single electron trap and a single hole center. Recombination of free electrons with trapped holes gives rise to the emission of light of wavelength λ. However, we also consider that the trapped electrons can absorb at this wavelength by undergoing a transition to the conduction band. The rate equations in this instance are

$$\frac{dn_c}{dt} = ns\exp\left\{-\frac{E_t}{kT}\right\} - n_c m A_{mn}\exp\{-\alpha n\} - n_c A_n\left(N-n\right), \qquad (2.133)$$

$$\frac{dn}{dt} = -ns\exp\left\{-\frac{E_t}{kT}\right\} + n_c A_n(N-n) - \left(1-\exp\{-\alpha n\}\right) n_c m A_m, \qquad (2.134)$$

and

$$\frac{dm}{dt} = -n_c m A_{mn} \qquad (2.135)$$

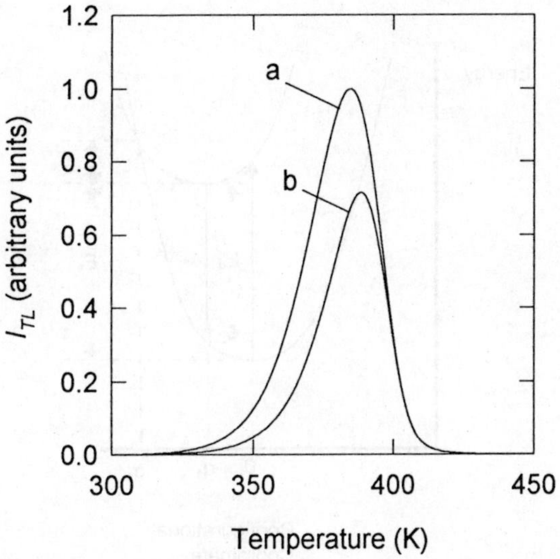

Figure 2.22 TL glow peaks obtained for $E_t = 1.0$ eV and $s = 10^{12}$ s^{-1} assuming (a) no self-absorption, and (b) self-absorption.

where α is the absorption probability of photons by the trapped electrons and has units of m^{-3}. In this case the measured TL intensity is given by that proportion of the photons generated during recombination which is not reabsorbed by the trapped electrons, i.e.

$$I_{TL} = n_c m A_{mn} - (1 - \exp\{-\alpha n\})\, n_c m A_{mn} = n_c m A_{mn} \exp\{-\alpha n\}. \quad (2.136)$$

One can see that the usual expression for the TL intensity (i.e. Eq. 2.135) is in this case modified by the term $\exp\{-\alpha n\}$. This modifier will be most prominent at the start of the TL peak when n is large. To illustrate the effect, numerical solutions to the above equations are given in Fig. 2.22 for the cases where $\alpha = 0$ and $\alpha = 1 \times 10^{-17}$ m^3. As expected, the two solutions diverge at lower temperatures with less light being emitted when self-absorption is strong. The solutions become the same on the high temperature side of the peak at which point self-absorption of the TL light, and the exponential term in Eq. 2.136, become less important.

In real materials one can imagine other possible situations - for example, the TL may be reabsorbed by deeper, thermally disconnected traps. In this case, the re-absorption will not only result in a reduced TL output but may also optically transfer electrons into the lower temperature, unstable traps. Clearly, one can imagine a wide variety of possible scenarios of this type. It is clear from

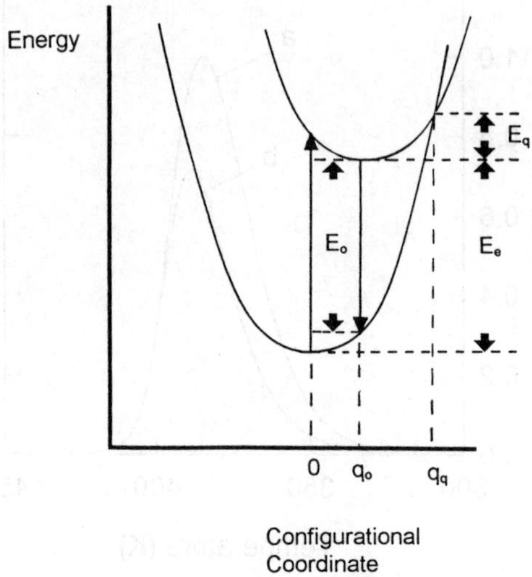

Figure 2.23 Configurational coordinate diagram of a luminescent center with potential energy curves for the excited and ground states as a function of the configurational coordinate.

these comments, however - as exemplified in Fig. 2.22 - that the result of self-absorption is to produce a distorted TL peak compared to that which would be expected if self-absorption is ignored. No studies have yet been carried out to examine the errors induced in the calculation of the trapping parameters from a TL peak under these circumstances.

2.5.3 Thermal Quenching

The emission of luminescence during TL occurs when electrons recombine with trapped holes (or vice-versa) leaving the site in an excited state. This is then followed by a relaxation to the ground state. Our discussions to this point have dealt only with the first part of this process which we have described using the energy band picture of a solid. The description of the second, light emitting, part of the process is accomplished using a configurational coordinate diagram, as illustrated in Fig. 2.23. Here the minima of the excited and ground states are displaced from each other by a distance q_o. The difference in the energies of these minima is the "excitation energy" E_e. A luminescence transition from the excited to the ground state, emitting a photon of energy E, is represented as a vertical downward arrow for which kinetic energy is conserved. E_o is the central energy of the emission band. At coordinate q_q, with vibrational energy E_q, the potential energy curves intersect and a non-radiative, thermal transition from the excited to the ground state is possible.

With this description the shape of a TL emission band, as a function of wavelength, at a given temperature T is given by

$$I(\lambda) = \frac{h^2c^2}{\lambda^3} \frac{A}{U\sqrt{2\pi}} \exp\left\{-\frac{(hc/\lambda - E_o)^2}{2U^2}\right\} \tag{2.137}$$

where A is a constant, c is the velocity of light, and

$$E_o = C_1 - C_2 \coth\left[\frac{\hbar\omega}{2kT}\right] \tag{2.138}$$

and

$$U^2 = S(\hbar\omega)^2 \coth\left[\frac{\hbar\omega}{2kT}\right]. \tag{2.139}$$

Here C_1 and C_2 are constants and S is the Huang-Rhys factor.

Thermal quenching of the luminescence is the phenomenon in which the luminescence efficiency decreases as the temperature increases. From the above description this can be understood by considering the increased vibration of the system as the temperature increases such that non-radiative transitions may occur at coordinate q_q. The probability Γ_{nr} that this occurs is thermally activated according to

$$\Gamma_{nr} = \Gamma_o \exp\left\{-\frac{E_q}{kT}\right\} \tag{2.140}$$

where Γ_o is a pre-exponential constant. If Γ_r is the probability of a radiative transition than one may write that the luminescence efficiency η (which earlier we had set equal to 1) at temperature T is

$$\eta = \frac{\Gamma_r}{\Gamma_r + \Gamma_{nr}}. \tag{2.141}$$

If kT is large compared to E_q then

$$\eta \approx \left(\frac{\Gamma_r}{\Gamma_o}\right) \exp\left\{\frac{E_q}{kT}\right\}. \tag{2.142}$$

The process described is not the only mechanism that can give rise to thermal quenching. In Section 2.4.4 we discussed the possibility that hole release may be occurring simultaneously with electron release. If the freed holes subsequently recombine non-radiatively with trapped electrons elsewhere in the system then the probability of there being holes available with which free electrons may radiatively recombine is also thermally activated. Using the notation of Section 2.4.4 we may then write that the efficiency is

$$\eta \approx c \exp\left\{-\frac{E_{tp}}{kT}\right\} \tag{2.143}$$

which is of identical form to Eq. 2.142. Here, c is a constant and we also assume that $kT << E_{tp}$.

The consequence of the above considerations is that the expression for the TL intensity I_{TL} has to be modified by the efficiency term η. For example, the expression for a first-order (Randall-Wilkins) peak (Eq. 2.28) would then become[76]

$$I_{TL} = Bs\exp\left\{-\frac{(E_t - W)}{kT}\right\}\exp\left\{-\left(\frac{s}{\beta}\right)\int_{T_0}^{T}\exp\left\{-\frac{E_t}{k\theta}\right\}d\theta\right\} \tag{2.144}$$

where $W = E_q$ or E_{tp}, and $B = (\Gamma_r/\Gamma_o)\, n_0$ or cn_0. The other expressions for the TL intensity would be similarly modified.

2.6 The Validity of the Arrhenius Law

The theory of thermoluminescence as discussed in this book is based on suggesting models consisting of localized energy levels within the forbidden gap of a solid, namely, trapping states and recombination centers, and writing rate equations governing the transitions taking place during excitation or during heating of the sample. One then either solves the equations numerically, or makes plausible approximations to arrive at analytical solutions. Providing the activation energies and the pre-exponential factors for the relevant trapping states are known, as well as the transition probabilities and occupancies of the relevant imperfections, one can arrive at descriptions of the TL glow curves which approximate experimental glow curves. The inverse problem, namely that of evaluating the trapping parameters from the obtained glow curve is a major goal of many TL studies, and will be discussed in detail in Ch. 3 and elsewhere.

At this point we would like to discuss the validity of the way the kinetic rate equations are written in the first place. We have assumed thus far that the thermal escape probability for a single electron per unit time is $s\exp(-E_t/kT)$. This kind of exponential dependence on the negative reciprocal of the absolute temperature T actually governs a variety of thermally dependent magnitudes such as diffusion coefficients, reciprocals of correlation times associated with dielectric, magnetic and viscoelastic relaxations, electronic and ionic mobilities, and rates of internal rotation around chemical bonds, and many other phenomena. This exponential dependence was first described by Arrhenius[77], and is usually termed the *Arrhenius Law*. A rather general account of the conditions leading to the occurrence of the Arrhenius law was given by Gibbs[78]. He assumed the system undergoing the thermally activated process consists of weakly interacting subsystems, each of which is subject to a potential of a double minimum type. Gibbs showed that the appearance of the Arrhenius exponential is a consequence of the dependence of the barrier climbing rate on the populations of reservoir subsystem levels which, in various aggregates, collectively possess and transfer the requisite energy for the barrier climbing process. Gibbs also discussed the frequency factor and showed that, usually, no significant temperature dependence of this factor is expected. This is to be compared with the strong temperature dependence of the exponential Arrhenius term.

Further work along similar lines was carried out by Peacock-López and Suhl[79]. In their notation, the reaction rate k (s^{-1}) is given by

$$k = A\exp(-\beta\epsilon_b) \tag{2.145}$$

where A (s^{-1}) is the pre-exponential frequency factor, $\beta = 1/k_BT$, k_B is Boltzmann's constant, T is the absolute temperature, and ϵ_b is the activation, or barrier, energy. The total probability of a reaction is the sum of the probabilities that the reactants absorb one, two, three.....etc. excitations with a total energy exceeding the barrier energy. The probability for absorption of exactly n excitations with energy ϵ is the square of the T-matrix element for this process, multiplied by $\rho_n(\epsilon)$, the density of states of the n excitations with total energy ϵ. As briefly explained below, it has been shown[79] that the sum over all n behaves like $e^{-(\beta-\beta_c)\epsilon}$, where $\beta = 1/k_BT$. $\beta_c = 1/k_BT_c$ is a function of a suitable coupling parameter and is a function of the way the single-excitation density $\rho_1(\epsilon)$ depends on ϵ. It is argued, however that in most cases, $\beta \gg \beta_c$ which brings us back to Eq. 2.145.

The sum of all possible processes leading to escape is written as

$$\frac{2\pi}{\hbar}\sum_{f,i} | < f\,|T|\,i > |^2 \frac{e^{-\beta E_i}}{z}\delta\left(E_f - E_i\right), \tag{2.146}$$

where $\{i\}$ is the complete set of quantum numbers of the system in the reactant well and $\{f\}$ is the set of quantum numbers of the reacted system. $\langle f\,|T|\,i\rangle$ is the T-matrix element. Eq. 2.146 assumes that quasiequilibrium conditions prevail during the reaction. $|i>$ and $|f>$ here are direct products of reactant states and bath excitation states. One can write

$$|i\rangle = |E_i^o, n_k + m_k\rangle,\ |f\rangle = \left|E_f^o, \{m_k\}\right\rangle, \tag{2.147}$$

$$E_i = E_i^o + \sum_k (m_k + n_k)\epsilon_k;\ E_f = E_f^o + \sum_k m_k\epsilon_k \tag{2.148}$$

where the bath, in its initial and final states, is characterized by the excitation number sets $\{m_k + n_k\}$ and $\{n_k\}$, respectively. Here, E_i^o is the reactant energy in the bound state of the reactant well, and E_f^o is the energy of the reaction products. k labels the excitations, and ϵ_k is the energy of the k-th of these.

Neglecting tunneling, one has $E_i^o < \epsilon_b$ and $E_f^o > \epsilon_b$. The transition occurs by absorption of the excess n_k excitations. Following this line of thought, the authors[79] reach the expression for the rate R, and for $\beta \gg \beta_c$

$$R = \mathcal{P}e^{-(\beta-\beta_c)\epsilon_b} \tag{2.149}$$

with $\mathcal{P}$, the pre-exponential factor, being a non-exponential function of β_c. This, indeed reduces to Eq. 2.145 since $\beta \gg \beta_c$. A further discussion is given, however, concerning cases in which β_c is not entirely negligible.

Tavgin and Stepanov[80] gave an independent activation process model. They also start by assuming a double potential well with a barrier, and make the calculations leading to the evaluation of the transition rate across the barrier as a result of subsequent transitions between a series of quantum subsystems which can be considered as identical harmonic oscillators.

In a series of papers, Maxia[51],[52],[81] explained the rate equations governing the TL process on the basis of non-equilibrium thermodynamics. By assuming the phosphor to be held in a cavity full of black-body radiation, and by applying the principle of minimum entropy production, the electron distribution in the conduction band was found to be in a steady quasiequilibrium with those in trapping states and recombination centers. Maxia also discussed entropy production in diffusion assisted thermoluminescence[52].

Böhm et al.[82] and Böhm and Scharmann[83] described a quantum mechanical treatment of the escape probability from traps in thermally stimulated processes and the evaluation of the pre-exponential factor. The authors discussed an atomistic model involving multiphonon processes for the calculation of the transition probability of a trapped charge carrier into an excited state in thermally stimulated processes. The displacement of lattice atoms from equilibrium positions gives rise to electron-lattice interactions. As a result, the temperature dependence was found to be of non-Arrhenius type in some cases. The difference in the probabilities for non-radiative transitions, as calculated by the quantum mechanical and by semi-classical methods, was also discussed in this work. The effects of non-radiative transitions on thermoluminescence was described, and the influence on the position and width of TL peaks was considered.

The authors used the time-dependent Schrödinger equation with a Hamiltonian which includes a system of electrons and the lattice atoms, as well as their mutual interactions. The static approximation was utilized and the lattice atoms were fixed in their equilibrium positions during the motion of the electrons. The approximation was considered as a first-order perturbation with the perturbed Hamiltonian

$$\hat{H}' = V_E(x, X) - V_E(x, X), \tag{2.150}$$

where x stands for the coordinates of the electrons and X for those of the lattice atoms, and V_E is the potential energy, including electron-electron interactions and the interaction of the electrons with the lattice atoms. The Hamiltonian describes that part of the interaction of the electrons with the lattice atoms which is caused by the displacement of the lattice atoms from their equilibrium positions, i.e., the electron-phonon interaction. Since the displacement of the lattice atoms is assumed to be small, using a Taylor series expansion to first-order, the Hamiltonian $\hat{H}'$ can be separated into two parts depending on the coordinates of the electrons and lattice atoms, respectively.

The probability for the transition of an electron from an initial state r to a final state r' can now be calculated using first-order peturbation theory. During the electronic transition all possible changes of the lattice states must be allowed. Thus, all initial and final states of the lattice are to be taken into account. Given the occupation probability p_v of the initial state, the transition probability is found to be

$$W_{r'r} = \frac{2\pi}{\hbar} \sum_{v'v} \left| H'_{r'v',rv} \right|^2 p_v \delta\left(E_{r'v'} - E_{rv}\right). \tag{2.151}$$

$H'_{r'v'}$ is the matrix element of $\hat{H}'$, and the occurrence of the δ-function guarantees the conservation of energy; E_{rv} and $E_{r'v'}$ are the energies of the system in the initial state rv and final state $r'v'$, respectively. The selection rule for the non-vanishing matrix element is $\Delta v = \pm 1$ where v is the vibration quantum number of the lattice. Therefore, only single-phonon processes should be allowed. To get an appreciable probability for multi-phonon processes, the potential energy of the lattice should depend on the energy state. If the electron is excited, both the equilibrium positions of the lattice atoms and the frequencies of the lattice vibrations are altered. Considering a particular matrix element, v and v' do not represent two different states of a single oscillator but, rather, states of different oscillators. This means that the selection rule disappears, and multi-phonon processes are allowed.

This calculation assumes that the equilibrium positions alter during the transition. As a result, the normal coordinates Q_ρ are changed by an amount a_ρ. The occupation probability p_v of the initial state of the lattice is a product of the occupation probabilities p_{v_ρ} of the normal vibrations which are characterized by Boltzmann statistics,

$$p_{v_\rho} = \exp\left(-\frac{\varepsilon_{v_\rho}}{kT}\right) / \sum_{v_\rho=0}^{\infty} \exp\left(-\frac{\varepsilon_{v_\rho}}{kT}\right) \tag{2.152}$$

where $\varepsilon_{v_\rho} = \hbar\omega_\rho(v_\rho + \frac{1}{2})$, and where v_ρ is the vibration quantum number of the normal modes. From Eq. 2.152, the transition probability is calculated under the mentioned assumption to be

$$W_{r'r} = W_o \cdot \left[-\frac{(\omega_e + S\omega_o + \alpha)^2}{2(H^2 + \gamma - \beta)} \right], \tag{2.153}$$

where $\hbar\omega_e = E_{r'} - E_r$; $S\omega_o = \sum_\rho \frac{M}{2\hbar}\omega_\rho^2 a_\rho^2$ and $H^2 = \sum_\rho \frac{M}{2\hbar}\omega_\rho^3 a_\rho^2 \coth\left(\frac{\hbar\omega_\rho}{2kT}\right)$. Here, ω_e is the frequency of the electronic transitions, S is the Huang-Rhys factor, ω_o is the average frequency of the lattice vibrations, ω_ρ are the frequencies of the normal modes, a_ρ is the displacement of the normal coordinates and M is the mass of the lattice atoms. At high temperatures, $x = \hbar\omega_\rho/2kT$; thus the argument of the coth is small and the hyperbolic cotangent can be approximated by $\coth x \approx 1/x$.

A relatively simple relation can be obtained in the strong coupling case, assuming large displacements of the normal coordinates and $S \gg 1$,

$$W_{r'r} = W_o \exp\left[-\frac{\tilde{C}}{\coth(\hbar\omega_o/2kT)} \right] \tag{2.154}$$

where $\tilde{C} = (\omega_e + S\omega_o)^2 / 3\frac{M}{\hbar}\omega_o^3 a_o^2)$. At high temperatures, the coth can be approximated as mentioned above, and this yields

$$W_{r'r} \approx W_o \exp\left(-\frac{C}{kT} \right) \tag{2.155}$$

where $C = \frac{1}{2}\tilde{C}\hbar\omega_o$. Here, the thermal activation energy is not necessarily equal to the trap depth. In fact, it depends on ω_e, ω_o and the number of participating phonons as expressed by the Huang-Rhys factor S. Böhm et al.[82] give examples of either a shift in the expected TL peak position or its width, where the presently described conditions apply.

In a second paper, Böhm and Scharmann[83] further described the quantum mechanical treatment which deals with non-radiative transitions due to electron-phonon coupling. This resulted in the possibility of getting an effective value of $s = 10^{15}\text{s}^{-1}$ for the frequency factor, which is 2-3 orders of magnitude higher than the highest expected from the lattice frequency. An alternative explanation, which accounts for even higher effective frequency factors is given in Ch. 10, which also has to do with non-radiative competition but in a way limited to the semi-classical treatment.

Swandic[84] presented a stochastic approach to recombination luminescence with retrapping in the steady state. A master equation for a stochastic model with one trapping state and one recombination center was derived. An expansion of this equation by van Kampen's Ω-expansion method allows one to recover the usual deterministic, macroscopic, kinetic, differential equations as well as to obtain the Fokker-Planck equation governing the fluctuations of the various particle numbers. In a later paper [85] the analysis was extended to the general case of several recombination centers and trapping states.

2.7 References

1. E. Wiedemann and G.C. Schmidt, *Ann. Phys. Chem. Neue Folge* **54** (1895) 604.
2. J.T. Randall and M.H.F. Wilkins, *Proc. Roy. Soc. London* **184** (1945a) 366.
3. J.T. Randall and M.H.F. Wilkins, *Proc. Roy. Soc. London* **184** (1945b) 390.
4. G.F.J. Garlick and A.F. Gibson, *Proc. Roy. Soc. London* **A60** (1948) 574.
5. R.H. Bube, *Photoconductivity of Solids*, Wiley and Sons, New York (1960).
6. S. Glasstone, K.J. Laidler and H. Eyring, *The Theory of Rate Processes*, McGraw-Hill, New York (1941).
7. N.F. Mott and R.W. Gurney, *Electronic Processes in Ionic Solids*, 2nd edition, Oxford Univ. Press, Oxford (1948).
8. F.E. Williams and H. Eyring, *J. Chem. Phys.* **15** (1949) 289.
9. D. Curie, *Luminescence in Crystals*, Methuen, London (1960).
10. E.I. Adirovitch, *J. Phys. Rad.* **17** (1956) 705.
11. P. Bräunlich, in *Thermally Stimulated Relaxation in Solids*, P. Bräunlich, editor, Springer-Verlag, Berlin (1979) 1.
12. R.R. Haering and E.N. Adams, *Phys. Rev.* **117** (1960) 451.
13. A. Halperin and A.A. Braner, *Phys. Rev.* **117** (1960) 408.
14. P.W. Levy, *Nucl. Tracks Radiat. Meas.* **10** (1985) 547.
15. P.W. Levy, in *The Encyclopedia of Physics,* R.G.. Lerner and G.L. Trigg, editors, 2nd. edition, VCH Publishers Inc., New York (1991) 1264.
16. C.E. May and J.A. Partridge, *J. Chem. Phys.* **40** (1964) 1401.
17. M.S. Rasheedy, *J. Phys.: Condens. Matter* **5** (1993) 633.
18. R. Chen, *J. Electrochem. Soc.* **116** (1969) 1254.
19. C. Christodoulides, *Phys. Stat. Sol. (a)* **118** (1990) 333.
20. G.A. Dussel and R.H. Bube, *Phys. Rev.* **155** (1967) 764.
21. I.J. Saunders, *J. Phys. C: Sol. St. Phys.* **2** (1969) 2181.
22. P.J. Kelly and P. Bräunlich, *Phys. Rev. B* **1** (1970) 1587.
23. M. Böhm and A. Scharmann, *Phys. Stat. Sol. (a)* **4** (1971) 99.
24. R. Chen, N. Kristianpoller, Z. Davidson and R. Visocekas, *J. Lumin.* **23** (1981) 293.
25. A. Opanowicz, *Phys. Stat. Sol. (a)* **116** (1989) 343.
26. R.K. Bull, S.W.S. McKeever, R. Chen, V.K. Mathur, J.F. Rhodes and M.D.

Brown, *J. Phys. D: Appl. Phys.* **19** (1986) 1321.

27. N.F. Mott and E.A. Davies, *Electronic Processes in Non-Crystalline Materials*, 2nd edition, Clarendon Press, Oxford (1979).
28. M.A. Lampert and P. Mark, *Current Injection in Solids*, Academic Press, New York (1970).
29. W.L. Medlin, *Phys. Rev.* **123** (1961) 502.
30. W.L. Medlin, *Phys. Rev.* **135** (1964) A1770.
31. G.G. Roberts, N. Aspley and R.W. Munn, *Phys. Rep.* **60** (1980) 59.
32. J. Grenet, C. Vautier, D. Carles and J.J. Chabrier, *Phil. Mag.* **28** (1973) 1265.
33. K.C. Kao and W. Hwang, *Electrical Transport in Solids,* Pergamon Press, Oxford (1981).
34. A. Bosacchi, B. Bosacchi, S. Franchi and L. Hernandez, *Sol. St. Commun.* **13** (1973) 1805.
35. L.F. Pender and R.J. Fleming, *J. Phys. C: Sol. St. Phys.* **10** (1977) 1571.
36. K. Hagekyriakou and R.J. Fleming, *J. Phys. D: Appl. Phys.* **15** (1982) 163.
37. Ph. Avouris and T.N. Morgan, *J. Chem. Phys.* **74** (1981) 4347.
38. T. Kikuchi, *J. Phys. Soc. Japan* **13** (1958) 526.
39. W.F. Hornyak, R. Chen and A.D. Franklin, *Phys. Rev. B* **46** (1992) 8036.
40. J.F. Fowler, *Proc. Roy. Soc.* **A276** (1956) 464.
41. R. Chen and Y. Kirsh, *Analysis of Thermally Stimulated Processes*, Pergamon Press, Oxford (1981).
42. W.F. Hornyak and A.D. Franklin, *Nucl. Tracks Radiat. Meas.* **14** (1988) 81.
43. W.F. Hornyak and R. Chen, *J. Lumin.* **44** (1989) 73.
44. G. Rudlof, J. Becherer and H. Glaefke, *Phys. Stat. Sol. (a)* **49** (1978) K121.
45. M. Schön, *Z. Physik.* **119** (1942) 463.
46. H.A. Klasens, *Nature* **158** (1946) 306.
47. J.P. Fillard, J. Gasiot and J.C. Manifacier, *Phys. Rev. B* **18** (1978) 4497.
48. P. Bräunlich and A. Scharmann, *Phys. Stat. Sol.* **18** (1966) 307.
49. S.W.S. McKeever, J.F. Rhodes, V.K. Mathur, R. Chen, M.D. Brown and R.K. Bull, *Phys. Rev. B* **32** (1985) 3835.
50. V. Maxia, *Phys. Rev. B,* **17** (1978) 3262.
51. V. Maxia, *Phys. Rev. B,* **21** (1980) 749.
52. V. Maxia, *Lett. Nuov. Cim.* **31** (1981) 27.
53. D. Shenker and R. Chen, *J. Comput. Phys.* **10** (1972) 272.
54. P.J. Kelly, M.J. Laubitz and P. Bräunlich, *Phys. Rev. B* **4** (1971) 1960.
55. A.C. Lewandowski and S.W.S. McKeever, *Phys. Rev. B* **43** (1991) 8163.
56. S.W.S. McKeever, B.G. Markey and A.C. Lewandowski, *Nucl. Tracks Radiat.*

Meas. **21** (1993) 57.

57. S.W.S. McKeever, A.C. Lewandowski and B.G. Markey, *Radiat. Prot. Dosim.*, **47** (1993) 9.
58. A.C. Lewandowski, B.G. Markey and S.W.S. McKeever, *Phys. Rev. B* **49** (1994) 8029.
59. C.J. Delbecq, Y. Toyozawa and P.H. Yuster, *Phys. Rev. B*, **9** (1974) 4497.
60. R. Visocekas, T. Ceva, C. Marti, F. Lefaucheux, and M.C. Robert, *Phys. Stat. Sol. (a)* **35** (1976) 315.
61. R. Visocekas, *Eur. PACT J.,* **3** (1979) 258.
62. R. Visocekas, *Nucl. Tracks,* **10** (1985) 521.
63. R. Visocekas, *Nucl. Tracks Radiat. Meas.*, **14** (1988) 163.
64. R.H. Templer, *Thermoluminescence Techniques for Dating Zircon Inclusions,* D.Phil. thesis, Univ. Oxford (1986).
65. R. Chen, *J. Mater. Sci.* **11** (1976) 1521.
66. D.G. Thomas, J.J. Hopfield and W.M. Augustiniak, *Phys. Rev.* **140** (1965) A202.
67. R.H. Templer, *Radiat. Prot. Dosim.* **17** (1986) 493.
68. G. Bonfiglioli, P. Brovetto and C. Cortese, *Phys. Rev.* **114** (1958) 951, 956.
69. B. Jassemnejad, R.J. Abbundi, M.D. Brown and S.W.S. McKeever, *Phys. Stat. Sol. (a),* **108** (1988) 753.
70. R.G. Fairchild, P.L. Mattern, K. Lengweiler and P.W. Levy, *J. Appl. Phys.* **49** (1978) 4512, 4523.
71. P.D. Townsend and Y. Kirsh, *Contemp. Phys.*, **30** (1989) 337.
72. J.R. Prescott, P.J. Fox, R.A. Akber and H.E. Jensen, *Appl. Opt.* **27** (1988) 3496.
73. B.J. Luff and P.D. Townsend, *Meas. Sci. Technol.* **4** (1993) 65.
74. T.M. Piters, W.H. Meulemans and A.J.J. Bos, *Rev. Sci. Instrum.* **64** (1993) 109.
75. P. Dorenbos, C.W.E. van Eijk, A.J.J. Bos and C.L. Melcher, *J. Phys.: Condens. Matter* **6** (1994) 4167.
76. A.G. Wintle, *Geophys. J.R. Astron. Soc.* **41** (1975) 107.
77. S. Arrhenius, *Z. Phys. Chem.* **4** (1889) 226.
78. J.H. Gibbs, *J. Chem. Phys.* **57** (1972) 4473.
79. E. Peacock-López and H. Suhl, *Phys. Rev. B* **26** (1982) 3774.
80. V.L. Tavgin and A.V. Stepanov, *Phys. Stat. Sol. (b)* **161** (1990) 123.
81. V. Maxia, *Lett. Nuov. Cim.* **20** (1977) 443.
82. M. Böhm, O. Erb and A. Scharmann, *Appl. Phys. A* **37** (1985) 165.
83. M. Böhm and A. Scharmann, *Appl. Phys. A* **43** (1987) 29.

84. J.R. Swandic, *Phys. Rev. B* **45** (1992) 622.
85. J.R. Swandic, *Phys. Rev. B* **53** (1996) 2352.

Chapter 3
ANALYSIS OF TL GLOW CURVES

3.1 General Considerations

In its most general sense, "analysis" of a physical phenomenon includes all inferences about the phenomenon, be they qualitative or quantitative, descriptive or mathematical. In this sense, practically all of this book deals with the analysis of TL glow curves. The scope of the present chapter, however, is more limited. Every model for TL includes a number of parameters which are rather interesting physical entities related to the trapping states responsible for the TL emission. Even in the simplest Randall-Wilkins[1] first-order model, presented in Eq. 1.1, one has three physical constants the evaluation of which is very important. These are the activation energy E_t (in eV), the frequency factor s (s^{-1}) and the initial filling of trapped charge carriers n_0 (m^{-3}). In fact, as suggested in Eq. 2.3, there is another important constant η which connects the TL intensity $I(t)$ with the rate of change of trapped carriers $-dn/dt$. This constant combines the efficiency of the recombination process (for those cases where the recombination transition is not 100% radiative, and providing the efficiency is constant) with the geometry of the device in which the sample is held, and with the efficiency of the light sensitive device (e.g. a photomultiplier tube). The constant, of course, can be dependent on the emission wavelength spectrum. In addition, η is a proportionality constant which depends on the units of the measured light (in photons or in units of energy). Setting the constant η to unity defines an arbitrary set of intensity units and retains the features of the TL glow curve, and, therefore, the information about the other parameters can still be extracted.

In the case of a simple first-order TL peak, as discussed here, the role of the initial concentration of trapped charge carriers is also merely that of a multiplicative constant (see Eq. 1.4) and, therefore, it can be lumped together with the constant η. The remaining expression, contains the two more "basic" physical quantities E_t and s and these can be evaluated from the peak shape and other methods. To complete this simple example, however, it should be noted that if the constant η *is* known, the value of n_0, which is an important parameter by itself, can be found as the total area under the TL peak. Another point to mention is that in non-first-order cases the initial filling of traps is not just a multiplicative constant. The advantage of being able to deal with fewer parameters (in the general-order kinetics, for example, three instead of four) does not exist any more, and n_0 should be explicitly treated as an independent parameter; on the other hand, if n_0 is of importance, information about it is still inherent in the glow curve.

If one wishes to consider a more realistic model than the Randall-Wilkins, first-order model, one can still consider one kind of trapping state and one kind of recombination center, but one can include the possibility of retrapping as well as recombination . The simultaneous differential equations governing the process are 2.17, 2.19, 2.20 and 2.22, re-written here in a slightly different way,

$$I = -dm/dt = A_m m n_c \tag{3.1}$$

$$dn/dt = -sn\exp\{-E_t/kT\} + n_c(N-n)A \tag{3.2}$$

$$dm/dt = dn/dt + dn_c/dt, \tag{3.3}$$

where we have dropped the subscript on I_{TL} and simplified the notation for the A constants. At this point we are not going to discuss the possible ways of solving this set of non-linear, simultaneous equations, but rather we are going to consider the parameters involved. In this discussion we take the constant η to be unity in line with the above explanation. Here, the emitted light is associated with the rate of change of the concentration of holes in centers $-dm/dt$, rather than with $-dn/dt$ as in the simplified Randall-Wilkins model. The constant η should, therefore, now multiply $-dm/dt$. A and A_m are the trapping and recombination probabilities, respectively, given in units of m^3s^{-1}. Although most researchers consider these as being constants, there is a slight temperature dependence of these magnitudes which will be ignored in the present context. E_t is the activation energy for releasing charge carriers from the trapping states and s is the associated frequency factor. In this framework, where the mobile charges are assumed to be electrons, n_0 and n_{c0} (both given in m^{-3}) are the initial concentrations of trapped and conduction band electrons respectively, and m_0 (in m^{-3}) is the initial concentration of holes in centers. In addition, N (m^{-3}) is the total concentration of trapping states (out of which $n(t)$ are occupied at time t).

We have just mentioned the magnitudes A_m, A, E_t, s, n_0, n_{c0}, m_0 and N, namely, eight independent parameters. These eight parameters determine the shape of the single glow peak in question. It is to be noted that, inasmuch as the above set of equations can be solved only numerically, the degree of difficulty in solving the equations can be considered as "intermediate". On the one hand, no simplifying assumptions have been made so far (e.g. quasiequilibrium). On the other hand, real-life situations can be, and often are, more complicated;

for example, more than one kind of luminescence center may be simultaneously involved. Also cases like localized transitions (see Ch. 2), as opposed to the present situation of carriers being raised into the conduction band, may occur at the same time. In addition, further complications such as a simultaneous thermal release of electrons and holes, or the possible transport of ions which may also result in some light emitting recombination, should sometimes be considered as well. Going back to the eight parameters in Eqs. 3.1-3.3, some general remarks should be made. As much as it is difficult to produce the glow curve numerically when the set of eight parameters is given, the inverse problem of extracting the eight parameters from an experimental curve is even more troublesome. Although it is obvious that all these eight parameters do influence the shape of the resulting glow peak, it is practically hopeless to expect the possibility of extracting all eight parameters from a single, smooth glow peak which, apart from its general shape, lacks special features. Putting this in other words, it is conceivable that two different sets of the eight parameters may result in two glow peaks which cannot be distinguished from each other within what can be considered as normal experimental error. Thus, there is a certain amount of degeneracy and it is obvious that finding the "correct" set of parameters is not straightforward.

Keeping this intrinsic difficulty in mind, we should attack the problem of parameter evaluation in different ways. Of course, in cases where, say, simple first-order kinetics is established, one can relatively easily calculate the two parameters E_t and s. In the more general case, however, one should differentiate between "important" parameters, which have a stronger influence on the shape of the curve (e.g. E_t and s) and can therefore be more easily found from an experimental peak, and the "less important" parameters. The best example here is n_{c0} which usually is close to zero at relatively low temperatures in insulators.

Taking these considerations into account, it appears that the task of evaluating *all* the physical quantities simultaneously is usually hopeless. We should therefore follow a more realistic route. Speaking in general terms, for evaluating a certain parameter, we are looking for a feature in the measured results which is most sensitive to this parameter and least sensitive to all the others. A good example here is the "initial rise" portion of the peak. As we discuss in full later, analysis of this region of a TL peak yields the activation energy, practically irrespective of the other details of the kinetics (and, therefore, independently of the other parameters; for an exception to this rule see Bräunlich[4]).

Of course, it is obvious that if a certain parameter has been evaluated (such as E_t as discussed here), the number of unknown parameters is reduced, and this may help in evaluating the remaining ones. The good news in this rather difficult situation is that TL and its related phenomena give a wealth of information that may help in evaluating the rest of the relevant physical parameters, as well as understanding the processes involved. This includes a series of measurements monitoring the glow peaks, say, in different spectral emission ranges, or after irradiation at various initial temperatures, or for different doses. Simultaneous measurements of TL, TSC, and TSEE can also be performed, as well as using information from these thermally stimulated phenomena along with information from other measurements such as ESR, optical absorption, photoluminescence, etc.

In summary, we can say that although the possibility of evaluating all the trapping parameters is far from being straightforward, we should try to utilize the complete arsenal of methods designed to analyze TL, other thermally stimulated processes, and other related phenomena. On the other hand we should understand that in some cases we will end up finding only some of the relevant parameters. In certain cases, we will have to accept the reality that some of the parameters dealt with are "heuristic", namely, we describe a peak with, say, seven physically relevant parameters by a model which includes only three. This is usually the situation, for example, when we evaluate the parameters E_t, s' and b in a "general-order" peak. Although s' and b are in most cases merely empirical in nature, the model gives a peak with variable symmetry features (the parameter b varying from below unity to above 2), and a physically significant value for E_t. This simplified model can serve at least for a characterization of the glow peak, and can help, among other things, the estimation of the stability of the glow peak at different ambient temperatures. This point is of great importance, for example, in applications in radiation dosimetry or dating.

3.2 Analysis of Phosphorescence

As pointed out in Ch. 2, phosphorescence is the decay of thermally stimulated luminescence as a function of time at a constant temperature. As mentioned already, phosphorescence differs from fluorescence in that the former is a temperature-dependent, long-time phenomenon measured in seconds, minutes or more, whereas the latter is much shorter and is essentially independent of temperature; phosphorescence is always associated with trapping of the carriers at rather shallow traps and their subsequent thermal release prior to their radiative recombination. It is in this sense that phosphorescence should be considered as thermoluminescence, or rather, thermally stimulated luminescence at constant temperature. It is to be noted that the validity of the requirement that the traps are shallow is very much temperature dependent. If the temperature at which the sample is held is, say, liquid nitrogen temperature (LNT), a shallow trap may be, say, an electron trap $0.1 - 0.2$ eV below the conduction band.

This, of course, is not a unique definition of a "shallow" trap, and this will depend on the other parameters involved, in particular the frequency factor s. The main point is that the total decay time will be between a few seconds and several minutes. Thus, if the ambient temperature is room temperature (RT, $\sim$300 K), the depth of the traps relevant to phosphorescence measurements will be substantially higher, say, 0.4 – 0.5 eV. If significantly higher energies are involved at a given temperature, the number of recombinations per second will be too small, and therefore, the light emission will not be measurable. If, on the other hand, the traps are too shallow, the decay time following excitation will be very small and may be difficult to measure with standard equipment. Also, this fast decaying phosphorescence may be confused with fluorescence which does not involve the intermediate state of trapping.

3.2.1 First-Order Decay

Phosphorescence usually exhibits a featureless luminescence decay curve and, as such, it may not be the best tool for evaluating kinetic parameters. Different kinds of decay do occur, however, and, as shown below, phosphorescence decay has been utilized to evaluate the particular process involved as well as the kinetic parameters.

The simplest kind of phosphorescence that can be expected is the exponential decay given by Eq. 2.4 as

$$I(t) = I_0 \exp\{-pt\} \tag{3.4}$$

where p^{-1} is the lifetime, and is given by

$$p = s \exp\{-E_t/kT\}\,. \tag{3.5}$$

This simple kind of behavior is associated with the first-order kinetics, as discussed in Ch. 2. The question of whether the kinetics are indeed first-order in TL as well as in any other thermally stimulated processes can easily be resolved by studying phosphorescence decay at constant temperature and performing a simple graphical (or, alternatively, numerical) analysis. It is obvious that if Eq. 3.4 describes the phosphorescence decay, one can write

$$\ln I - \ln I_0 = -pt. \tag{3.6}$$

In other words, a plot of $I(t)$ as a function of time on a semilog scale should yield a straight line with a slope of $-p$. An example is shown in Fig. 3.1, taken from Acharya and Mukherje[5]. The figure shows the phosphorescence decay of KCl:Sn at different temperatures. It is clear that any other behavior appearing in the decay curve indicates that either the kinetics are not of first order, or that more than a single trapping state is involved in the process. Some of these possibilities, and the ways to handle them and extract trapping parameters,

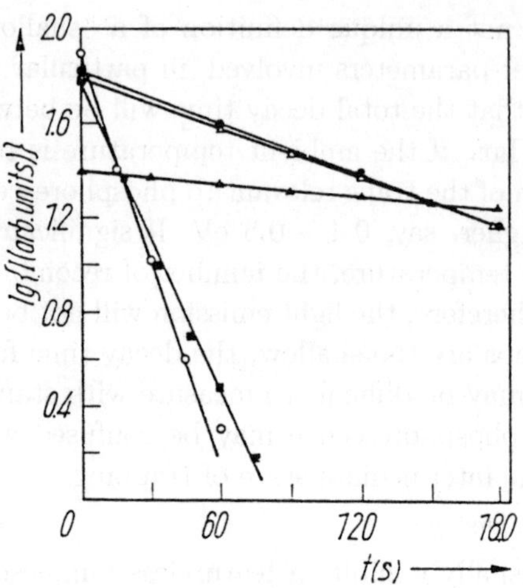

Figure 3.1 Exponential decay of phosphorescence and photostimulated decay in KCl. Full triangles indicate the results of phosphorescence decay at 75 °C, full circles at 151 °C, and open circles at 180 °C. Full squares give the results of KCl at 190 °C and open triangles give the photostimulated decay of KCl:Sn at 180 °C. (After Acharya and Mukherje[5].)

will be discussed in a later section. If, however, the first-order property is established, one gets p as the (negative) slope at the given temperature as shown in Eq. 3.5. This is now a simple expression which includes the two important parameters E_t and s. If we wish to evaluate these parameters, we can repeat the measurement at a somewhat different temperature and get a second equation with E_t and s, derived from the slope at the second temperature. Solving the two equations will yield the values of the two unknowns E_t and s. Better still, one can perform a series of phosphorescence decay measurements at different temperatures thus getting a set of values for $p(T_i)$ for the slope as given by Eq. 3.5 as a function of temperature. Taking the logarithm on both sides of Eq. 3.5 one gets

$$\ln p = \ln s - E_t/kT_i \ . \tag{3.7}$$

Thus, plotting p as a function of $1/T_i$ on a semilog scale should now yield a straight line with a slope of $-E_t/k$, from which the activation energy E_t can readily be evaluated. The value of $\ln s$ can be easily found from the intercept of the line with the y axis, and thus the value of the frequency factor s is immediately known. An example of such a plot is shown in Fig. 3.2, taken from Mariani et al.[6].

It seems rather tempting to utilize a broad range of temperatures in

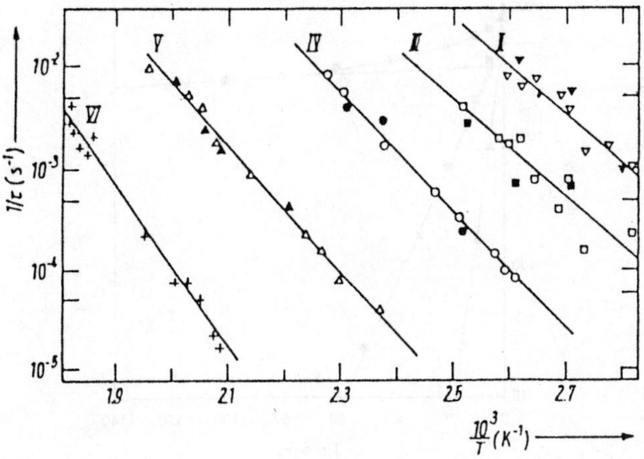

Figure 3.2 Inverse temperature dependence of the inverse of phosphorescence lifetimes. Empty symbols indicate γ-irradiated samples whereas full symbols are for photostimulated samples.

(After Mariani et al.[6].)

which the phosphorescence decay is monitored so as to get more reliable results for E_t and s. This, however, has to be prescribed with caution. For given values of E_t and s, if the temperature is too low, the emission may be too weak to be detected, and certainly its variation with time cannot be accurately determined. If the temperature is too high, the decay is very fast and the standard TL equipment may not be capable of following it. Of course, equipment for shorter time decay is easily available but, as pointed out already, the measured decay may be confused with fluorescence which is basically independent of the trapping of the charge carriers. Furthermore, the broader the temperature range, the larger is the risk that another trapping state will be involved. Sometimes this is unavoidable, in particular where continua of trapping states are concerned, but as long as discrete levels are studied, it is advisable to utilize a limited temperature range for the evaluation of a certain pair of activation energy and frequency factor. It should also be mentioned here that if the temperature range chosen is too broad so that the effect of a neighboring trap (say, deeper) starts to show up, this will be revealed on the plot of $\ln p$ vs. $1/T_i$ as a change in slope. The situation is even more complicated when the trapping states in question have nearly equal activation energies. More often than not this is the reason for pure exponential decays not showing up even when first-order kinetics are concerned. This is due to several exponential decays occurring at the same temperature. Different researchers have used various methods for the

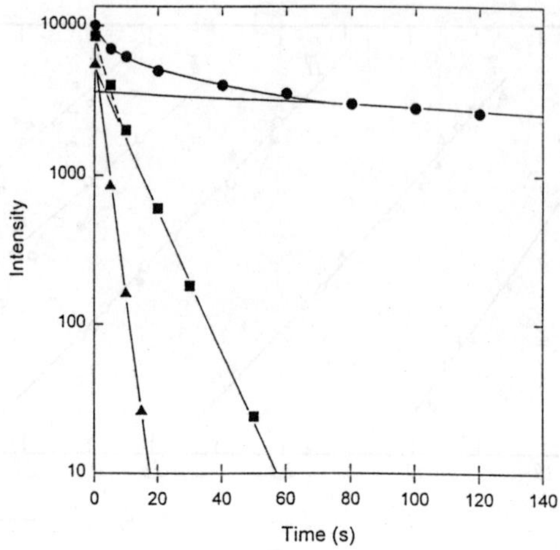

Figure 3.3 A plot of log intensity of phosphorescence decay vs. time in CaS:Pd and its deconvolution into three exponential components.
(Redrawn from Patil et al.[7].)

deconvolution of these decay curves. Figure 3.3 shows an example by Patil et al.[7] in which the decay curve is split into three components, each of which shows as a straight line on the semilog scale. Care must be taken with this type of analysis, however, since almost any curve can be approximated by the sum of multiple straight line components and false conclusions may easily be obtained.

3.2.2 Second-Order Decay

As mentioned a number of times already, first-order kinetics are by no means the only possible behavior expected even in the relatively simple situation in which one trapping state and one kind of recombination center are involved. To demonstrate the fact that entirely different behaviors can sometimes be found, let us concentrate on second-order kinetics. The governing equation in this situation is

$$I = -dn/dt = p'n^2 \tag{3.8}$$

where $p' = s' \exp\{-E_t/kT\}$. The solution of this equation is

$$I(t) = I_0/(1 + p'n_0t)^2. \tag{3.9}$$

The validity of this expression can be checked by plotting $\sqrt{I_0/I}$ as a function of the time t. In this second-order situation, a straight line is expected with a slope of $n_0p' = n_0s' \exp(-E_t/kT)$. Here too, if this is established, the temperature may be varied and the parameters E_t and $s'n_0$ can be evaluated. An example of such a behavior is shown in Fig. 3.4, taken from Tatumi et al.[8].

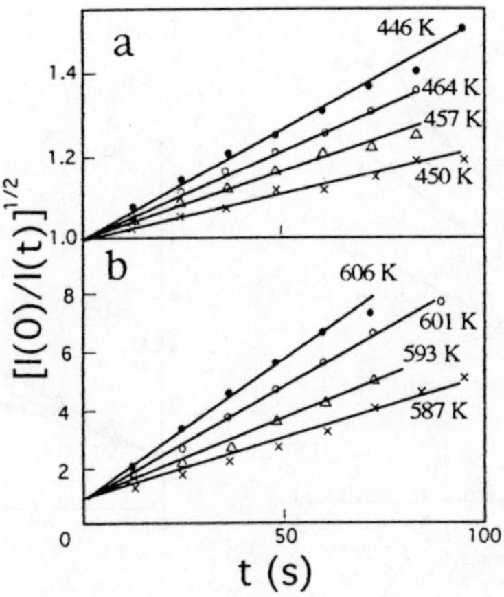

Figure 3.4 Plot of $[I(0)/I(t)]^{1/2}$ vs. time from phosphorescence decay results in natural dolomite.

(After Tatumi et al.[8].)

3.2.3 General-Order Decay

Continuing along the same line, we can consider the expected decay curve in cases where the kinetics are of "general order", as described in Ch. 2. The following discussion is limited to situations in which b is a constant, although, as discussed in Ch. 2, b may be temperature dependent to some extent. The governing differential equation is

$$I(t) = -dn/dt = p'n^b \tag{3.10}$$

where b attains various values between, say, 0.7 and 2.5 (see Chen[9]). At a constant temperature T, the solution of the equation is

$$I(t) = n_0^b p' / \left[1 + (b-1)p' n_0^{b-1} t\right]^{b/(b-1)} . \tag{3.11}$$

This also is a decaying function of time. The general-order feature of the phosphorescence decay can be demonstrated if a plot of $I^{(1-b)/b}$ vs. t yields a straight line. An example given by Nagpal and Kathuria[10] is shown in Fig. 3.5 . Here, the plot of $I^{(1-b)/b}$ is shown as a function of time for phosphorescence measured at 75 °C and 85 °C in MgO. Since the value of b is not known beforehand, the authors show the curve as calculated for different values of b, namely: 1.5, 1.7 and 1.9. Since the plot closest to a straight line is found for $b = 1.7$, it is determined that this is the order of the kinetics.

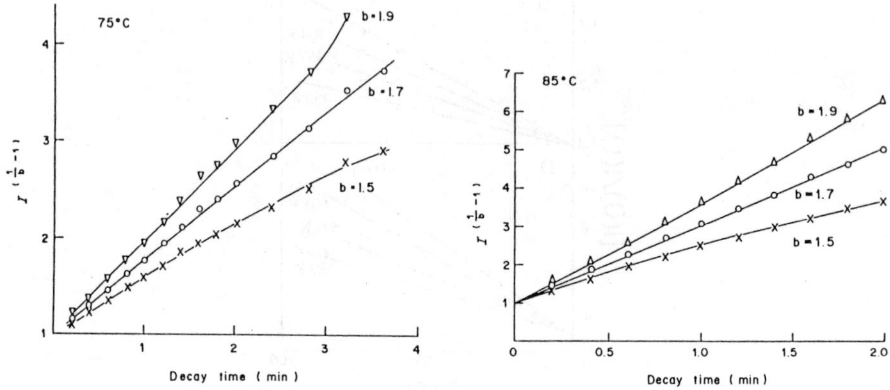

Figure 3.5 Plots of $I^{(1-b)/b}$ vs. decay time for γ-irradiated MgO at 75°C (left) and 85°C (right).

(After Nagpal and Kathuria[10].)

γ	0.33	0.5	1.0	2.0	3.0	5.0	10.0	25.0	50.0	100.0	500.0
r	2.5	2.3	2.0	1.75	1.63	1.47	1.34	1.15	1.01	0.97	0.8

Table 3.1 Relation between the power r and the ratio γ

3.2.4 Becquerel's Decay Law

We should remember that Eq. 3.10 represents an empirical approach using four parameters, s', n_0, b and E_t instead of the eight original parameters in Eqs. 3.1-3.3. The result of Eq. 3.11 should, therefore, be accepted with appropriate caution. In this context, it is of interest from the historical point of view that Eqs. 3.1-3.3 were first presented by Adirovitch [11] while studying phosphorescence decay. Starting with these equations and using some approximations, including that of quasiequilibrium, Adirovitch arrived at Becquerel's decay law

$$I(t) = I_0\left[(c+t)/c)\right]^{-r} \tag{3.12}$$

where c and r are simply related to the parameters in Eq. 3.11, c having dimensions of time and r dimensionless. Adirovitch found a quasi-empirical relation between the power r and the ratio $\gamma = A_m/A_n$ between the recombination and retrapping probabilities. Values are given in Table 3.1. In fact, Adirovitch showed that r was rather strongly dependent on γ and very weakly dependent on the other trapping parameters.

Medlin[12] studied phosphorescence decay from UV-irradiated calcite. An example of his results is shown in Fig. 3.6. Here, the results are depicted

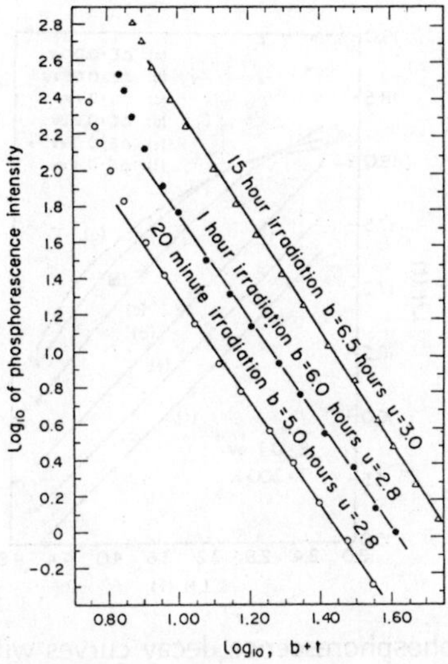

Figure 3.6 Results of phosphorescence decay of UV-irradiated calcite. (After Medlin[12].)

as the logarithm of phosphorescence intensity vs. the logarithm of $c/(c+t)$ (or, which is practically equivalent, $\log(c+t)$) which, when the best values of c are chosen, indeed yield straight lines.

The technique used is that of plotting the phosphorescence intensity vs. time. After long enough time on a log – log scale, t is large enough as compared to c in Eq. 3.12, and a straight line is expected on the log – log scale; the slope is $-r$. Hoogenstraaten[13] reported experimental results for ZnS crystals which confirmed the Becquerel-type decay. The value of r was found, however, to be slightly temperature dependent. In ZnS it was experimentally found to be $r = 1 + 2.3 \times 10^{-3}T$ (T in degrees absolute).

3.2.5 Continua

Some authors studied the phosphorescence decay associated with a continuum of trapping states. Randall and Wilkins[3] assumed a uniform distribution of trapping levels, stretching to infinity, and, for first-order kinetics, wrote the expression

$$I(t) = \int_0^{\infty} N_{E_t} s \exp\left\{-\frac{E_t}{kT}\right\} \exp\left[-st \exp\left\{-\frac{E_t}{kT}\right\}\right] dE_t = \frac{nkT}{t}\left[1 - \exp\{-st\}\right]. \tag{3.13}$$

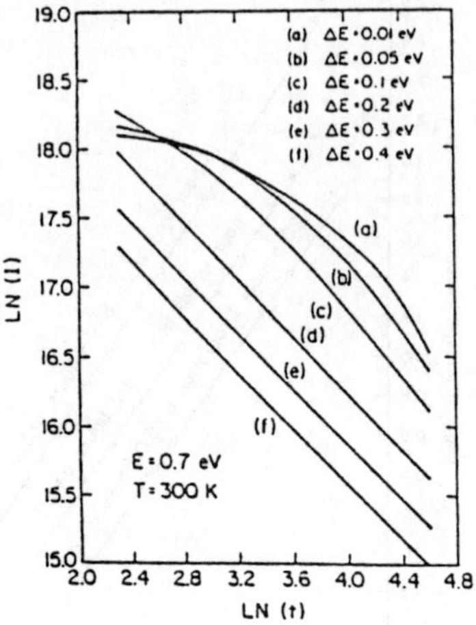

Figure 3.7 Calculated phosphorescence decay curves with different widths of the uniform energy range ΔE_t.
(After Hornyak and Chen[17].)

For $st \gg 1$, which normally occurs after a very short time, this yields

$$I(t) = nkT/t. \tag{3.14}$$

It should be noted that various authors consider the continuous distribution to be the normal situation in "quasi-disordered" materials (Bosacchi et al.[14]), amorphous materials(Pender and Fleming[15]),and glassy materials(Kikuchi[16]). Hornyak and Chen[17] made the point that, of course, the activation energy distribution ranges over a finite energy interval only. They conducted some numerical calculations, based on first-order kinetics, a typical result of which is shown in Fig. 3.7. The calculated decay curves are shown in curves (a-f) on a ln-ln scale for a case with the frequency factor $s = 10^{10}\text{s}^{-1}$, $T = 300$ K and a uniform finite distribution of ΔE_t centered at $E_o = 0.7$ eV. Whereas for narrow ΔE_t the result is curved, indicating a behavior similar to that of a discrete level, for $\Delta E_t = 0.2$ eV (i.e. an energy distribution from 0.6 eV to 0.8 eV) the decay curve looks straight as if the energies indeed stretch over an infinite range. Hornyak and Chen[17] pointed out that a similar behavior can be expected for any symmetric distribution, not necessarily uniform, as long as it is broad enough.

Wünsche et al.[18] gave a slightly different version of Eq. 3.14, stating that for many organic compounds, isothermal luminescence measurements at low temperature behave like

$$I(t) = I_0/(1 + \alpha t). \tag{3.15}$$

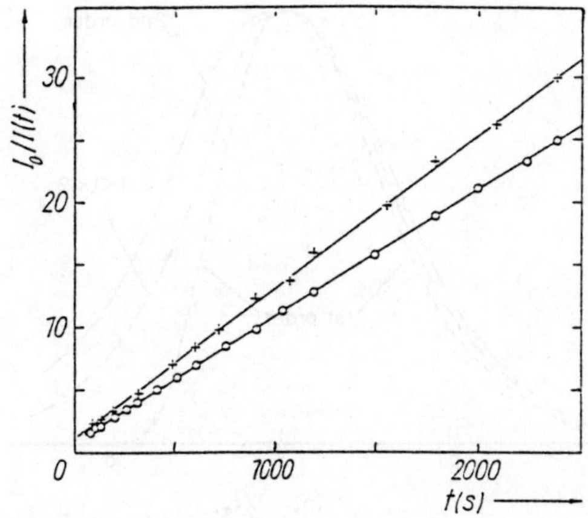

Figure 3.8 Isothermal luminescence of PMMA(○) and P(MMA-MA)(+) after electron irradiation at $T = 80$ K. Given in terms of $I_o/I(t)$ vs. time.
(After Wünsche et al.[18].)

This can, of course be seen as Eq. 3.12 with $r = 1$, or Eq. 3.14 with $\alpha t \gg 1$. In order to check this behavior, they plot $I_0/I(t)$ vs. t as shown in Fig. 3.8, in Poly-Methylmethacrylate-Methacrilic Acid (PMMA and PMMA-MA). The straight line indicates that, indeed, $\alpha t \gg 1$, even during the first seconds of the measurement.

3.2.6 TL-like Presentation of Phosphorescence

The behavior of the phosphorescence intensity vs. time is a rather featureless decay function and, therefore, it is rather difficult to extract trapping parameters from it. Moreover, the methods mentioned consist of plotting the results on either log – log or semi-log scales which may further decrease the accuracy of the parameter evaluation from the specific features of the decay curve. Randall and Wilkins[3], the pioneers of the study of TL, suggested to plot It as a function of $\ln t$ for a phosphorescence decay curve. This plot should certainly yield $It =$ *constant* in the situation of a broad-enough, symmetric distribution of activation energies, as seen from Eq. 3.14. However, in the somewhat more common first-order, second-order and general-order cases, it has been pointed out by Randall and Wilkins, and later elaborated by Visocekas[19] and Chen and Kristianpoller[20], that a peak-shaped curve is seen on this plot which resembles in shape the TL glow peak with the same order of kinetics. Figure 3.9 depicts three calculated curves of this kind as given by Visocekas.

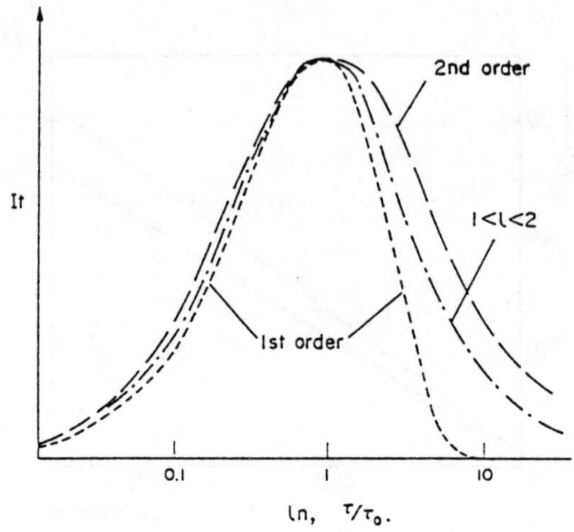

Figure 3.9 Isothermal decay curves for first-, second- and "general"-order as shown on It vs. ln$t$ coordinates.

(After Visocekas[19].)

In the first-order case, the appropriate expression is

$$It = sn_0 t \exp\{-E_t/kT\} \exp[-st \exp\{-E_t/kT\}] \tag{3.16}$$

which, if one writes $y = It$ and $x = \ln t \Rightarrow t = \exp\{x\}$, becomes

$$y = (n_0/\gamma) \exp\{x\} \exp[-\exp\{x\}/\gamma] \tag{3.17}$$

where $\gamma = s^{-1} \exp\{E_t/kT\}$. The peak is asymmetric, and looks very much like a first-order TL peak under a constant heating rate and with the same values of E_t and s. In fact, it can be shown that the similarity becomes an identity if a hyperbolic heating function is utilized for TL instead. By finding the derivative of y with respect to x and equating it to zero, we get

$$\exp\{x_{\max}\} = \gamma \Rightarrow t_{\max} = \gamma$$

where the subscript max indicates the maximum point of the curve. This is the reason t is given in units of γ in Fig. 3.9, so that the maximum occurs at $t = 1$ on this scale. The outer curve in Fig. 3.9 represents a second-order peak on the same scale. Defining here $\gamma = (n_0 s') \exp\{E/kT\}$, and x and y as before, we get

$$y = \frac{n_0 \exp\{x\}/\gamma}{[1 + \exp\{x\}/\gamma]^2}. \tag{3.18}$$

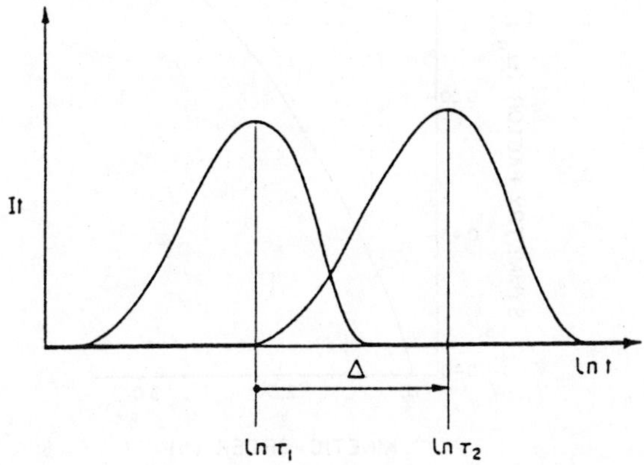

Figure 3.10 It vs. $\ln t$ presentation of isothermal decay at two different temperatures T_1 and T_2. The parameter Δ is shown.
(After Visocekas[19].)

In this case, the peak is perfectly symmetric, in resemblance to a second-order TL peak which is nearly symmetric with a linear heating function and is precisely symmetric with a hyperbolic heating function. Here too, equating the derivative (with respect to x) to zero yields $\exp\{x_{\max}\} = \gamma \Rightarrow t_{\max} = \gamma$. The curve in between those of first and second order in Fig. 3.9 represents an intermediate "general-order" case with $1 < b < 2$. The same considerations can be used to prove that $t_{\max} = \gamma$ and that the peak is similar (with a linear heating function) or identical (with a hyperbolic function) to the general-order TL peak with the same value of b.

As shown by Visocekas[19], repeating this procedure at a different temperature yields a peak shifted on the $\ln t$ scale. Figure 3.10 depicts two curves of It vs. $\ln t$, calculated at two temperatures, T_1 and T_2; t is in regular time units, say, seconds. Two peaks are seen having maxima of $\ln \gamma_1$ at T_1 and $\ln \gamma_2$ at T_2. The shift $\Delta = \ln \gamma_2 - \ln \gamma_1$ is thus defined. Using these two maxima, one gets

$$\Delta = \ln \gamma_2 - \ln \gamma_1 = (E_t/k)\,(1/T_2 - 1/T_1) \tag{3.19}$$

in which all the magnitudes except for E_t are measurable, and E_t can therefore be evaluated. Chen and Kristianpoller[20] have studied the shape of the TL-like curve for different values of the kinetic order b. The original differential equation here is Eq. 3.10, the solution of which is Eq. 3.11. Defining a (concentration dependent) frequency factor $s = s' n_0^{b-1}$(having units of s^{-1}), they gave the

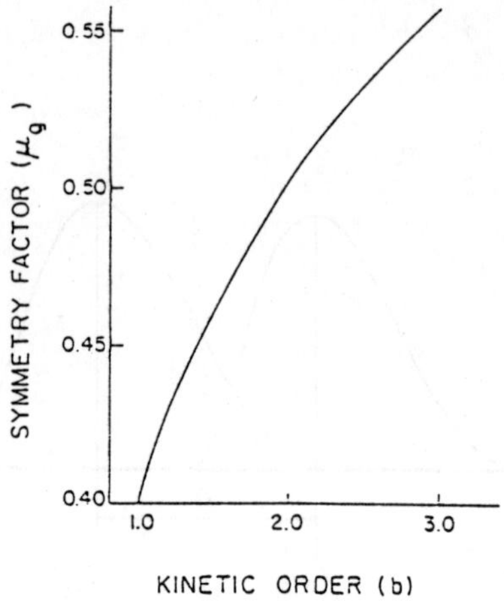

Figure 3.11 The symmetry factor as a function of the kinetic order b for the TL-like presentation of phosphorescence.
(After Chen and Kristianpoller[20].)

expression for phosphorescence decay

$$I(t) = sn_0 \exp\{-E_t/kT\} / [1 + (b-1)s \exp\{-E_t/kT\} t]^{b/(b-1)} . \tag{3.20}$$

Using the same definitions for x and y as before, they found

$$y = \frac{n_0 \exp\{x\}/\gamma}{[1 + (b-1)\exp\{x\}/\gamma]^{b/(b-1)}} \tag{3.21}$$

where, again, $\gamma = \exp\{x_{\max}\}$. As mentioned, this is very similar to the general-order TL peak with a linear heating function, and the same values of E_t, s', n_0, and b. Chen and Kristianpoller[20] found a one-to-one correlation between the kinetic order b and the "symmetry factor" μ_g defined as $\mu_g = \delta/\tau$, with $\delta = y_2 - y_m$; $\tau = y_2 - y_1$ and where y_m, y_1, y_2 are, respectively, the maximum value and the two half-intensity values on any of the peaks of the form shown in Fig. 3.9. A plot of μ_g vs. b is given in Fig. 3.11.

A method which, in principle, does the opposite - namely, transforms thermograms to equivalent isothermal curves - has been suggested by Ozawa[21]. Ozawa defined a reduced time

$$\tau = \int_0^t \exp\{-E/kT\}\, d\theta \tag{3.22}$$

where θ is the time dummy variable and T is a function of time. He showed that, for first-order kinetics, one has

$$-\ln(1-c) = z\tau \quad (3.23)$$

where c is the fraction of material reacted upon (analogous to the instantaneous concentration $n_0 - n$ in TL theory), and z is the pre-exponential (frequency) factor. The first-order property is associated with Eq. 3.23 by the fact that the decay of the amount of material not reacted upon, $1-c$, decays exponentially with the reduced time,

$$1 - c = \exp\{-z\tau\}\,. \quad (3.24)$$

Ozawa[21] suggested to plot $-\ln(1-c)$ vs. τ, and the first-order attribute is established if a straight line is seen. The identification of n-th order reactions, as well as other kinds of kinetics relevant to thermoanalytical measurements, are also discussed by Ozawa.

3.3 The Initial Rise Methods

The simplest, and most generally applicable method for evaluating the activation energy E_t of a single TL peak, as well as practically all the other thermally stimulated processes, is the initial rise method. The basic premise upon which this method is based is that at the low temperature end of the peak, all the relevant occupancies of the states in question - the trap, the recombination center and, in some cases, other interactive states, can be considered as being approximately constant. The rise of the measured intensity as a function of temperature in this region is, therefore, very close to exponential, thus

$$I(T) = C\exp\{-E_t/kT\} \quad (3.25)$$

where the constant C includes all the dependencies on the other parameters and occupancies as exemplified below for a number of different cases.

Following Garlick and Gibson[22] we can write Eq. 3.25 as

$$\ln(I) = C_1 - E_t/kT. \quad (3.26)$$

Thus, if I is plotted as a function of $1/T$, a straight line is expected in the initial rise range, with the slope of $-E_t/k$, from which the activation energy E_t is readily found. The equivalent of numerically finding the best fit between the initial rise range of the glow peak and the experimental function is obvious (see e.g. Chen and Haber[23]), from which the parameter to be found is, of course, the activation energy E_t. The parallel method of finding the value of E in a chemical thermally stimulated processes (e.g. DTA or DTG), is usually termed the "Arrhenius plot".

In order to illustrate the applicability of the initial rise method, let us return to the simple Randall-Wilkins[2] first-order equation

$$I = sn \exp(-E_t/kT). \tag{3.27}$$

As long as n varies only slightly, $n \cong n_0$, Eq. 3.25 holds, and the utilization of the initial rise method is straightforward. In a different way, considering the solution of the Randall Wilkins equation

$$I = sn_0 \exp\{-E_t/kT\} \exp\left[-(s/\beta)\int_{T_0}^{T} \exp\{-E_t/k\theta\}\, d\theta\right], \tag{3.28}$$

we can say that, as long as $T \cong T_0$, the integral in the second exponent is nearly zero, which makes the second exponent close to unity and practically constant in this region, and therefore we get again

$$I \cong sn_0 \exp\{-E_t/kT\}. \tag{3.29}$$

The situation is very much the same in the more general situation of the "GOT" equation resulting from the three original simultaneous equations 3.1-3.3 with the quasiequilibrium assumption,

$$I = sn \exp\{-E_t/kT\}\, A_m m/\left[A_m m + A(N-n)\right]. \tag{3.30}$$

As long as only the initial region is concerned, $n \cong n_0$ and $m \cong m_0$ can be considered to be constants, and therefore, again, $I(T) \propto \exp\{-E_t/kT\}$ and the initial rise techniques can be utilized to find E_t.

Similar considerations can be made directly in reference to the three simultaneous differential equations 2.17, 2.19 and 2.20 governing a single TL peak. From $I = A_m m n_c$ one can see that as long as m is practically constant, the emitted light intensity $I(T)$ is proportional to $n_c(T)$. From Eq. 2.17, however, $dn_c/dt = sn \exp\{-E_t/kT\} - n_c[mA_m + (N-n)A]$ and it can be seen that since dn_c/dt is substantially smaller than each of the two terms on the right hand side, one gets $n_c(T) \propto \exp\{-E_t/kT\}$ and therefore, $I(T) \propto \exp\{-E_t/kT\}$. This analysis does not differ substantially from the discussion of Eq. 3.30 except that, as an important by-product, it is established that $n_c(T)$ also behaves exponentially with temperature initially, and, therefore, the initial rise method can be utilized with TSC peaks as well. The same is true, with some reservations, with a TSEE peak as well as the other thermally stimulated processes.

An example of the simple use of the initial rise method is shown in Fig. 3.12, taken from Lörincz et al.[24]. Here, a plot of TL intensity is given, on a

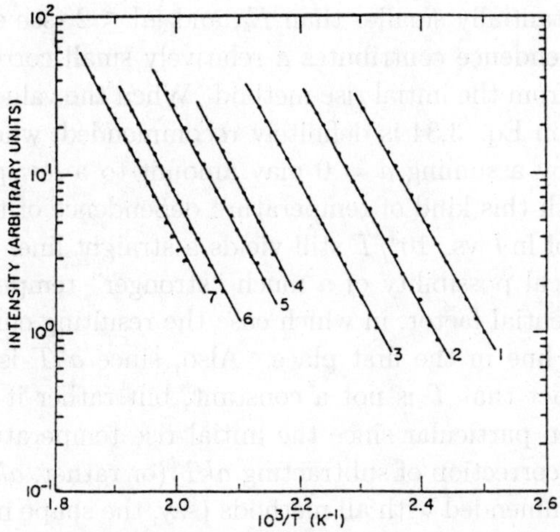

Figure 3.12 Initial rise of the 710 nm emission of TL intensity in Spinel, preheated to (1) 164 ^{o}C, (2) 187 ^{o}C, (3) 204 ^{o}C, (4) 234 ^{o}C, (5) 241 ^{o}C, (6) 256 ^{o}C, (7) 276 ^{o}C.

(After Lörincz et al.[24].)

logarithmic scale vs. $10^3/T$, as measured from the initial rise range of TL glow curves in γ- and x-irradiated Al_2O_3:Cr^{3+} samples, previously heated to various temperatures between 140 °C and 290 °C. The activation energy resulting from the straight lines is 1.5 ± 0.1 eV. A modification of the initial rise method, suggested first by Aramu et al.[25], is related to the possibility that the pre-exponential factor depends on temperature like T^a. In the initial rise range, the intensity of TL (or any other thermally stimulated phenomenon) behaves like

$$I(T) \propto T^a \exp\{-E_t/kT\}. \tag{3.31}$$

From Eq. 3.25 it is obvious that evaluating E_t by the initial rise method is equivalent to finding the energy

$$E_{ir} = -k\,[d\ln I/d(1/T)]. \tag{3.32}$$

With the temperature dependent pre-exponential factor, Eq. 3.32 yields

$$E_{ir} = E_t + akT. \tag{3.33}$$

from which E_t can be found once the power of temperature dependence a is known,

$$E_t = E_{ir} - akT. \tag{3.34}$$

Since kT is substantially smaller than E_t, and $|a| \leq 2$ (see e.g. Chen[26]), this temperature dependence contributes a relatively small correction to the value directly derived from the initial rise method. When the value of a is known, the correction given in Eq. 3.34 is definitely recommended; when it is not known, the error made by assuming $a = 0$ may amount to a few percent. The main point is that, with this kind of temperature dependence of the pre-exponential factor, the plot of $\ln I$ vs. $10^3/T$ still yields a straight line. This is in contrast to the hypothetical possibility of a much "stronger" temperature dependence of the pre-exponential factor, in which case the resulting curve is not expected to be a straight line in the first place. Also, since akT is a relatively small correction, the fact that T is not a constant, but rather it is a variable, is of no significance, in particular since the initial rise temperature range is rather small. A similar correction of subtracting akT (or rather, akT_m) will be shown below to be recommended with all methods (say, the shape methods) where the pre-exponential factor goes like T^a.

A discussion of the consequence of the quasiequilibrium (QE) approximation on the results of the initial rise analysis of TL and TSC peaks has been given by Lewandowski and colleagues[27]-[29]. This discussion can be understood by examining the general equation for first-order kinetics as described by Eq. 2.110, re-written here thus:

$$I_{TL} = \frac{n_0}{Q} \exp\left\{-\frac{E_t}{kT}\right\} \exp\left[-\frac{s}{\beta}\int_{T_0}^{T} \exp\left\{-\frac{E_t}{k\theta}\right\} d\theta\right]. \tag{3.35}$$

where all the terms have been previously defined. In the initial rise region, when $T \approx T_0$, Eq. 3.35 produces:

$$\ln(I_{TL}) = \ln(n_0) + \ln(s) - \ln(Q) - \frac{E_t}{kT}. \tag{3.36}$$

Since n_0 is a constant, independent of T, differentiating Eq. 3.36 with respect to $\delta = 1/T$ yields:

$$\frac{d\ln(I_{TL})}{d\delta} = \left(-\frac{1}{k}\right)\left(-\frac{k}{s}\frac{ds}{d\delta} + \frac{k}{Q}\frac{dQ}{d\delta} + E_t\right). \tag{3.37}$$

from which we see not only the correction term necessary for the temperature-dependence of the frequency factor ($d\ln(s)/d\delta) = aT$, when $s = T^a$), but also an additional term due to the temperature-dependence of Q. Thus, the conditions for the initial rise analysis to be valid are not only that $T \approx T_0$ (i.e. $n \approx n_0$), but also that $d\ln(Q)/d\delta \approx 0$. Note that the QE approximation is not necessary for the initial rise method to be valid since $Q = 1$ is just a special condition of

$d\ln(Q)/d\delta \approx 0$. Hence, the initial rise method is independent of the closeness to QE, as well as the order of kinetics. Indeed, in all of the theoretical situations examined by Lewandowski and colleagues[27],[28] - e.g. the data illustrated in Fig. 2.18 - the slope of the Q-function over the initial rise region was small compared to the value of Q itself - i.e. $(1/Q)\,dQ/d\delta \approx 0$. Thus, the initial rise method is indeed seen to have a wide generality.

As mentioned already, it is quite common to have in TL, as well as in other thermally stimulated processes, several overlapping peaks. If, for example, the glow curve consists of a main peak accompanied by *satellites* at lower temperatures, an effective way of measuring the activation energy of the main peak is by a thermal bleaching of the satellites. In this very simple experimental procedure, the sample is heated up to, say, half of its maximum intensity and held at that temperature for a certain period of time. Although this would decrease the overall intensity, it would effectively eliminate the lower-temperature accompanying peaks. This method has been further developed by Nahum and Halperin[30], in such a way that the sample was subsequently heated and cooled several times. For each cycle, $\log I$ vs. $1/T$ is plotted and, therefore, the activation energies of a series of overlapping peaks, can be found. The energies thus found can be plotted as a function of the temperature which is, say, the middle of the limited range of each heating. An example of this method, termed by Gobrecht and Hoffman[31] the "fractional glow technique" (FGT) is shown in Fig. 3.13. This figure, taken from Kirsh et al.[32], shows the fractional glow curves from albite and microcline. The resulting activation energies as a function of temperature are seen in Fig. 3.14. For some additional work on FGT see the papers by Firszt and Oczkowski[33] and Chruścińska[34].

It is interesting to note that exactly the same method has been used by Bell and Sizmann[35] for determining activation energies from thermal annealing curves, and by Creswell and Perlman[36] for studying thermally stimulated currents.

Halperin et al.[37] suggested an important improvement of the initial rise method which extends the usable range by plotting $\log(In^{-b})$ vs. $1/T$ when the kinetic order b is known. This method is based on the use of Eq. 2.32 (repeated as Eq. 3.10 in this chapter), according to which

$$I/n^b = s' \exp\{-E_t/kT\}\,. \tag{3.38}$$

It is quite obvious that whereas the plot of $\ln I$ vs. $1/T$ normally yields a straight line only up to ~5% of the maximum intensity[23], using the correct value of b and plotting $\ln(I/n^b)$ vs. $1/T$ would extend this range substantially, thus enabling the evaluation of the activation energy for low intensity peaks. In order to be able to use this method, one has to evaluate the value of n for different points along the glow peak. Suppose T_i is a temperature point

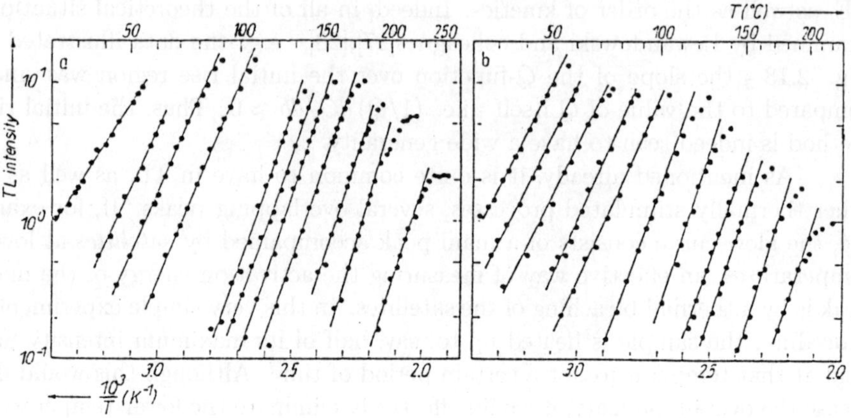

Figure 3.13 Fractional glow curve for (a) albite and (b) microcline measured in the blue emission region. The dots represent experimental results and the straight lines are the least-square best fit of the linear part.

(After Kirsh et al.[32].)

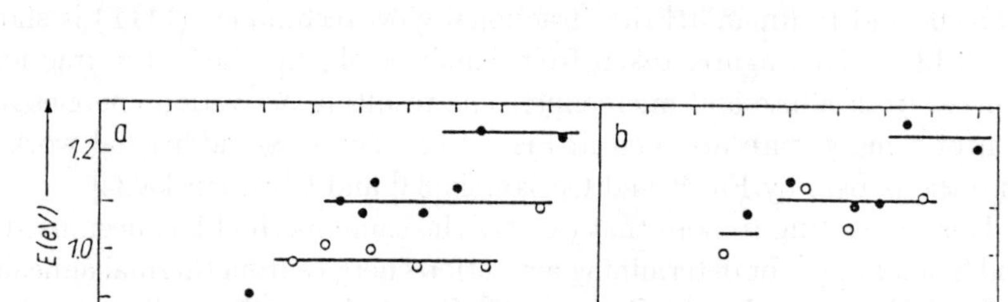

Figure 3.14 Activation energies from the FGT as a function of temperature. The lines represent the average activation energies found by the FGT for (a) albite and (b) microline. The circles represent individual measurements of blue emission (full circles) and red emission (empty circles).

(After Kirsh et al.[32].)

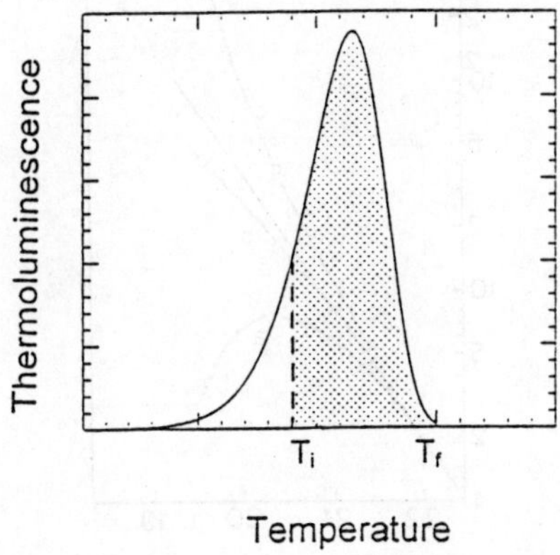

Figure 3.15 A schematic isolated glow peak. The shaded area is proportional to the concentration of carriers in the trap at the temperature T_i.
(Redrawn from Chen and Kirsh[38].)

somewhere along the TL peak and T_f is the temperature at the end of the peak (theoretically $T_f = \infty$, however any point at the high temperature side of the peak where TL is negligibly small can be used). Assume that t_i and t_f are the corresponding times. Let us consider the shaded area in Fig. 3.15 , taken from Chen and Kirsh[38]. From the basic relation $I = -dn/dt$, we immediately get

$$n(T_i) = \int_{t_i}^{t_f} I dt = (1/\beta) \int_{T_i}^{T_f} I dT \tag{3.39}$$

where β is the constant heating rate. Thus, $n(T_i)$ is evaluated for different temperatures and, therefore, $\ln(I_i/n_i^b)$ can be plotted as a function of $1/T_i$. In case b is not known beforehand, the process can be repeated a number of times with different values of b, and the best straight line searched for. An example is shown in Fig. 3.16, taken from Halperin et al.[37] . The results here are for a glow peak of NaCl at 490 K. The best straight line is that of $\ln(I/n)$ which gives the best activation energy and yields the order of kinetics b as well. Similar results, taken from Muntoni et al.[39] of TL peaks in ZnS (~80 K) and NaCl (590 °C = 863 K) are shown in Fig. 3.17 Here too, different values of the kinetic order are tried, and the best straight lines are found with $b = 1.35$ in the former and $b = 1.5$ in the latter case.

A third example, taken from Dražič and Trontelj[40], is given in Fig. 3.18, which shows the results for a TL peak at 313 °C in $CaSO_4$:Dy. Here, trial

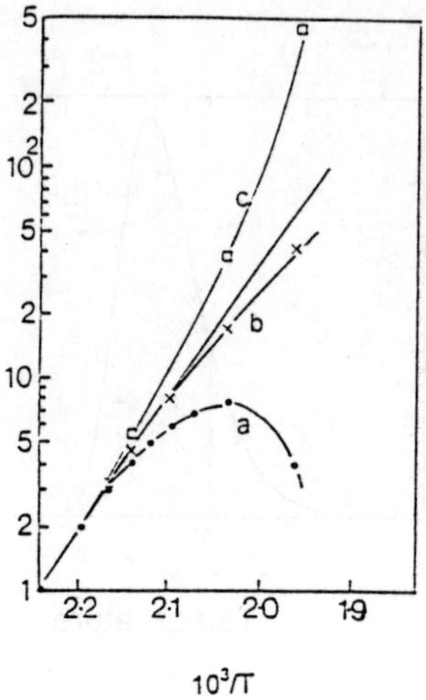

Figure 3.16 A glow peak of NaCl at 490 K plotted vs. $1/T$ with (a) log$I$, (b) log(I/n) and (c) log(I/n^2) in the ordinate. The best results are seen for first-order kinetics.

(Redrawn from Halperin et al.[37].)

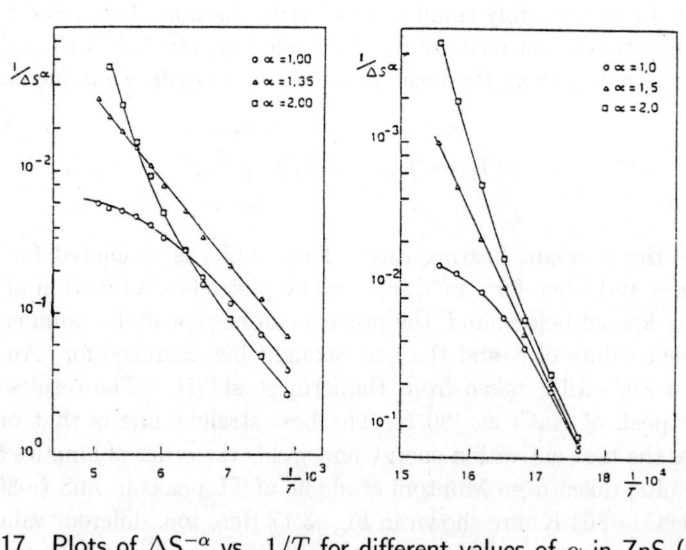

Figure 3.17 Plots of $\Delta S^{-\alpha}$ vs. $1/T$ for different values of α in ZnS (left hand side) and NaCl (right hand side).

(After Muntoni et al.[39].)

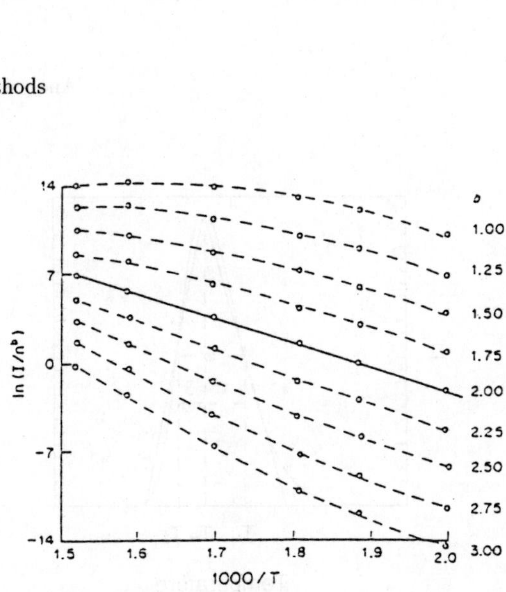

Figure 3.18 Plots of $\ln(I/n^b)$ against $1000/T$, for several assumed values of b, used to calculate of the activation energy from peak III in $CaSO_4$. (After Dražič and Trontelj[40].)

values of b between 1.0 and 3.0 were taken, and the best straight line was found for $b = 2$, establishing the second-order kinetics for this case. It is interesting to note that practically the same method for evaluating the activation energy from thermogravimetric data was suggested by Farmer as reported by Sharp and Wentworth[41]. If w denotes here the reaction mass fraction, they suggested to plot $\ln[-(dw/dt)w^{-b}]$ as a function of $1/T$, and from the slope of the straight line, to evaluate the activation energy, and the frequency factor.

Apart from the experimental difficulty related to the fact that the measured points are taken at the low intensity portion of the peak, Bräunlich[42] pointed out another limitation to the use of the initial rise method for TL peaks. He showed that if the relevant trapping state is initially filled to saturation and if retrapping is much stronger than recombination, $A_n \gg A_m$, the activation energy evaluated by the initial rise method may be too low by up to a factor of 2 (for $A_n/A_m = 10^3$) as compared to the real value. This difficulty can, however, be overcome by repeating the measurement using a significantly lower dose, so that the trap filling is farther from saturation in cases where the situation discussed by Bräunlich prevails.

An important advantage of the initial rise method as compared to some of the other ways of determining the activation energy is that the measured glow depends only on the temperature and not on the heating rate in the initial rise range. This calls for the use of a relatively small heating rate since it eliminates or, at least substantially reduces, the experimental difficulties related

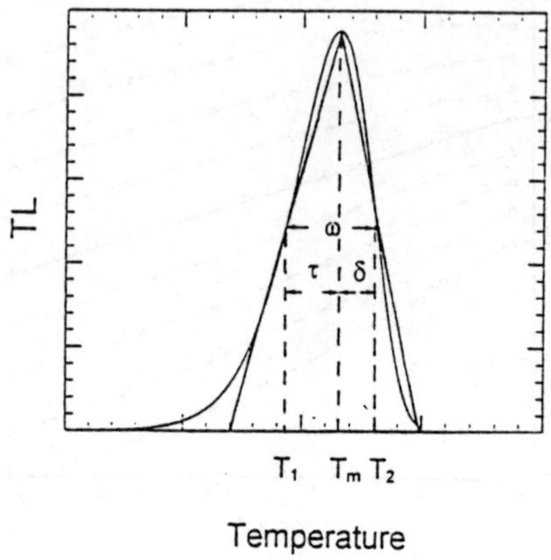

Figure 3.19 A schematic thermoluminescence peak showing the maximum temperature T_m, the low and high temperatures of half intensity T_1 and T_2 and the derived half widths $\omega = T_2 - T_1$, $\delta = T_2 - T_m$ and $\tau = T_m - T_1$.
(Redrawn from Chen and Kirsh[38].)

to temperature differences between the sample and the temperature reading device, as well as problems of temperature gradients within the sample.

3.4 Peak Shape Methods

3.4.1 Half Maximum Intensity Methods

Both the initial rise method described in the previous section and the curve fitting methods to be discussed in section 3.6 can qualify as being peak shape methods. However, in the generally accepted terminology in the field, the term *peak shape method* is reserved to methods in which the kinetic parameters are evaluated using a small number of points extracted from the glow curve. Usually, the temperature of maximum intensity T_m is used, as well as the two temperatures of half maximum intensity T_1 and T_2. A schematic TL peak is shown in Fig. 3.19 in which T_1, T_m and T_2 are shown, as well as the derived parameters $\omega = T_2 - T_1$, $\delta = T_2 - T_m$ and $\tau = T_m - T_1$. The triangles to which different parts of the area under the glow peak are compared in developing the expression for the activation energy are also shown. A very simple, although not very accurate, method which depends only on the maximum temperature T_m has been given by Urbach[43]. From TL experiments on KCl, he deduced empirically that $E_t = T_m/500$ where T_m is the absolute temperature in K and E_t the activation energy in eV. Later authors showed that the activation energy thus found can be inaccurate by a factor of up to 2, and therefore, the frequency factor determined from this may be wrong by several orders of magnitude.

The first method in this family was developed by Grossweiner[44], who dealt with first-order TL peaks, and developed the equation

$$E_t = 1.41kT_mT_1/\tau. \tag{3.40}$$

In fact, Grossweiner gave a coefficient of 1.51 in Eq. 3.40 which he found using only the first term in the asymptotic series utilized for evaluating the exponential integral (see Section 6.1). Dussel and Bube[45] changed the coefficient to 1.41 by the use of more terms in the series.

Another method in the same family was developed by Lushchik[46] who dealt with both first- and second-order TL peaks observed with a constant heating rate. For the sake of demonstration, we shall show here the method for developing Lushchik's equation for first-order TL peaks. Lushchik assumed that the area under the curve for $T \geq T_m$ is the same as that of a triangle with the same height and the same high-temperature half-width $\delta = T_2 - T_m$. Fig. 3.19 depicts schematically a TL peak as well as the triangle the area of which is assumed by Lushchik to be equal to that of the high temperature half of the peak. The *triangle assumption* can be expressed as

$$\delta I_m = \beta n_m, \tag{3.41}$$

where I_m is the maximum intensity, β is the (constant) heating rate and n_m is the concentration of carriers at the maximum. This implies that

$$n_m = \int_{t_m}^{\infty} I dt. \tag{3.42}$$

From Eq. 3.27 one gets, for first-order kinetics, $I/n = s\exp\{-E_t/kT\}$ which reads at the maximum

$$I_m/n_m = s\exp\{-E_t/kT_m\}. \tag{3.43}$$

Rewriting the maximum condition for a first-order TL peak, Eq. 1.5,

$$s\exp\{-E_t/kT_m\} = \beta E_t/(kT_m^2), \tag{3.44}$$

one immediately gets

$$I_m/n_m = \beta E_t/(kT_m^2). \tag{3.45}$$

Utilizing the assumption in Eq. 3.41 gives

$$\beta/\delta = \beta E_t/(kT_m^2) \tag{3.46}$$

from which the Lushchik formula for first-order peaks emerges,

$$E_t = kT_m^2/\delta. \tag{3.47}$$

In a similar way Lushchik found the following expression for second-order peaks

$$E_t = 2kT_m^2/\delta. \tag{3.48}$$

Chen[26] questioned the triangle assumption and tested its validity for a wide range of activation energies and pre-exponential factors for both first- and second-order peaks. Instead of the Lushchik assumption expressed in Eq. 3.41, which can be written as $\delta I_m/\beta n_m = 1$, one can write

$$\delta I_m/\beta n_m = c_\delta, \tag{3.49}$$

and test:

1. To what extent c_δ is constant, namely invariant with variations of E_t and s.

2. By how much c_δ (or, rather, its average over the ranges of interest of E_t and s) is different from unity.

For both first and second order, the variations of c_δ with s (or s') and E_t were confirmed by numerical calculations to be very small, although the values of c_δ were very much different for first and second order. The (nearly constant) value of c_δ for first-order peaks was found to be $c_\delta = 0.976$. For second-order peaks, c_δ was determined to be 0.85, again, with only minor variations with the activation energies and the pre-exponential factors. For the second-order case Eq. 3.48 should, therefore, be multiplied by 0.85 so that the amended, more accurate formula for second order is

$$E_t = 1.7kT_m^2/\delta. \tag{3.50}$$

In the first-order case, once E_t is evaluated, s can immediately be determined by substituting the value of E_t into Eq. 3.44. When E_t is given by an expression such as Eq. 3.47, preferably with the correction of multiplication by the coefficient 0.976, one can get as an explicit function for s, thus

$$s = (0.976\beta/\delta) \exp\{0.976 T_m/\delta\}. \tag{3.51}$$

A similar use can be made of expressions like Eq. 3.40.

As mentioned already, a dependence of the pre-exponential factor on the temperature like T^a where $-2 \leq a \leq 2$ is rather common. Keating[47] was the first who developed a method, based on the shape of a first-order peak, which takes into account this temperature dependence. He derived the equation

$$kT_m/E_t = (1.2\gamma - 0.54) + 0.0055 - \left[\frac{1}{2}(\gamma - 0.75)\right]^2 \tag{3.52}$$

with $\gamma = \delta/\tau$. Although this equation was expected to be workable for any value of a, the approximations made in developing it introduced large errors in the results. Also, the method is too tedious to be used routinely, in particular since methods are available which are both easier to use and are more accurate.

Halperin and Braner[48] developed a method based on the measurement of T_m and τ for TL peaks. They distinguished between first- and second-order cases and between cases of excitation of carriers into a band (where $s \propto T^2$) and into excited states within the forbidden gap prior to recombination, in which case the pre-exponential factor is independent of the temperature. They utilized Lushchik's assumption, along with some other approximations. As shown later by Chen[26], their assumptions are equivalent to the single one that the *low temperature* half of the peak fulfils the condition that the area under it is equal to that of the triangle with the same height and half width (see, again, Fig. 3.19). This means in our notation that

$$\beta(n_0 - n_m) \approx \tau I_m. \tag{3.53}$$

By using the first two terms in the asymptotic series (see Ch. 6), Halperin and Braner[48] found that for first-order kinetics and s independent of T,

$$E_t = 1.72\left(kT_m^2/\tau\right)(1 - 1.58\Delta) \tag{3.54}$$

where $\Delta = 2kT_m/E_t$. They also found similar equations for the other three cases, one in which $s \propto T^2$ and the corresponding two for second-order kinetics. For example, the first-order case with $s \propto T^2$ yields a similar equation with 2.58 replacing 1.58. Although any method based upon the use of τ has certain advantages over those making use of δ or ω, Eq. 3.54 has two shortcomings. First, it should be used iteratively since the evaluation of Δ includes the knowledge of E_t. One should, say, start with assuming $\Delta = 0$, get a first value for E_t, find thus an approximation of Δ and re-insert it into Eq. 3.54. One should now repeat this procedure a number of times and get a self consistent solution of the equation. Alternatively, one can consider Eq. 3.54 as a quadratic equation in E_t. When trying to solve the quadratic equation, one faces the problem of having to choose between two solutions. The other problem is related to the fact that the triangle approximation (Eq. 3.53) may not be very accurate. Using similar considerations to those discussed above regarding the Lushchik equation 3.47, Chen[26] gave a better version of Halperin and Braner's formula, namely

$$E_t = 1.52kT_m^2/\tau - 1.58(2kT_m), \tag{3.55}$$

which is non-iterative and more accurate than the original version. For the case of $s \propto T^2$, the coefficient 1.58 is replaced, like in the Halperin and Braner version, by 2.58. This amounts to a reduction of $2kT_m$ from the value given

	First Order			Second Order		
	τ	δ	ω	τ	δ	ω
c_α	1.51	0.976	2.52	1.81	1.71	3.54
b_α	1.58+$a/2$	$a/2$	1+$a/2$	2+$a/2$	$a/2$	1+$a/2$

Table 3.2 Coefficients appearing in Eq.3.59 for various shape methods of calculating activation energies. a in this table is the power of the temperature dependence of the pre-exponential factor; usually, $-2 \leq a \leq 2$.

by Eq. 3.55. Chen[26] also developed similar methods based on the full width at half maximum $\omega = T_2 - T_1$ along with the maximum temperature T_m. The *triangle assumption* for this case is $\omega I_m/\beta n_0 \approx 1$. To complete the analogy with the previous treatment, we can define

$$c_\omega = \omega I_m/\beta n_0 \tag{3.56}$$

and check numerically, for broad ranges of E_t and s, whether c_ω is practically a constant, and, if so, by how much this constant differs from unity. This has been tested numerically for activation energies between 0.1 and 2.0 eV and for pre-exponential factors in the range of $10^5 - 10^{13}$ s^{-1}. The value found for first-order peaks was $c_\omega = 0.92 \pm 0.8\%$[26]. It is quite surprising, although very useful, that the variation ($\pm 0.8\%$) is so small for such a broad range of E_t and s, which means, in other words, that c_ω is constant for all practical purposes. Moreover, the value of c_ω was found to be practically independent of a in the power dependence of the frequency factor, $s \propto T^a$(with $-2 \leq a \leq 2$). For second-order peaks, a value of $c_\omega = 0.88$ was found with a slightly larger variation with the activation energies and frequency factors. The resulting first-order equation is

$$E_{\omega 1} = 2.52kT_m^2/\omega - \left(1 + \frac{1}{2}a\right) 2kT_m, \tag{3.57}$$

and for second order

$$E_{\omega 2} = 3.54kT_m^2/\omega - \left(1 + \frac{1}{2}a\right) 2kT_m. \tag{3.58}$$

Equations 3.47, 3.48, 3.50, 3.55, 3.57 and 3.58 can be summarized as

$$E_\alpha = c_\alpha \left(kT_m^2/\alpha\right) - b_\alpha(2kT_m) \tag{3.59}$$

where α stands for δ, τ or ω. The values of c_α and b_α for the three methods and for first- and second- order processes are given in Table 3.2.

For an example of the use of these formulae for a first-order TL peak in a KCl:Ag crystal see the work by Takeuchi et al.[49]. A method similar to that by Grossweiner[44] has been given by Maeta and Sakaguchi[50], utilizing temperature points at $1/m$ of the maximum intensity with m being in the range of $1.1 - 5.0$ (including $m = 2$, dealt with by Grossweiner).

Concerning the distinction between first-order and second-order peaks, Halperin and Braner[48] suggested the use of the *symmetry factor* $\mu_g = \delta/\omega$. As further discussed by Chen[26], a first-order peak is characterized by $\mu_g \cong 0.42$ and a second-order peak by $\mu_g \cong 0.52$, where μ_g is only slightly dependent on the particular activation energy and pre-exponential factor involved.

A somewhat different treatment of the same measured quantities has been suggested by Balarin[51]. He also utilized the symmetry properties of the peak to distinguish between first- and second-order kinetics, but preferred to check the value of δ/τ which should be in the range of $0.7 - 0.8$ for first-order peaks, and $1.05 - 1.2$ for second-order peaks. Balarin gave a monogram to aid in the evaluation of the parameters from the measured glow curve. Another monogram was given by Zsakó[52] for the evaluation of kinetic parameters from the shape of thermogravimetry (TG) curves; this can be utilized as it stands for TL glow peaks also. It should also be mentioned at this point that Fleming[53] tabulated the temperatures at which the intensities of first- and second-order peaks equal one half, two thirds and four fifths of the peak intensity, for given activation energies and maximum temperatures.

For the case of a temperature independent pre-exponential factor ($a = 0$), Chen[26] suggested another variant of Eq. 3.57, namely

$$E_t = 2.29kT_m^2/\omega. \tag{3.60}$$

This is somewhat easier to use, though slightly less accurate, than Eq. 3.57. A similar formula has been developed for annealing curves of first order by Wenzl[54] with a less accurate coefficient of 2.5 instead of 2.29.

It is readily seen that the difference between the temperature independent case ($a = 0$) and those in which $a \neq 0$ results in the subtraction of akT_m from the formula in the latter case. This is the same as described in Eq. 3.34 with respect to the initial rise method as related to the temperature dependent pre-exponential factor. Also to be mentioned are very similar methods for evaluating activation energies of first-order thermal desorption peaks by the use of the total half peak, described by Edwards[55] and Chan et al.[56]. Chen[9] also investigated the general-order case in which $I = -dn/dt \propto n^b$(see Eq. 2.32) for b not necessarily 1 or 2. In this work, the correlation between the geometrical shape factor $\mu_g = \delta/\omega$ defined by Halperin and Braner[48], and the three parameters involved, i.e. the activation energy, E_t, the pre-exponential factor s' (or s) and the order of kinetics b, was studied. Repeating Eq. 2.33, the solution

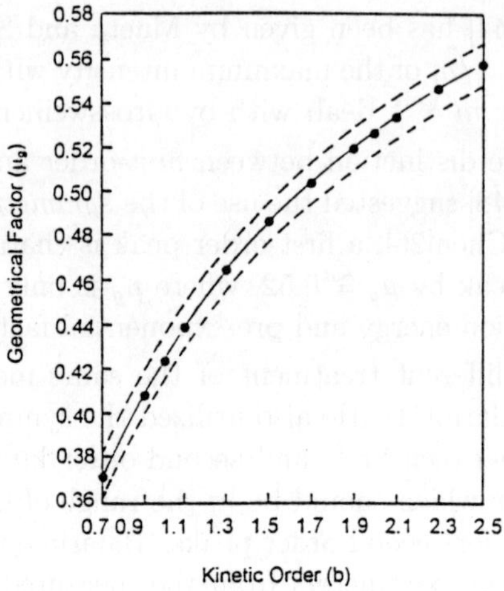

Figure 3.20 Calculated geometrical factor factor μ_g as a function of the given order b. The central curve shows the average values, the upper and lower dashed lines give the largest possible deviations for different values of E_t and s'' for the given order b.

(Redrawn from Chen[9].)

of Eq. 2.32

$$I = s''n_0 \exp\{-E_t/kT\}\left\{1 + \left[(b-1)\frac{s''}{\beta}\right]\int_{T_0}^{T} \exp\{-E_t/k\theta\}\,d\theta\right\}^{-b/(b-1)} \tag{3.61}$$

where $s'' = s'n_0^{b-1}$, we have the basic equation from which the symmetry properties of the general-order peak can be investigated. Chen[9] found that μ_g is practically independent of E_t (for $0.1 \le E_t \le 1.6$ eV) and s'' ($10^5 \le s'' \le 10^{13}$ s^{-1}), and is very strongly dependent on b (for $0.7 \le b \le 2.5$). The results are shown in Fig. 3.20 where the solid line indicates *average* values of μ_g as a function of b for various values of E_t and s'' and the dashed lines indicate the minimum and maximum μ_g for a given b and for extreme values of E_t and s'' within the mentioned ranges. Figure 3.20 can be used to evaluate the effective order from the measured value of μ_g in a single TL peak. Chen[9] has also suggested a method for finding c_α and b_α (in Eq. 3.59) for the non-first- or non-second-order cases, by using the known values of $\mu_g = 0.42$ for first-order and $\mu_g = 0.52$ for second-order, and using the measured value of μ_g as an interpolation (or extrapolation) parameter. The equations for this general case are still based on Eq. 3.59, with the coefficients being

$$c_\tau = 1.51 + 3(\mu_g - 0.42);\ b_\tau = 1.58 + 4.2(\mu_g - 0.42) + a/2, \tag{3.62}$$

$$c_\delta = 0.976 + 7.3(\mu_g - 0.42);\ b_\delta = a/2, \tag{3.63}$$

$$c_\omega = 2.52 + 10.2\,(\mu_g - 0.42);\ b_\omega = 1 + a/2. \tag{3.64}$$

The applicability of these formulae, attained by linear interpolation over μ_g was first checked for numerically generated general-order peaks and gave very satisfactory results. Furthermore, these equations were used for the more general case given by the solution of the three simultaneous equations 3.1-3.3 by Shenker and Chen[57]. The activation energies thus calculated were within 5% of the known value of E_t, initially introduced into the original simultaneous equations.

The methods associated with the general-order kinetics have been successfully used by various authors working on different materials. These include Kristianpoller and Kirsh[58], Spurný and Hobzovà[59], De-Blasi et al.[60], Azorin and Guitérrez[61], Yeh[62] and Cooke et al.[63].

An extension to this kind of formula was given by Singh et al.[64]. They suggest to utilize temperature points at $\frac{2}{3}$ and $\frac{4}{5}$ of the maximum intensity, in addition to the half-maximum intensity discussed above.

3.4.2 Inflection Points Methods

In addition to these methods, based on temperature points at which the glow intensity attains a specified fraction of the maximum intensity, some authors preferred to utilize the two points of inflection, i.e. where $d^2I/dT^2 = 0$, in addition to the maximum temperature. Land[65] studied first-order peaks and developed the formula

$$E_t = kT_mT_i \ln(A/\varphi)/\,|T_i - T_m| \tag{3.65}$$

where $\varphi = (T_i/T_m)^2$ and $A = [3 \pm (5 + 8/X_i)^{1/2}]/2$, and where $X_i = E_t/kT_i$. Land realized the obvious difficulty in using his method, namely, the accurate evaluation of the inflection point. To overcome this, he suggested taking the approximation that the intensity at the first and second inflection points are given by $0.685I_m$ and $0.534I_m$ respectively. This introduces, however, an inaccuracy which is added to those inherent in the development of Eq. 3.65. Furthermore, it turns the method into one which relies on points of predetermined height, rather than the genuine inflection points.

A similar method has been suggested by Yang and Steinberg[66] for the analysis of DTA curves. Here, general-order kinetics are assumed, and the resulting formulae are meant to evaluate all three parameters, namely, the activation energy, the pre-exponential factor and the kinetic order. Gartia et al.[67] studied general-order TL peaks and offered a method for evaluating the three parameters, using the two inflection points. These authors also mention the inherent difficulty of an accurate determination of the inflection point.

3.4.3 Hyperbolic Heating Functions

The discussion in this section, so far, has assumed the use of linear heating functions, namely, constant heating rates, $T(t) = T_0 + \beta t$ where β is a constant. Of course, the use of a different heating function $T = T(t)$ would cause changes in the peak shape. As long as a linear heating function with a given heating rate is used, time dependent TL properties such as $I(t)$ and $n(t)$ can trivially be changed into temperature dependent ones, $I(T)$, $n(T)$, etc. If the heating rate is changed, and, more so, if the heating function is varied, magnitudes like $I(T(t))$ and $n(T(t))$ can rather easily be evaluated. However, the general dependence on the heating function should be considered as a functional of the heating function $T = T(t)$, as suggested by Fillard and Gasiot[68]. These can be written as $I\{T(t)\}$ and $n\{T(t)\}$. The main point here is that these quantities are not anymore unique functions of the temperature even if one considers only monotonically increasing heating functions, which is an obvious requirement. Thus, $I(t)$ and $n(t)$ depend not only on the given values of t and T, but rather on the preceding temperature program $T(t)$, namely, on the *history*. Of course, the use of a certain heating function $T(t)$ would cause changes in the peak shape as compared, say, to a given linear function. The exception to this is the position of the maximum T_m which, as shown in the next section depends only on the *instantaneous* heating rate and, therefore, does not depend on the history, at least in the first-order case.

Having this in mind, one would usually prefer the simplest linear heating function as an accepted standard for TL as well as all other thermally stimulated phenomena. However, several investigators, covering practically all thermally stimulated processes, have pointed out the mathematical advantage of using an apparatus for hyperbolic heating functions. This is a scheme that goes like $T = T_0/(1 - \beta' t)$ where β' (s^{-1}) is a constant. This can be written as $1/T = 1/T_0 - at$ where $a = \beta'/T_0$. With modern computer controlled equipment, this kind of heating can be achieved by rather simple programming. Written in a differential manner, the same heating function can be expressed as

$$-d\,(1/T)\,/dt = \left(1/T^2\right)(dT/dt) = a\,. \tag{3.66}$$

The use of this expression changes the integral appearing in the different TL expressions from

$$\int_0^t \exp\{-E_t/kT\}\,dt \tag{3.67}$$

into

$$a\int_{T_0}^{T} \left(1/\theta^2\right)\exp\{-E_t/k\theta\}\,d\theta \tag{3.68}$$

which is an elementary integral, as opposed to the "usual" expression for constant heating rate

$$\frac{1}{\beta}\int_{T_0}^{T} \exp\{-E_t/k\theta\}\, d\theta\,. \tag{3.69}$$

This approach has been employed by various investigators who studied TL as well as other thermally stimulated processes. For example, we can mention Arnold and Sherwood[69], Kelly and Laubitz[70], Stammers[71] and Christodoulides[72],[73]. Also, the same technique has been utilized for the study of thermal desorption peaks by Ehrlich[74] and for Ionic Thermocurrents (ITC) by Müller and Teltow[75].

The maximum condition for a hyperbolic heating function, (which replaces Eq. 3.44 for the linear heating function) and first-order kinetics is

$$E_t/a = s\exp\{-E_t/kT_m\}\,. \tag{3.70}$$

This can be written as an explicit function of T_m in terms of E_t and s

$$T_m = E_t/\left[k\ln(ska/E_t)\right]. \tag{3.71}$$

A very simple expression is found here for the maximum intensity

$$I_m = n_0 a E_t/ke \tag{3.72}$$

where e is the basis of the natural logarithm, and n_0 is the initial trap concentration. Müller and Teltow[75] found a simple explicit formula for the activation energy by the use of the half-width t_h when the glow curve is measured on a time scale, namely

$$E_t = 2.447/(at_h)\,. \tag{3.73}$$

From the last two equations 3.72 and 3.73, one immediately gets

$$n_0 = eI_m t_h/2.447 = 1.111 I_m t_h. \tag{3.74}$$

This shows that on the time scale and with a hyperbolic heating function, the area of the triangle with the same height and width as the TL peak (or any other TSP peak) is exactly 90% of that under the curve itself. Kelly and Laubitz[70] developed formulae for the evaluation of the activation energies from glow curves measured under hyperbolic heating functions, both for first- and second-order kinetics, based on the measured width of the peak on a temperature scale. Their equations are similar to that by Grossweiner[44]; for first-order they found

$$E_t = 1.46kT_mT_1/\tau \tag{3.75}$$

where T_1 is the low-temperature, half-intensity temperature and τ is the low-temperature half-width. Another formula given by these authors was

$$E_t = 0.985kT_mT_2/\delta \tag{3.76}$$

where T_2 is the high-temperature, half-intensity temperature, and δ the high-temperature half-width.

Second-order peaks, which are nearly symmetrical under linear heating functions, turn out to be exactly symmetrical in reciprocal temperature when hyperbolic heating rates are utilized. Since a "normal" TL glow peak is rather narrow as compared to its temperature of occurrence, $\omega \ll T_m$, these peaks are also nearly symmetrical in the "regular" temperature scale. The two analogous equations for E_t in this case are

$$E_t = 1.763kT_1T_m/\tau = 1.763kT_mT_2/\delta. \tag{3.77}$$

This implies that $T_1/\tau = T_2/\delta$, and using the definitions $\tau = T_m - T_1$, and $\delta = T_2 - T_m$, this readily yields

$$T_m^{-1} = 0.5\left(T_1^{-1} + T_2^{-2}\right) . \tag{3.78}$$

Further work along these lines has been described by Christodoulides[72],[73] who added to this kind of study the possibility of general-order kinetics, i.e. $b \neq 1$, $b \neq 2$.

It is interesting to note that the discussion above concerning the TL-like presentation of phosphorescence gave expressions which are identical to those shown here for TL under hyperbolic heating rates. Thus, for example, the curve shown above concerning the relation between the shape factor μ_g and the kinetics order b would be identical here, in the hyperbolic heating function case, to Fig. 3.11 dealing with the TL-like presentation of phosphorescence. This relation is a unique, one-to-one correspondence between μ_g and b, as opposed to the similar relation with the constant heating rate shown in Fig. 3.20 where there is a slight variation due to different values of E_t and s.

3.4.4 Comparison of the Different Shape Methods

To finalize this section, we compare the relative advantages of the methods included in the general expression - Eq. 3.59 - where α is replaced by τ, δ or ω and the appropriate coefficients are used accordingly. Chen[26] found that the possible theoretical error due to slight variations of c_τ, c_δ and c_ω, within the plausible range of E_t and s, for first-order peaks is largest for the τ method and smallest for the δ method, with the ω method being of intermediate accuracy. For the second-order case, however, the ω formula yields the smallest possible theoretical error - namely, smallest variations of $E_t = E_\omega$ for different pre-exponential factors - and the τ method is still the worst in this sense. The situation is, however, entirely different when one is considering the possible experimental error. The amended Lushchik formula (for E_δ), which should be recommended on theoretical grounds, would actually be useful only when extremely clean peaks, measured under ideal conditions, are observed. The small difference $\delta = T_2 - T_m$ entering the calculation is subject to relatively large errors if considerable inaccuracies are involved in the measurement of the relevant temperatures T_m and T_2. The main factor in the possible error $\Delta\delta = \Delta T_2 + \Delta T_m$ is ΔT_m, since the possible error in determining the temperature is much bigger at a point where the derivative dI/dT is zero (T_m), than in points of high absolute value of the derivative (T_1 and T_2). Thus, $\Delta\delta$ and $\Delta\tau = \Delta T_1 + \Delta T_m$ are expected to be much larger than $\Delta\omega = \Delta T_1 + \Delta T_2$. Moreover, since ω is bigger than τ and δ, the relative errors $\Delta\tau/\tau$ and $\Delta\delta/\delta$ are significantly larger than $\Delta\omega/\omega$. The term T_m (or rather, T_m^2) appearing in the equations contributes only a small additional inaccuracy since $\Delta T_m/T_m$ is very small due to the high value of T_m as compared to τ, δ and ω. In summary, from this point of view, the method using ω is judged to be the best.

Another experimental aspect that should be taken into account is that in many cases a thermally stimulated peak is not "clean" in the sense that it is often accompanied by satellites - namely, smaller TL peaks which may occur adjacent to the main one. As mentioned already, if such a satellite appears at the low temperature half, it can be thermally bleached quite efficiently by heating the sample up to beyond the temperature of the satellite, leaving the main peak little altered. If, however, the satellite appears at the high temperature half of the peak, it cannot be thermally cleaned. From this point of view, the method using the value of τ is expected to be preferred whereas that utilizing δ would probably give inaccurate results. The performance of the ω method will be intermediate in this respect.

When an indeterminacy exists about the order of kinetics, methods using the low-temperature half of the peak, namely, the amended Halperin-Braner and Grossweiner formulae, are preferable since the value of τ is the least sensitive to the order of kinetics of the process. The poorest method in this regard is that using the high-temperature half of the peak, since δ is very sensitive to the order of the kinetics.

3.5 Methods of Various Heating Rates

As discussed, changing the heating function or even changing the heating rate while keeping the heating function linear, alters the shape of the TL peak and, in particular, the maximum temperature T_m. Several methods have been developed, all based on the repeated measurement of a certain TL (or other TSP) glow peak at different heating rates, while keeping all the other parameters, such as the initial trap occupancy n_0, constant. Such methods have independently been developed in the research of different thermally stimulated phenomena. Repeating Eq. 3.44 in a slightly different form we get

$$\beta = (sk/E_t)T_m^2 \exp\{-E_t/kT_m\} \ . \tag{3.79}$$

While going from one measurement in a series to another, increasing the heating rate β on the left-hand side of this equation would necessarily result in an increase in the right-hand side. Since $T_m^2 \exp\{E_t/kT_m\}$ is an always increasing function of T_m, a growth in the value of the function necessarily implies an increase in the argument T_m. Booth[76], Bohun[77] and Parfianovitch[78], all dealing with TL peaks, have independently suggested the use of this shift for the evaluation of E_t and s. Using two different heating rates, β_1 and β_2, one gets two maximum temperatures, T_{m1} and T_{m2}, respectively. Writing Eq. 3.44 twice, for β_1 and β_2, dividing one by the other and rearranging, one gets

$$E_t = [kT_{m1}T_{m2}/(T_{m1} - T_{m2})] \ln[(\beta_1/\beta_2)(T_{m2}/T_{m1})^2] \ . \tag{3.80}$$

Either this expression or the value of E_t thus obtained can be inserted into Eq. 3.44 for either β_1 or β_2, and the value of s can be calculated. Hoogenstraaten[13] proposed the use of several linear heating rates. From Eq. 3.44 it is obvious that with this series of results, a plot of $\ln(T_m^2/\beta)$ vs. $1/T_m$ should yield a straight line with the slope of $-E_t/k$, from which E_t can readily be evaluated. Extrapolation to $1/T_m \rightarrow 0$ gives the value of $\ln(sk/E_t)$ from which s can be calculated by the use of E_t/k found from the slope.

Osada[79] was the first to extend the method to non-linear heating functions. He proved that Eq. 3.44 is valid for an exponential heating function

$$T = T_\infty - (T_\infty - T_0)\exp\{-\alpha t\} \tag{3.81}$$

where α is a constant (with units of s^{-1}) and T_∞ is the final temperature approached asymptotically with time. This heating scheme is what we get "naturally" if we let a "cold" sample warm up while being in thermal contact with an infinite thermal bath at temperature T_∞. The instantaneous heating rate at T_m, which we denote by β_m should now replace β in Eq. 3.44, which leads directly to the validity of Eq. 3.80 for various exponential heating functions (different values of α and/or T_∞), as well as to the use of Hoogenstraaten's method in this case. Chen and Winer[80] showed that Eqs. 3.44 and 3.79 are correct for any first-order peak measured under *any monotonically increasing* heating function. To show this let us rewrite Eq. 1.3 in terms of temperature (rather than time) for a general heating function $T = T(t)$

$$I(T) = n_0 s \exp\{-E_t/kT\} \exp\left[-s\int_{T_0}^{T} \frac{1}{(d\theta/dt)} \exp\{-E_t/k\theta\}\, d\theta\right] \ . \tag{3.82}$$

The assertion of an increasing heating function ensures that dT/dt (i.e. $d\theta/dt$) $\neq 0$. (In fact, $dT/dt > 0$.) Setting the derivative of Eq. 3.82 to zero yields

$$\beta_m E_t / \left(kT_m^2\right) = s \exp\{-E_t/kT_m\} \tag{3.83}$$

where $\beta_m = (dT/dt)_{T_m}$. Equation 3.83 is, of course, a generalization of Eq. 3.44 for any heating function. Here too, by using two heating functions with $\beta_{m1} \neq \beta_{m2}$, one can use Eq. 3.80 for evaluating E_t, and by utilizing various heating functions, one can repeat Hoogenstraaten's method, namely, plotting $\ln(\beta_m/T_m^2)$ vs. $1/T_m$, and readily evaluate the activation energy from the slope and the frequency factor from the intercept with the vertical axis.

As described so far, this method has been developed for first-order kinetics only. However, many investigators have applied it to various TL peaks as well as other thermally stimulated phenomena, without testing whether the first-order condition occurs. Nevertheless, the method turns out to be a very good approximation for practically any kinetics of the process. For a detailed discussion of this point, see Singh et al.[81]. Chen and Winer[80] showed that the methods of various heating rates are applicable for general-order kinetics which includes the second-order case. It should be noted, however, that as opposed to the first-order case, in which Eq. 3.83 (on which these methods rely), is strictly correct, the equivalent equation for other kinds of kinetics are only a good approximation. To demonstrate this point, let us consider the more general equation

$$-dn/dt = K_o F(n) \exp\{-E_t/kT\} \tag{3.84}$$

written by Damask and Dienes[82] and further discussed by Bell and Sizmann[35] in relation to thermal annealing. The expression given by these authors for getting the activation energy by the use of two heating rates β_1 and β_2 is

$$E_t = \frac{kT_{m1}T_{m2}}{T_{m2} - T_{m1}} \ln\left[\frac{\beta_1}{\beta_2}\left(\frac{T_{m1}}{T_{m2}}\right)^2 \frac{(dF/dn)_{T_{m1}}}{(dF/dn)_{T_{m2}}}\right]. \tag{3.85}$$

The function $F(n)$ is assumed to be "normal" in the sense that its first derivative with respect to its argument n is only slowly varying, which results in $(dF/dn)_1/(dF/dn)_2 \cong 1$. Under these circumstances, Eq. 3.85 reduces to Eq. 3.80. Of course, a more general situation is possible, namely, that the TL peak results from the solution of the three differential equations 3.1-3.3, which is not covered by Eq. 3.84. The broad practical success of the method, however, indicates that it is a very good approximation even under such general conditions. Chen and Winer[80] examined this method and found it very successful for computer generated, second-order and general-order TL peaks, as well as for measured TL peaks in $ZnS{:}Er^{3+}$. An example of the latter results is shown in curve b of Fig. 3.21, showing $\ln(\beta_m/T_m^2)$ vs. $1/T_m$ which yields an activation

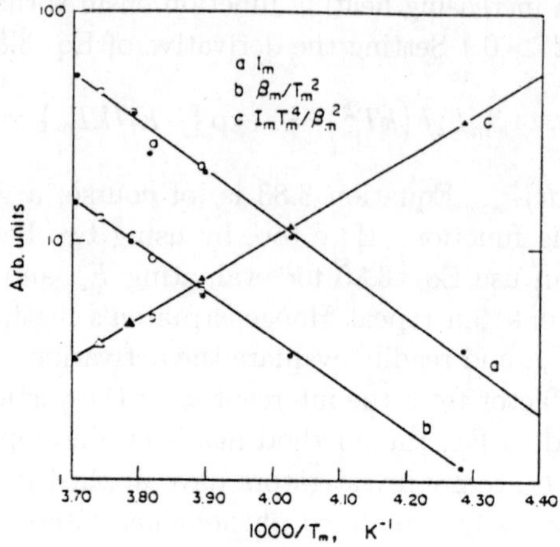

Figure 3.21 Experimental results of $ZnS{:}Er^{3+}$ TL peak at 260 K. Plots Of I_m(curve a), β_m/T_m^2 (curve b) and $I_m T_m^4/\beta_m^2$ (curve c) vs. $1000/T_m$. Some of the points were observed using non-linear heating functions. (After Chen and Winer[80].)

energy of 0.43 eV, in good agreement with the initial rise result of 0.41 eV for the same peak. Similar results of straight lines on this kind of plot for different kinetic orders are given in Fig. 8 in the paper by Townsend et al.[83] for TL in LiF, by Panchenko et al.[84] for TSC in $Bi_{12}SiO_{20}$, by Kissinger[85] for DTA measurements, by Maciejewski[86] for thermogravimetry measurements, and by Gravelle[87] for thermal desorption experiments. The applicability of the various heating rates method for different kinetic orders has been studied by Singh et al.[81].

Examining, again, Eq. 3.79, one can see that the main T_m-dependence of the right hand side is that of the exponent rather than the T_m^2 factor. This fact is the basis for an approximate version of the various heating rates method, namely getting E_t from the slope of $\ln\beta$ vs. $1/T_m$ (or $\ln(\beta_m)$ vs. $1/T_m$ in cases of a nonlinear heating function). This has been done with rather good results by Dussel and Bube[45] for TL peaks and by Ozawa[21] for DTA curves. Ozawa has also developed a method which depends on the shift of the whole peak (rather than that of T_m only).

In a way similar to that discussed with respect to the initial rise and the peak shape methods, one can evaluate the activation energy by subtracting $2kT_m$ from the value obtained from the slope of $\ln(\beta) = f(1/T_m)$. To prove this, let us consider again Eq. 3.79. The approximate activation energy found from the slope of this nearly straight line is

$$E_a = -k\left[d\ln\beta/d(1/T_m)\right] \; . \tag{3.86}$$

By a straightforward differentiation of Eq. 3.79 we get

$$E_a = E_t - 2k\frac{d\,(\ln T_m)}{d(1/T_m)} = E_t + 2kT_m \tag{3.87}$$

from which we find immediately

$$E_t = E_a - 2kT_m \; . \tag{3.88}$$

In the methods described so far only the variation of T_m with varying values of β (or β_m) have been utilized. It will now be shown that the variations of the maximum intensity with the heating rate can also be used for the evaluation of the activation energies. In their investigation of TSC peaks, Haering and Adams[88] showed that for two extreme cases - *slow retrapping* and *fast retrapping* - one has approximately $\sigma(T_m) \propto \exp\{-E_t/kT_m\}$, where $\sigma(T)$ is the conductivity. Therefore, plotting $\ln\sigma(T_m)$ as a function of $1/T_m$ for various heating rates would yield, again, a straight line, from the slope of which E_t can readily be evaluated. Chen and Winer[80] showed that actually, in any first-order peak (the TSC fast- and slow-retrapping cases included) one has

$$I_m \cong sn_0 \exp\{-E_t/kT_m\} \exp\{\Delta_m\} \tag{3.89}$$

where $\Delta_m = 2kT_m/E_t$. Since Δ_m is normally 0.1 or less, $\exp\{\Delta_m\} \cong 1+\Delta_m$, and the small variations in T_m with changing heating rates cause negligible changes in the additional term, leaving the approximation $I_m \propto \exp\{E_t/kT_m\}$. The formula of plotting $\ln(I_m)$ vs. $1/T_m$ should result in a nearly straight line with a slope of $-E_t/k$. In a slightly different manner, differentiation of Eq. 3.89 yields

$$E_t = -k\left[d\ln(I_m)/d(1/T_m)\right] - 2k^2T_m^2/E_t \tag{3.90}$$

which is a better approximation than before. It is obvious, however, that the second term does not contribute more than 1 – 2% of the evaluated value of E_t and, therefore, one gets good results when plotting $\ln(I_m)$ vs. $(1/T_m)$. In principle, this should be limited to various *linear* heating functions, but it seems that if one uses a slowly varying heating rate, the additional error introduced should be tolerable. Curve *a* of Fig. 3.21 depicts an example of a plot of I_m vs. $1/T_m$ on a semilog scale. The slope is seen to be practically the same as in curve *b*.

While dealing with the *general-order* case ($-dn/dt \propto n^b$), Chen and Winer[80], used similar arguments and showed that

$$I_m^{b-1}\left(T_m^2/\beta_m\right)^b = \left(E_t/bks''\right)^b \exp\{E_t/kT_m\} \tag{3.91}$$

which suggests the use of a plot of the logarithm of the left hand side as a function of $1/T_m$ for different heating rates. The apparent advantage of Eq. 3.91 is that it is rigorously correct for any heating function; however, its potential use is very limited since one has to know the kinetic order b in advance. The limiting case of $b \to 1$ brings this expression back to Eq. 3.79, with $s'' = s$. Where the peak is known to be of second-order, one should plot $\ln[I_m(T_m^2/\beta)^2]$ as a function of $1/T_m$ in order to evaluate E_t from the slope. Curve c of Fig. 3.21 shows such a plot. The slope here is very much the same as in curves a and b (except for the sign). This is so in spite of the fact that the peak in question is not of second order. It is obvious however, that the expression which yields line c can be obtained by dividing the expression for I_m in line a by the square of β_m/T_m^2 in line b. Since the latter is strictly accurate for first-order peaks, the approximation implied in line c is the same as that in line a. All in all, the fact that the slopes (in absolute values) of the three curves in Fig. 3.21 are practically the same indicates that all three methods for evaluating E_t are not very sensitive to the kinetic order or to the heating function employed.

As a by-product of the derivation of this method, Chen and Winer[80] suggested a method for evaluating E_t based on the shift of non-first-order peaks as a function of the initial concentration of trapped carriers as shown, for example, in Fig. 2.8(a). (See also the discussion by Kessler[89] related to non-first-order thermally stimulated depolarization peaks.) The right hand side of Eq. 3.91 can be written as $(1/s')(E_t/bk)^b \exp\{E_t/kT_m\}$, where $s' = s'' n_0^{1-b}$ and s' is a constant independent of the initial concentration n_0. Therefore, one can plot $\ln\left[I_m^{b-1}(T_m^2/\beta_m)^b\right]$ vs. $1/T_m$ for various initial concentrations, and get a straight line with slope E_t/k. While going from one measurement to another, one can change both n_0 and the heating rate, and a straight line is still expected. If one wishes to concentrate on variations of n_0 only, the same heating rate can be used in subsequent measurements. One can then plot $\ln\left[I_m^{b-1}T_m^{2b}\right]$ vs. $1/T_m$ for cases of $b \neq 1$. This condition is important since in first-order peaks, T_m is not expected to depend on n_0 at all; actually, only if b is substantially different from unity, would one expect accurate results. For the second-order case, $b = 2$ and the plot will be of $\ln[I_m T_m^4]$ vs. T_m. Figure 3.22 depicts such a plot for computer-generated, second-order peaks. The value of the activation energy used in the simulation was 0.4 eV and the energy evaluated from the slope came out as 0.398 eV. An important point is that the values of n_0 are not directly utilized in this method and, therefore, it is applicable even in cases where the initial concentrations n_0 are not linear with the applied dose (see Ch. 4), provided that the kinetics of the process are not altered.

The pre-exponential factor is, in certain cases, proportional to T^a where $-2 \le a \le 2$. Chen and Winer[80] showed that in this case plotting $\ln(\beta_m/T_m^{a+2})$ vs. $1/T_m$ should give a straight line from the slope of which E_t is found; a

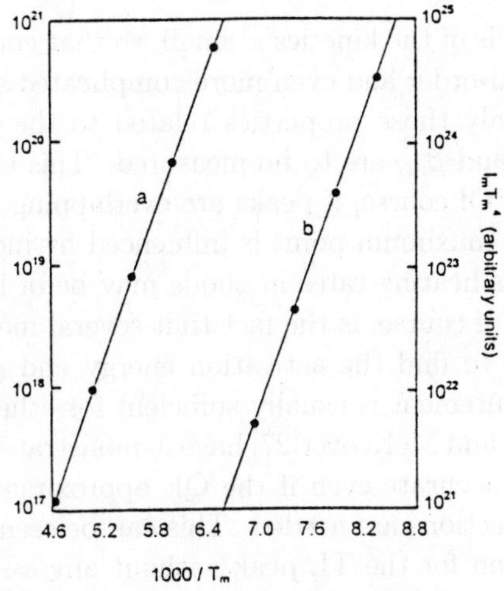

Figure 3.22 $I_m T_m^4$ vs. $1000/T_m$ for various concentrations n_o, for a second-order peak. (b) is a continuation of (a).

(Redrawn from Chen and Winer[80].)

particular case of $a = 3/2$ had previously been discussed by Schön[90] as related to the two-heating-rates version of the method. Equation 3.80 is written for this case as

$$E_t = [kT_{m1}T_{m2}/(T_{m1} - T_{m2})] \ln \left[(\beta_1/\beta_2)(T_{m2}/T_{m1})^{7/2} \right] . \tag{3.92}$$

Alternatively, it can be shown that if one plots $\ln(\beta_m/T_m^2)$ vs. $1/T_m$ for the temperature-dependent (T^a) pre-exponential factor case, the resulting energy would be $E_t + akT_m$ rather than E_t. As prescribed above for the initial rise and shape methods, a better value of E_t can thus be found by subtracting akT_m from the result directly derived from the slope.

3.5.1 *Advantages and Limitations*

These methods of various heating rates transpire from the above discussion as being very general and easy to use in order to determine the basic kinetic parameters of TL as well as other thermally stimulated phenomena. From the experimental point of view it is an advantage that practically any heating scheme can be utilized, including the linear, hyperbolic and exponential functions mentioned here. It has been shown that even if the heating function is not known, the only thing required is the instantaneous heating rate at the maximum point. The main importance of this point is the mentioned fact that the intensity function $I(T)$ is a functional of the heating function $T(t)$, which adds some complication to the study of the whole curve, but remains rather simple as long as only T_m is concerned. It has also been shown that the sensitivity of these methods to the details of the kinetics is small, so that good results are found for second-order, general-order and even more complicated situations. Yet another advantage is that only those properties related to the maximum point itself, namely T_m (or I_m) and β_m, are to be measured. This eliminates the difficulty caused by satellites. Of course, if peaks are overlapping in such a way that the measurement at the maximum point is influenced by more than one peak, the results of the various heating rates methods may be of limited value. Another slight disadvantage, of course, is the fact that several measurements have to be carried out in order to find the activation energy and pre-exponential factor, while only one measurement is usually sufficient for other methods.

Lewandowski and McKeever[27] have demonstrated that the various heating rates method is accurate even if the QE approximation is invalid. In this case, only small corrections are needed. This can be seen from Eq. 2.109 which is a general expression for the TL peak without any assumption of quasiequilibrium, kinetic order, or temperature dependence of s. At the peak maximum we have from Eq. 2.109

$$\ln\left(\frac{T_m^2}{\beta}\right) = \frac{E_t}{kT_m} + \ln\left(\frac{E_t}{ks_m}\right) - \ln\left(\frac{Q_m}{Q_m + P_m} + \frac{d\left\{\ln\left[(Q+P)/s\right]\right\}_m/dT}{(s_m/\beta)\exp\{-E/kT_m\}}\right) \tag{3.93}$$

where the subscript m refers to the values of these parameters at the peak maximum. The third term on the left-hand side is the correction term when non-quasiequilibrium, general-order kinetics and a temperature dependence for s all apply. If one imposes the constraint s = a constant, Q = a constant (a good approximation near $T = T_m$) and $P \ll 1$ (i.e. slow retrapping) then the correction term equates to zero and this generalized form reduces to the standard form (viz. Eq. 3.86).

From the theoretical point of view, the range of heating rates utilized should be as broad as possible. This is limited by the following experimental conditions:

I. At too small heating rates, the peaks are broad and of low intensity, which makes a precise evaluation of the maximum temperature rather difficult.

II. At too high heating rates, there could be some delay between the temperature of the sample and the thermometric device (e.g. a thermocouple) which results, of course, in errors in the temperature reading. While changing the heating rate, the size of the error varies. Moreover, an important complication that normally cannot be avoided, in particular with high heating rates, is the possibility of temperature gradients in the measured sample. This has a bearing on all the methods of extracting information and should, therefore, be discussed in some detail. If this is the case, a given peak would appear as if a separate curve arises from any isothermal region in the sample; the apparently different peaks, shifted from one another, would superimpose to give a

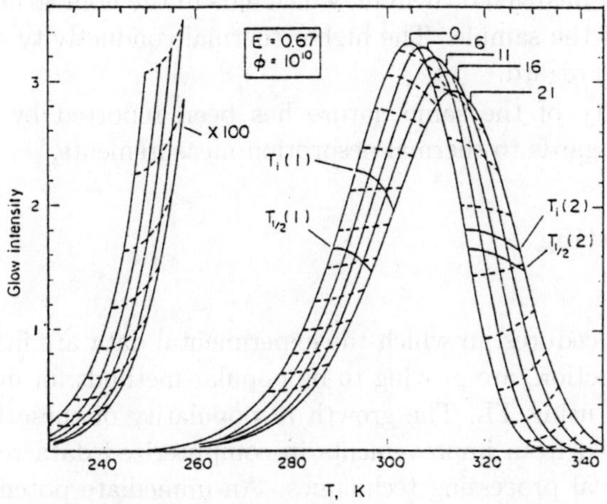

Figure 3.23 The change from a first-order TL peak resulting from a linear temperature gradient in the sample.
(After Land[65].)

thermally stimulated curve which usually does not look like a single peak even if the basic process is, say, of first-order. An example of such a TL peak taken from Land[65] is shown in Fig. 3.23. As pointed out by Land[65], Chvoj[91] and Chvoj and Pokorný[92], the result is a broadening of the given single peak. These authors have also given some detailed calculation as to the shape of the expected curve under these circumstances. For further work along the same lines see Betts et al.[93], Betts and Townsend[94], Kitis et al.[95] and Piters and Bos[96]. In many cases the main aim is to extract the trapping parameters from the measured curve, and this can be done only by eliminating the effect of temperature gradients as much as possible. The obvious precautions which reduce this effect are using a thin sample and relatively slow heating rates. As just mentioned, the latter would reduce the measured signal in all the thermally stimulated processes. As for the use of thin samples, it usually reduces the signal in TL measurements, but may not do so in some of the other phenomena, and may be even of advantage in some of them. A somewhat better way to cope with the problem is to use, whenever possible, the sample in powder form which is placed, or preferably glued, directly on the oven plate which is heated in a pre-determined manner. This is done, for example, in TL measurements of quartz and other minerals in dating of archaeological samples (see Ch. 7). Of course, in other thermally stimulated processes such as TSC, this is not applicable. Another method routinely utilized in the reading of TL from archaeological and geological samples is to fill the sample compartment with an inert gas, usually nitrogen or helium, which aids in the homogeneous heating of different parts of the sample. The higher thermal conductivity of helium is an advantage in this regard.

A difficulty of the same nature has been reported by Chiarena and Gillet[97] with regards to thermal desorption measurements.

3.6 Curve Fitting

3.6.1 Principles

Curve fitting procedures, in which the experimental data are fitted to a previously defined function, are proving to be popular methods for evaluating trapping parameters using TL. The growth in popularity of these techniques has probably stemmed from improvements in computerized data recording methods, and numerical processing techniques. An immediate potential advantage that these methods possess over those described thus far is the fact that the whole curve, or almost the whole curve, is utilized in the analysis, rather than just a few points (as in the peak shape, peak position, or initial rise methods). Clearly, the more data that are used to define the glow curve, the greater the potential for accurate determination of the glow curve parameters.

Unfortunately, this advantage is often offset in real cases by the requirement for "clean" glow peaks with accurately determined shapes. Small distortions in the glow curve can yield erroneous parameter values as the computerized fitting routine attempts to define the "best fit" to the numerical data. For this reason it is essential to define experimentally the error associated with the recording of the glow curve. Any fit to the data which then falls within these error limits should be admitted as acceptable. This in turn defines the uncertainty associated with the evaluated parameters (e.g.[98]). Obtaining undistorted glow curves requires careful consideration of experimental details. It is essential to adopt a small heating rate when performing such analyses in order that the temperature recorded (usually the temperature of the heating planchet) is actually the same as the temperature of the sample, as noted in the previous section. Likewise, great care should be taken to avoid temperature gradients across the sample. (See references[93] and [94].) The number of data points used in the analysis should be sufficient to define the (sometimes) large number of fitting parameters involved in the procedure. A good rule of thumb is to use at least three data points for each parameter. Fortunately, with modern computerized data handling apparatus this is not usually a difficulty, but this consideration can sometimes be overlooked.

Other possibilities for causing distortions in the glow curve shape relate to the topics discussed in Ch. 2 - namely the variation of the emission spectrum with temperature, self-absorption and thermal quenching. Ideally, what one wishes to plot on the vertical axis of a "TL" glow curve is $dn/dT = (1/\beta)dn/dt$, with T on the horizontal axis. In practice, however, one always plots TL intensity in the assumption that this is proportional to dn/dt. In Fig. 2.21 we showed how a changing emission spectrum during TL readout can give rise to a distorted glow curve due to the fact that dn/dt is not proportional to the TL intensity. The figure demonstrated the fact that if one monitors one particular wavelength, a certain glow peak is obtained, whereas if one monitored a different wavelength, a separate peak is obtained. A good experimental example of this was given by Hornyak et al.[99]. Thus, if one analyzed either peak separately, one would get erroneous answers. However, if one added both curves together (i.e. recorded the glow curves without the use of filters) one should get a much better fit. The difficulty here comes from the fact that the sensitivity of most photon detectors is not uniform with respect to the photon energy. Thus, the accurate way to obtain data suitable for analysis in this case is to monitor TL-versus-wavelength-versus-temperature, correct for the detector wavelength response, and add all "slices" along the wavelength axis. These precautions had already been noted by earlier authors[100],[101] and have been stressed again more recently[102].

Self-absorption of the emitted TL can cause further distortions in the glow curve, as was demonstrated in Fig. 2.21. Regrettably, if such events are occurring there is little that can be done to correct for the problem without a detailed knowledge of the emission spectrum and the absorption spectrum over the wavelength and temperature range of interest.

Additional problems may occur as a result of thermal quenching. A decreasing luminescence efficiency as a function of increasing temperature can give rise to gross distortions of the recorded glow curve, making analysis meaningless without proper correction. Wintle[103] first highlighted the potential difficulties posed by thermal quenching during glow curve analysis in an examination of the trapping parameters of quartz. Similar discussions have centered around the analysis of the glow curve from α-Al_2O_3. Kitis et al.[104] and Kortov et al.[105] both discuss the effect of thermal quenching on the shape of the main dosimetric TL peak from single crystal α-Al_2O_3 for which the lower temperature part of the glow curve is favored over the higher temperature region.

All of the above experimental considerations have particular importance for curve fitting procedures since small deviations of the data from the real glow curve shape can lead to the curve fitting minimization routine converging at the wrong minimum, with subsequent large errors in the calculated parameters. (However, similar considerations also apply to all of the analytical methods discussed in this chapter. Whatever the analysis, the value of it is only as good as the original data.)

3.6.2 Equations and Fitting Procedures

Having obtained data suitable for fitting it is then necessary to choose the appropriate equation to fit to the data. The most popular procedure is to either assume a kinetic order (perhaps using data obtained from a study of the variation in the peak position with absorbed dose) or to allow the kinetic order to vary as a free parameter. In the former case one would fit either the Randall-Wilkins first-order equation (Eq. 2.28), or the Garlick-Gibson second-order equation (Eq. 2.31) to the data, using E_t, s and n_0 as fitting parameters (a 3-parameter fit). Alternatively, by using the kinetic order as a fitting parameter one can use the May-Partridge equation (Eq. 2.33) in a 4-parameter fit (E_t, s, n_0 and b). As discussed in Ch. 2, Eqs. 2.28, 2.31 and 2.33 assume no interaction between the trapping centers. Analysis of complex glow curves involving, say, p overlapping peaks usually proceeds by using a simple linear superposition of these equations. Each of the $4p$ parameters is varied independently until a predefined "best fit" criterion is reached. The "best fit" is usually recognized by the minimization of some error function between the experimental data and the linear superposition of the assumed functions. The most popular error function is a simple "chi-squared" (weighted or unweighted), although a "figure-or-merit" function is also popular. Convergence to the minimum of these functions is then taken as indicating the "best fit". Minimization routines, such as Marquardt, or Levenberg-Marquardt, are normally used. The error functions and the minimization routines are discussed more fully in Ch. 6.

Clearly, for p peaks of the above type one is working in $3p$- or $4p$-dimensional parameter space and locating the global minimum, as opposed to a local minimum, is not certain. To aid in this process, estimates for each of the $3p$ or $4p$ parameters should first be obtained using the analytical techniques described in earlier sections of this chapter, and the estimates then used as starting values in the fitting process. Tests to determine if the true minimum has been found should include varying these starting values to verify that the minimization calculation always converges to the same minimum.

Mohan and Chen[106] described the curve fitting method for computer-simulated, isolated first- or second-order peaks, using Eqs. 2.28 and 2.31, while Shenker and Chen[107] discussed the application of the method to general-order kinetics using Eq. 2.33. Bull et al.[108] showed further that even when there are multiple peaks in the glow curve and the trap levels interact, treatment of the glow curve as the superposition of several individual glow peaks is valid for first-order kinetics, but becomes progressively less valid for higher-order kinetics. Thus, curve fitting a first-order glow curve by the linear superposition of first-order Randall-Wilkins equations is unlikely to lead to gross errors.

Piters[109] and Bos and Piters[110] also examined the possibility of fitting the simple Randall-Wilkins expression to computer simulated glow peaks, but in this case the simulated curves were generated by a numerical solution to the rate equations which were modified by allowing the individual defect species to interact via ionic diffusion processes (a situation suggested to be relevant to TL from LiF doped with Mg impurities). These authors concluded that even this more complex case, not considered during the derivation of the Randall-Wilkins expression, could be very well fitted by the Randall-Wilkins equation. The observance of a good fit, therefore, should not be taken to indicate the validity of the chosen model.

McKeever at al.[111] examined the kinetics of Schön and Klasens, previously described in Ch. 2, and analyzed the simulated glow curves produced using this kinetic scheme. Among the analytical techniques used was curve fitting using either the Randall-Wilkins or Garlick-Gibson equations. Once again, it was observed that these simple expressions for TL can be made to fit this more complex case, and again one must warn against accepting a "good fit" as being proof of the validity of the chosen model and equations.

Additional complexity to the fitting process arises if one avoids the assumption of quasiequilibrium and, for example, uses the Eq. 2.116 for first-order kinetics instead of the 3-parameter Randall-Wilkins equation. As pointed out by Lewandowski et al.[28], this equation may be used in a fitting protocol if the lifetime τ_n is parametrized; for example:

$$\tau_n = C_\tau T^\alpha \tag{3.94}$$

where C_τ and α are constants. However, such a procedure leads to 5 fitting parameters per peak - namely, E_t, s, n_0, C_τ, and α. Alternatively, τ_n could be considered constant, reducing the number of parameters to 3. The inner integral in Eq. 2.116 is evaluated using one of several approximations (see Ch. 6) while the outer integral may be evaluated using an adaptive Simpson's rule formalism. Computation times, however, will be necessarily lengthy.

Lewandowski and colleagues also examined the errors that are introduced if non-QE, first-order kinetics TL peaks are evaluated using the simple Randall-Wilkins equation (Eq. 2.28) instead of the computationally more difficult Eq. 2.116. In the array of cases examined by these authors, assuming an input value for E_t of 0.300 eV, the range of observed errors in this parameter was -0.005 to +0.010 eV, which may be considered favorable if this were a real experimental situation.

Expressions to describe TL assuming a distribution of trap energies have been derived, as discussed in Ch. 2. However, the expressions for Gaussian or exponential distributions contain an integral that needs to be solved numerically and involve long computational times. However, Hornyak and Chen[17] derived a simple expression for the case when the distribution of energy states is rectangular, as already given in Eq. 2.74. This expression contains no integrals and can therefore be used for fitting.

It is clear from the above discussion that the adoption of curve fitting as an analytical method can lead to erroneous conclusions regarding models and parameter values unless great care is taken in data collection and computational analysis. The ability of the simple expressions for TL (e.g. the Randall-Wilkins equation) to fit glow curves generated from models more complex than those from which these simple expressions were derived, leads one to recommend extreme caution in interpreting results. Nevertheless, the validity of the superposition principle for first-order kinetics, the lack of sensitivity of first-order kinetics to quasiequilibrium, and the apparent ability of curve fitting to return the accurate values for the trapping parameters makes this an attractive technique for analyzing first-order glow curves. The ability of the technique to analyze non-first-order curves correctly is less clear.

3.6.3 Some Experimental Examples

Perhaps the glow curve which has been analyzed most often using curve fitting is that of LiF:Mg, especially in its dosimetric form of TLD-100. An example fit of the glow curve to a linear superposition of first-order Randall-Wilkins equations is shown in Fig. 3.24, from the work of Bos et al.[112]. Here we see the best fit of the TLD-100 glow curve to four first-order peaks. The fitting parameters for each peak are E_t, s and n_0. Also shown in the figure is the residue, i.e. the difference between the experimental data and the sum of the four component peaks. The almost random nature of the residue, centered on a value of zero for temperatures less than 480 K, is evidence of the excellence of this particular fit. Exercises of this type on this material have consistently given similar values for the activation energies (e.g.[98], [101],[110]) and this is taken as further evidence that confidence can be given to the validity of the fit.

Example fits of the "100 °C" peak in quartz, also to the first-order Randall-Wilkins equation, are shown in Fig. 3.25. Again, an apparently satisfactory fit is obtained. Analysis of the high temperature TL from this material, however, is more difficult. Hornyak et al.[99] showed that the TL peak shape from this material at high temperature is wavelength dependent and, furthermore, that the glow curve shape can best be represented by assuming a uniform distribution of trapping energies, rather than a single energy. Example data are illustrated in Fig. 3.26.

The major difficulty with applying curve fitting to real TL glow curves is that the actual model appertaining in the sample under study is usually unknown. One may be able to determine the kinetic order from the application of the analytical techniques suggested in the preceding sections, and from an observation of how the peak shape and position depend upon absorbed dose, but verifying that, say, the Randall-Wilkins equation is a sufficiently exact equation for analytical purposes may be non-trivial. We have already noted that

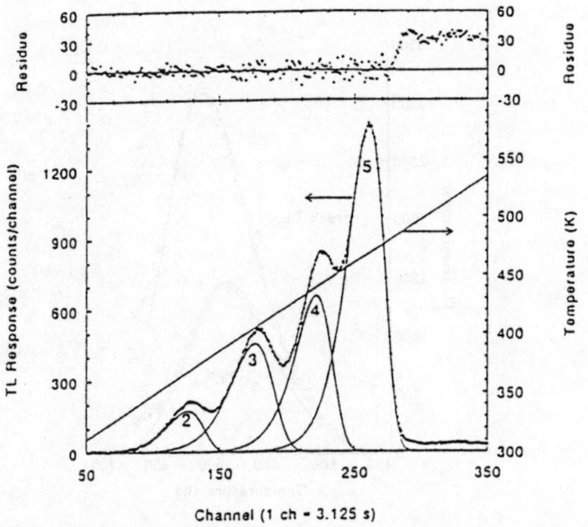

Figure 3.24 Fit of the glow curve from LiF TLD-100 to four first-order peaks using the Randall-Wilkins equation. The upper figure gives the residue between the data and the sum of the four component peaks.

(After Bos et al.[112].)

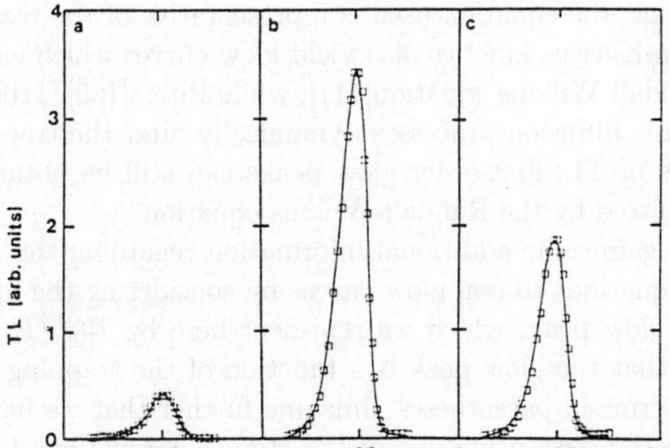

Figure 3.25 Fit of the 100^oC TL peak from quartz to the Randall-Wilkins equation. Three examples are shown, each giving the same E_t value.

(After McKeever et al.[113].)

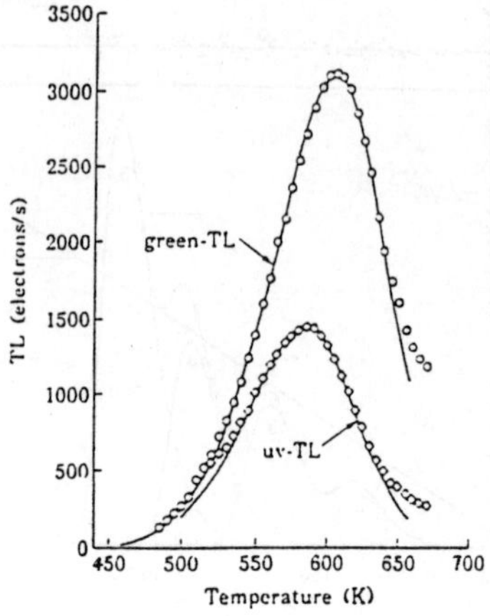

Figure 3.26 Experimental glow curves from quartz, recorded at two different wavelengths, and fitted to a TL peak assuming a uniform distribution of trapping states.

(After Hornyak et al.[99].)

real data may be distorted by non-quasiequilibrium, or thermal quenching, or self-absorption, or other reasons. We have also demonstrated that the ability to obtain what appears to be a good fit to the data should not be taken as conclusive evidence that the equation used is representative of the real material. For example, Schön-Klasens kinetics also yield glow curves which can be easily fitted by the Randall-Wilkins equation[111], while Piters[109],[110] demonstrated that even if ionic diffusion processes dynamically alter the trap structure during the process of TL, first-order glow peaks can still be obtained which can apparently be fitted by the Randall-Wilkins equation.

We can gain some additional information regarding the applicability of the standard equations to real glow curves by considering the following: Imagine a real TL glow peak, which we represent here by $I_{TL}^{real}(E_t, s, n_0, b)$ where we emphasize that the glow peak is a function of the trapping parameters by placing these terms in parentheses. Imagine further that we have obtained an excellent fit to the data using one of the standard equations for TL, denoted by $I_{TL}\left(E_t', s', n_0', b\right)$, for which the evaluated parameters E_t', s', and n_0' do not necessarily equal the real parameters. We may thus write:

$$I_{TL}^{real}(E_t, s, n_0, b) \approx I_{TL}\left(E_t', s', n_0', b\right). \tag{3.95}$$

Using the Randall-Wilkins equation as an example for $I_{TL}\left(E_t', s', n_0', b\right)$

(with $b = 1$) we have:

$$I_{TL}^{real}(E_t, s, n_0, b) \approx n_0^{'} s' \exp\left\{-\frac{E_t^{'}}{kT}\right\} \exp\left[-\frac{s'}{\beta}\int_{T_0}^{T} \exp\left\{-\frac{E_t^{'}}{k\theta}\right\} d\theta\right] \tag{3.96}$$

where the rest of the terms in Eq. 3.96 have the same meaning as in Eq. 2.28. It follows that

$$\frac{dI_{TL}^{real}(E_t, s, n_0, b)}{dT} \approx -\frac{s'}{\beta}\exp\left\{-\frac{E_t^{'}}{kT}\right\} I_{TL}^{real}(E_t, s, n_0, b) + I_{TL}^{real}(E_t, s, n_0, b)\frac{E_t^{'}}{kT^2} \tag{3.97}$$

which may be re-written

$$\frac{d\ln I_{TL}^{real}(E_t, s, n_0, b)}{dT} \approx \frac{E_t^{'}}{kT^2} - \frac{s'}{\beta}\exp\left\{-\frac{E_t^{'}}{kT}\right\}. \tag{3.98}$$

Solving for s' at $T = T_m$ gives

$$s' = \exp\left\{\frac{E_t^{'}}{kT_m}\right\}\frac{\beta E_t^{'}}{kT_m^2} \tag{3.99}$$

which, when substituted in Eq. 3.98, yields

$$\frac{E_t^{'}}{k}\left[\frac{1}{T^2} - \frac{1}{T_m^2}\exp\{\frac{E_t^{'}}{k}\left(\frac{1}{T_m} - \frac{1}{T}\right)\right] \approx \frac{d\ln I_{TL}^{real}(E_t, s, n_0, b)}{dT}. \tag{3.100}$$

The left side of Eq. 3.100 is such that it cannot be solved analytically for $E_t^{'}$, but it can be solved numerically quite easily. The main point of this analysis is that the parameter $E_t^{'}$ should be interpreted as that energy value which must be inserted into the Randall-Wilkins equation in order to force an exact fit between the Randall-Wilkins equation and the real data. It can be observed from an examination of Eq. 3.100 that $E_t^{'}$ must be temperature dependent. Thus the interpretation is that $E_t^{'}(T)$ is that temperature-dependent energy *function* that one must use in the simple Randall-Wilkins expression to obtain an exact fit to the data. Thus, one may proceed experimentally by obtaining the natural logarithm of $I_{TL}^{real}(E_t, s, n_0, b)$, differentiating, and numerically solving the left side of Eq. 3.100 at each value of T. Clearly, if the data really are accurately described by the Randall-Wilkins equation one will find that $E_t^{'}(T) = E_t^{'} =$ a constant. Deviations from this will indicate the extent of the failure of the chosen equation to describe the data.

Lewandowski et al.[28] applied this analysis to computer-generated data for which the QE approximation was invalid and their results are illustrated in figure 3.27. An example of the application of the analysis to real data is shown

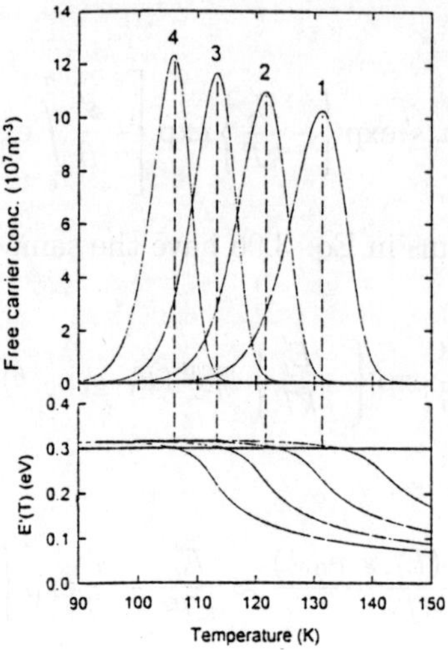

Figure 3.27 $E_t'(T)$ (lower figure) for four different TSC (i.e. free carrier concentration) peak (upper figure). Each peak is computer simulated and deviates from the Randall-Wilkins equation because of the invalidity of the QE approximation. (After Lewandowski et al.[28].)

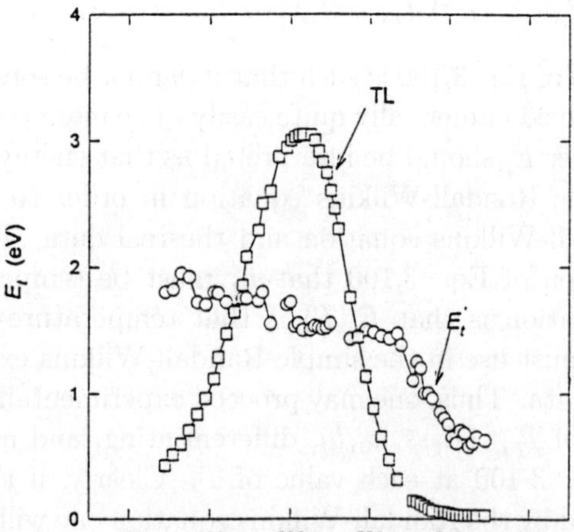

Figure 3.28 $E_t'(T)$ for TL from LiF:Mg,Cu,P.

in figure 3.28. Previous attempts to fit the glow curve from LiF:Mg,Cu,P to the Randall-Wilkins equation have resulted in apparent success[114]. However, figure 3.28 suggests that this might not be the most appropriate equation to use in this case.

Another way that one can get information from a glow *peak* is by embedding it in the broader context of the entire glow *curve*, as described by Maxia et al.[118]. Following Bonfiglioli et al.[119], Maxia et al. assumed a given series of TL peaks to result from electrons excited from a single trapping state with recombination into a series of hole centers. Of course, an entirely analogous situation is that of a single hole trapping state and a series of electron centers. The equations governing the recombinations, making the quasiequilibrium assumption are,

$$\frac{dn_i}{dt} = \frac{s\exp(-E_t/kT)B_i m_i n}{\left(A(N-n) + \sum\limits_{j=-p}^{q} B_{i+j}m_{i+j}\right)} \tag{3.101}$$

where m_i is the concentration of holes in the i-th luminescence center, and A and B_i are the probabilities for retrapping and recombination, respectively. N is the concentration of available electron traps and n is the concentration of trapped electrons. The neutrality condition is given by

$$n - \sum_{j=-p}^{q} m_{i+j} = n_0 - \sum_{j=-p}^{q} m_{i+j,0} \tag{3.102}$$

where $m_{i+j,0}$ denotes the initial concentration of m_{i+j}. These equations deal with $p+q+1$ overlapping peaks, but can also be applied when some transitions, or even all but one, are radiationless. One can start with the case of non-overlapping peaks. Assuming a constant heating rate, $\beta = dT/dt$, we can write the equation for the intensity $I(T)$ as

$$-dm_i/dt = \alpha I(T) \tag{3.103}$$

where α is a constant depending on the light collection efficiency and the units used. By integrating, Eq. 3.103 one gets

$$m_i(T) = \alpha \int_T^{T_f} I(T')d(T') = \alpha S(T) \tag{3.104}$$

and

$$m_i(T_0) = \alpha \int_{T_0}^{T_f} I(T)dT = \alpha S_o, \tag{3.105}$$

where T_0 and T_f are, respectively, the temperature at the beginning and the end of the glow curve. Defining the areas σ and Σ by using α as given in Eqs. 3.104 and 3.105, we have

$$n_0 = \alpha\sigma, \text{ and } N = \alpha\Sigma. \tag{3.106}$$

Using these definitions, Maxia et al.[118] developed the expression

$$y = \ln\left\{I(T)\frac{\cos\theta + \sin\theta \cdot S(T)}{S^2(T) + \Delta \cdot S(T)}\right\} = -\frac{E_t}{kT} + \mu \tag{3.107}$$

where

$$\tan\theta = \frac{B - A}{A(\Sigma - \sigma + S_o)}, \tag{3.108}$$

$$\Delta = \sigma - S_o \tag{3.109}$$

and

$$\mu = 0.5\ln\left\{(\beta/BS)^2\left[A^2(\Sigma - \sigma + S_o) + (B - A)^2\right]\right\}. \tag{3.110}$$

All the quantities appearing on the left hand side of Eq. 3.107 can be determined from the recorded glow curve, except from the "angle" θ which depends on unknown parameters. Had θ been known, the plot of y vs. $1/T$ would be a straight line, the slope of which yields the activation energy E_t. In order to evaluate θ, Maxia et al.[118] suggested a curve fitting technique which chooses the best value of θ in the sense that the correlation coefficient

$$\rho = \frac{cov(x, y)}{[v(x)v(y)]^{1/2}} \tag{3.111}$$

is closest to unity, where $x = 1/T$. This means that one performs a linear regression of y on x. Once the best θ is found, the ratio A/B can be evaluated by

$$A/B = [1 + (\Sigma - \Delta)\tan\theta]^{-1}, \tag{3.112}$$

and the frequency factor s by

$$s = (1 - A/B)\beta\exp\mu/\sin\theta. \tag{3.113}$$

One of the difficulties in using this method is that Σ, the evaluation of which is crucial in the last two equations, cannot be obtained directly from the recorded glow curve. An estimate of the value of Σ can be derived from the fact that $\sigma < \Sigma$ and that σ approaches Σ when the sample is excited to saturation.

Maxia et al.[118] state that the method can give reasonable results even when slight overlapping of peaks occurs, but it cannot be used when strongly overlapping peaks are involved. They applied the method to TL peaks in ZnS and NaBr. For a peak at 180 K in ZnS, a value of $E_t = 0.146$ eV was found, with $\mu = 7.2$ and $A/B = 0.24$. The frequency factor determined was $s = 6 \times 10^2$ s^{-1}, which is suspiciously low. For a peak at 423 K in NaBr they found $E_t = 0.95$ eV, with $\mu = 23.8$ which yields $A/B = 0.25$ and $s = 10^{10}$ s^{-1}.

Onnis and Rucci[120] developed a similar method for the case of a series of trapping levels and a single recombination center. The method involves finding the best value of ξ_i which makes linear the plot of

$$y = \ln\left\{I(T)\frac{1+\xi_i+S_i(T)}{S_i(T)+\Delta_i S_i(T)}\right\} \text{ vs. } 1/T \tag{3.114}$$

where

$$\Delta_i = \sigma_o - S_{io}\ ;\ \xi_i = \frac{1-A_i/B}{(A_i/B)\Sigma_i+\Delta_i};\ \mu_i = \ln\frac{S_i/\beta}{(A_i/B)\Sigma_i+\Delta_i}, \tag{3.115}$$

and where

$$N_i = \alpha\Sigma_i;\ n_{io} = \alpha\sigma_{io};\ m_0 = \alpha\sigma_o. \tag{3.116}$$

This variant is also applicable only when overlapping of adjacent peaks is practically negligible. Onnis and Rucci[120] applied the method to TL peaks of x-irradiated NaCl and found, for a peak at $T_m = 360$ K, an activation energy of 0.75 eV, $s = 10^{10}$ s^{-1} and $A/B = 0.04$ - i.e. very small retrapping as compared to recombination. For a peak at 570 K they found $E_t = 0.95$ eV, $s = 10^8$ s^{-1} and $A/B = 0.07$. They conclude that the two peaks are nearly of first order. These methods were later used by Aramu et al.[121] for TL peaks in AgCl and AgBr. The obvious shortcoming of these methods is that one has to know beforehand whether a given series of peaks is related to several trapping states or to several kinds of centers. Aramu and Maxia[122] improved the method by avoiding some of the approximations involved so that it can be used for practically any peak. They applied this method to a peak at 144 K in AgCl and found $E_t = 0.13$ eV and $s = 3 \times 10^{-4}$ s^{-1}.

3.7 Dealing with a Given Glow Curve

The performance comparison between the members of a particular group of methods for evaluating kinetic parameters has already been discussed in the different sections of this chapter. For example, the advantages and disadvantages of different peak shape methods have been compared in section 3.4. Here, we add some more general remarks.

All the analytical methods described are rather sensitive to the occurrence of *satellites* accompanying the peak of interest. All methods will also suffer to some degree from low intensity signals and from poor temperature control. Apart from these very practical, and real problems, all the methods generally yield the information with similar levels of accuracy, even though some do so with significantly less computational effort than others. In general practice, however, it is a very good idea to evaluate the parameters using a number of methods, and to compare the results.

While utilizing any of the formulae appearing in the this chapter, one should be familiar with the approximations incurred in their development, and to make sure that they are valid for the case under study. For example, one needs to assess whether the model chosen (the most important of the assumptions used) is in fact a reasonable, albeit simplified, description of the system under examination. One needs to be certain that the kinetic order assumed is in fact the correct kinetic order for the particular system of interest. Experimental methods to check this have been described. How well does the quasiequilibrium approximation describe the kinetics under study? (See, for example, the paper by Opanowicz and Przybyszewski[115].) What mathematical approximations are used in the analysis chosen? For example, the initial rise technique relies upon the assumption that the integral appearing in the TL equation is approximately equal to 1. This assumption will only hold when $T \simeq T_0$ and to extend the initial rise analysis too far up the leading edge of the glow curve will undoubtedly lead to erroneous conclusions.

As another example we anticipate here a method described in Ch. 8 for analysis of other thermally stimulated phenomena - namely the method by Coats and Redfern[116], which is cited and used in many works. Most of the investigators do not mention the fact that the method is based upon the approximation $\int_o^T \exp(-E/R\theta)d\theta \cong (RT^2/E)\exp(-E/RT)$ which can be inaccurate by up to 10% (see Ch. 6 for more details). In a better approximation, one can take two terms in the asymptotic series, but then, the linearity upon which the method is based will be perturbed. One should check, therefore, once a first approximation for E is available, if the magnitude $(1-2RT/E)$ changes appreciably along the peak, and try to correct the first estimate.

Understanding the details of the development of a formula is of importance for another, rather trivial reason. For example, in the mentioned paper by Coats and Redfern[116], one of the equations includes an error, which was later pointed out by Ozawa[117]. Some investigators used the original equation as is and, of course, ended up with inaccurate results.

As mentioned with relation to practically all thermally stimulated processes, the pre-exponential factor, which is normally assumed to be temperature independent as a first approximation, is actually rather often dependent on temperature like T^a. For example, values of a between -2 and 2 were considered in both TL and the other thermally stimulated processes. This temperature dependence is rather weak as compared to the exponential dependence on temperature $\exp(-E_t/kT)$. However, it is obvious that if one disregards the T^a

dependence while evaluating the kinetic parameters, one gets results of the parameters with some error. This may amount to a few percent in the evaluation of the activation energy E_t, and of course, to a much larger error in the evaluation of the frequency factor s. As mentioned in this chapter with relation to some of the methods, various correction procedures can be made and in general, it seems to be always valid, as is shown in detail for some of the shape methods by Chen[26] (see also section 3.4), that if one gets a value of the activation energy by ignoring the temperature dependence of the pre-exponential factor, a better estimate will be reached by subtracting akT where T is the average temperature in the relevant range.

The theory of TL can easily explain the occurrence of a single peak related to a given trap with given trapping parameters, its dependence on the heating rate and on the initial filling conditions which are related to the excitation dose. By using the quasiequilibrium approximation, the theory can also easily explain the possibility of a single peak to be asymmetrical in the *first-order* case or to be nearly symmetrical in the *second-order* case. The theory permits us to state explicitly the conditions that would lead from the general situation to the first- and second-order cases, or to intermediate situations such as the *mixed-order*. The theory can also explain the occurrence of a series of peaks related to either a series of trapping states or to a series of recombination centers, or to both. The unfortunately common situation in which the observed peaks are not single peaks, can quite easily be explained as well. One can think in this regard of two possible degrees of complication.

1. Two or more peaks are not interrelated as far as the process itself is concerned, but they occur at the same temperature region and thus, the measured result is a superposition of the two. This is in particular a common situation when the sample investigated is a mixture of various materials, which is very often the case in the study of archaeological and geological specimens (as mentioned in more detail in Ch. 7). The assumption of linear superposition of glow peaks is valid in this case.

2. The processes bringing about the two or more peaks are interrelated. For example, during the heating, electrons released from one kind of trap may retrap into another one which may later participate in the process by releasing electrons in the same temperature range. We should consider this as a combined effect and not a mere superposition. This should be kept in mind while using any method of deconvolution.

By using existing theories one can explain peaks which are accompanied by conductivity counterparts, as well as those which are not. Also, one can usually distinguish between peaks related to the transport of electrons as compared to those related to the transport of holes, e.g. by using TSEE which, as explained, generally reveals only the thermally released electrons.

The more important problem, which is also the more difficult is the inverse one, namely, given a glow curve, how can one extract the relevant parameters? We will distinguish between three possible situations.

a. The curve consists of a single peak or a series of distinct peaks.

b. The curve consists of overlapping peaks related to distinct energy levels.

c. The glow curve results from a continuum of levels in the forbidden gap.

When the glow curve consists of a series of separate peaks, one can, at least as a first approximation, try to analyze each of them separately. The beginning of the discussion concerns, therefore, the analysis of one peak. We suggest the following guidelines in the study of a single peak.

1. As a first step one should ensure that the temperature of the sample is the actual temperature being monitored. Since in all TL measurements the temperature actually measured is that of the planchet, not the sample, this is not necessarily a trivial aspect. Heating within an inert gas atmosphere, a low rate (to reduce thermal gradients) and with the sample directly adhering to the planchet (rather than within a container) all help to ensure that the temperature monitored is the real sample temperature.

2. The second step should be to ensure that the peak is "clean" - i.e. free from overlapping, nearby peaks. Thermal bleaching of the low temperature side of the peak can be performed by heating the sample to a temperature below the maximum, and cooling it back before reheating. This should usually eliminate, or at least reduce, the low temperature satellites. One should check that thermal quenching of the luminescence is not occurring over the temperature range of interest or, if it is occurring, one should make corrections for it. Emission spectra data should also be taken and corrections made for the wavelength response of the detection system. It is important that all the photons being emitted are counted, not just those of a particular wavelength. To avoid the influence of experimental noise, the peak should be excited to a high level.

3. If the peak appears to be clean, one can use any of the methods based on three parameters, E_t, s and b. Even if the underlying situation is more complicated, being dependent on more parameters, the three parameter presentation provides a characterization of the peak as far as its shape is concerned. This approach is more reasonable if the three parameters remain the same when the intensity of the peak varies due to a change in the applied dose, and is more questionable otherwise. If a shape method is utilized, for evaluating the activation energy, methods based on the measurement of the low temperature half $\tau = T_m - T_1$ should be preferable, due to the possibility of cleaning effectively this portion of the peak, as well as other reasons discussed in detail in section 3.4. Otherwise, methods that utilize the complete data set

- preferably the whole glow peak - should be preferred over methods that rely upon just two or three points at specific temperatures while ignoring the rest of the data. It is obvious that neither first- nor second-order kinetics should be presumed beforehand unless, of course, the shape of the peak shows that clearly. Otherwise tests such as the variation in the shape and position of the peak as a function of trap filling should be carried out. As argued at different points within this chapter, analysis of first-order peaks is generally more reliable than non-first-order peaks.

4. A number of methods for finding the parameters should be used. Agreement between results from different methods is always somewhat reassuring. In case of a discrepancy, a more in-depth study is required, e.g. as described in #5.

5. The peak should be measured under various conditions. These may include a change in the dose of excitation and the source of excitation and a change in the heating rate. In some cases such as in quartz samples, variations may occur between bulk and powder samples, or within the latter, between different grain sizes. In TSC and TSD (see Ch.8), different kinds of contacts should also be used.

One can certainly try to analyze a glow peak in the framework of the full number of relevant parameters. As stated already, this may turn out to yield ambiguous results since it may happen that different sets of parameters may be fitted to the same given curve within the given experimental accuracy. Instead of just examining the glow peak, one should use a broader arsenal of additional experiments. These may include isothermal decay, dose dependence, excitation and emission spectra and simultaneous measurements, such as with TSC.

Let us consider the case of a series of distinct peaks through an example. Suppose that we have such a series of TL peaks, all having the same emission spectrum, but different activation energies. This would mean that there exists in the crystal a series of different traps and suggests (but is not a guarantee of) a single kind of recombination center. The occurrence of accompanying TSC peaks would indicate that the charge carriers have been raised to the band before recombining with carriers of opposite sign in recombination centers. Of course, the absence of TSC peaks will strongly suggest localized transitions. (Again, this is not guaranteed; one can imagine charge carriers with low mobilities such that the TSC is not easily detected.) The sign of the mobile carriers may possibly be determined independently by the Hall-effect, or by thermoelectric power measurements, or by time-of-flight techniques. If TSEE occurs in parallel with TL and TSC it indicates, in most cases, that the mobile carriers are electrons rather than holes. Müller[123] suggested a method based on a TSD measurement to find the sign of the mobile carriers involved.

All of this can enable us to determine the location of the relevant levels within the forbidden gap. Let us assume for this discussion that the mobile carriers are electrons. The area under any of the peaks represents the total concentration of electrons in a certain trapping state, n_0, whereas the total area under the series of the peaks represents the concentration of holes in centers, m_0. The total concentration N of traps of a certain kind can be evaluated by bringing the sample to saturation when $n_0 \approx N$. Generally, however, only relative, not absolute, values for the concentrations can be determined since, in TL, not all of the emitted photons are collected. One possibility for estimating the absolute values was described by Bube for the case of TSC[124]. Here one illuminates the sample with a known photon flux and determines the flux necessary to produce a photocurrent which is the same order of magnitude as the TSC signal. With the assumption that one absorbed photon yields one photoelectron, one can estimate the number of electrons necessary to yield the observed current. This procedure was described for TSC from BSO and BGO by Bloom and McKeever[125]. The occupancy of thermally disconnected traps can be found as the difference between the total area, representing m_0 at saturation, and N. A more elaborate treatment of this case was given by Onnis and Rucci[120] for a series of peaks. This followed an earlier report by Maxia et al.[118] dealing with the opposite situation, namely, that of a series of peaks having all the same activation energy and different emission spectra due to transitions into different centers.

Within the framework of the above guidelines, the analysis of TL, TSC, or other thermally stimulated phenomena may proceed.

3.8 References

1. J.T. Randall and M.H.F. Wilkins, *Proc. Roy. Soc.* **A184** (1945) 347.
2. J.T. Randall and M.H.F. Wilkins, *Proc. Roy. Soc.* **A184** (1945) 366.
3. J.T. Randall and M.H.F. Wilkins, *Proc. Roy. Soc.* **A184** (1945) 390.
4. P. Bräunlich, *J. Appl. Phys.* **38** (1967) 2516.
5. B.S. Acharya and M.L. Mukherje, *Phys. Stat. Sol. (a)* **53** (1979) 373.
6. D.F. Mariani, G. Gonzáles, J. Vignolo, M. Jiménez de Castro and I. Ibarra, *Phys. Stat. Sol. (a)* **112** (1989) K53.
7. M.G. Patil, S.H. Pawar and R.D. Lawangar, *Sol. St. Comm.* **34** (1980) 105.
8. S.H. Tatumi, M. Matsuoka and S. Watanabe, *Phys. Stat. Sol. (a)* **129** (1992) K57.
9. R. Chen, *J. Electrochem. Soc.* **116** (1969) 1254.
10. J.S. Nagpal and S.P. Kathuria, *Int. J. Appl. Radiat. Isot.* **35** (1984) 495.
11. E.I. Adirovitch, *Phys. Rev.* **17** (1956) 705.
12. W.L. Medlin, *Phys. Rev.* **135** (1964) A1770.

13. W. Hoogenstraaten, *Philips Res. Repts.*, **13** (1958) 515.
14. A. Bosacchi, B. Bosacchi, S. Franchi and L. Fernandez, *Sol. St. Comm.* **13** (1973) 1805.
15. L.F. Pender and R.J. Fleming, *J. Phys. C: Sol. St. Phys.* **10** (1977) 1571.
16. T. Kikuchi, *J. Phys. Soc. Japan* **13** (1958) 526.
17. W.F. Hornyak and R. Chen, *J. Lumin.* **44** (1989) 73.
18. P. Wünsche, R. Kunze and M. Wäsche, *Phys. Stat. Sol. (a)* **103** (1987) 145.
19. R. Visocekas, *Thèse*, Université Pierre et Marie Curie, Paris 6 (1978).
20. R. Chen and N. Kristianpoller, *Radiat. Prot. Dosim.* **17** (1986) 443.
21. T. Ozawa, *J. Therm. Anal.* **2** (1976) 369.
22. G.F.J. Garlick and A.F. Gibson, *Proc. Phys. Soc.* **60** (1948) 574.
23. R. Chen and G.A. Haber, *Chem. Phys. Lett.* **2** (1968) 483.
24. A. Lörincz, M. Puma, F.J. James and J.H. Crawford, *J. Appl. Phys.* **53** (1982) 927.
25. F. Aramu, P. Brovetto and A. Rucci, *Phys. Lett.* **23** (1966) 308.
26. R. Chen, *J. Appl. Phys.* **40** (1969) 570.
27. A.C. Lewandowski and S.W.S. McKeever, *Phys. Rev. B* **43** (1991) 8163.
28. A.C. Lewandowski, B.G. Markey and S.W.S. McKeever, *Phys. Rev. B* **49** (1994) 8029.
29. S.W.S. McKeever, A.C. Lewandowski and B.G. Markey, *Radiat. Prot. Dosim.* **47** (1993) 9.
30. J. Nahum and A. Halperin, *J. Phys. Chem. Solids* **24** (1963) 823.
31. H. Gobrecht and D. Hoffman, *J. Phys. Chem. Solids* **27** (1966) 509.
32. Y. Kirsh, S. Shoval and P.D. Townsend, *Phys. Stat. Sol. (a)* **101** (1987) 253.
33. F. Firszt and H.L. Oczkowski, *Phys. Stat. Sol. (a)* **111** (1989) K113.
34. A. Chruścińska, *J. Lumin.* **62** (1994) 115.
35. F. Bell and R. Sizmann, *Phys. Stat. Sol.* **15** (1966) 369.
36. R.A. Creswell and M.M. Perlman, *J. Appl. Phys.* **41** (1970) 2365.
37. A. Halperin, A.A. Braner, A. Ben-Zvi and N. Kristianpoller, *Phys. Rev.* **117** (1960) 416.
38. R. Chen and Y. Kirsh, *Analysis of Thermally Stimulated Processes*, Pergamon Press, Oxford (1981).
39. C. Muntoni, A. Rucci and A. Serpi, *La Ricc. Sci.* **38** (1968) 762.
40. D. Dražič and M. Trontelj, *J. Appl. Radiat. Isot.* **37** (1986) 337.
41. J.H. Sharp and S.A. Wentworth, *Anal. Chem.* **41** (1969) 2060.
42. P. Bräunlich, *J. Appl. Phys.* **38** (1967) 2516.
43. F. Urbach, *Wiener Berich. IIa*, **139** (1930) 363.

44. L.I. Grossweiner, *J. Appl. Phys.* **24** (1953) 1306.
45. G.A. Dussel and R.H. Bube, *Phys. Rev.* **155** (1967) 764.
46. C.B. Lushchik, *Sov. Phys. JETP* **3** (1956) 390.
47. P.N. Keating, *Proc. Phys. Soc.* **78** (1961) 1408.
48. A. Halperin and A.A. Braner, *Phys. Rev.* **117** (1960) 408.
49. N. Takeuchi, H. Ohta and S. Nakamura, *Phys. Stat. Sol. (a)* **130** (1985) K185.
50. S. Maeta and K. Sakaguchi, Japan. *J. Appl. Phys.* **19** (1980) 519.
51. M. Balarin, *Phys. Stat. Sol. (a)* **31** (1975) K111.
52. J. Zsakó, *J. Therm. Anal.* **15** (1979) 369.
53. R.J. Fleming, *Can. J. Phys.* **46** (1968) 1569.
54. H. Wenzl, *Ang. Phys.* **15** (1961) 286.
55. D. Edwards Jr., *Surf. Sci.* **69** (1977) 301.
56. C.M. Chan, R. Aris and W.H. Weinberg, *Appl. Surf. Sci.* **1** (1978) 360.
57. D. Shenker and R. Chen, *J. Comput. Phys.* **10** (1972) 272.
58. N. Kristianpoller and Y. Kirsh, *Phys. Stat. Sol. (a)* **21** (1974) 87.
59. Z. Spurný and L. Hobzovà, *Radiochem. Radioanal. Lett.* **29** (1977) 287.
60. C. De-Blasi, S. Gallassini, C. Manfredotti, G. Miccoci, L. Ruggiero and A. Tepore, *Sol. St. Comm.* **25** (1978) 149.
61. J. Azorin and A. Guitérrez, *Nucl. Tracks* **11** (1986) 167.
62. S.H. Yeh, *Radiat. Eff.* **88** (1986) 39.
63. D.W. Cooke, E.P. Gravathas and M.D. Brown, *J. Appl. Phys.* **54** (1983) 1165.
64. S.J. Singh, R.K. Gartia and P.S. Mazumdar, *J. Phys. D: Appl. Phys.* **22** (1989) 467.
65. P.L. Land, *J. Phys. Chem. Solids* **30** (1969) 1681.
66. R.T. Yang and M. Steinberg, *Anal. Chem.* **49** (1977) 998.
67. R.K. Gartia, S.J. Singh, T.S.C. Singh and P.S. Mazumdar, *J. Phys. D: Appl. Phys.* **24** (1991) 1451.
68. J.P. Fillard and J. Gasiot, *J. Electrostat.* **3** (1977) 37.
69. W. Arnold and H. Sherwood, *J. Phys. Chem.* **63** (1959) 2.
70. P.J. Kelly and M.J. Laubitz, *Can. J. Phys.* **45** (1967) 311.
71. K. Stammers, *J. Phys. E* **12** (1979) 637.
72. C. Christodoulides, *Phys. Stat. Sol. (a)* **94** (1986) 251.
73. C. Christodoulides, *J. Phys. D: Appl. Phys.* **19** (1986) 1555.
74. G. Ehrlich, *Adv. in Catalysis and Related Subjects* **14** (1963) 255.
75. P. Müller and J. Teltow, *Phys. Stat. Sol. (a)* **12** (1972) 471.
76. A.H. Booth, *Canad. J. Chem.* **32** (1954) 214.

77. A. Bohun, *Czech. J. Phys.* **4** (1954) 91.

78. I.A. Parfianovitch, *J. Exp. Theor. Phys. SSSR*, **26** (1954) 696.

79. K. Osada, *J. Phys. Soc. Japan* **15** (1960) 145.

80. R. Chen and S.A.A. Winer, *J. Appl. Phys.* **41** (1970) 5227.

81. T.S.C. Singh, P.S. Mazumdar and R.K. Gartia. *J. Phys. D: Appl. Phys.* **23** (1990) 562.

82. A.C. Damask and G.J. Dienes, *Point Defects in Metals*, Gordon and Breach (1963).

83. P.D. Townsend, C.D. Clark and P.W. Levy, *Phys. Rev.* **155** (1967) 908.

84. T.V. Panchenko, A.Ya. Kudzin and N.A. Truseva, *Sov. Phys. Sol. Stat.* **22** (1980) 1077.

85. H.E. Kissinger, *Anal. Chem.* **29** (1957) 1702

86. M. Maciejewski, *Rocz. Chem. Anal. Soc. Chim. Polon.* **51** (1977) 535.

87. P.C. Gravelle, *J. Therm. Anal.* **14** (1978) 53.

88. R.R. Haering and E.N. Adams, *Phys. Rev.* **117** (1960) 451.

89. A. Kessler, *J. Electrochem. Soc.* **123** (1976) 1236.

90. M. Schön, *Tech. Wiss. Abhandl. der Osram Studiengesellschaft* **7** (1958) 175.

91. Z. Chvoj, Czech. *J. Phys.* **B27** (1977) 957.

92. Z. Chvoj and P. Pokorný, *Czech. J. Phys.* **B28** (1978) 446.

93. D.S. Betts, L. Couturier, A.H. Khayrat, B.J. Luff and P.D. Townsend, *J. Phys. D: Appl. Phys.* **26** (1993) 843.

94. D.S. Betts and P.D. Townsend, *J. Phys. D: Appl. Phys.* **26** (1993) 849.

95. G. Kitis, M. Spiropulu, J. Papadopoulos and S. Charalambous, *Nucl. Instrum. Meth.* **73** (1993) 367.

96. T.M. Piters and J.J. Bos, *J. Phys. D: Appl. Phys.* **27** (1994) 1747.

97. J.C. Chiarena and E. Gillet, *J. Appl. Phys.* **48** (1977) 1712.

98. S.W.S. McKeever, *Phys. Stat. Sol. (a)* **62** (1980) 331.

99. W.F. Hornyak, R. Chen and A.D. Franklin, *Phys. Rev. B* **46** (1992) 8036.

100. P.D. Townsend, in *Thermoluminescence of Geological Materials*, D.J. McDougall, editor, Academic Press, London (1968) 51.

101. R.G. Fairchild, P.L. Mattern, K. Lengweiler and P.W. Levy, *J. Appl. Phys.* **49** (1978) 4512.

102. B.J. Luff and P.D. Townsend, *Meas. Sci. Technol.* **4** (1993) 65.

103. A.G. Wintle, *Geophys. J. Roy. Astr. Soc.* **41** (1975) 107.

104. G. Kitis, J. Papadopoulos, S. Charalambous and J.W.N. Tuyn, *Radiat. Prot. Dosim.* **55** (1994) 183.

105. V.S. Kortov, I.I. Milman, V.I. Kirpa and J. Lesz, *Radiat. Prot. Dosim.* **55**

(1994) 279.

106. N.S. Mohan and R. Chen, *J. Phys. D: Appl. Phys.* **3** (1970) 243.
107. D. Shenker and R. Chen, *J. Phys. D: Appl. Phys.* **4** (1971) 287.
108. R.K. Bull, S.W.S. McKeever, R. Chen, V.K. Mathur and M.D. Brown, *J. Phys. D: Appl. Phys.* **19** (1986) 1321.
109. T.M. Piters, *Ph.D. Thesis*, Delft University of Technology (1993).
110. A.J.J. Bos and T.M. Piters, *Radiat. Prot. Dosim.* **47** (1993) 41.
111. S.W.S. McKeever, J.F. Rhodes, V.K. Mathur, R. Chen, M.D. Brown and R.K. Bull, *Phys. Rev. B.* **32** (1985) 3835.
112. A.J.J. Bos, R.N. M. Vijverberg, T.M. Piters and S.W.S. McKeever, *J. Phys. D: Appl. Phys.* **25** (1992) 1249.
113. S.W.S. McKeever, C.Y. Chen and L.E. Halliburton, *Nucl. Tracks* **10** (1985) 489.
114. J. McKeever, F.D. Walker and S.W.S. McKeever, *Nucl. Tracks Radiat. Meas.* **21** (1993) 179.
115. A. Opanowicz and K. Przybyszewski, *Proc. SPIE Int. Soc. Opt., Eng.* **2** (1995) 236.
116. A.W. Coats and J.P. Redfern, *J. Polym. Sci.* **3B** (1965) 917.
117. T. Ozawa, *J. Therm. Anal.* **5** (1973) 499.
118. V. Maxia, S. Onnis and A. Rucci, *J. Lumin.* **3** (1971) 378.
119. G. Bonfiglioli, P. Brovetto and C. Cortese, *Phys. Rev.* **114** (1958) 956.
120. S. Onnis and A. Rucci, *J. Lumin.* **6** (1973) 404.
121. F. Aramu, V. Maxia and G. Spano, *J. Lumin.* **9** (1974) 104.
122. F. Aramu and V. Maxia, *J. Lumin.* **10** (1975) 77.
123. P. Müller, *Phys. Stat. Sol. (a)* **33** (1976) 543.
124. R.H. Bube, *Photoelectronic Properties of Semiconductors*, Cambridge University Press, Cambridge (1992).
125. D. Bloom and S.W.S. McKeever, *J. Appl. Phys.* **77** (1995) 6511, 6521.

Chapter 4
DOSE DEPENDENCE OF TL; NONLINEARITIES

4.1 General Considerations and Experimental Evidence

For our purposes we can define the dose either as the accumulated amount of irradiation incident on the sample during the excitation period at low temperature, or the total amount absorbed prior to the heating. In most cases, there is no direct way to relate the released TL to the amount of radiation exposure in absolute units (see e.g. Cameron et al.[1]) and, therefore, a proportionality factor will always be included or implied in the considerations of response to dose. In addition, no variations in the absorption of the irradiation are usually observed during the excitation, and therefore, we can alternatively speak of incident and absorbed doses, and only another proportionality factor is to be added. As long as the applied radiation intensity (or dose rate), X, measured either in incident photons or particles per units of time and area, is kept constant, the dose is given by the product Xt where t is the time of excitation. In cases where the applied dose varies during the excitation like $X(t)$, the total dose is given by $\int_0^t X(\tau)d\tau$. In the application of TL for the measurement of the dose of irradiation and for archaeological and geological dating (see Ch. 7), one usually hopes that, at least in certain dose ranges, the measured effect would be proportional to the applied dose. This is indeed so for certain materials and in certain dose ranges. In many other materials or dose ranges, however, different degrees of nonlinearities of different sorts are observed. These are very important to comprehend, both from the applications point of view and the theoretical aspects of understanding the processes taking place in the thermoluminescent material during its excitation and heating.

A number of ways for the classification of the nonlinear dose dependence effects are to be considered. One can distinguish between *sublinear* and *superlinear* (or *supralinear)* behavior. One can discuss the difference between various models explaining the nonlinear effects, e.g., models related to processes taking place during the excitation stage and those occurring during the heating phase. One should also differentiate between nonlinearities associated with diverse radiations. For example, Moscovitch and Horowitz[2] discuss the different dose responses attained from γ rays and electrons (β particles) on one hand and

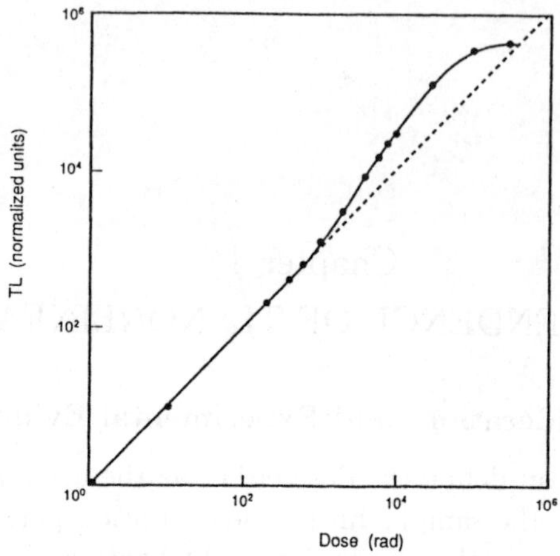

Figure 4.1 TL output vs. γ dose in LiF.
(After Zimmerman[7].)

heavy charged particles (e.g. α particles) on the other. Whereas in the former case (which should include UV excitation as well), the energy deposition may be characterized as a spatially uniform low density sea of initially free electrons, holes and excitons. For heavy charged particles, on the other hand, ion tracks are being produced and the immediate vicinity of the ion track is characterized by a very high instantaneous density of electrons and holes combined with rather high density of secondary electrons (δ rays) penetrating to larger radial distances from the track. Yet another kind of classification has to do with the dose range at which the nonlinearity occurs.

We will start here with some experimental examples taken from the literature which will touch upon the subjects of characterizations, and classifications, and will move on to some basic definitions before going into the different theoretical models in some detail.

Nonlinear growth of TL as a function of the absorbed β or γ dose , mainly in LiF(TLD-100), has been discussed in some detail in the book by Cameron et al.[1]. This was based on previous work by Cameron et al.[3],[4],[5],[27], and Zimmerman[7]. In these studies the term *supralinearity* was utilized. The characteristic behavior, as shown in Fig. 4.1, is that of an initial linear dependence on the dose, followed by a "more than linear" range after which, an approach to saturation is observed. A similar behavior has been reported by various other authors under varying experimental circumstances, e.g. Mische and McKeever[8] and Sunta et al.[9]. A similar dose dependence has been reported in other materials such as Mg_2SiO_4 by Mikado et al.[10] and fluorite by

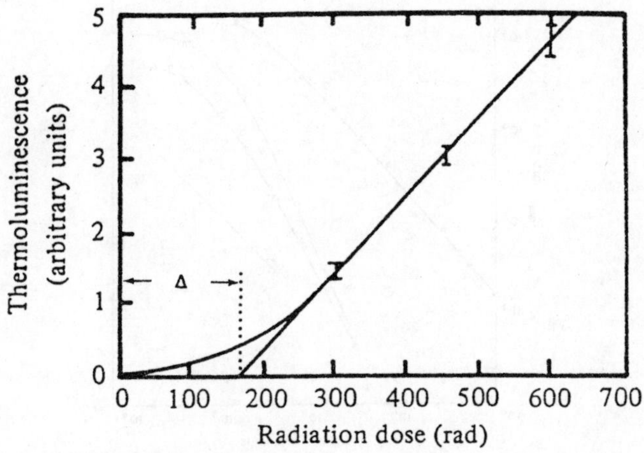

Figure 4.2 Dose dependence of the TL from pottery quartz. The dose dependence is seen to be more than linear at low doses, and becomes linear at higher doses. (After Aitken[13].)

Jain[11].

Another kind of nonlinearity was discovered in the mid-sixties by Tite[12] in pottery quartz (see also Aitken[13]). Here, the dose dependence curve is linear in a broad range, except for the lowest doses where the TL starts increasing at a small rate with respect to dose and the rate increases gradually until it reaches a steady value which prevails over a broad dose range. An example taken from Aitken is shown in Fig. 4.2. Tite and Aitken were mainly concerned with the possible error thus introduced into the evaluation of the dose (which in their case, was translated into an archaeological age) and defined the *supralinearity parameter* Δ which is also shown in Fig. 4.2.

In the same period of the mid-sixties another sort of *superlinearity* was reported. Halperin and Chen[14] studied the dose dependence of UV irradiated semiconducting diamonds (type IIb). UV light of 225 nm (5.5 eV), which enables a band-to-band electron transition to occur, produced a strong TL peak at ~250 K which was linear with the dose. UV light at longer wavelengths, however, caused a different dose dependence, as shown in Fig. 4.3. When excitation was performed with UV light in the range of 300 – 400 nm (4 - 3 eV), a dose dependence as seen in curves *b* and *c* of Fig. 4.3 was found. This is the plot of the maximum TL intensity (representing, in fact, the total TL) as a function of the dose. Since it is plotted on a log – log scale, those slopes which are larger than unity represent *superlinearity* (the term was suggested by Halperin and Chen[14]). The behavior here is that the dependence is more

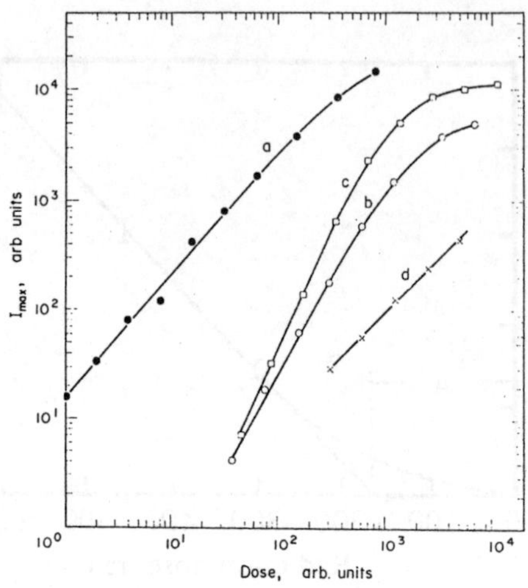

Figure 4.3 Dependence of the blue TL peak at 250 K in semiconducting diamond on the dose of UV light. Excitation wavelengths were: a - 225 nm, b - 300 nm, c - 375 nm, d - 425 nm.

(After Halperin and Chen[14].)

than linear in a broad dose range, and as of the lowest dose which enables the measurement of TL. Halperin and Chen[14] assumed that the dose dependence goes like

$$I_{\max} = aD^k \tag{4.1}$$

where $I_{\max}$ is the maximum TL intensity and D the applied dose. They found values of k between 2 and 3 in the mentioned range of 300 - 400 nm. When plotted on a $\log - \log$ scale this resulted in straight lines (up to the dose range where saturation effects occur) with slopes equal to k - namely, between 2 and 3. It is interesting to note that Otaki et al.[15] have reported a similar behavior (the slope derived from their Fig. 2 is $\sim$3.3) in CaF_2:Tb_4O_7 exposed to sunlight. Halperin and Chen[14] explained the observed superlinear dose dependence as being the result of a multistage transition of electrons from the valence band into the conduction band. This can explain why, once the excitation is with high enough energy photons of 5.5 eV (225 nm), the dependence is linear. It stands to reason that a similar interpretation can be given to the results by Otaki et al.[15] of strong superlinearity under UV excitation in CaF_2:Tb_4O_7. A similar effect of two stage excitation of TL of $CaSO_4$ was described by Yaskolko[16].

Before going into more details concerning the different kinds of superlinearity (or supralinearity), we shall give some definitions for the characterization of nonlinearities as described by Chen and McKeever[17]. So far, we have used the term "more than linear", or mentioned that the TL intensity may depend on the dose like D^k with $k > 1$. It is obvious that the former is too vague whereas the latter does not cover cases like that in which the dose dependence is linear at low doses and goes superlinear at higher ones, or cases where $I_{\max} = \alpha D^{k_1} + \beta D^{k_2}$ in which k_1, $k_2 > 1$. Chen and McKeever[17] suggested the use of the two terms *superlinearity* and *supralinearity* to describe two related, but not identical properties of the growth curve.

4.1.1 Superlinearity

Following the work of Chen and Bowman[18], Chen and McKeever[17] suggested to reserve the term *superlinearity* for the increase in the derivative of the dose dependence function. Let us denote the measured TL signal by $S(D)$, be it either the maximum intensity or the total area under the peak, as a function of the dose D. The derivative of this function at any point D is dS/dD (or $S'(D)$), and an increase of the derivative at a certain point is expressed by stating that d^2S/dD^2 (or $S''(D)$) is positive. Thus, $d^2S/dD^2 > 0$ represents ranges of superlinearity; $d^2S/dD^2 < 0$ characterizes ranges of sublinearity, and of course, $d^2S/dD^2 = 0$ signifies a range (or point) of linearity. The difficulty with this definition is that it is merely qualitative since it has to do with the sign of the second derivative only. Chen and McKeever[17] proposed a normalized quantity which not only describes the qualitative feature of being super- or sub-linear, but which is also a quantitative measure of these properties. Moreover, as we have noted, one of the ways of detecting superlinearity is by plotting the results on a log – log scale. In this presentation of the function $S = \alpha D^k$, the slope is k and is a direct measure of the degree of superlinearity (or sublinearity). Chen and McKeever[17] wished to define a magnitude which evaluates numerically the amount of superlinearity, and which would attain the value of k in the special case in which $S \propto D^k$. They defined the *superlinearity index* $g(D)$ as

$$g(D) = \left[\frac{DS''(D)}{S'(D)}\right] + 1 \,. \tag{4.2}$$

As long as one is dealing with a range of an increasing dose dependence ($S'(D) > 0$, which is nearly always true), it is obvious that the condition $g(D) > 1$ indicates superlinearity since $S''(D) > 0$. Furthermore, $g(D) = 1$ means a range (or point) of linearity and $g(D) < 1$ signifies sublinearity. It can readily be seen that $DS''(D)/S'(D)$ is a dimensionless quantity, which makes the addition of unity legal, and obviously makes $g(D)$ a dimensionless function. It is immediately seen that taking $S(D) = \alpha D^k$ results in $g(D) = k$, as initially requested. In fact, even for $S(D) = \alpha D^k + \beta$ one gets $g(D) = k$.

In order to understand better the significance of the quantity $g(D)$, Chen and McKeever[17] considered some particular cases:

1) As suggested already, when plotting $\log S$ vs. $\log D$, ranges of slope larger than unity occur in which the slope varies with the dose in a certain dose range, say, from 2 to 3. (See Halperin and Chen[14] and Fig. 4.3, and Chen et al.[19], Fig. 4.7.) Let us consider the function

$$S(D) = \alpha D^2 + \beta D^3. \tag{4.3}$$

One readily gets

$$g(D) = (2\alpha D + 6\beta D^2)/(2\alpha D + 3\beta D^2) + 1. \tag{4.4}$$

This yields $g(D) = 2$ for the low doses, and $g(D) = 3$ for high doses, and of course, a transition from 2 to 3 in between.

2) Another interesting situation is that of dose ranges in which there is an approach to saturation. One example of such behavior is the function

$$S = S_0 (1 - \exp\{\gamma D\}) \tag{4.5}$$

for $\gamma > 0$. Using Eq. 4.2, one gets

$$g(D) = 1 - \gamma D \tag{4.6}$$

which may be near unity at low doses (where the dose dependence of S is nearly linear), positive but smaller than unity at higher doses, zero at a dose $D = 1/\gamma$ and negative at higher doses. Of course, $g(D) < 1$ always indicates sublinearity.

3) Let us consider the hypothetical case of the strong superlinearity

$$S(D) = S_0 \exp\{\gamma D\}. \tag{4.7}$$

where $\gamma > 0$. This yields

$$g(D) = \gamma D + 1. \tag{4.8}$$

The main feature of this is that the superlinearity starts as being ~ 1 at low doses, which represents a nearly linear behavior at low doses, where $\exp\{\gamma D\} \cong 1 + \gamma D$. At higher doses, the amount of superlinearity $g(D)$ grows steadily as more terms of higher order are to be included in the series expansion of $\exp\{\gamma D\}$. In other words, the dependence of $g(D)$ on D (linearly) demonstrates the very rapid increase of $S(D) = S_0 \exp\{\gamma D\}$ and its derivatives at high doses.

Figure 4.4 presents an example, taken from Chen and Hag-Yahya[20] of $g(D)$ derived from numerical results of dose dependence of TL under competition between trapping states (for the details, see below). Here, $g(D)$ starts at 2 (namely, a quadratic dose dependence), grows to much higher values at a certain dose range, before reducing to low values when saturation effects become dominant.

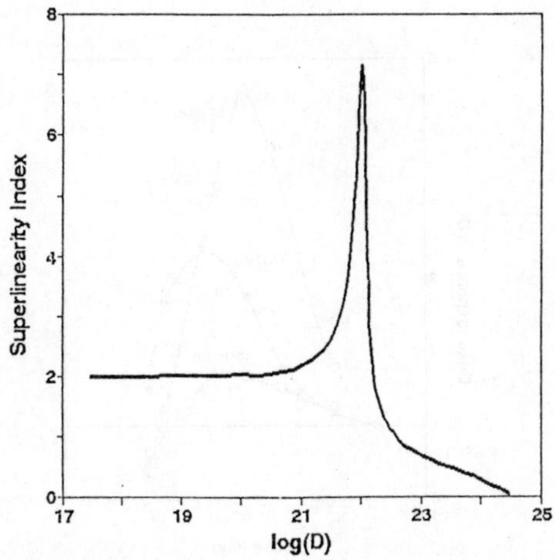

Figure 4.4 The superlinearity index $g(D)$ as a function of D numerically calculated for a model of two competing traps. $g(D)$ starts at the value of 2, attains higher values and then goes down to below unity.

(After Chen and Hag-Yahya[20].)

4.1.2 Supralinearity

As we have pointed out, a number of materials have the characteristic property shown in Fig. 4.1, namely, an initial linear range which is followed by a nonlinear (i.e. superlinear) region before saturation effects set in. Chen and McKeever[17] suggested the term *supralinearity* to describe this property of the measured quantity being above the continuation of the initial linear range. In fact past authors, starting with Cameron et al.[1], used the same term but without differentiating between the presently described attribute and superlinearity as defined above. Some authors, such as Horowitz[21] and Mische and McKeever[8] quantified this property by defining the dimensionless function

$$f(D) = [S(D)/D] / [S(D_1)/D_1] \tag{4.9}$$

termed by these authors the *supralinearity index* or *dose response function.* Here, D_1 is a normalization dose in the initial linear range. Of course, this definition is applicable only if such a linear range exists at the low dose range. Figure 4.5 gives an example of the dose response as a function of D derived from a TL dose dependence measured from LiF:Mg. The value of unity at the left hand side is related to D_1, at the end of the initial linear range. It is obvious that $f(D) > 1$ means values of $S(D)$ above the initial linear range, which is the numerical expression for supralinearity.

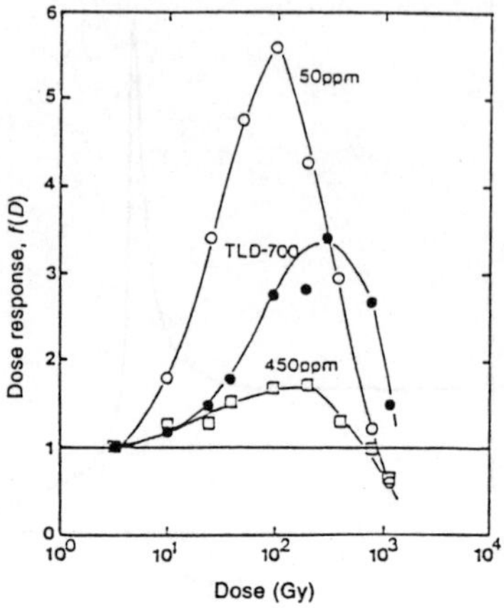

Figure 4.5 Plot of $f(D)$ vs. D for various samples of LiF:Mg. The samples used for these data were TLD-700 as well as LiF doped with either 50 or 450 ppm Mg. (After Mische and McKeever[8].)

Chen and McKeever[17] proposed a modified definition of $f(D)$, namely,

$$f(D) = \left[\frac{S(D) - S_0}{D}\right] / \left[\frac{S(D_1) - S_0}{D_1}\right] \tag{4.10}$$

where S_0 is the intercept on the TL intensity axis of the extrapolation of the linear region in curves like that shown in Fig. 4.2. For cases such as those shown in Fig. 4.1, it is clear that $S_0 = 0$, and thus the new, more general definition of $f(D)$ is identical to the older definition of Eq. 4.9. However, the advantage of the new definition is that it can be applied to the two cases in which a broad linear range occurs either preceded by or followed by a range of nonlinearity. Thus, in cases like that of Fig. 4.2 in which the superlinear region precedes the linear range, it is impossible to define $f(D)$ according to Eq. 4.9. Equation 4.10 may be used, however, noting that for a dose response of this type we have $S_0 < 0$. A negative value of S_0 yields $f(D) > 1$ in the low dose region, i.e., that part of the dose dependence in which the TL intensity is greater than that inferred from a backwards extrapolation of the linear range.

With this definition of $f(D)$, we see that as long as the function $S(D)$ is proportional to the dose, then obviously $f(D) = 1$, and in this case we see that $g(D) = 1$ (see previous subsection) and $f(D) = 1$ mean the same thing. However, for other cases, $g(D) = 1$ does not necessarily mean $f(D) = 1$ and vice-versa. Values of $S(D)$ above the extrapolated linear portion (e.g. in either Fig. 4.1 or Fig. 4.2) cause $f(D)$ to be larger than 1. It is obvious that in

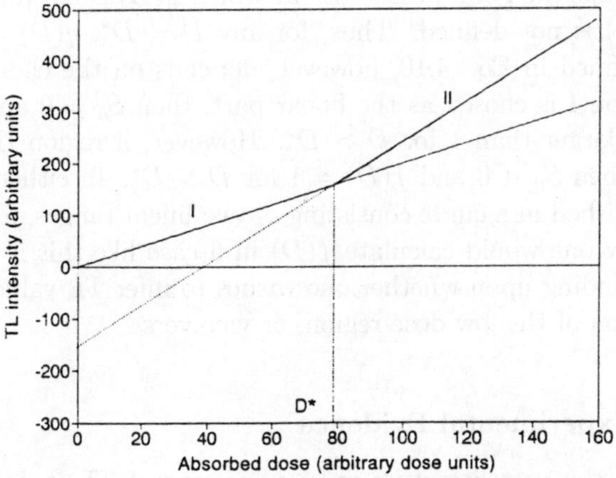

Figure 4.6 Hypothetical dose dependence curve consisting of two linear segments (I and II) with different slopes. D^* is 79 arbitrary dose units. S_0 for region I is 0, while S_0 for region II is -158, in arbitrary units.
(After Chen and McKeever[17].)

those dose ranges in which the function $S(D)$ "takes off" above linearity, the dose dependence is both superlinear and supralinear. At higher doses, however, when saturation effects start to occur, it may very well happen that the curve is not superlinear anymore ($d^2S/dD^2 \leq 0$ or $g(D) \leq 1$) but is still above the extrapolation range and therefore $f(D) \geq 1$, i.e. it is supralinear. Of course, $f(D)$ monitors exactly the property it is supposed to monitor, namely the amount to be considered for extrapolation purposes. In cases (1) and (3) of the previous subsection, $f(D)$ cannot be defined since there is no linear region at all. In case (2), $f(D) = 1$ at low doses, and becomes less than 1 as saturation is approached.

To summarize this discussion of the definitions of super- and supralinearity, one should understand that the former property, namely superlinearity, is a measure of the rate of change with the dose of the dose-dependent function, described quantitatively by the index $g(D)$. The latter property of supralinearity is related more to the applications of TL in dosimetry and archaeological and geological dating, and is presented quantitatively by the supralinearity index $f(D)$. Basically, this has to do with the correction to be made in extrapolation in cases where supralinearity occurs following an initial linear dose range, or prior to such a linear range. Figure 4.6 depicts a hypothetical dose dependence curve consisting of two straight segments intersecting at the dose D^* (79 units in the example shown). Obviously, no superlinearity occurs, excluding perhaps the point at $D = D^*$ at which $S(D)$ is not differentiable, and therefore, $g(D)$ is not defined. Thus, for any $D \neq D^*$, $g(D) = 1$. The value of $f(D)$ as defined in Eq. 4.10, however, depends on the choice of the linear region. If region I is chosen as the linear part, then $S_0 = 0$ and $f(D) = 1$ up to D^*, and is larger than 1 for $D > D^*$. However, if region II is chosen as a linear range, then $S_0 < 0$ and $f(D) = 1$ for $D > D^*$. In either case, supralinearity is established in a curve consisting of two linear ranges, for each of which $g(D) = 1$. How one would calculate $f(D)$ in a case like this is really a matter of choice, depending upon whether one wishes to infer TL values at high doses by extrapolation of the low dose region, or vice versa.

4.2 More Experimental Evidence

In addition to the vast literature on nonlinearity of TL in LiF (see also e.g. Jain[22] and Lakshmanan et al.[23]), high temperature peaks of pottery quartz and semiconducting diamonds, other materials exhibited the effect as well. Brynkar et al.[24] reported superlinearity of aluminophosphate glass detectors under γ ray excitation. Rodine and Land[25] studied the UV excited TL of ThO_2 crystals and found an initial quadratic dose dependence of some of the peaks whereas others depended linearly on the dose. Their findings and qualitative explanation were the basis for the theory of competition during heating developed later. An even stronger superlinearity was reported in the $\sim$ 110 °C TL peak in synthetic quartz irradiated with β rays, by Chen et al.[19]. In this case, under certain conditions, the slope on the log – log scale was 3, indicating a cubic dose dependence. An example is shown in Fig. 4.7. For more evidence of this superlinear dose dependence in synthetic quartz see the work by Kitis et al.[26]. These results are also further discussed below in relation to the model of competition during heating.

With the exclusion of the case of semiconducting diamond, all the explanations of superlinearity include the element of competition, be it during excitation or during the heating phase, or both. As a preliminary intuitive explanation (which is partially implied in the work of Rodine and Land[25]), let us consider a glow curve consisting of a series of TL peaks, the total area under which grows linearly with the excitation dose. If, at a certain dose range, the intensity of one (or more) peaks starts to approach saturation, the linearity of the total area means that at least one of the other peaks should grow faster than linearly. Here, the superlinearity of some of the peaks comes at the expense of some of the others, which grow sublinearly. This concept can be extended to cases in which some of the peaks are invisible, being related to radiationless transitions, or otherwise unmeasurable due to limitations of the measuring system. In this way superlinearity will be seen without a direct way of observing

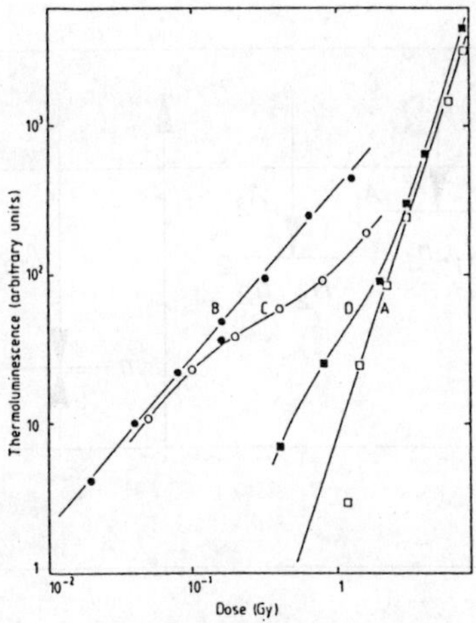

Figure 4.7 Growth curves of synthetic quartz.
(After Chen et al.[19].)

the peaks at the expense of which it occurs.

In what follows we discuss in detail the concepts of competition, both during excitation and during heating.

4.3 Competition During Excitation

The following explanation of superlinearity has been given by Suntharalingam[6], Suntharalingam and Cameron[28] and Aitken at al.[29], and further elaborated by Chen and Bowman[18]. The underlying assumption in this model is that the TL system has one recombination center, level 3, with m being the instantaneous occupancy of this center, and two trapping states 1 and 2, of concentrations N_1 and N_2 and occupancies n_1 and n_2, respectively, as shown in Fig. 4.8. The theoretical considerations discussed here follow the filling of the *active trap* n_1 in the presence of the competitor occupancy n_2, and *assumes* that the resulting TL intensity is proportional to this filling. The validity of this assumption under different conditions will be examined in a later section.

To begin with, we shall give an intuitive explanation to the superlinear filling of the active trap under these circumstances. Suppose that the concentration N_2 of the competitor is smaller than that of the active trap N_1, but the trapping probability A_2 of the competitor is larger than that of the active trap A_1. Qualitatively, at low doses, the excitation fills both traps linearly. At a certain dose, however, the competing trap N_2 saturates, hence more electrons are made available to the trap of interest N_1. This causes a faster, though linear, fill-

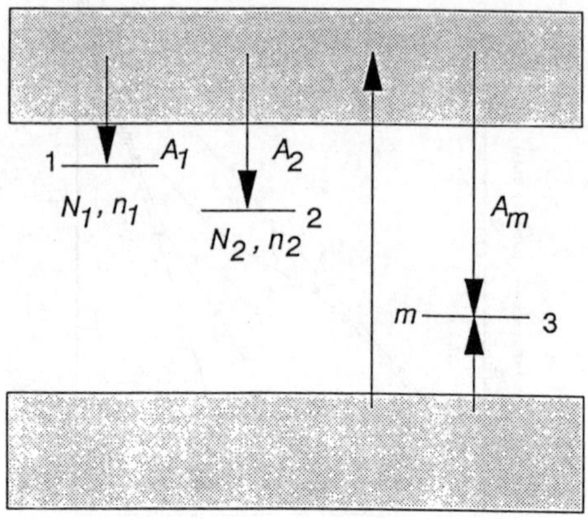

Figure 4.8 Energy levels involved in the competition during excitation. m - concentration of recombination centers, N_1 and N_2 - concentrations of active and competing traps, of which n_1 and n_2, respectively, are occupied. n_c - instantaneous concentration of electrons in the conduction band, A_1, A_2 and A_m are the retrapping, competing trapping and recombination probabilities, respectively.

ing of this trap. The transition region from one linear range to the other would, however, appear to be superlinear since this transition obviously occurs continuously. It is to be noted that such competition-related superlinearity bears some resemblance to superlinear luminescence dependence as described by Nail et al.[30] and superlinearity of photoconductivity as explained by Duboc[31], Cardon and Bube[32] and Dussel and Bube[33], the main difference being that in luminescence and photoconductivity the superlinear behavior is of signal vs. excitation intensity rather than vs. the dose.

Concerning the main assumption of the model, namely, that the TL intensity is proportional to the final filling of n_1, in the context of this model it is obvious that $n_1 < m$ at the end of the excitation and therefore, the peak intensity S (area under the curve) or I_m (maximum intensity) is indeed proportional to n_1. A further discussion on this point and the circumstances for its validity will be given in the next section.

Before entering a more mathematical presentation of the model, we would like to mention that the condition to be considered regarding superlinearity is, as explained above, $d^2n_1/dD^2 > 0$, and thus this will be analytically studied. However, in a section to follow we shall derive an expression with D

as a function of n_1 and, therefore, it is easier to investigate d^2D/dn_1^2. Since

$$d^2n_1/dD^2 = -\left(d^2D/dn_1^2\right)/\left(dD/dn_1\right)^3 \tag{4.11}$$

and since $dn_1/dD > 0 \Rightarrow dD/dn_1 > 0$, the condition $d^2n_1/dD^2 > 0$ is equivalent to $d^2D/dn_1^2 < 0$.

We shall start with a somewhat simplified version of the case of electrons being raised by the irradiation from the valence into the conduction band, and being trapped in levels 1 and 2. We thus neglect the possibility of band-to-band or band-to-center recombination during excitation. The equations governing the electron transitions are (see also Paige[34])

$$dn_1/dt = A_1(N_1 - n_1)n_c \tag{4.12}$$

$$dn_2/dt = A_2(N_2 - n_2)n_c \tag{4.13}$$

$$dn_c/dt = X - dn_1/dt - dn_2/dt \tag{4.14}$$

where n_c is the concentration of electrons in the conduction band, A_1 and A_2 are the transition probabilities ($\mathrm{m^3s^{-1}}$) into 1 and 2, respectively, and X is the rate of creation of electron-hole pairs by the irradiation. It is implicitly assumed here that the excitation of the whole sample is homogeneous, as is the case with irradiation by x-rays, γ rays, etc. Also, the above assumptions lead to the conclusion that all the created holes are accumulated in the recombination center m.

By eliminating n_c from Eqs. 4.12 and 4.13 we get

$$n_2 = (N_2 - n_{20})\left[(N_1 - n_1)/(N_1 - n_{10})\right]^{A_2/A_1} \tag{4.15}$$

where n_{10} and n_{20} are the initial values of n_1 and n_2. Using now the "usual" assumptions that the rate of change of electron concentration in the conduction band is much smaller than that of the charge in traps, i.e. quasiequilibrium, where $|dn_c/dt| \ll |dn_1/dt|$ and $|dn_c/dt| \ll |dn_2/dt|$, Eq. 4.14 becomes

$$X = dn_1/dt + dn_2/dt. \tag{4.16}$$

By integration we immediately get

$$D = \int_0^t X dt = n_1 + n_2 - n_{10} - n_{20}. \tag{4.17}$$

Substituting n_2 from Eq. 4.15 yields

$$D = n_1 - n_{10} + N_2 - n_{20} - (N_2 - n_{20})\left[(N_1 - n_1)/(N_1 - n_{10})\right]^{A_2/A_1}. \tag{4.18}$$

The superlinearity feature can be shown by differentiating twice, which gives

$$\frac{d^2D}{dn_1^2} = -\frac{A_2}{A_1}\left(\frac{A_2}{A_1} - 1\right)\frac{N_2 - n_{20}}{(N_1 - n_{10})^2}\left[\frac{N_1 - n_1}{N_1 - n_{10}}\right]^{A_2/A_1 - 2} . \tag{4.19}$$

The condition for this to be negative, thus yielding superlinear behavior, is $A_2 > A_1$. This condition is explicitly that which has been previously stated, namely, that the trapping probability of the competitor should be larger than that of the trap directly involved in the TL process. It is to be mentioned that filling of level 2 follows the same Eq. 4.18, with the subscripts 1 and 2 interchanged, and the same is true for Eq. 4.19. This shows, immediately that for $A_2 > A_1$, n_2 grows sublinearly or, in other words, is on its way to saturation. It is also to be mentioned that Eq. 4.18 is expected to yield a superlinear dependence for $A_2 > A_1$ in the whole range of $0 < n_1 < N_1$, whereas according to the qualitative description above as well as much experimental evidence, a linear region of growth occurs at low doses, and certainly, at high doses, saturation of level 1 occurs. However, observe that in the low dose range, d^2n_1/dD^2 is positive but very small, which means that the dependence is practically linear. As for the high dose range, Eq. 4.19 shows that once level 2 is very close to saturation, $N_2 - n_{20} \approx 0$ which results in $d^2D/dn_1^2 \approx 0$, and in turn, using Eq. 4.11, $d^2n_1/dD^2 \approx 0$, which still means a linear increase of $n_1(D)$. The fact that no saturation effect is predicted here is related to the assumption that recombination during excitation is negligible. This may approximately be the case at the low dose range, including the superlinearity region, but it probably ceases to be so at higher doses. One addition to the model which attempted to take care of this situation was considered by Aitken et al.[29] who examined the case in which band-to-band recombination is allowed. Equation 4.14 is replaced by

$$dn_c/dt = X - dn_1/dt - dn_2/dt - A_p n_c p_v \tag{4.20}$$

where p_v is the concentration of free holes in the valence band and A_p the band-to-band transition probability. In addition, neutrality has to be fulfilled, namely,

$$p_v = n_1 + n_2 + n_c \tag{4.21}$$

which implies the possibility of accumulating holes in the valence band. However, such an accumulation seems rather unlikely, and Chen and Bowman[18] modified this model as follows. Remembering that we denoted by m the accumulated concentration of holes in centers, Eq. 4.20 becomes

$$m + p_v = n_1 + n_2 + n_c. \tag{4.22}$$

However, p_v can often be neglected as compared with the accumulated concentration m (except for the very beginning of the excitation; a more accurate treatment would require a fifth differential equation to be fulfilled simultaneously with the other four) and we arrive at

$$m \approx n_1 + n_2 + n_c. \tag{4.23}$$

This is the same as Eq. 4.21 where the left hand side has now changed its meaning to be the holes trapped in the center which, as opposed to p_v, can accumulate during excitation. At the same time, we can change the last term on right hand side in Eq. 4.20 to be $A_m n_c m$, which is the recombination of free electrons with holes in centers. It is implicitly assumed here that the trapping probability of free holes into m is relatively fast. All in all, Chen and Bowman[18] changed the model of Aitken et al.[29] in such a way that now, band-to-band transitions are neglected, whereas band-to-center recombination during the excitation is allowed, and, in fact, is quite important. This is in line with the statement by Pilkuhn[35] that, indeed, band-to-band transitions in the model have much smaller probability than band-to-center transitions and is a statement of the Schokley-Read recombination processes. Also, not allowing band-to-center transitions during excitation, would not be consistent with the crucial role this transition plays during the heating. We thus take the result by Aitken et al.[29], replacing p_v by m and A_p by A_m, and get the solution of the revised equations 4.12, 4.13 and 4.20 as

$$\begin{aligned} D \;=\; & N_2 \frac{A_m - A_2}{A_2}\left[\left(1 - \frac{n_1}{N_1}\right)^{\frac{A_2}{A_1}} - 1\right] + \frac{A_1 - A_m}{A_1} n_1 \\ & - \frac{N_1 A_m}{A_1}\left(\frac{N_2}{N_1} + 1\right) \ln\left(\frac{N_2}{N_1} + 1\right). \end{aligned} \tag{4.24}$$

For the sake of simplicity, it has been assumed that $n_{10} = n_{20} = 0$, otherwise n_1 has to be replaced by $n_1 - n_{10}$, N_1 by $N_1 - n_{10}$ and N_2 by $N_2 - n_{20}$.

The first derivative of Eq. 4.24 is

$$\frac{dD}{dn_1} = -\frac{N_2}{N_1}\frac{A_m - A_2}{A_1}\left(1 - \frac{n_1}{N_1}\right)^{A_2/A_1 - 1} + \frac{A_1 - A_m}{A_1} + \frac{A_m}{A_1}\left(\frac{N_2 + N_1}{N_1 - n_1}\right). \tag{4.25}$$

For very high doses, we may expect $n_1 \approx N_1$ (nearly all the traps are filled), thus the first term in Eq. 4.25 goes to zero provided $A_2 > A_1$. The latter condition is shown below to be an important part of the conditions for superlinearity. The third term tends to infinity and so does the sum; in fact, the sum goes to infinity even if $A_2 \leq A_1$. Therefore, $dn_1/dD \to 0$ which demonstrates the expected saturation behavior at high doses. The second derivative will now be

$$\frac{d^2 D}{dn_1^2} = \frac{N_2}{N_1^2}\frac{A_m - A_1}{A_1}\left(\frac{A_2}{A_1} - 1\right)\left(1 - \frac{n_1}{N_1}\right)^{A_2/A_1 - 2} + \frac{A_m}{A_1}\frac{N_2 + N_1}{(N_1 - n_1)^2}. \tag{4.26}$$

Having illustrated that saturation occurs at high doses, we now consider the low dose range for which

$$\frac{d^2D}{dn_1^2}|_{n_1\approx 0} \cong \frac{1}{A_1N_1}\left[\frac{A_mA_2N_2}{A_1N_1} + \frac{A_2N_2}{N_1} + A_m - \frac{A_2^2N_2}{A_1N_1}\right]. \quad (4.27)$$

The size of the range which can be considered as the "low dose" region depends on the various parameters involved. Thus, a necessary and sufficient condition for superlinearity at low doses is

$$N_2A_2^2/(A_1N_1) > A_m + N_2A_2/N_1 + A_mA_2/N_1 + A_mA_2N_2/(A_1N_1). \quad (4.28)$$

Since each of the terms both on the right and left hand side of expression 4.28 are positive, we can deduce three necessary conditions for the superlinearity at low doses, namely

$$N_2A_2^2/(N_1A_1) > A_m;\ A_2 > A_m;\ \text{and}\ A_2 > A_1. \quad (4.29)$$

The first two conditions are fulfilled for small values of A_m. This seems to point to the fact that the superlinearity of the filling of the active trap is dependent on the competing recombination during excitation and, as a consequence, during heating. When the recombination probability is strong, no superlinearity of this kind is expected whereas when it is small, superlinearity of the trap filling is expected. The third condition $A_2 > A_1$ is the same as that mentioned above in the framework of the simpler model, namely, that trapping into the active trap is weaker than retrapping into the competitor. The first condition can be re-written as

$$A_2 > (A_mA_1N_1/N_2)^{1/2}, \quad (4.30)$$

and, therefore, the three necessary conditions can be summarized by saying that A_2 has to be large enough as compared to the relevant magnitudes of the other parameters in the system. The necessary condition for this kind of superlinearity can be summarized thus

$$A_2 > \max\left(A_m, A_1, \sqrt{A_1A_mN_1/N_2}\right). \quad (4.31)$$

Obviously, a simple, sufficient condition for the superlinear filling of the active trap in the presence of a competitor is

$$A_2 > 3\max\left(A_m, A_1, \sqrt{A_1A_mN_1/N_2}\right). \quad (4.32)$$

It is important to note here that the full picture of the superlinear dose dependence of *TL* (as opposed to trapped charge concentrations) is more complicated

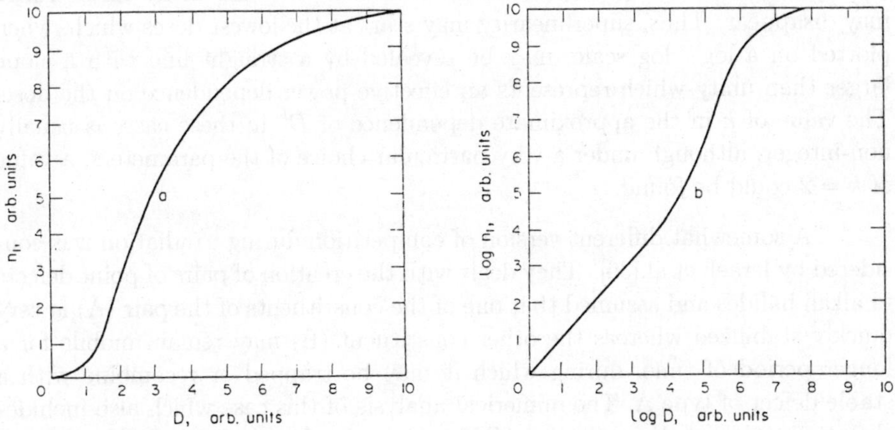

Figure 4.9 The linear-superlinear-linear-saturation dose dependence as numerically calculated from Eq.4.24, on a linear scale (a) and on a log-log scale (b). The parameters used are $N_1 = N_2 = 10^{23}$ m^{-3}, $A_1 = A_m = 10^{-22}$ m^{3}s^{-1} and $A_2 = 3 \times 10^{-21}$ m^{3}s^{-1}.

(After Chen and Bowman[18].)

since the same energy levels are involved in the processes during the following read-out (heating) stage.

It is obvious that when the relation given in expression 4.28 is reversed, which can easily occur for a high value of A_m, the dependence of n_1 on the dose will be sublinear starting at the lowest doses. It is to be mentioned that the analysis would be identical to that shown if one assumed filling conditions under which competition of the recombination center occurred, provided that the final concentration of holes in the active recombination center is smaller than that of electrons in the trap associated with the TL process, i.e., in the case where TL intensity depends on the center population. This point will be further discussed in the following sections along with the possible complication of the combined effect of competitions during excitation as well as heating. The two "mirror image" cases are certainly also possible, namely, the cases dealing with hole traps and electron centers, and their filling during the irradiation.

A demonstration of the dose dependence of n_1 as calculated from Eq. 4.24 is shown in Fig. 4.9, taken from Chen and Bowman[18]. As seen in curve (a), the dependence is linear at low doses, it goes superlinear at high doses and goes linear again before approaching saturation at high doses. As shown by Chen and Bowman, this typical behavior may alter if the parameters are chosen such that the capacity of the competing trap N_2 is relatively small (or, rather, that of $N_2 - n_2$ is small), in which case the initial linear range may disappear. Thus, superlinearity may start at the lowest doses which, when plotted on a log – log scale, may be revealed by a straight line with a slope larger than unity which represents an effective power dependence on the dose. The value of k in the approximate dependence of D^k in these cases is usually non-integer, although under a very particular choice of the parameters, a value of $k = 2$ could be found.

A somewhat different version of competition during irradiation was considered by Israeli et al.[36]. They dealt with the creation of pairs of point defects in alkali halides and assumed that one of the constituents of the pair (A) is very quickly stabilized whereas the other constituent (B) may remain mobile for a longer period of time, during which it may be trapped or recombine with a stable defect of type A. The numerical analysis of this case which also includes the attenuation of the excitation UV beam in the depth of the sample, yielded ranges of superlinear dependence on the dose.

4.4 Competition During Heating

As noted, Rodine and Land[25] reported the dose dependence of some of the TL peaks in β irradiated ThO_2 to be quadratic with the dose as of the lowest doses, and explained the effect qualitatively as being the result of competition during heating. We will start by giving an intuitive and more comprehensive explanation to this kind of superlinearity. Let us assume a situation in which the trap and center directly involved in the creation of a TL peak, are linearly dependent on the dose, at least at low doses. One may think that since two entities, the trapping state and the recombination center are involved, with initial fillings of n_0 and m_0 respectively following the excitation (see e.g. Fig. 1.1), then the emerging TL should be dependent on the product $n_0 m_0$. If we are dealing with a dose range in which both n_0 and m_0 are linearly dependent on the dose, one would expect a quadratic dependence on the dose. As is well known, the normal dependence of TL on the dose is linear or nearly linear, which suggests that the above explanation is incorrect in most cases. In fact, under most circumstances, if S denotes the measured TL (be it the area under the peak or the maximum intensity), we expect to have $S \propto \min(n_0, m_0)$ rather than $S \propto m_0 n_0$. The explanation for this is that the total number of recombinations should be determined by an integration over the TL intensity $I(T)$ which can be written also as $I(t)$ for a certain heating function $T = T(t)$. With the appropriate choice of units, we have $I(t) = -dm/dt$. For the sake of the present discussion, let us assume the existence of only one TL peak. We can integrate over $I(t)$ from an initial time t_0 (which we can set to zero) to infinity, and get

$$m_0 - m_\infty = \int_0^\infty (-dm/dt)dt = \int_0^\infty I(t)dt = S. \tag{4.33}$$

If $m_0 < n_0$, the peak terminates because the center (m) is exhausted, therefore $m_\infty = 0$ and thus

$$m_0 = \int_0^\infty I(t)dt = S, \tag{4.34}$$

and the right hand side is, in the appropriate units, the area S under the glow peak. If, however, $m_0 > n_0$, the peak terminates because the trap (n) is exhausted, thus we remain with $m_\infty = m_0 - n_0$ holes in the centers and therefore,

$$S = \int_0^\infty I(t)dt = m_0 - m_\infty = n_0. \tag{4.35}$$

Equations 4.34 and 4.35 can be summarized as

$$S = \min(n_0, m_0). \tag{4.36}$$

Thus, if $n_0 \propto D$, and $m_0 \propto D$ we still have $S \propto D$, which agrees with the behavior usually seen in the experiments. As for the maximum intensity I_m, it is normally close to being proportional to the total area S, and therefore, one usually gets $I_m \propto D$. This situation may change substantially in the presence of a competitor.

Let us consider again the energy level scheme of one kind of recombination center and two trapping states, this time concentrating on the transitions during heating, as shown in Fig. 4.10. Continuing the intuitive, qualitative argument, we can slightly change the statement that $S \propto \min(n_0, m_0)$ by saying that the area under the peak should be proportional to the quantity n_0 or m_0 which expires first. When these two levels are the only active ones in existence this indeed should be $\min(n_0, m_0)$, however, in the presence of a strong competitor this may not be the case anymore. We may very well have a situation in which we start with $n_0 > m_0$, but since during the heating some of the electrons are trapped in the competitor rather than recombining with holes in the centers, the peak may terminate when n_0 is exhausted whereas some holes still remain in the center. We therefore have here $S \propto n_0$, where n_0 is assumed to depend linearly on the dose. In addition, if we consider the electrons in the conduction band during the heating, they will distribute so that some recombine with trapped holes m and others go into the competitor. Roughly speaking, we can say that if the concentration of empty competitors $N_2 - n_2$ does not

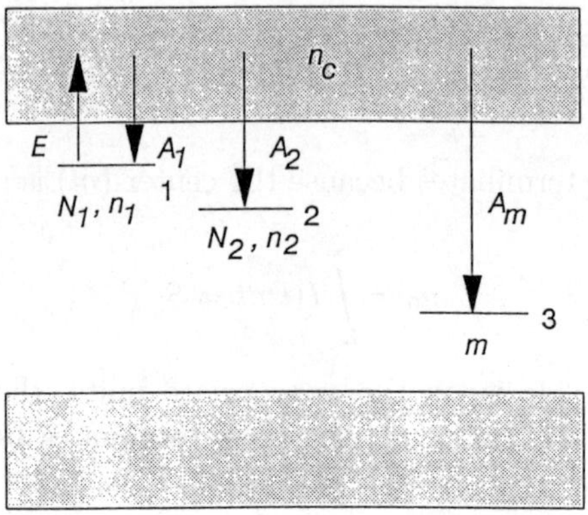

Figure 4.10 Energy levels involved in competition during heating (read-out). m - concentration of holes in centers (level 3); N_1, N_2 - concentrations of active TL traps, and competing traps, respectively (levels 1 and 2); n_1, n_2 - concentrations of electrons in these traps, respectively; A_1, A_2 - trapping probabilities into 1 and 2, respectively; A_m - the recombination probability; and n_c - the concentration of free electrons in the conduction band.

depend strongly on the initial dose (that is, if N_2 is far from saturation, or, $N_2 \gg n_2$), then the larger the value of m, the more recombinations into the center are expected. Therefore, if m_0 is proportional to the dose, this is an additional dependence on the dose of the TL intensity, and therefore, under these circumstances, $S \propto n_0 m_0$ and thus, $S \propto D^2$. If the competitor is in the range of an approach to saturation, the dependence of the TL intensity on the excitation dose may be even stronger than D^2.

This hand-waving argument for explaining the superlinear dose dependence of TL under competition during heating has been given a more convincing mathematical formulation by Kristianpoller et al.[37]. With reference to Fig. 4.10, the equations governing the processes occurring during heating are

$$dn_1/dt = -\gamma n_1 + A_1 n_c (N_1 - n_1), \tag{4.37}$$

$$dn_2/dt = A_2 (N_2 - n_2) n_c, \tag{4.38}$$

$$I = -dm/dt = A_m m n_c, \tag{4.39}$$

$$dm/dt = dn_1/dt + dn_2/dt + dn_c/dt, \tag{4.40}$$

where $\gamma = s \exp(-E_t/kT)$, in which E_t is the trap depth and s is the frequency factor. From Eqs. 4.38 and 4.39 one gets

$$n_c = \frac{1}{A_2}\frac{d}{dt}[\ln(N_2 - n_2)] = \frac{1}{A_m}\frac{d}{dt}(\ln m) \tag{4.41}$$

which can be integrated to yield

$$N_2 - n_2 = (N_2 - n_{20})(m/m_0)^{A_2/A_m}, \tag{4.42}$$

where m_0 and n_{20} denote the initial values of m and n_2, respectively. With the quasiequilibrium assumption

$$|dn_c/dt| \ll |dm/dt|, \tag{4.43}$$

Eq. 4.40 can be written as

$$dm/dt \cong dn_1/dt + dn_2/dt. \tag{4.44}$$

By substituting Eqs. 4.37, 4.38 and 4.39 into Eq. 4.44 we get an explicit expression for n_c

$$n_c = \gamma n_1 / \left[A_1(N_1 - n_1) + A_2(N_2 - n_2) + A_m m\right], \tag{4.45}$$

and by inserting Eq. 4.45 into Eq. 4.41 we have

$$I(t) = -dm/dt = A_m \gamma n_1 m / \left[A_1(N_1 - n_1) + A_2(N_2 - n_2) + A_m m\right]. \tag{4.46}$$

The three terms appearing in the denominator represent the probabilities (s^{-1}) for retrapping at the original trap (level 1), trapping at the competitor (level 2) and for recombination (at level 3) respectively. Integration of Eq. 4.44 yields

$$m - m_0 = n_1 - n_{10} + n_2 - n_{20} \tag{4.47}$$

where n_{10} is the initial value of n_1. By substituting Eqs. 4.42 and 4.47 into Eq. 4.46, one gets the differential equation

$$I(t) = -dm/dt = \gamma A_m m F(m) \tag{4.48}$$

with

$$F(m) = \frac{(n_{10} + n_{20} - m_0 - N_2) + m + (N_2 - n_{20})(m/m_0)^{\frac{A_2}{A_m}}}{A_1(N_1 + N_2 - n_{10} - n_{20} + m_0) + (A_m - A_1)(N_2 - n_{20})(m/m_0)^{\frac{A_2}{A_m}}}. \tag{4.49}$$

This equation can be numerically solved for various sets of given parameters.

It can be shown that Eq. 4.48 includes the quadratic dose dependence of TL (see Rodine and Land[25]) as a special case. If we assume that trapping in the competitor is much faster than both retrapping and recombination, namely,

$$A_2(N_2 - n_2) \gg A_1(N_1 - n_1) + A_m m, \tag{4.50}$$

then Eq. 4.46 becomes

$$I(t) = -dm/dt = A_m \gamma n_1 m / \left[A_2(N_2 - n_2)\right]. \tag{4.51}$$

If we assume, in addition, that the retrapping into level 1 is very small as compared with the rate of release of trapped carriers,

$$\gamma n_1 \gg A_1 n_c (N_1 - n_1), \tag{4.52}$$

Eq. 4.37 becomes

$$dn_1/dt = -\gamma n_1, \tag{4.53}$$

the solution of which is simply

$$n_1 = n_{10} \exp\left\{-\int_0^t \gamma d\tau\right\}. \tag{4.54}$$

Substituting Eqs. 4.42 and 4.54 into Eq. 4.51 yields

$$I(t) = -dm/dt = A_m \gamma n_{10} \exp\left\{-\int_0^t \gamma d\tau\right\} m / \left[A_2(N_2 - n_{20})(m/m_0)^{A_2/A_m}\right]. \tag{4.55}$$

Integration of this equation gives

$$\int_0^t \gamma n_{10} \exp\left\{-\int_0^\tau \gamma d\tau'\right\} d\tau = (N_2 - n_{20})\left[1 - (m/m_0)^{A_2/A_m}\right]. \tag{4.56}$$

Since

$$\int_0^\infty \gamma \exp\left\{-\int_0^t \gamma d\tau\right\} dt = 1, \tag{4.57}$$

Eq. 4.56 transforms into

$$n_{10} = (N_2 - n_{20})\left[1 - (m_\infty/m_0)^{A_2/A_m}\right] \tag{4.58}$$

for $t = \infty$, where m_∞ is the final value of the filling of level 3. From Eq. 4.58 we immediately get

$$m_\infty = m_0\left[1 - n_{10}/(N_2 - n_{20})^{A_m/A_2}\right]. \tag{4.59}$$

Assuming further that $n_{10} \ll N_2 - n_{20}$, we now get

$$m_\infty \cong m_0 \left[1 - (A_m/A_2)(n_{10}/(N_2 - n_{20})\right]. \tag{4.60}$$

As pointed out in Eq. 4.35, the area S under the glow peak equals $m_0 - m_\infty$, which leads, therefore to

$$S = m_0 - m_\infty \cong [A_m/A_2(N_2 - n_{20})]\, m_0 n_{10}. \tag{4.61}$$

Following Kristianpoller et al.[37] we can add that if level 2 is far from saturation, $N_2 \gg n_{20}$, we get

$$S \cong [A_m/(A_2 N_2)]\, m_0 n_{10}, \tag{4.62}$$

which yields the quadratic dose dependence reported by Rodine and Land[25] if both m_0 and n_{10} are in their linear range. Some additional theoretical discussion on the quadratic dose dependence under similar circumstances can be found in the work by Zavt and Savikhin[38] and by Kantorovich et al.[39]. However, as pointed out by Chen et al.[19], if n_{20} in Eq. 4.61 is dose dependent, the overall dose dependence of S can be more than quadratic, which may be very significant in ranges where the competitor is coming close to saturation.

We can thus distinguish between two elements of superlinear dependence of TL due to competition during heating. When the competitor N_2 is far from saturation, we can expect a quadratic dependence on the dose in agreement with Eq. 4.62 and assuming linear dependencies on the dose of the filling of the recombination center and the active trap (level 1). In addition, at the range of approach to saturation of the competitor, an extra element of superlinearity, which may be quite strong, may occur. Kristianpoller et al.[37] simulated this situation by assuming the approach to saturation of n_{20} as

$$n_{20} = N_2[1 - \exp\{-\alpha D\}] \tag{4.63}$$

where α is a positive constant. Curve A of Fig. 4.11 shows the dose dependence of the area under the curve S under these circumstances. At low doses, S grows quadratically with the dose up to a point where the approach to saturation of the competitor gets noticeable, at which point the dose dependence function becomes steeper. At still higher doses S grows linearly with D as the competitor is practically full and is therefore ineffective as a competitor. The fact that the dose dependence in curve A is linear all the way up to the filling of level 2 is related to the non-physical assumption that no saturation effect appears in the filling of the active trap. A more accurate treatment of the model should take into account the real dose dependence of n_{10} and m_0. As for the initially quadratic behavior, it is quite obvious that if the measuring system is not capable of detecting below the level of ~ 11.5 on the log S scale, the apparent initial

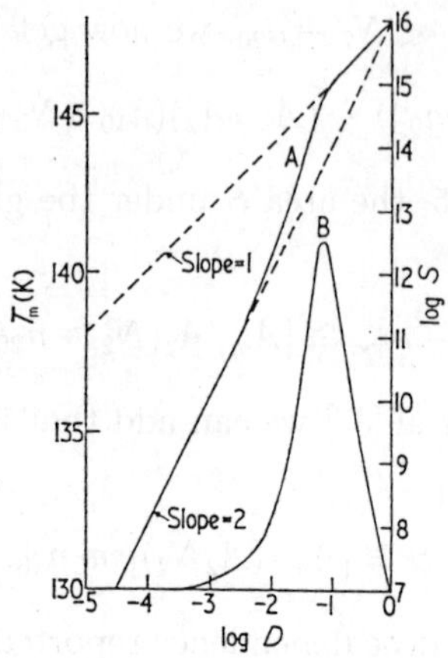

Figure 4.11 The area S (curve A) and the maximum temperature T_m (curve B) as functions of the dose D numerically calculated from Eqs. 4.37-4.40. The chosen parameters are: $N_1 = 10^{23}$ m^{-3}, $N_2 = 10^{22}$ m^{-3}, $A_1 = A_2 = A_m = 10^{-22}$ m^{3}s^{-1} and the dose dependence filling functions are $n_{10} = n_{20} = 10^{22} D$ m^{-3} and $n_{20} = 10^{22}[1 - \exp\{-50D\}]$ m^{-3}. The dose is normalized so that its maximum value is unity.

(After Kristianpoller et al.[37].)

dose dependence will be more than quadratic, and the "real" initial quadratic behavior at lower doses will not be observed. Note also curve B which shows the resulting dose dependence of the maximum temperature T_m. This starts at a constant value of 130 K at low doses, goes up by more than 10 K in the critical range of approach to saturation of the competitor, and goes down back to 130 K at the higher dose range. The idea of assuming an exponential approach to saturation of all three relevant trap and center occupancies was further discussed in the above-mentioned work by Sunta et al.[9], and results concerning the expected superlinearity of the dose dependence have been discussed. Of course, the disadvantage of this treatment is in assuming the simplest kind of approach to saturation rather than using the real results of the same energy level model during the excitation stage.

A rather similar situation which, at a first glance may appear to lead to a similar kind of superlinearity, is a model with one trapping state, concentrtaion n, and two kinds of centers, concentrations m_1 and m_2, as shown in Figs. 4.12 and 4.13. As far as the filling of the center is concerned, the general considerations made in section 4.3, are indeed expected to hold. This means that one of the centers, say m_2, which we consider as being the radiationless competitor, starts linearly with the dose, and tends to saturation when it is about to be filled

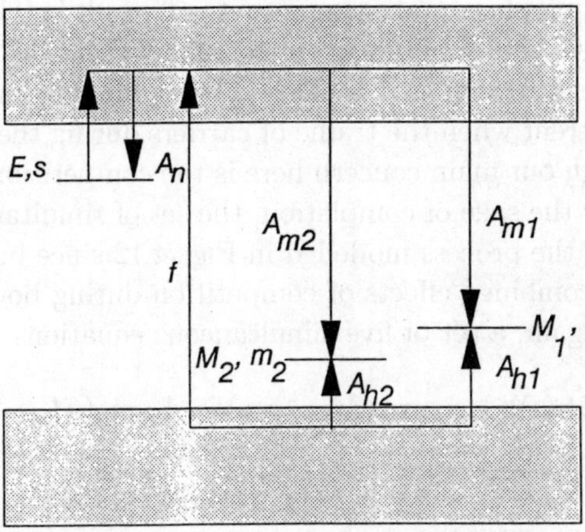

Figure 4.12 An energy level scheme for the model of two competing centers and one trapping state. The meaning of the different parameters is given in the text. The transitions shown are those occurring during the excitation period. (After Chen et al.[40].)

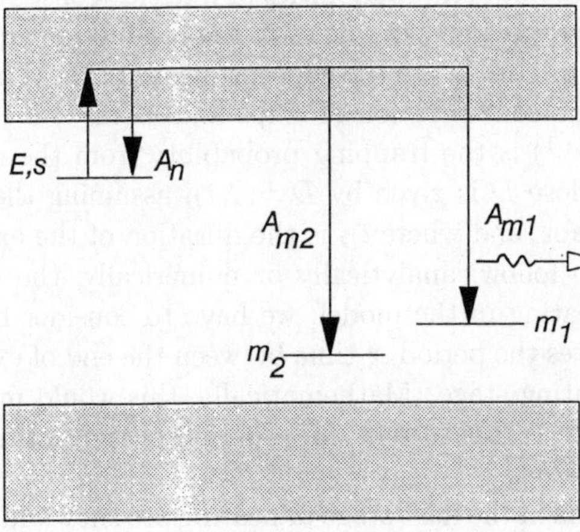

Figure 4.13 The transitions occurring in the model shown in Figure 4.12 during heating. It is assumed that the measured emission is only from m_1, and therefore, the TL intensity I is directly related to $-dm_1/dt$. (After Chen et al.[40].)

to capacity. The center m_1, which is considered to be the *active* one assuming that during the heating phase transition into it will result in TL emission, starts linearly with the dose, gets superlinear in the range where m_2 is saturated, and finally goes itself to saturation as it is close to being filled to capacity. The situation is entirely different when the traffic of carriers during the heating phase is considered. Although our main concern here is the competition during heating, let us write first, for the sake of completion, the set of simultaneous differential equations governing the process modelled in Fig. 4.12 since in the next section we will discuss the combined effects of competition during both excitation and heating. We have, again, a set of five simultaneous equations

$$dm_h/dt = X - A_{h1}m_h(M_1 - m_1) - A_{h2}m_h(M_2 - m_2) \quad (4.64)$$

$$dm_1/dt = A_{h1}m_h(M_1 - m_1) - A_{m1}m_1n_c \quad (4.65)$$

$$dm_2/dt = A_{h2}m_h(M_2 - m_2) - A_{m2}m_2n_c \quad (4.66)$$

$$dn/dt = A_n(N - n)n_c \quad (4.67)$$

$$dm_1/dt + dm_2/dt + dm_h/dt = dn/dt + dn_c/dt. \quad (4.68)$$

Here, M_1 and M_2 (m^{-3}) are the concentrations of the two centers and m_1, m_2 (m^{-3}) are their occupancies, respectively. A_{h1} and A_{h2} (m^3s^{-1}) are the probabilities for capturing free holes from the valence band, and A_{m1}, A_{m2} (m^3s^{-1}) are the recombination probabilities of free electrons. X ($m^{-3}s^{-1}$) is the rate of creation of free electrons and holes in the conduction and valence bands, respectively. n_c and m_h (m^{-3}) are the concentrations of these free electrons and holes. N (m^{-3}) is the concentration of traps and n (m^{-3}) is its instantaneous occupancy. A_n (m^3s^{-1}) is the trapping probability from the conduction band into the trap. The dose D is given by $D = Xt_D$ assuming that the excitation intensity X is constant, and where t_D is the duration of the excitation.

If we wish to follow, analytically or numerically, the whole procedure of excitation and heating in the model, we have to consider here a relaxation period which simulates the period of time between the end of excitation and the beginning of the heating stage. Mathematically, this would mean the solution of Eqs. 4.64-4.68 with X set to zero, for a period of time such that n_c and m_h decay to negligible values.

The transitions in the next stage of heating are shown in Fig. 4.13. Here, E_t (eV) and s (s^{-1}) are the trap depth and the frequency factor of the trap, respectively. The governing set of equations is now

$$I = -dm_1/dt = A_{m1}m_1n_c \quad (4.69)$$

$$-dm_2/dt = A_{m2}m_2n_c \quad (4.70)$$

$$-dn/dt = sn \exp\{-E_t/kT\} - A_n(N-n)n_c \tag{4.71}$$

$$dn/dt + dn_c/dt = dm_1/dt + dm_2/dt \ . \tag{4.72}$$

The numerical solution of the simultaneous sets of equations 4.64-4.68 and subsequently 4.69-4.72 will be discussed in the next section. However, approximate considerations, similar to those made above in the case of competition between traps, show that competition between centers leads to an entirely different result. Making the analogous assumptions to those leading to Eq. 4.61, we get a very similar result with m_{20} replacing $(N_2 - n_{20})$ where m_{20} is the final filling of M_2 following the first stages of excitation and relaxation, i.e. the initial concentration for the heating stage. The result is

$$S \cong [A_{m1}/(A_{m2}m_{20})]\, m_{10}n_0 \ . \tag{4.73}$$

As much as the expression is analogous to Eq. 4.61, the consequences concerning the dose dependence are entirely different. If, for example, n_0 and m_{20} behave in a certain dose range in a similar manner, say, both grow linearly with the dose, the measured TL dependence on the dose will behave in the same way as m_{10}. If, due to competition during excitation, m_{20} is sublinear with the dose, the superlinearity of TL will be stronger in this range. As opposed to the case of competition between traps, however, this kind of trap and center filling may account for a behavior that starts linearly, gets superlinear at higher doses and approaches saturation. This is opposed to the previous case where the initial dose dependence was expected to be quadratic, and at higher doses became even more superlinear before starting its approach to saturation. The reason for the fundamentally different behavior between the competing trap and the competing center cases lies in the fact that the number of competing traps $(N_2 - n_2)$ *decreases* with dose, whereas the number of competing centers m_2 *increases* with dose.

Since these conclusions are based on simplifying assumptions, it is advisable to support them by numerical calculations demonstrating the two kinds of possible behaviors, as was done by Chen et al.[40]. In this latter work, particular sets of parameters were chosen for each of the two different models. Of course, the advantage is that no simplifying assumptions were required, and the calculated results were the direct outcome of the properties of the physical situation embodied by the relevant model. At the same time, it is obvious that results based on the choice of specific sets of parameters are just special cases that may demonstrate that certain behavior is possible, but does not prove a general attribute. In this sense, each of the two approaches, namely, numerical solutions of specific cases vs. quasi-general results based on approximations are complementary methods that can be used to understand trends such as the dose dependence in certain models, as discussed here.

4.5 Combined Competition in Excitation and Heating

It is obvious that the complete dichotomy between competition during excitation and competition during heating, as discussed so far, is rather artificial, and can point mainly to trends in the expected behavior rather than be used to explain details. The obvious reason is that following a given irradiation of a sample, one measures the resulting TL without being able to observe directly the filling of the traps. Also, the assumptions of linear dose dependence of m_0 and n_{10} as well as an exponential approach to saturation of n_{20} on the dose, do give an idea of the possible expected dose dependence of the measured TL, but in doing so, we are assuming externally given dose dependencies rather than solving the model equations from scratch. In fact, in principle one should start with the set of Eqs. 4.12 - 4.14 and 4.23, and solve them during the excitation period. In order to mimic the experimental results, one should then consider at this point a period of relaxation during which time, the free electrons and holes left in the conduction and valence bands respectively, relax and become trapped, or recombine, prior to the heating. As noted earlier the equations governing this process are the same with X, the rate of production of electron-hole pairs, set to zero. The contribution of this relaxation may or may not be significant, depending on the set of parameters being used. More often than not, the addition to the trapped carriers during this time is negligible. Finally, one has to solve the set of equations 4.37-4.40 with a chosen heating function $T = T(t)$, be it linear or not. Of course, the same energy level scheme with exactly the same parameters are used in the heating stage as are utilized in the excitation and relaxation stages. The procedure explained so far appears to be rather difficult, and in most cases it indeed is. As mentioned before, since the differential equations under consideration are not linear, we cannot expect to be able to get an analytical solution. In principle, we can choose one of two routes to tackle the problem. Unfortunately, each of them has its own serious pitfalls. One way is to use assumptions which may lead to analytical, albeit approximate expressions. Taking this route, it may happen that the combination of the assumptions made will lead the analytical results astray, and in this sense, the "general" result obtained is questionable. The alternative is to solve numerically the set of differential equations during the three stages. With modern numerical methods, this can be done very accurately. Although such results may be very accurate and instructive, the difficulty here is that we get each time a solution corresponding to a particular set of parameters, and even if we repeat this for a number of sets, it is very difficult to draw general conclusions. We do maintain, however, that even the demonstration that certain dose dependencies are possible is of great importance. Also, we believe that the combination of the two mentioned routes can bring about a better understanding of the possible dose dependencies. Some works dealing with the combined effect of competition dur-

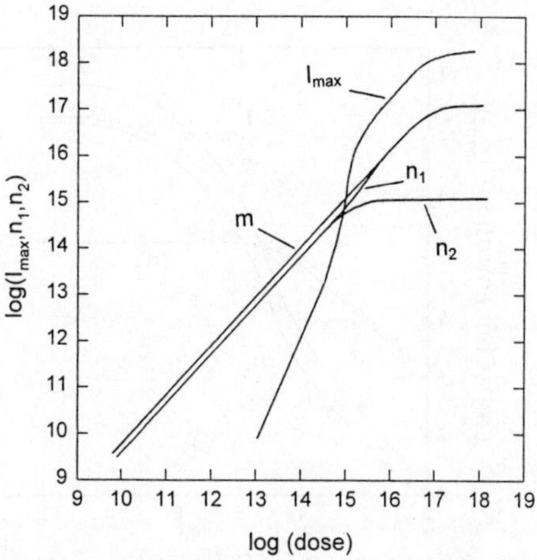

Figure 4.14 Representative results of the numerical calculations related to a model with two traps and one center. n_1, n_2 and m are the values attained for given doses of irradiation following the first two stages of excitation and relaxation. I_{max} is the resulting maximum of TL, calculated during the heating stage.
(Redrawn from Chen and Fogel[42].)

ing excitation and readout are to be mentioned here. Lee and Chen[41] solved the two sets of equations sequentially, using the quasiequilibrium assumption, and found the implicit expression for the TL signal S

$$S + (N_2 - n_{20}) \left[1 - (1 - S/m_0)^{A_2/A_m}\right] = n_{10}, \tag{4.74}$$

which in turn, could easily be solved numerically to yield the dose dependence. If retrapping is much stronger than recombination, a quadratic dose dependence results whereas if the competing trap is close to saturation, S depends linearly on the dose. In other cases, however, Eq. 4.74 was numerically solved and yielded results such as linear-superlinear-linear-saturation, which is sometimes found in experiments.

The alternative numerical solution of the simultaneous differential equations during the three stages, points to interesting results. Chen and Fogel[42] performed such calculations, using the above-mentioned model of two competing traps and one recombination center. These authors chose the following set of parameters: $N_1 = 10^{23}$ m^{-3}, $N_2 = 10^{21}$ m^{-3}, $A_1 = A_m = A_n = 10^{-21}$ m^{3}s^{-1}, and $A_2 = 10^{-19}$ m^{3}s^{-1}. A result which is typical of this model (more than merely a result obtained with this specific set of parameters only) is shown in Fig. 4.14 on a log – log scale. The basic feature here is that n_1 and n_2 start linearly with the dose, and at a certain dose range at which n_2 approaches saturation, n_1 goes

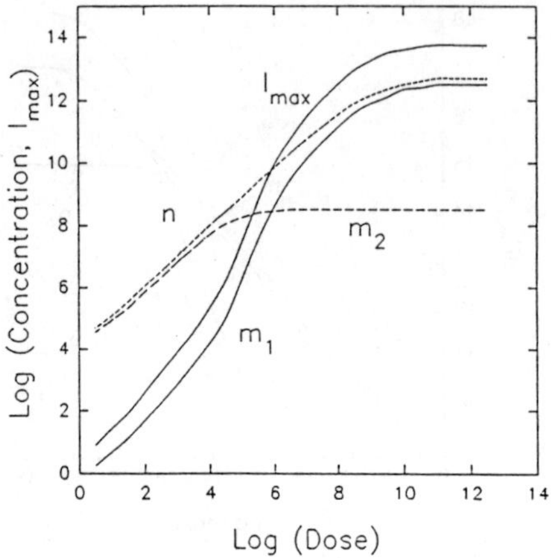

Figure 4.15 A sample result of the model with two centers and one trap. The parameters chosen are given in the text.
(After Chen et al.[40].)

slightly superlinear, before approaching saturation. The emitted TL, however, measured here by $I_{\max}$, starts quadratically, becomes more superlinear in the range where n_2 approaches saturation, then goes linear and finally to saturation at higher doses. These results resemble those of the approximate solution by Kristianpoller et al.[37] and, anyway, point to the fact that the governing effect is that of competition during heating. To some extent, using different parameters could change the superlinear range of n_1, and its contribution to the final TL, but in all the results reported this was a minor contribution and the main effect was that of competition during heating with the basic result that the dose dependence was superlinear as of the lowest doses.

The alternative effect of a model with two competing centers and one trapping state was studied by Chen et al.[40]. A sample result is shown in Fig. 4.15. The given parameters here are $E_t = 1.0$ eV, $s = 10^{13}$ s^{-1}, $X = 10^{21}$ m^{-3}s^{-1}, $A_{m1} = A_{m2} = 10^{-21}$ m^{3}s^{-1}, $A_n = 2 \times 10^{-21}$ m^{3}s^{-1}, $M_1 = 9 \times 10^{23}$ m^{-3}, $M_2 = 10^{23}$ m^{-3}, $A_{h1} = 10^{-21}$ m^{3}s^{-1}, $A_{h2} = 10^{-20}$ m^{3}s^{-1}, and $N = 10^{24}$ m^{-3}. The x-axis here gives t_D, the duration of irradiation, to which the excitation dose Xt_D is proportional. The curves show the dose dependencies of m_{10}, m_{20}, n_0 and $I_{\max}$. Here, m_1, which is considered to be the active center, grows linearly as long as m_2 is far from saturation. It then goes up superlinearly in the range where m_2 approaches saturation, and finally m_1 approaches saturation at higher doses. In the heating stage, transitions into m_2 are considered to be radiationless or, which is equivalent for the present purposes, to emit light in a range to which the light detecting device is "blind". The trap filling n is seen to grow linearly until it approaches saturation at about the same dose range as m_1. The resulting TL is seen to follow nearly exactly the dose dependence of the active trap m_1. It is obvious that the dominating effect here is that of competition during excitation whereas competition during heating adds only a minor contribution. Here too, varying the relevant parameters does not seem to change substantially the results. This appears to be at least one of the ways to explain the known dose dependence of linear-superlinear-linear-saturation.

To conclude this section, the following points are to be made (see references [40],[8],[43]).

1. In general, it is inadequate to consider the two kinds of competition (i.e. during excitation or during heating) separately. In this respect, the arguments on which kind of competition occurs in specific cases should include both possibilities since both types may contribute to the observed superlinearity in the same material and under the same experimental conditions.

2. The superlinear effect due to competition during heating has two elements. One has to do with the final result being proportional to the concentrations of charge carriers in both traps and centers in the presence of a strong competitor. Thus, if each of them is proportional to the dose, the total dose dependence is quadratic. In addition, in the dose range where the competitor approaches saturation, the reduction in competition causes some extra superlinearity (more than quadratic behavior).

3. The numerical solutions of the sets of simultaneous differential equations enable us to bypass the possible problems of artifacts due to the approximations made. Thus, if a certain behavior is observed, we can be sure that it is a net effect of the model under examination and, therefore, can be compared with the relevant experimental results. On the other hand, the numerical results are associated with the chosen sets of parameters, and it is rather difficult to draw general conclusions. It is felt, however, that it is of prime importance to be able to demonstrate that the experimentally observed behaviors can, indeed, be associated with these models. It is also important, however, to mention the approach typified in the mentioned paper by Lee and Chen[41]. These authors used the opposite way, namely, they reached some conclusions concerning superlinearity by making approximations, thus getting semi-analytical results.

4. As opposed to what might have been thought, it has been found, both from analytical considerations and numerical results that there is a big difference between a situation with a competition of two trapping states and a competition between two centers. In the former, a typical behavior is an initially quadratic dose dependence followed by stronger superlinearity before saturation effects set in. In the latter, a typical dependence is an initial linear range, followed by moderate superlinearity, and in turn, going back to linearity

and saturation. It appears that the main difference between a trap competitor and a center competitor is that, in the former case, the concentration $N_2 - n_2$ which competes with the concentration m_{10} for released electrons, decreases with the dose. In the latter case, m_{20} which is the concentrtaion of the relevant competitor, increases with the dose.

5. The apparent discrepancy between the theoretical dependence which starts quadratically and gets more superlinear at higher doses, and the more than cubic initial dose dependence in the 110 °C peak in synthetic quartz can be explained easily. It is very plausible that the initial expected quadratic dependence cannot be observed due to an insufficient sensitivity of the measuring system in the low dose range. As stated by Mische and McKeever[8], under the circumstances of a competing luminescence center, no superlinearity is expected for competition during heating if the center is far from saturation. This is, more or less, confirmed in the work by Chen et al.[40]. In this case the only superlinearity possible seems to be that related to the superlinear filling of the centers involved, in which case the measured TL would follow practically the same pattern.

4.6 Dose-Rate Effects

In the discussion so far we have addressed the dose imparted on a sample as a basic independent variable, uniquely defined by the physical circumstances of the irradiation of the sample. It should be remembered, however, that the excitation of the sample is performed, in fact, by utilizing a radiation intensity $X(t)$, and the applied dose should be defined as $D = \int_0^t X(\tau)d\tau$. Even if we limit the discussion to the rather common case in which $X(t)$ is a constant X, we should remember that since the dose D is given by $D = Xt$, one can get the same dose by using a high intensity (also termed *dose rate*) and short time or low dose-rate and long time. We will define *dose-rate effects* as the cases in which the application of a given dose using different dose rates result in different TL intensities. An obvious situation in which dose-rate effects are expected are cases where the sample is excited at a temperature below, but rather close to that of the expected TL peak. Under these circumstances, it is obvious that thermal decay of the TL peak occurs simultaneously with the build-up, which influences the final measured signal. As a result of this dynamical competition, it is clear that more TL will be emitted with high dose rates and short irradiation times than with low dose rates imparted at longer times. A similar situation occurs when the peak appears at a significantly higher temperature than that of irradiation, but anomalous fading is taking place.

In the present section we discuss some experimental evidence found in the literature for the dose-rate effect and mention some theoretical work explaining the possibility of dose-rate effects where no simultaneous fading (normal or anomalous) occurs.

To begin with, most research workers in the fields of TL dating and dosimetry, assume that the TL response is independent on the dose rate (see, for example, Cameron et al.[1] and Aitken[13]). Justification for this assumption stems predominately from the experimental work by Karzmark et al.[44] and Tochilin and Goldstein[45] who observed a dose-rate independent TL response from LiF for dose rates ranging from 5 to 1.7×10^8 Gy/s. This, of course, does not prove that dose-rate effects are impossible in other materials and/or other dose-rate ranges. This is important, in particular in the study of archaeological samples in which the natural dose rate to which the sample is exposed in antiquity may be as low as 10^{-3} Gy/y (3×10^{-11} Gy/s), whereas the laboratory dose rate, applied for calibration may be as high as several Gy/s (see Ch. 7). It is to be noted that a dose-rate effect such as that reported by Facey[46] is not the "real" effect discussed here since it has to do with the thermal decay during the excitation. However, Groom et al.[47] reported a significant effect where no thermal decay takes place. Here, a decrease of TL up to a factor of 5, was reported with increased dose rate in powdered samples of Brazilian crystalline quartz when irradiated with ^{60}Co γ rays at dose rates ranging from 1.4×10^{-3} to 3.3 Gy/s. A smaller effect of the same sort was reported by Hsu and Weng[48]. Some additional experimental evidence of a similar nature was given by Wintersgill and Townsend[49] who described a dose rate effect in the excitation of TL in LiF. An opposite effect of higher TL for larger dose rate was reported by Kvasnička[50],[51] who found the effect in Brazilian and milky quartz excited by ^{60}Co γ rays, using dose rates of 2×10^{-5} Gy/s to 2×10^{-2} Gy/s. The apparent discrepancy between the two sets of results may very well be due to the different (though somewhat overlapping) ranges utilized.

In a theoretical work by McKeever et al.[52], it was demonstrated that while using a simple model of one trap and one center, and taking into consideration the recombination occurring during the excitation as well as the relaxation time following the excitation and prior to the heating of the sample, a dose-rate dependence of the filling of the relevant trap can be found. In this model, and with specific sets of parameters chosen, it was shown that whereas at high total dose, the filling of the trap increases with the dose rate, at lower total doses, the effect goes in the opposite way and at higher dose rates, the final filling of the trap is lower, which should result in a lower measured TL.

A further work along the same lines was given by Chen et al.[53], a sample result is shown in Fig. 4.16. The model utilized here is similar to that in Fig. 4.8, but with one trapping state only, denoted by concentrtaion N, having an instantaneous occupancy of n and a trapping probability of A. The probability of capturing a hole in the valence band into the hole center is denoted by A_h.

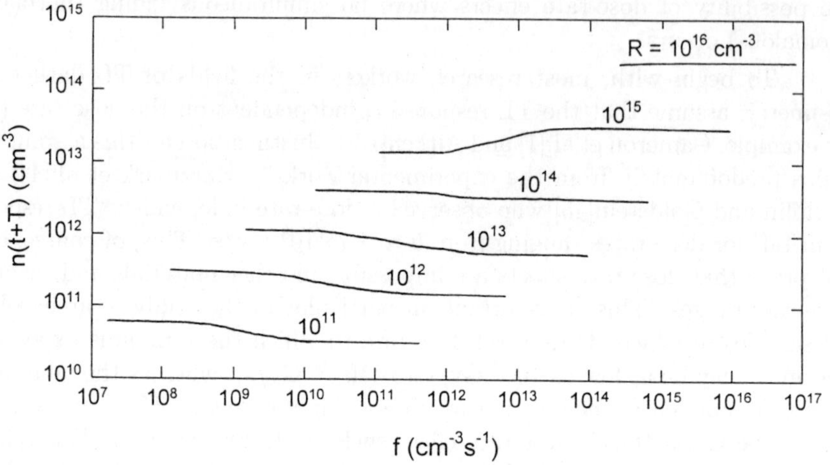

Figure 4.16 The dependence of the final filling of the trap n upon the intensity (dose-rate) f for different values of the total dose D.
(Redrawn from Chen et al.[53].)

The values of the parameters used in the figure are: $M = 3 \times 10^{20}$ m^{-3} where M is the total concentration of the hole center. $N = 10^{21}$ m^{-3}; $A = 10^{-23}$ m^3s^{-1}; $A_m = 10^{-19}$ m^3s^{-1} and $A_h = 10^{-21}$ m^3s^{-1}. The x-axis denotes, on a logarithmic scale the intensity (dose rate) f ($m^{-3}s^{-1}$) and the total dose $D = ft_D$ is shown next to the different curves (note that $f = X$ in our notation). In some cases, n is seen to decrease with increasing f, but as D is increased, this behavior changes to one of increasing n with increasing f. Here, in addition to the dose-rate dependencies shown, it is also seen that in certain dose-rate ranges, the results are dose-rate independent. Finally, in a later work, Chen et al.[54] showed that while considering models of either one trap and two centers or one center and two traps, different dose-rate dependencies may be predicted for two TL peaks in the same sample and for the same total dose.

4.7 Non-Isotropic Distribution of Traps and Centers

In previous sections we have discussed how, in principle, competition during excitation or during heating will result in the non-linear growth of TL as a function of absorbed dose, although, experimentally, it is generally observed that competition during heating dominates. The analyses of the previous sections have each assumed a spatially isotropic distribution of traps and recombination centers throughout the sample and have proceeded on the basis of kinetic analyses of the appropriate rate equations. With the imposition of a non-isotropic distribution of traps and recombination centers, this type of analysis is called into question. Although some authors (e.g. Sunta et al.[9],[55]-[57]) have tried, nevertheless, to modify the rate equations to account for this, others have attempted to move away completely from the rate equation picture and to describe the competition processes entirely from considerations of the spatial distribution of the traps and centers and the probability of "interaction" (i.e. charge exchange) between them. Mische and McKeever[8] examined this question for the particular case of TL from Mg-doped LiF. Here the evidence supporting the dominance of competition during heating is unequivocal. In particular, it is observed that experimental monitors of the trapped charge concentration - namely, optical absorption and/or electron spin resonance - show that this concentration is linear-sublinear with dose, whereas the TL signal resulting from this trapped charge concentration is linear-superlinear-sublinear (see, for example, the evidence described by Horowitz[58], Mische and McKeever[8], and McKeever[59]). Thus, the challenge with LiF:Mg is to explain how a linear growth of trapped charge concentrations can lead to a linear-superlinear TL growth.

The analysis of the TL process described by Mische and McKeever proceeds as follows: Consider a material containing only one species of TL-producing electron trap (concentration n_1), one species of hole trap (recombination center; concentration m), and one species of (competing) non-TL-producing trap (concentration n_2'). At this point we make no stipulation as to the origin of n_2' (i.e. we do not say whether these are trapped holes or empty electron traps). They are, however, considered to be thermally disconnected and, therefore, are thermally stable over the temperature range in which the TL-producing electrons are released. Consider further one electron trap, out of the n_1 trapped electrons per unit volume, at the center of a spherical volume of radius r. If we denote the thickness of a spherical shell of radius r to be dr, then the number of trapped holes within the volume of this shell is $4\pi r^2 m dr$. If the capture cross-section of a trapped hole is σ_m then the probability of an electron released from the trap at the center of the sphere being captured by a hole trapped in the shell volume is $(\sigma_m/4\pi r^2)\,4\pi r^2 m dr$. If the mean free path of an electron between trapping events is $\mu = 1/(\sigma_m m + \sigma_2 n_2')$ then the probability that an electron will be released from the center, travel a distance r and recombine with the hole in the shell volume is

$$dP(r) = \left(\frac{\sigma_m}{4\pi r^2}\right) 4\pi r^2 m dr \exp\{-r/\mu\} = \sigma_m m \exp\{-r/\mu\}\, dr. \tag{4.75}$$

In the above expression for the mean free path σ_2 is the capture cross-section of the competing center. Integrating over a spherical volume of radius R yields

$$P(R) = \mu\sigma_m m\,[1 - \exp\{-R/\mu\}]\,. \tag{4.76}$$

This is the probability that the electron will recombine, producing a TL photon, somewhere in the spherical volume of radius R. If R is of the dimensions of the crystal d, then $d \approx R \gg \mu$ and so $P_d = \mu\sigma_m m$, where $P_d = P(R)$ at $R = d$.

For the case of a small concentration of competitors it is trivial to see that $n_2' \ll m$ and, therefore, $P_d = 1$. Thus, the released electron must recombine somewhere in the crystal eventually because there are no alternatives. This will be true for each electron, and so for n_1 electrons we have $P_d^{tot} = n_1 P_d = n_1$. Assuming an intrinsic efficiency η the TL intensity will be given by

$$S = \eta n_1 \tag{4.77}$$

and the TL intensity can be seen to follow n_1. Thus, if n_1 is linear with dose, so too is S.

On the other hand, if there are many competitors then $n_2' \gg m$ and $\mu \approx 1/\sigma_2 n_2'$ and now $P_d = \sigma_m m/\sigma_2 n_2'$. Thus, for n_1 electrons we have $P_d^{tot} = n_1 P_d = n_1 \sigma_m m/\sigma_2 n_2'$, yielding a TL signal

$$S = \frac{\eta n_1 \sigma_m m}{\sigma_2 n_2'}. \tag{4.78}$$

We can gain further insight by stipulating that the competitors are trapped holes - i.e. non-radiative recombination centers. For this we may write $n_2' = m_2$ and we have the charge neutrality condition $n_1 = m + m_2$. Applying this to Eq. 4.78 gives

$$S = \eta \frac{\sigma_m}{\sigma_2} m \tag{4.79}$$

since we have already stipulated that $n_2' = m_2 \gg m$. In section 4.5 we have already examined this case and shown that as long as the competing center is far from saturation (i.e. $n_2' = m_2 \ll N_2' = M_2$) then m will grow linearly with dose, and thus S will also be linear. However, should the competing trap approach saturation ($n_2' = m_2 \approx N_2' = M_2$) then the growth of m with dose can become strongly superlinear and so too will the growth of S. Thus, in this way a linear-superlinear growth pattern may be obtained and we have succeeded in verifying the results of the previous section, but by an entirely different line of argument.

For the case of LiF:Mg, however, the above case is not an option since *all* the observed absorption bands and ESR signals exhibit linear growth with dose, and, furthermore, there is strong evidence to suggest that the competing centers are, in fact, empty electron traps. To accommodate this latter condition we write $n_2' = N_2 - n_2$, and the charge neutrality condition should be re-written $n_1 + n_2 = m$ where it should be recognized that $N_2 - n_2$ is the concentration of *available* competing electron traps. Thus, Eq. 4.78 remains unaltered, but is here written in a slightly different way as

$$S = \eta \frac{\sigma_m}{\sigma_2} \frac{n_1 m}{(N_2 - n_2)}. \tag{4.80}$$

Examination of this equation reveals it to be of the same form as Eq. 4.60, with $A_m = \sigma_m v$, $A_2 = \sigma_2 v$ (v is the electron thermal velocity) and $\eta = 1$. In Eq. 4.60 the concentrations are explicitly defined at the end of the irradiation cycle and the start of the heating cycle (n_{10}, n_{20} and m_0) which is implicit in Eq. 4.80. Thus, once again we have verified the analysis of the previous sections using an entirely different line of argument.

As pointed out in previous sections, however, Eq. 4.80 predicts a superlinear growth of TL, even assuming linear growth of n_1, n_2 and m, from even the lowest doses. To explain the initial linear growth of TL from LiF:Mg (Fig. 4.1), Mische and McKeever introduced the concept of spatial association between the electron traps and the recombination centers. For this, consider that the TL-producing electron trap is spatially associated with the recombination center, within a distance s (defined to be small enough so that no other defects of any type - trap or center - exist within the volume $(4/3)\,\pi s^3$). In reality there is likely to be a distribution of trap-center distances, such that a fraction K are $\leq s$ and $(1 - K)$ are $> s$. Implicit in what follows is the assumption that both families of electron trap contribute to the same TL peak - i.e. Mische and McKeever assumed that s is not small enough to significantly distort the potential around the electron trap.

Within the volume $(4/3)\pi s^3$ the probability that the released electron will recombine with its trapped-hole partner (nearest-neighbor recombination) is simply $\sigma_m/4\pi s^2$. The probability that the electron will avoid nearest-neighbor recombination, travel a distance R, and recombine elsewhere, is

$$P(R) = \left(1 - \frac{\sigma_m}{4\pi s^2}\right) \mu \sigma_m m \left[1 - \exp\left\{-\frac{R}{\mu}\right\}\right]. \tag{4.81}$$

Thus, the total probability of an electron recombining in a nearest-neighbor center, or recombining elsewhere within the crystal is

$$P_K^{tot} = n_1 K \left[\frac{\sigma_m}{4\pi s^2} + \left(1 - \frac{\sigma_m}{4\pi s^2}\right) \mu \sigma_m m\right] \tag{4.82}$$

and the total recombination probability, including all electron traps is

$$P^{tot} = n_1 K \left[\frac{\sigma_m}{4\pi s^2} + \left(1 - \frac{\sigma_m}{4\pi s^2}\right) \mu \sigma_m m\right] + (1 - K)\, n_1 \mu \sigma_m m. \tag{4.83}$$

The TL intensity is then

$$S = \eta n_1 \left[K \frac{\sigma_m}{4\pi s^2} + \left(1 - K \frac{\sigma_m}{4\pi s^2}\right) \mu \sigma_m m\right]. \tag{4.84}$$

For the case of few competitors ($m \gg N_2 - n_2$) Eq. 4.84 simplifies to $S = \eta n_1$ - i.e. the same as Eq. 4.77.

For the case of many competitors $\mu = 1/\sigma_2(N_2 - n_2)$ and Eq. 4.84 becomes

$$S = \eta n_1 \left[K\frac{\sigma_m}{4\pi s^2} + \left(1 - K\frac{\sigma_m}{4\pi s^2}\right) \frac{\sigma_m m}{\sigma_2(N_2 - n_2)} \right]. \tag{4.85}$$

This equation is seen to consist of the sum of a linear term ($\eta n_1 K\sigma_m/4\pi s^2$) and a nonlinear term $\eta n_1 \left(1 - K\sigma_m/4\pi s^2\right)\left(\sigma_m m/\sigma_2\left(N_2 - n_2\right)\right)$. As the dose increases, the nonlinear term will begin to dominate, dependent upon the value of K. An example data set calculated from Eq. 4.85 is shown in Fig. 4.17. To generate these data it was assumed that n_1 and m varied with dose like

$$n_1 = N_1\left[1 - \exp\{-\alpha D\}\right] \tag{4.86}$$

and

$$m = M\left[1 - \exp\{-\beta D\}\right] \tag{4.87}$$

in accordance with the experimental optical absorption and ESR results. Here N_1 and M are the total available concentrations of electron traps and hole traps and α and β are the dose constants. $N_2 - n_2$ is determined from the charge neutrality condition $n_1 + n_2 = m$. The parameters used in the calculation are $s = 10^{-8}$ m, $\alpha = 0.5 \times 10^{-4}$ dose units^{-1}, $\beta = 3 \times 10^{-5}$ dose units^{-1}, $K\sigma_m/4\pi s^2 = 0.8 \times 10^{-3}$, $\sigma_m/\sigma_2 = 1$, $M = 10^{23}$ m^{-3}, $N_1 = 2 \times 10^{22}$ m^{-3} and $N_2 = 3 \times 10^{24}$ m^{-3}. It is clear from Fig. 4.17 that this analysis can produce the linear-superlinear-sublinear behavior characteristic of LiF:Mg.

Also shown in Fig. 4.17 is the dose response that would result from this model if there were no competitors ($N_2 - n_2 \ll m$). Comparison of the two curves leads to the realization that superlinear behavior is a result of an under-response at low doses, rather than an over-response at high doses (as the phenomenon is sometimes described).

Sunta and colleagues[9],[55]-[57] use an almost identical model to describe non-linear behavior, but attempted to use a kinetic analysis similar to that of Kristianpoller et al.[37]. The important modification used by Sunta and colleagues, however, is to introduce two sub-species of TL-producing trap, depending upon whether the trap is spatially associated with a luminescence center, or is separated from it. Thus, they describe *non-interacting* and *interacting* traps, respectively. Electrons freed from the "interacting" traps (concentration N_{in}) can either be localized in competing traps, or captured by the luminescence centers. Electrons freed from the "non-interacting" traps (concentration N_n) can be captured only by the spatially associated luminescence centers. This situation is slightly different from that described by Mische and McKeever who allowed electrons to "escape" from the spatially associated pair

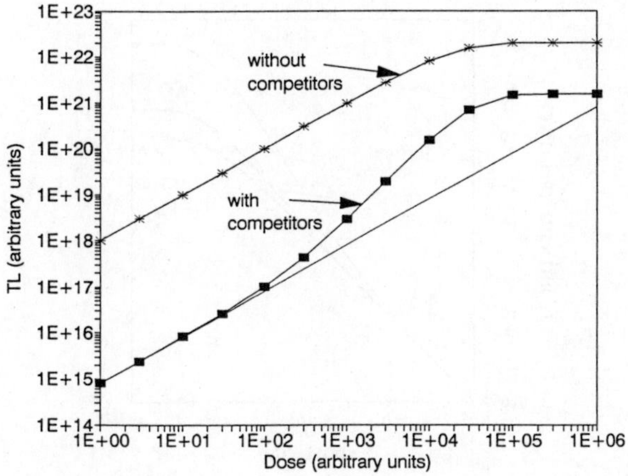

Figure 4.17 Simulated non-linear growth of TL versus dose, as predicted from the Mische and McKeever model, assuming spatial association between traps and centers. The linear-sublinear growth predicted when competitors are absent is also shown.

and to be captured at any other site; the analysis of Sunta et al. does not allow this. In accordance with Mische and McKeever, however, Sunta et al. also presume that the "non-interactive" and the "interactive" traps both give rise to the same TL signal.

The analysis of Sunta et al. shows that the TL intensity, as a function of absorbed dose, for the interacting traps follows the non-linear relationship

$$S_{in} = \frac{n_{in} m \sigma_m}{(N_2 - n_2)\,\sigma_2 + m\sigma_m} \tag{4.88}$$

where the other terms have the same meaning as before and the subscript "in" denotes "interactive". This equation is identical to Eq. 4.80 with the stipulation that $(N_2 - n_2)\,\sigma_2 \gg m\sigma_m$, and $\eta = 1$. Writing $N_2 = aN_{in}$ along with $\sigma_2 = \sigma_m$, and assuming both n_{in} and n_2 grow with dose in the manner of Eqs. 4.86 and 4.87, but with the same dose constant α', Eq. 4.88 becomes:

$$S_{in} = \frac{N_{in}\,(a+1)\,(1 - \exp\{-\alpha' D\})^2}{(a + 1 - \exp\{-\alpha' D\})}. \tag{4.89}$$

For the "non-interacting" traps we have

$$S_n = n_n = N_n\,[1 - \exp\{-\alpha' D\}] \tag{4.90}$$

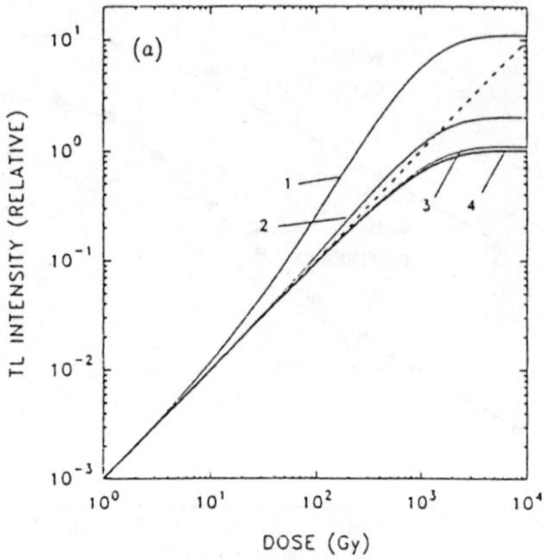

Figure 4.18 TL intensity as a function of dose for the case $a = 1$. Curves 1-4 refer to $K = 10, 1, 0.1$, and 0.01, respectively.
(After Sunta et al. [55].)

where the same dose constant α' is assumed for n_{in} as for n_n, and the subscript n refers to non-interacting. The total TL is then $S_{tot} = S_{in} + S_n$, thus

$$S_{tot} = N_n \left[1 - \exp\{-\alpha' D\}\right] \left[1 + K \frac{(a+1)(1 - \exp\{-\alpha' D\})}{(a + 1 - \exp\{-\alpha' D\})}\right]. \tag{4.91}$$

In this expression K is the ratio of "interacting" to "non-interacting" traps (i.e. $K = N_{in}/N_n$).

Some solutions of Eq. 4.91 are shown in Fig. 4.18, for various values of K. It is clear that for large values of K a linear-superlinear-sublinear behavior results.

4.8 Track Interaction Theory

Both of the above sets of analyses (i.e. that of Mische and McKeever[8] and of Sunta et al.[9],[55]-[57]) seek to explain the linear-superlinear-sublinear behavior of TL versus dose in essentially the same way, namely by invoking (i) competition during heating, and (ii) spatial association between traps and centers. Implicit in the analyses is the assumption of an isotropic production of free charge carriers (electrons and holes) during the irradiation phase. That is, the above analyses are strictly applicable only to low linear energy transfer (LET) irradiation - e.g. gamma irradiation. Irradiation with heavy charged particles (HCP) of high LET produces a non-isotropic distribution of densely ionized

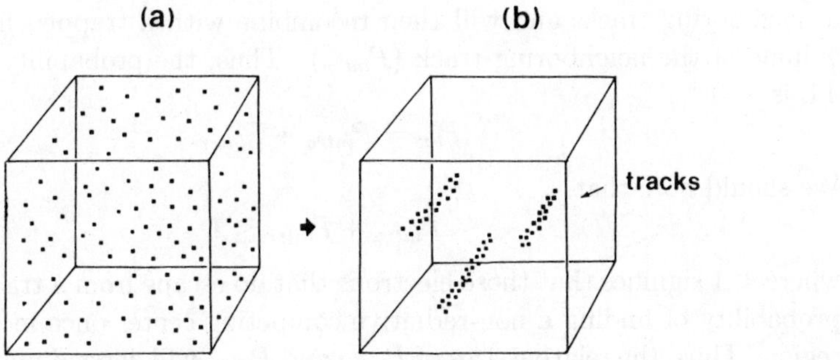

Figure 4.19 (a) The isotropic defect distribution assumed for low LET (e.g. gamma) irradiation. (b) The non-isotropic defect distribution assumed for high LET (e.g. HCP) irradiation.
(After Mische and McKeever[8].)

regions along linear tracks. Calculations ([60],[61]) indicate that the dose is deposited within a few hundred angstroms of the track center and that the tracks have approximately cylindrical geometry. For example, a 4 MeV alpha particle deposits 90% of its dose within 100 μm of the track center, and 98% within 200 μm of the track center (Horowitz and Rosenkrantz[62]).

The non-isotropic distribution of trapped charge in an HCP-irradiated specimen is represented in Fig. 4.19 in which we compare schematically the isotropic distribution assumed for gamma-irradiated samples, with the trapped charge distribution assumed for HCP irradiation. Clearly, in such a defect configuration as that shown in Fig. 4.19(b), we must consider two entirely different environments - namely, those regions *within* each track where the trapped charge populations are high and the competing trap population is low (since we consider the competitors to be empty electron traps), and those regions *between* traps, where the trapped charge populations are low and the competing trap population is high. Indeed, the latter regions can be considered as zero dose regions in which the filled trap concentrations are approximately always zero and, therefore, independent of dose.

We can make some general statements by way of introduction to our consideration of the dose response function under high LET irradiation. To evaluate the probability that a released electron will find a luminescence center and produce a TL photon, one now has to consider the probability that this will occur *within* a track (P_{intra}) and the probability that an electron will escape from the track, will successfully migrate through the inter-track region to reach a neighboring track, and will then recombine with a trapped hole within the volume of the neighboring track (P_{inter}). Thus, the probability for producing TL is

$$P_{tot} = P_{intra} + P_{inter}. \tag{4.92}$$

We should note that

$$P_{intra} + P_{inter} \leq 1 \tag{4.93}$$

where < 1 signifies that those electrons that do escape from a track have a finite probability of finding a non-radiative competing center once in the inter-track region. Thus, the relative sizes of P_{intra} and P_{inter} will depend upon such parameters as the mean distance between tracks (i.e. the track distribution function), the radius of the track (i.e. the ionization density within each track), the concentration of competing traps, and the temperature (since the charge carrier mobility will be temperature dependent). In other words, the TL sensitivity will be dose dependent, LET dependent, sample dependent, and temperature dependent - all of which are observed experimentally.

Examining first the intra-track recombination term, we consider a cylindrical track of length $L >>$ radius R. Mische and McKeever[8] consider an electron starting from the center of the track and moving in a direction θ with respect to the track axis. These authors show that the net probability of the electron recombining within the track is given by:

$$P_{intra}^{tot} = (2/\pi) \int_0^{\pi/2} P(\theta) d\theta \tag{4.94}$$

where $P(\theta)$ is the probability that an electron starting from the center of the track and moving away at an angle θ will either recombine with a luminescence center, or be trapped at a competitor, before reaching the edge of the track, and is described by

$$P(\theta) = \mu_i m_i \sigma_m \left[1 - \exp\{-R/\mu_i\} \sin\theta\right]. \tag{4.95}$$

Here we define the mean free path and recombination center concentrations, μ_i and m_i, respectively, *inside* the track. As before, the definition of μ_i is

$$\mu_i = 1/\left(\sigma_m m_i + \sigma_2 n'_{2i}\right) \tag{4.96}$$

where n'_{2i} is the empty competitor concentration inside the track. Note that with this definition of μ_i, P_{intra}^{tot} is interpreted as the probability that the released electron will *either* recombine *or* be retrapped in a competitor before leaving the track. Thus, this is to be distinguished from P_{intra} which is the probability that only recombination will occur (and a TL photon will be released). Clearly, for a high enough ionization density inside the track n'_{2i} may approach zero, or $n'_{2i} << m_i$, in which case $P^{tot}_{intra} = P_{intra}$.

As an example, for a 4 MeV alpha particle we have $L \approx 10^5$ Å and $R \approx 200$ Å (thus, $L >> R$). For a typical defect concentration of 100 ppm and $\sigma_m \approx 10^{-16}$ m^2, we have $\mu_i \approx 2 \times 10^{-9}$ m (assuming $n'_{2i} \approx 0$), and therefore $R \approx 10\mu_i$. With these values, Eq. 4.94 yields $P_{intra} = 1$. In fact, for any value of $R > 8\mu_i$, it is certain that the electron will recombine or retrap within the track, independent of angle θ. Since in such circumstances $P_{inter} = 0$, then the TL intensity becomes

$$S = \eta n_{1i} P_{tot} = \eta n_{1i} \tag{4.97}$$

and again the TL dose response curve is linear, i.e. $S(D)$ follows $n_{1i}(D)$. The point here is that HCP irradiation tends to produce a linear dose response curve. The key feature, however, is the ionization density - i.e. the LET of the charged particle. For certain HCPs, one can imagine that a finite fraction of the freed electrons will in fact escape from the track and enter the inter track region. The key to producing a TL photon after the escape thus lies with the possibility that the electron will survive the inter track region and reach an adjacent track where radiative recombination again becomes probable. For this we need to consider the distribution of tracks within the specimen and the probability that the free electron will reach one of these tracks. This was examined by Horowitz and colleagues[2],[61]-[64] who considered the probability of a released electron migrating from one track to another to be given by the product of a term $\exp\{-r/\mu_o\}$ and a solid angle, geometric configuration factor $g(r,R)$ between two neighboring tracks, where r is the distance between tracks. For a first approximation, $g(r,R)$ was taken to be $g(r,R) \approx 2R/r$[62]. Note that the mean free path used here is that *outside* the tracks, given by $\mu_o = 1/\sigma_2 n'_{20}$, where the subscript o signifies the inter track region, for which we have assumed $n_{10} = m_0 = 0$.

The probability that a given track has its j-th nearest neighbor at a distance r is denoted by $P_j(\phi, r)$, for a given HCP fluence ϕ. The probability of a freed electron finding a neighboring track is found by integrating from the radius of the original track to infinity (i.e. to the crystal dimension) and is given by $\int_R^\infty g(r,R)\exp\{-r/\mu_o\}\left[\sum_j P_j(\phi,r)\right] dr$. Hence,

$$P_{inter} = \left(1 - P^{tot}_{intra}\right)\left[\int_R^\infty g(r,R)\exp\{-r/\mu_o\}\sum_j P_j(\phi,r)\,dr\right] P_{intra}. \tag{4.98}$$

The first $(1 - P^{tot}_{intra})$ term represents the probability that the electron escapes from the first track, and the last P_{intra} term represents the probability that having reached the second track, a TL photon will result. Horowitz and Rosenkrantz [62] evaluated $P_j(\phi, r)$ for $j = 1$, 2, and 3 (first-, second-, and third-nearest neighbors).

Thus, at low HCP doses when the distance between adjacent tracks is large, $P_{inter} \ll 1$ and (from Eq. 4.92) $P_{tot} \approx P_{intra}$, where $P_{intra} \leq 1$. Since P_{intra} depends only upon LET, and not upon dose (i.e. fluence) we find that the TL response is linear over this region. At higher fluences, when the distance between tracks decreases, P_{inter} increases and thus P_{tot} increases - i.e. superlinearity occurs. The extent of the superlinearity for a given dose (fluence) depends upon the LET of the particle. If the LET is large enough, P_{intra} is always $\gg P_{inter}$ and the dose response is always linear. As the LET decreases, so P_{inter} increases and superlinearity becomes more probable.

The supralinearity index for HCP irradiation can now be calculated, using Eq. 4.9 , thus:

$$f(D) = 1 + \frac{n_e}{n_{1i}} \sum_j \int_R^{\infty} g(R, r) \exp\{-r/\mu_o\} P_j(\phi, R)\, dR \tag{4.99}$$

where the 1 indicates linear growth and the second term represents the enhanced TL efficiency due to track interaction. Here $n_e = n_{1i}(1 - P_{intra}^{tot})$ is the number of electrons which escape from the track. Obviously, if $P_{inter} \ll 1$, then $f(D) \approx 1$. As P_{inter} increases, so $f(D)$ increases. Eq. 4.99 differs slightly from that derived by Horowitz and colleagues[2],[61]-[64] who replace the term n_e/n_{1i} with n_e/n_w, where $n_w = n_{1i}P_{intra}$ is the number of electrons which recombine inside the track. However, in doing this Horowitz and colleagues implicitly assume that once an electron reaches a new track a TL photon is always produced - i.e. they set the last P_{intra} term in Eq. 4.98 equal to 1. Also implicit in the derivation of Eq. 4.99 is the assumption that $n_{1i} = \alpha\phi$ over all ranges of ϕ considered (where α is the dose (fluence) constant).

Fig. 4.20 illustrates solutions to Eq. 4.99 for 1st-, 2nd-, and 3rd-nearest neighbors, and Fig. 4.21 shows a comparison of a fit of Eq. 4.99 to data from LiF:Mg,Ti following 4 MeV alpha particle irradiation. The best fit yields a track radius of 105 Å and a mean free path (μ_o) of 3.7 nm.

One should note finally that the treatment described by Mische and McKeever[8] for gamma irradiated specimens and the Track Interaction Model of Horowitz and colleagues[2],[61]-[64] for HCP irradiated specimens are in fact very similar, with the local trap-center complex of the Mische-McKeever analysis being replaced by the charged particle track of the Horowitz et al. analysis. This small difference leads to some corresponding differences in definitions, but the basic treatment is the same in both instances. This was noted initially by

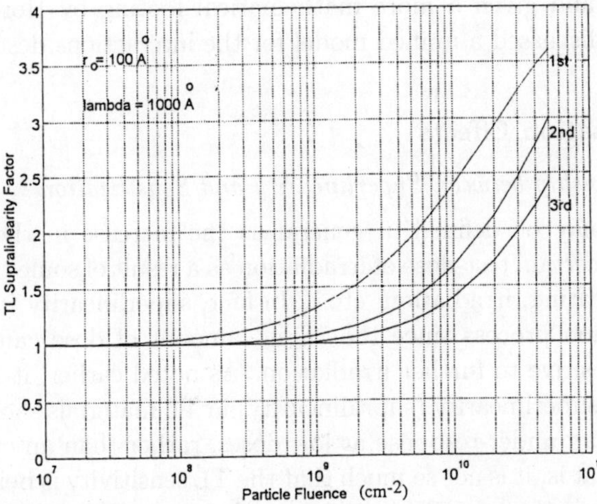

Figure 4.20 Supralinearity index $f(D)$ versus particle fluence for 1st-, 2nd- and 3rd-nearest neighbors.

(After Horowitz et al.[64].)

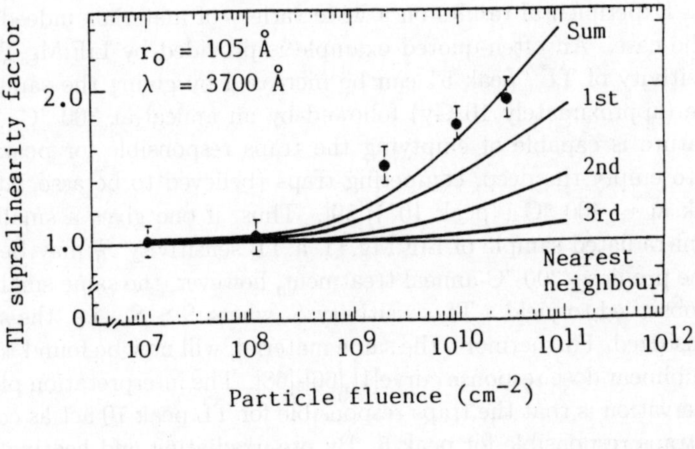

Figure 4.21 The supralineraity factor as a function of 4 MeV alpha particle fluence for peak 8 in LiF:Mg,Ti. The dots are the data (with error bars). The solid lines represent calculations for 1st-, 2nd-, and 3rd-order nearest neighbors, and the sum of all three, as obtained by fitting the expression for $f(D)$ to the data.

(After Horowitz and Rosenkrantz[62].)

McKeever[65] and given a more mathematical footing by Horowitz et al.[64] who formally discussed a unified model for the interactions described above.

4.9 Sensitization Effects

4.9.1 Relationship between Superlinearity and Sensitization

Sensitization may be defined in general as the increase in the response (TL intensity) to a certain test-dose of irradiation as a result of some treatment of the sample (i.e. heating, irradiation, etc.). In fact, superlinearity itself is evidence of a sensitization process since the "first" amount of dose causes the sample to be more sensitive to further irradiation. As noted earlier, it is important to recognize that superlinearity (supralinearity) in TL materials should properly be interpreted as an under-response at low doses, rather than an over-response at high doses. That is, it is not so much that the TL sensitivity is being enhanced at high doses as it is that the TL sensitivity is being depressed at low doses. That this is so is a result of the competition effects noted in earlier sections whereby the competitors become less and less effective as the dose increases. If all the competitors could be removed - by, say, thermal annealing, or by administering a large pre-dose to the sample in order to pre-fill the competitors - then one might expect that the super(supra)linearity would be removed, and, furthermore, that the sensitivity would be enhanced across the entire dose range (until sublinearity sets in). Experimental results on a wide variety of materials indeed show this to be the case. An often-quoted example is provided by LiF:Mg,Ti in which the sensitivity of TL "peak 5" can be increased by giving the sample a large pre-dose (approximately 10 Gy) followed by an anneal at 300 °C. The latter temperature is capable of emptying the traps responsible for peak 5, but is unable to empty the deep, competing traps (believed to be associated with a TL peak at $\sim$ 400 °C ("peak 10"))[59]. Thus, if one gives a small test-dose to an unirradiated sample of LiF:Mg,Ti, a TL sensitivity S_i may be obtained. After the pre-dose/300 °C-anneal treatment, however, the same small test-dose will be observed to yield a TL sensitivity S, where $S > S_i$ - i.e. the sample has been sensitized. Furthermore, the same material, will now be found to display a linear-sublinear dose response curve[1],[66]-[68]. The interpretation placed upon this observation is that the traps responsible for TL peak 10 act as competitors for the traps responsible for peak 5. By pre-irradiating and heating to 300 °C the peak 5 traps are left empty, but the peak 10 traps are left full. In this way the competitors are removed and a subsequent irradiation of the sample leads to an enhanced TL signal and a linear-sublinear dose response[59].

In a similar manner, the sensitivity of the 110 °C TL peak in quartz is found to be extremely superlinear in as-received samples (see curve A in Fig. 4.7). However, in quartz samples which have been annealed at high temperatures (e.g. 900 °C) the same TL peak is found to be significantly sensitized, and the dose response curve is linear (curve B in Fig. 4.7)[19]. The interpretation placed upon this observation by Chen et al.[19] was that the high temperature annealing destroyed the centers which were acting as competitors to the traps responsible for the 110 °C TL peak, although no independent evidence exists to support this assertion.

In each of the above cases, the competing traps are effectively removed from the TL process. In the LiF case the competitors are pre-filled, but not destroyed. In the quartz case, the competitors are thought to be destroyed. In either case no competitors are available for competition effects - during either the subsequent excitation or the heating phase.

Sensitization can be modeled using the same kinetic models as presented earlier to describe superlinearity. To demonstrate this we use the same energy band model as shown in Fig. 4.8 consisting of a main TL trap, of concentration N_1, a competing trap of concentration N_2, and a luminescent recombination center, of concentration M. Following the earlier analysis, the TL from such a system can be represented by Eq. 4.88, with $n_{in} = n_1$. Re-writing this equation, with the added simplifying criterion that $\sigma = \sigma_m$ and using $m = n_1 + n_2$, gives

$$S = \frac{n_1 (n_1 + n_2)}{N_2 + n_1}. \tag{4.100}$$

In section 4.3 we demonstrated how the model of Fig. 4.8 can lead to a linear-superlinear-sublinear growth of the concentration n_1, typified in Fig. 4.9, depending upon the relative values of the various system parameters. In particular, an important aspect is the relative size of N_2 compared with N_1 - i.e. the dose at which the competing traps saturate is an important parameter dictating the extent of the linear region observed in Fig. 4.9. Accordingly, we define here a fraction $a = N_2/N_1$. Although the shapes of the $n_1(D)$ and $n_2(D)$ curves critically depend upon these factors, we shall make a first approach to the problem, for the purposes of simplification, by assuming that n_1 and n_2 both grow with dose according to

$$n_1 = N_1 (1 - \exp\{-\alpha D\}) \tag{4.101}$$

and

$$n_2 = N_2 (1 - \exp\{-\alpha D\}) = a N_1 (1 - \exp\{-\alpha D\}). \tag{4.102}$$

Thus, for a small test-dose D_t, we may define a fraction ζ_t of each type of trap to be filled according to

$$\zeta_t = \frac{n_1}{N_1} = \frac{n_2}{N_2} = (1 - \exp\{-\alpha D_t\}). \tag{4.103}$$

In this way Eq. 4.100 may be written

$$S_i = \frac{\zeta_t N_1 (\zeta_t + \zeta_t a)}{a + \zeta_t} \tag{4.104}$$

where S_i signifies the initial TL sensitivity for a test-dose D_t.

For a large pre-dose D ($\geq D_t$) the traps will fill according to Eqs. 4.101 and 4.102. If a pre-dosed sample is then heated so as to completely empty the level 1 traps, but to leave the competing traps untouched, then a subsequent test-dose D_t will yield a TL signal

$$S' = \frac{\zeta_t N_1 \left[\zeta_t + (\zeta_t + \zeta)a\right]}{a + \zeta_t} \tag{4.105}$$

where ζ is the fraction of filled competing traps for pre-dose D, and $\zeta \geq \zeta_t$. In writing Eq. 4.105 we have implicitly assumed that the traps fill in the same way after the pre-dose as before. This, along with Eq. 4.101, is tantamount to assuming that $a > 1$ since if the competing traps do not approach saturation during filling, then the filling of level 1 is unaffected by the actual extent to which level 2 is populated. Furthermore, we should also remark that during the process of heating to empty level 1 some electrons will be transferred from level 1 to level 2. Thus, the fraction of level 2 actually occupied at the time the second test-dose is administered will be greater than that immediately after irradiation, i.e. $\eta > 1 - \exp\{-\alpha D\}$. Nevertheless, in the calculation below we assume that $\eta = 1 - \exp\{-\alpha D\}$.

With Eqs. 4.104 and 4.105 we may define a *sensitization factor* $F(D)$, thus

$$F = \frac{S'}{S_i} = \frac{\zeta_t + (\zeta_t + \zeta)\, a}{\zeta_t + \zeta_t a} = \frac{1 + a\left[(\zeta_t + \zeta)/\zeta_t\right]}{1 + a} > 1. \tag{4.106}$$

In Fig. 4.22(a) is plotted the sensitization factor as a function of dose (using Eq. 4.106) for the dose response curve shown in Fig. 4.22(b) (using Eqs. 4.100 and 4.101). The values used for this calculation were: $N_1 = 2 \times 10^{22}$ m^{-3}, $a = 1.5$, $\alpha = 0.5 \times 10^{-4}$ (dose units)$^{-1}$, and $D_t = 10$ dose units. Since we do not consider spatial association between centers, and since the competitors are all considered to be electron traps (i.e. level 2 in Fig. 4.8) the dose response is observed to be superlinear from the lowest dose levels used, until saturation. For reference, the line of linearity is also shown.

In this analysis we have assumed that Eq. 4.101 is an accurate description of the growth of the trapped charge populations. As pointed out in earlier sections, the only way to confirm this for any given set of parameters is to numerically solve the appropriate rate equations during the irradiation phase, as well as the heating phase, and to allow the rate equations to dictate the filling functions without any assumptions as to their shape. This is done for a two-trap/one-center system in which parameter $a = 1.5$, as above. (The value of N_1 was assumed to be 2×10^{17} m^{-3}, rather than 2×10^{22} m^{-3} for the previous example, for speed of computation). The results are displayed in Fig. 4.23 in

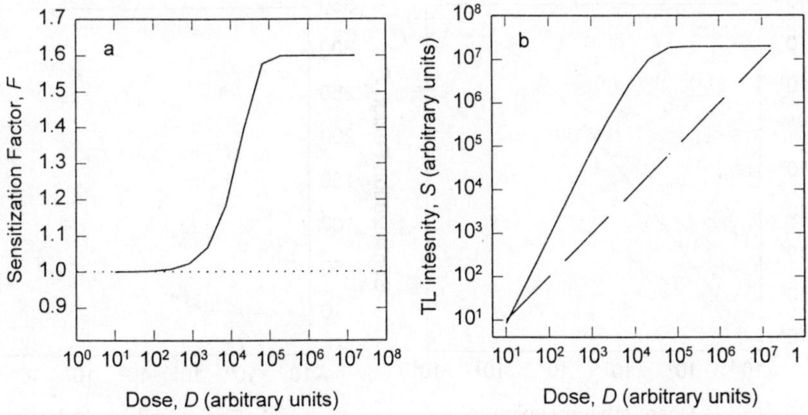

Figure 4.22 (a) The sensitization factor F versus dose, for the dose response function S shown in (b), for the case of a competing trapping center and the linear-sublinear growth of the trapped electron concentrations. The dashed line in (b) indicates the line of linearity.

which we show the TL growth curve as a function of dose (a), and the sensitization factor $F(D)$ as a function of pre-dose (b). The curve shapes are the same as those shown in Fig. 4.22, thus illustrating the validity of the approach. The value of F is much greater in this example, due primarily to the actual value for α being different in the numerical calculations compared with that assumed in the previous analysis.

As noted earlier, Sunta et al.[57] include the concept of spatial association between the traps and centers in order to explain the observed linear-superlinear behavior of LiF:Mg,Ti. For this, these authors included the non-interacting traps, as well as the interacting traps. Following Eq. 4.91 the TL resulting from a test-dose D_t, administered *before* the pre-dose D, is thus:

$$S_i = \zeta_t N_n \left[1 + K\frac{(\zeta_t + \zeta_t a)}{(a + \zeta_t)}\right] \tag{4.107}$$

and that obtained *after* the pre-dose is

$$S' = \zeta_t N_n \left[1 + K\frac{[\zeta_t + (\zeta_t + \zeta)\, a]}{(a + \zeta_t)}\right]. \tag{4.108}$$

The sensitization factor now becomes

$$F = \frac{S'}{S_i} = 1 + K\zeta(a/b) \tag{4.109}$$

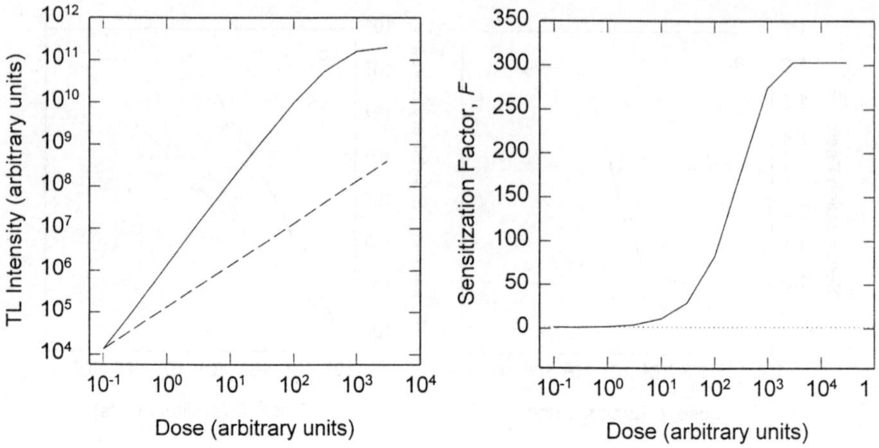

Figure 4.23 Numerical solutions to the rate equations describing the two-trap/two-center system, showing (a) the growth of TL with dose, and (b) the sensitization factor $F(D)$ as a function of the pre-dose. The dashed line in (a) shows the line of linearity.

where $b = a + \zeta_t[1 + K(1 + a)]$. This is more general than the result of Sunta et al. who assumed $a >> \zeta_t$ and $\zeta >> \zeta_t$. Clearly the latter assumption is questionable at low pre-doses.

Fig. 4.24(a) shows an example of the TL which results from such a model with $K = 10$, and the values of the other parameters as described for Fig. 4.22. Also shown in the figure (b) are the supralinearity factor $f(D)$, using the definition of Eq. 4.10 with $S_o = 0$, and the sensitization factor $F(D)$, as expressed by Eq. 4.109. An important point to note is that $F(D)$ and $f(D)$ do not reach their maximum values at the same dose, despite being caused by the same physical mechanism (i.e. filling of the deep, competing centers). On several occasions arguments have been forwarded in the literature to the effect that since the sensitization and supralinearity factors generally do not have a maximum at the same dose they cannot be caused by the filling of the same competing traps ([69]). The present analysis indicates that this line of argument is false reasoning[55],[57].

4.9.2 Sensitivity Changes due to Annealing

A consequence of the effects mentioned is that repeated reuse of a TL sample (i.e. cycles of irradiation and readout) mean that the sensitivity will evolve and be non-constant over the usage period. This is because reuse repeatedly places charge into the deep, competing traps, and this charge accumulates with each

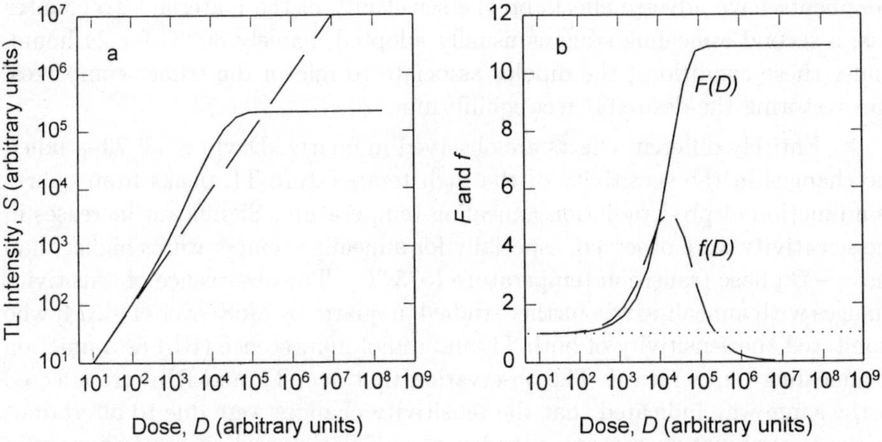

Figure 4.24 (a) TL growth curve for the case of a degree of spatial association between traps and centers. (b) Resulting supralinearity $f(D)$ and sensitization $F(D)$ factors.

reuse. This behavior is observed for most TL materials, *unless* the doses used are such that the level of trap filling is very much less than the available trap concentrations. Under these circumstances the competitors fill negligibly and the effect of the competitors remains unaltered. Thus, the repeated reuse of LiF:Mg,Ti in personal dosimetry, where the dose levels used are very small, has been shown to be possible without the need for a high temperature annealing of the material (to empty the deep traps) between uses. In contrast, for use in clinical dosimetry, a 400 °C anneal for one hour is required to remove the residual charge in the competitors and thereby avoid sensitization[70]. Similarly, a high temperature anneal of α-Al_2O_3:C is unnecessary if the doses are kept at low personal dosimetry levels. Higher doses, however, require a high temperature anneal at $T > 800$ °C to ensure that the deep traps in this material are emptied[71].

High temperature annealing can have disadvantageous side effects. In LiF:Mg,Ti the defects responsible for the main TL signal in this material ("peak 5") are believed to be trimer clusters of Li_{vac}-Mg_{Li} dipoles[72]. (These are Mg impurities which have substituted for Li ions in the LiF lattice (Mg_{Li}); charge compensation is provided by Li vacancies (Li_{vac}) and coulombic attraction between the two components results in the formation of Li_{vac}-Mg_{Li} dipolar defects. Clusters of three such defects (trimers) are considered to be the source of the traps responsible for the peak 5 TL signal.) High temperature annealing at 400 °C causes the dissociation of the trimers into dipoles and in this way the thermal treatments have adverse effects on the sensitivity of the material. To counter this, a second annealing stage is usually adopted, namely 80 °C for 24 hours. Under these conditions, the dipoles associate to reform the trimer complexes, thus restoring the desired defect equilibrium.

Entirely different effects are observed in quartz. David et al.[73] studied the changes in the sensitivity of the high temperature TL peaks from quartz as a function of pre-irradiation annealing temperature. Significant increases in the sensitivity were observed, especially for annealing temperatures higher than the $\alpha - \beta$ phase transition temperature (573 °C). The observance of sensitivity changes with annealing was further studied in quartz by McKeever et al.[74] who monitored the sensitivity of both TL and radioluminescence (RL) as a function of annealing temperature. The observation that *both* TL and RL were affected in the same way indicated that the sensitivity changes were due to alterations to the recombination centers. Rendell et al.[75] observed spectral changes in the TL of quartz as a function of cooling rate after annealing at 1000 °C for 15 hours. Changes to the emission spectra were also observed by Hashimoto et al.[76] in a study of TL color images from natural quartz after annealing at 1000 °C for 100 hours. In particular, the latter authors observed a large increase in red emission from the samples. Similar observations have been reported for the Optically Stimulated Luminescence (OSL) and phototransferred TL (PTTL) signals from these materials[77]-[79]. In the latter work, not only was a large enhancement in sensitivity observed after annealing, but the OSL versus dose curve became superlinear, having been linear before annealing. These effects are shown in Fig. 4.25(a) and (b). This is a particularly challenging observation to explain since, as we have discussed at length in the foregoing sections, one usually expects that an increase in sensitivity is accompanied by the removal of superlinearity. At least, that is what we observe in sensitization caused by pre-irradiation annealing. Here the sensitization is the result of annealing and the opposite effect is observed.

The emission spectral changes support the conclusions of McKeever et al.[74] that the sensitivity increases in the high temperature TL of quartz after high temperature annealing are related to changes in the recombination center concentrations. From these observations alone, however, it is not possible to state whether these changes are the result of the creation of additional radiative recombination sites, or the destruction of non-radiative recombination sites. In an attempt to describe the sensitivity changes for OSL in quartz, Bøtter-Jensen et al.[79] examined the model of Fig. 4.26 in which there is a shallow trap (level 1) which represents all those shallow traps that yield TL peaks at temperatures less than the TL peak of interest. Level 2 represents that trap under study during TL (or OSL) measurement. Level 3 represents all those deep traps which are thermally stable over the temperature range in which level 2 thermally

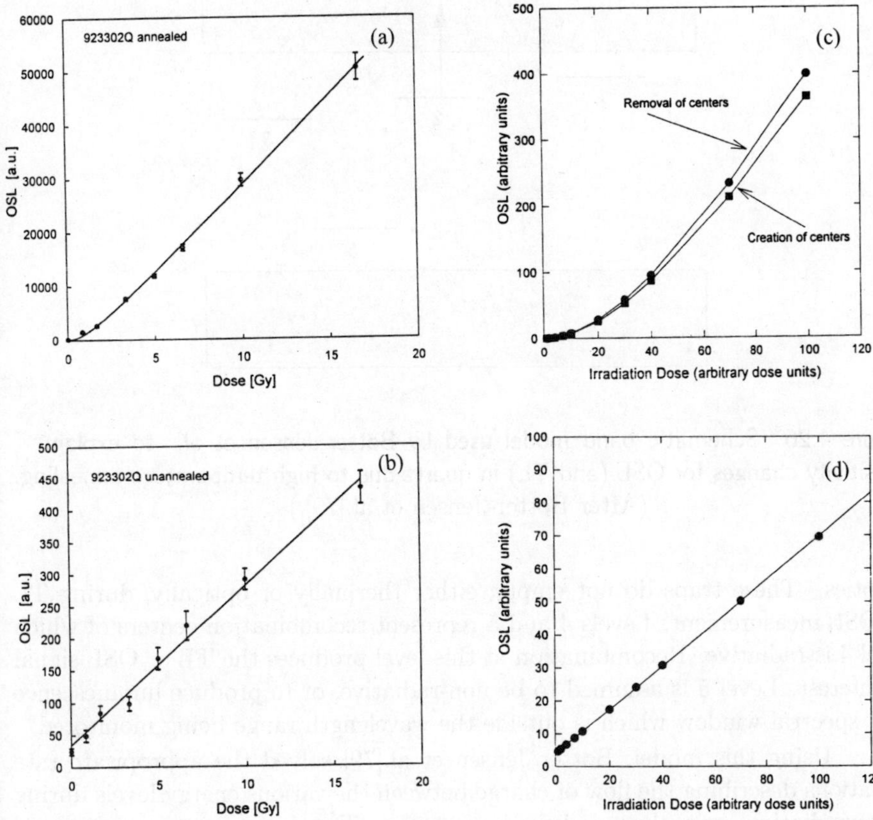

Figure 4.25 Experimental OSL versus dose curves for two sets of natural sedimentary quartz: (a) after annealing at 800 ^oC for 1 hour prior to irradiation; (b) without annealing. The two curves are plotted using the same arbitrary OSL units and indicate an approximately two orders of magnitude increase in sensitivity along with the inducement of a non-linear dose response after annealing. In curves (c) and (d) are shown, respectively, the computer simulated OSL dose response, using the model described in the text, for "annealed" and "unannealed" samples. During the annealing it was assumed that either radiative centers were created, or non-radiative centers were removed; qualitatively the same result is obtained in either case.

(After Bøtter-Jensen et al.[79].)

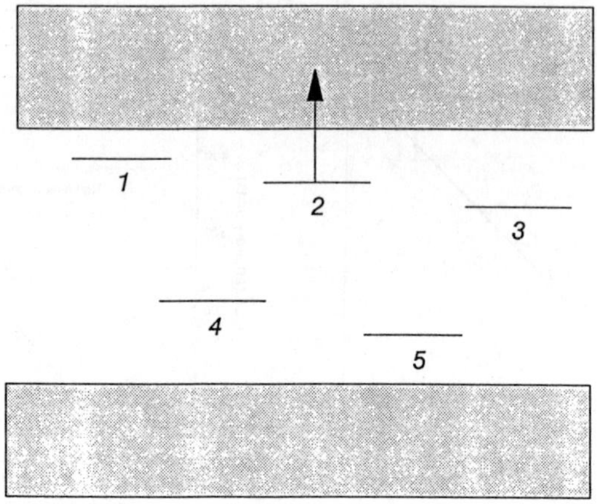

Figure 4.26 Schematic band model used by Bøtter-Jensen et al. to explain sensitivity changes for OSL (and TL) in quartz due to high temperature annealing. (After Bøtter-Jensen et al.[79].)

empties. These traps do not empty, either thermally or optically, during TL or OSL measurement. Levels 4 and 5 represent recombination centers of which level 4 is radiative. Recombination at this level produces the TL or OSL signal of interest. Level 5 is assumed to be non-radiative, or to produce luminescence in a spectral window which is outside the wavelength range being monitored.

Using this model, Bøtter-Jensen et al.[79] solved the appropriate rate equations describing the flow of charge between the various energy levels during the irradiation, annealing and readout stages. The dose response results are shown in Fig. 4.25(c) and (d). For these calculations it was assumed that either the high temperature annealing creates additional radiative recombination sites, or removes non-radiative recombination centers. The two assumptions produce similar conclusions since in either case the ratio of the radiative to non-radiative centers is increased and an enhanced sensitivity results. The inducement of superlinearity is explained in the following way. The samples under study are natural specimens for which there has already been a long radiation history prior to excavation. As a result, the deep centers are full, or almost full. (It was found by Bøtter-Jensen and colleagues that the simulation of this pre-excavation history was essential to obtain the desired effects.) Thus, when the specimen is irradiated in the laboratory, the deep centers are not effective as competing centers and a linear dose response is obtained. During annealing, however, the deep centers are emptied making them effective competing centers. Normally, as discussed in the preceding sections, this would also result in a *reduction* in luminescence sensitivity. However, in this case the expected reduction is countered by a sensitivity increase due the noted changes to the recombination center concentrations. McKeever et al.[80] used the same model to simulate the OSL sensitivity changes expected to occur during cycles of irradiation and luminescence readout when reusing the same aliquot, while Chen et al.[93] used a similar model to explain TL sensitivity changes expected with similar cycles of irradiation and readout.

4.9.3 The Pre-Dose Effect

The above discussion of the TL from quartz illustrates the complex changes that can occur during annealing and/or pre-irradiation. The lower temperature TL from this material (specifically, the so-called 110 °C peak), however, exhibits ostensibly more complex sensitivity changes as a combined result of irradiation and annealing. Known as the *Pre-Dose Effect*, the sensitivity changes in the 110 °C TL peak result only when the sample has been (a) pre-exposed to a dose of irradiation (the "pre-dose") and subsequently heated to a given temperature (the "activation temperature"). Without either one of these components the sensitization effect is not observed. To observe the pre-dose effect on a sample which has previously been irradiated with a pre-dose D (e.g. the natural, or archaeological dose), one gives the quartz sample a small test-dose d ($\ll D$) in the laboratory and records the intensity of the TL emission at 110 °C (S_o). (Alternatively, if the pre-dose D is given in the laboratory, the sample should be heated to a temperature just above that of the 110 °C peak (say, 150 °C) to remove the induced TL. This is not required for a natural dose D since this will have been delivered over a very long period, during which time the 110 °C peak is unstable.) The sample is then heated to a temperature T_a (the activation temperature; typically 500 °C) and upon cooling to room temperature and re-irradiating with the same test-dose one obtains an enhanced sensitivity (S_D). A second heating to T_a and irradiation with the test-dose will yield the same TL sensitivity ($S'_D = S_D$). However, if the sample is now given a larger dose β ($d < \beta < D$) and is again heated to T_a, then, after an additional test-dose a new, larger sensitivity is obtained ($S_{D+\beta}$). The important observation is that $S'_D = S_D$; thus, heating to T_a alone does not induce the sensitivity increase. With the assumption that the increase from S'_D to $S_{D+\beta}$ is proportional to the dose β, and the increase from S_o to S_D is proportional to the dose D, we may write

$$D = \beta\,(S_D - S_o)\,/\,(S_{D+\beta} - S'_D)\,. \tag{4.110}$$

Fleming[81] suggested the use of this effect as a means of determining the unknown dose D and refinements of the procedure have led to reliable methods of dose determination in both archaeological dating[13] and retrospective radiation dosimetry[82]-[86].

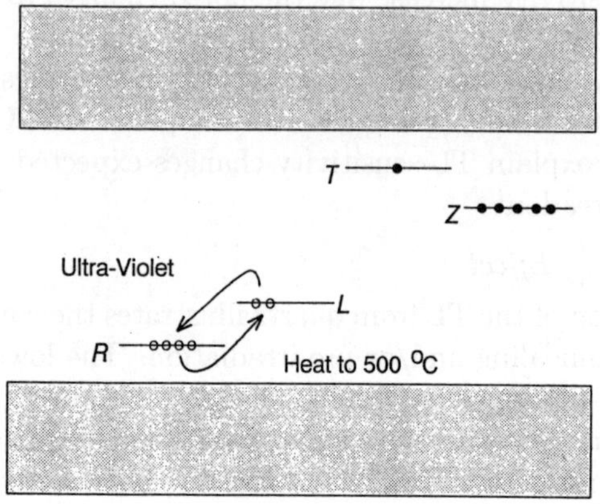

Figure 4.27 Zimmerman's model for the Pre-Dose Effect. (After Zimmerman[87].)

Zimmerman[87] was the first to suggest a model for the effect, based on observations that both the TL and RL sensitivities were affected in the same way after "activation". This suggested that the sensitization process is related to changes in the probability of radiative recombination. An additional observation was that illumination of the sample with UV light (230 - 250 nm) after sensitization will reverse the effect (although not completely[91]). The model that Zimmerman proposed is represented in Fig. 4.27 in which is shown the 110 °C peak trap (T), a population of deep traps (Z), a set of non-luminescent hole centers (R; termed the "reservoir" centers), and radiative recombination centers (L). R is assumed to possess a larger hole capture cross-section than L, such that the number of holes in L is assumed to be proportional to the pre-dose D. Charge neutrality is maintained by a similar large number of electrons in Z.

Administering a small test-dose in the laboratory puts a small number of electrons into T (and Z), and holes into L and R, such that the number of electrons in T is less than the number of holes in L. Heating releases the electrons from T, giving the original sensitivity, S_o. Heating the material to an activation temperature T_a is then assumed to release holes from R into the valence band and to transfer them to L. Upon cooling to room temperature more holes now exist in L than previously, such that a second test-dose and heating will produce a larger TL signal, S_D. Since the enhancement from S_o to S_D is caused by the transfer of holes from R to L, and since the number transferred will depend upon the number already in R, then one can expect the sensitivity increase $S_D - S_o$ to be dependent upon pre-dose D. It is supposed

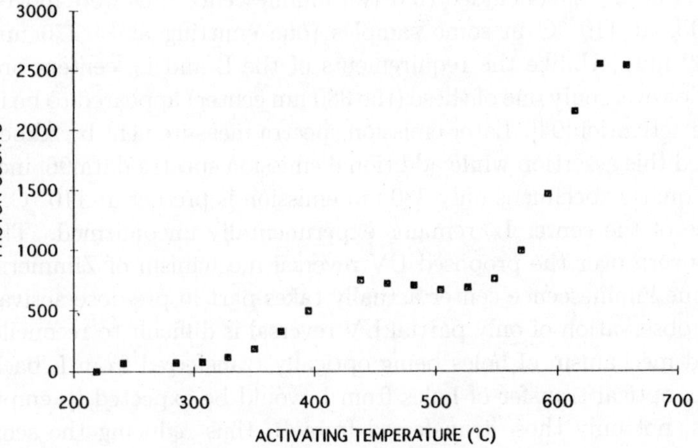

Figure 4.28 Thermal activation curve for a sample of quartz extracted from a porcelain specimen, showing multiple-stage activation. (After Stonehan[88].)

that UV illumination causes the transfer of holes from L back into R, via the valence band.

In the Zimmerman model the transfer of holes from R to L is stimulated by heat and thus the sensitivity increase is thermally activated. This is illustrated in Fig. 4.28 in which we see the thermal activation curve (TAC) for quartz removed from a sample of porcelain collected in the vicinity of the Chernobyl accident. This is a typical example and illustrates that the activation process is multi-staged, suggesting, within the confines of the Zimmerman model, that several different R-type sites are available. The second stage increase observed in the figure coincides with the $\alpha - \beta$ phase transition in this material, as previously discussed by David et al.[89]. Often a slight decrease in the sensitivity may be observed at certain values of the activation temperature (e.g.[90]). Known as "deactivation", the cause of this is unclear in terms of the Zimmerman model.

Several modifications have been introduced into the Zimmerman model in order to account for various aspects of the experimental results. Firstly, Bailiff's observation[91] that the sensitization cannot be completely reversed by UV illumination was explained by him in terms of an additional luminescence site L_o which contributes to the 110 °C TL peak and to the pre-dose effect, but does not take part in the UV reversal process. Bailiff's data indicate that L_o can account for a major fraction of the 110 °C TL signal, and the pre-dose effect. In attempts to identify the defects taking part in the pre-dose effect

McKeever et al.[92] concluded that two luminescence sites were indeed involved in the TL at 110 °C in some samples (one emitting at $\sim$ 470 nm and one at $\sim$ 380 nm). Unlike the requirements of the L and L_o centers proposed by Bailiff, however, only one of these (the 380 nm center) appeared to be involved in pre-dose activation[94]. Later emission spectra measurement by Akber et al.[95] confirmed this assertion while additional emission spectra data[96] indicate that in some quartz specimens only 380 nm emission is present at 110 °C. Thus, the existence of the center L_o remains experimentally unconfirmed. This in turn casts concern over the proposed UV reversal mechanism of Zimmerman, since if only one luminescence center actually takes part in pre-dose activation, then Bailiff's observation of only partial UV reversal is difficult to reconcile with the proposed mechanism of holes being optically transferred from L back to R. In any case, optical transfer of holes from L would be expected to empty L of all its holes, not only those transferred from R, thus reducing the sensitivity to zero - a phenomenon which is not observed.

Chen[97] pointed out that in the Zimmerman model the population of electrons in T must be less than that of holes in L in order that the signal be linear with the test-dose. This conflicts with the requirement that the TL sensitivity is dependent upon the number of trapped holes, which implies that the population of holes in L is less than the population of electrons in T. (This is because, as explained in earlier sections of this chapter, one would normally expect that the TL signal would be dependent upon the smaller of the two concentrations (vis. Eq. 4.36).) To remove this conflict, Chen suggested the addition of a competing trap, T_2, which competes with the L centers for the electrons released from T. Alternatively, Z may fulfill this role. In either situation, the TL intensity will depend upon both the number of trapped electrons, and the number of trapped holes, as required. The growth of the 110 °C TL signal as a function of dose is expected to be non-linear in this case (see Section 4.4).

The simple description noted above of how the pre-dose effect may be used in dosimetry (vis. Eq. 4.110) assumed a linear relationship between sensitivity and pre-dose. Chen[97] accounts for the possibility of a non-linear growth of sensitivity with dose by assuming a relationship of the type

$$S = S_\infty \exp\{-N_R/B\} \tag{4.111}$$

in which N_R is the number of holes trapped at R and B is a constant. The pre-exponent factor S_∞ is the maximum sensitivity attained at high doses. With the assumption $N_R \propto D$, a plot of $\ln(S_\infty - S)$ versus D will yield a straight line from which the value of D for $S = S_o$ can be obtained by extrapolation.

One should note emission spectra measurements[96] which indicate that some of the higher temperature TL peaks may share the same emission centers

(L) as the 110 °C peak. This is particularly true for the so-called "325 °C peak". Activation of the pre-dose effect takes place over the same temperature range as that which is required to produce these higher temperature peaks. Thus, one is left concluding that the L centers are being filled with holes due to thermal transfer from R centers, at the same time that they are being emptied of holes due to recombination with electrons released from the 325 °C traps. Thus, unusual kinetics should be expected, if one accepts the Zimmerman model.

We finish this chapter by noting that the Zimmerman model for the pre-dose effect has not yet received independent verification. Despite its general acceptance possible alternative views of the effect should not be ruled out. For example, we may consider the fact that during the thermal activation stage one is actually emptying the deeper traps which themselves yield TL - e.g. the "325 °C" trap and the "375 °C" traps (so-called because of the position of the corresponding TL peaks when heated at $\sim$ 20 °C/s). These freed electrons may recombine (producing the observed TL signals) or be retrapped in deeper, competing traps. The resulting change in the trapped charge distribution may cause a sensitivity change when the sample is irradiated a second time. Indeed, this scenario was exactly that used to describe the OSL sensitivity changes during single aliquot reuse - as described earlier[80] and was also used by Chen et al.[93] to describe sensitivity changes in the 470 nm emission from the 110 °C peak due to repeated cycles of irradiation and heating. By extension of this logic one might reasonably expect that the 380 nm 110 °C TL sensitivity would also be affected in a similar way. The sensitivity changes would be expected to be dependent upon pre-dose D, and the temperature of thermal activation, T_a. To our knowledge, however, detailed examination of these ideas has not yet been forthcoming.

4.10 References

1. J.R. Cameron, N. Suntharalingam and G.N. Kenney, *Thermoluminescent Dosimetry*, The University of Wisconsin Press, Madison (1968).
2. M. Moscovitch and Y.S. Horowitz, *J. Phys. D: Appl. Phys.* **21** (1988) 804.
3. J.R. Cameron, D.W. Zimmerman, G.N. Kenney, R. Buch, R. Bland and R. Grant, *Health Phys.* **10** (1964) 25.
4. J.R. Cameron, D.W. Zimmerman and R.W. Bland, *Report COO-1105-102, USAEC* (1965).
5. J. Wagner and J.R. Cameron, *Report COO-1105-117, USAEC* (1966).
6. N. Suntharalingam and J.R. Cameron, *Report COO-1105-130, USAEC* (1967).
7. J. Zimmerman, *J. Phys. C: Sol. St. Phys.* **4** (1971) 3277.
8. E.F. Mische and S.W. McKeever, *Radiat. Prot. Dosim.* **29** (1989) 159.
9. C.M. Sunta, E. Okuno, J.F. Lima and E.M. Yoshimura, *J. Phys. D: Appl.*

Phys.. **27** (1994) 2636.

10. T. Mikado, T. Tomimasu, T. Yamazaki and M. Chiwaki, *Nucl. Instrum. and Meth.* **157** (1978) 109.
11. V.K. Jain, *Radiat. Eff.* **54** (1981) 99.
12. M.S. Tite, *Archaeometry* **9** (1966) 155.
13. M.J. Aitken, *Physics and Archaeology*, 2nd edition, Oxford University Press (1974).
14. A. Halperin and R. Chen, *Phys. Rev.* **148** (1966) 839.
15. H. Otaki, H. Kido, A. Hiratsuka, Y. Fukuda and N. Takeuchi, *J. Mater. Sci. Lett.* **13** (1994) 1267.
16. V. Ya. Yaskolko, *Phys. Stat. Sol. (a)* **157** (1996) 507.
17. R. Chen and S.W.S. McKeever, *Radiat. Meas.* **23** (1994) 667.
18. R. Chen and S.G.E. Bowman, *Eur. PACT J.* **2** (1978) 216.
19. R. Chen, X.H. Yang and S.W.S. McKeever, *J. Phys. D: Appl. Phys.* **21** (1988) 1452.
20. R. Chen and A. Hag-Yahya, *Radiat. Prot. Dosim.* **21** (1996) 17.
21. Y.S. Horowitz, *Phys. Med. Biol.* **26** (1981) 765.
22. V.K. Jain, *Nucl. Instrum. and Methods* **180** (1981) 195.
23. A.R. Lakshmanan, B. Chandra, R.C. Bhatt, W. Hoffman and R. Spallek, *J. Phys. D: Appl. Phys.* **18** (1985) 1673.
24. Z. Brynkar, J. Plichta and J. Kratil, *Kernenergie* **31** (1988) 162.
25. E.T. Rodine and P.L. Land, *Phys. Rev. B* **4** (1971) 2701.
26. G. Kitis, M. Spiropulu, S. Charalambous and E.R. Weber, *Nucl. Inst. Methods B* **101** (1995) 255.
27. N. Suntharalingam, *Ph.D. Thesis* (1967) University of Wisconsin.
28. N. Suntharalingam and J.R. Cameron, *Phys. Med. Biol.* **14** (1969) 397.
29. M.J. Aitken, J. Thompson and S.J. Fleming, Proc. 2nd Int. Conf. Lumin. Dosim., Gattlinburg, Tenn., *US At. Ener. Comm., CONF-680920* (1968) 364.
30. N.R. Nail, F. Urbach and D. Pearlson, *J. Opt. Soc. Am.* **39** (1949) 690.
31. C.A. Duboc, *Brit. J. Appl. Phys.* **6** (1955) 5107.
32. F. Cardon and R.H. Bube, *J. Appl. Phys.* **35** (1964) 3344.
33. G.A. Dussel and R.H. Bube, *J. Appl. Phys.* **37** (1966) 13.
34. F.G. Paige, *Phil. Mag.* **2** (1957) 864.
35. M.H. Pilkuhn, *Proc. 13th Int. Conf. Semicond.* (1976) 61.
36. M. Israeli, N. Kristianpoller and R. Chen, *Phys. Rev. B* **6** (1972) 4861.
37. N. Kristianpoller, R. Chen and M. Israeli, *J. Phys. D: Appl. Phys.* **7** (1974) 1063.

38. G.S. Zavt and F.A. Savikhin, *Trud. IFAN Estonian SSR* **43** (1975) 8.
39. L.N. Kantorovich, G.M. Fogel and V.I. Gotlib, *J. Phys. D: Appl. Phys.* **21** (1988) 1008.
40. R. Chen, G. Fogel and C.K. Lee, *Radiat. Prot. Dosim.* **65** (1996) 63.
41. C.K. Lee and R. Chen, *J. Phys. D: Appl. Phys.* **28** (1995) 408.
42. R. Chen and G. Fogel, *Radiat. Prot. Dosim.* **47** (1993) 23.
43. M. Rosenkrantz and Y.S. Horowitz, *Radiat. Prot. Dosim.* **47**, (1993) 27.
44. C.J. Karzmark, J. White and J.F. Fowler, *Phys. Med. Biol.* **9** (1964) 273.
45. E. Tochilin and N. Goldstein, *Health Phys.* **12** (1966) 1705.
46. R.A. Facey, *Health Phys.* **12** (1966) 715.
47. P.J. Groom, S.A. Durrani, K.A.R. Khazal and S.W.S. McKeever, *Eur. PACT J.* **2** (1978) 200.
48. P.C. Hsu and P.S. Weng, *Nucl. Inst. and Meth.* **147** (1977) 453.
49. M.C. Wintersgill and P.D. Townsend, *Radiat. Effects* **38** (1978) 113.
50. J. Kvasnička, *Health Phys.* **36** (1979) 521.
51. J. Kvasnička, *Int. Appl. Rad. Isot.* **34** (1983) 713.
52. S.W.S. McKeever, R. Chen, P.J. Groom and S.A. Durrani, *Nucl. Inst. Meth.* **175** (1980) 43.
53. R. Chen, S.W.S. McKeever and S.A. Durrani, *Phys. Rev. B* **24** (1981) 4931.
54. R. Chen, S.W.S. McKeever and S.A. Durrani, *Eur. PACT J.* **6** (1982) 295.
55. C.M. Sunta, E.M. Yoshimura and E. Okuno, *J. Phys. D: Appl. Phys.* **27** (1994) 852.
56. C.M. Sunta, E.M. Yoshimura and E. Okuno, *J. Phys. D: Appl. Phys.* **27** (1994) 1337.
57. C.M. Sunta, E.M. Yoshimura and E. Okuno, *Radiat. Meas.* **23** (1995) 655.
58. Y.S. Horowitz, *Radiat. Prot. Dosim.* **6** (1984) 17.
59. S.W.S. McKeever, *J. Appl. Phys.* **68** (1990) 724.
60. J.J. Butts and R. Katz, *Radiat. Res.* **122** (1967) 855.
61. J. Kalef-Ezra and Y.S. Horowitz, *Radiat. Isot.* **33** (1982) 1085.
62. Y.S. Horowitz and M. Rosenkrantz, *Radiat. Prot. Dosim.* **31** (1990) 71.
63. Y.S. Horowitz, M. Rosenkrantz, S. Mahajna and D. Yossian, *J. Phys. D: Appl. Phys.* **29** (1996) 205.
64. Y.S. Horowitz, S. Mahajna, M. Rosenkrantz and D. Yossian, *Radiat. Prot. Dosim.* **65** (1996) 7.
65. S.W.S. McKeever, *Radiat. Prot. Dosim.* **33** (1990) 83.
66. M.R. Mayhugh and G.D. Fullerton, *Health Phys.* **28** (1975) 297.
67. M.N. Charles, H.D. Mistry and Z.U. Khan, *Nucl. Instrum. Methods* **175** (1980)

51.

68. B. Chandra, A.R. Lakshmanan and R.C. Bhatt, *J. Phys. D: Appl. Phys.* **15** (1982) 1803.
69. A.R. Lakshmanan, R.C. Bhatt and K.G. Vohra, *Phys. Stat. Sol. (a)* **53** (1979) 617.
70. Y.S. Horowitz, *Radiat. Prot. Dosim.* **47** (1993) 135.
71. M.S. Akselrod, V.S. Kortov and E.A. Gorelova, *Radiat. Prot. Dosim.* **47** (1993) 159.
72. S.W.S. McKeever, *Radiat. Prot. Dosim.* **8** (1984) 81.
73. M. David, C.M. Sunta and A.K. Ganguly, *Ind. J. Pure Appl. Phys.* **15** (1977) 277.
74. S.W.S. McKeever, J.A. Strain, P.D. Townsend and R.A. Wood, *Eur. PACT J.* **9** (1983) 123.
75. H.M. Rendell, P.D. Townsend and R.A. Wood, *Radiat. Meas.* **23** (1994) 441.
76. T. Hashimoto, S. Sakaue and M. Ichino, *Radiat. Meas.* **23** (1994) 293.
77. M.J. Aitken and B.W. Smith, *Quat. Sci. Rev.* **7** (1988) 387.
78. L. Bøtter-Jensen, H. Jungner and V. Mejdahl, *Radiat. Prot. Dosim.* **47** (1993) 643.
79. L. Bøtter-Jensen, N. Agersnap Larsen, V. Mejdahl, N.R.J. Poolton, M.F. Morris and S.W.S. McKeever, *Radiat. Meas.* **24** (1995) 535.
80. S.W.S. McKeever, L. Bøtter-Jensen, N. Agersnap Larsen, V. Mejdahl and N.R. J. Poolton, *Radiat. Prot. Dosim,* **65** (1996) 49.
81. S.J. Fleming, *Archaeometry* **15** (1973) 13.
82. E.H. Haskell, P.L. Kaipa and M.E. Wrenn, *Nucl. Tracks Radiat. Meas.* **14** (1988) 114.
83. I.K. Bailiff and E.H. Haskell, *Radiat. Prot. Dosim.* **6** (1984) 245.
84. E.H. Haskell and I.K. Bailiff, *Nucl. Tracks Radiat. Meas.* **10** (1985) 503.
85. E.H. Haskell and I.K. Bailiff, *Radiat. Prot. Dosim.* **34** (1990) 195.
86. E.H. Haskell, *Radiat. Prot. Dosim.* **47** (1993) 297.
87. J. Zimmerman, *J. Phys. C: Sol. St.* **4** (1971) 3265, 3277.
88. D. Stoneham, *Radiat. Meas.* **24** (1995) 499.
89. M. David, *Ind. J. Pure Appl. Phys.* **19** (1981) 1048.
90. M.J. Aitken, *Eur. PACT J.* **3** (1979) 319.
91. I.K. Bailiff, *Eur. PACT J.* **3** (1979) 345.
92. S.W.S. McKeever, C.Y. Chen and L.E. Halliburton, *Nucl. Tracks Radiat. Meas.* **10** (1985) 489.
93. R. Chen, G. Fogel and N. Kristianpoller, *Radiat. Meas.* **23** (1994) 277.
94. X.H. Yang and S.W.S. McKeever, *Radiat. Prot. Dosim.* **33** (1990) 27.

95. R.A. Akber, G.B. Robertson and J.R. Prescott, *Nucl. Tracks Radiat. Meas.* **14** (1988) 21.
96. R.B. Scholefield, J.R. Prescott, A.D. Franklin and P.J. Fox, *Radiat. Meas.* **23** (1994) 409.
97. R. Chen, *Eur. PACT J.* **3** (1979) 325.

Chapter 5
OPTICAL PHENOMENA

5.1 Optical Processes and TL

Thus far, the entire discussion has concerned thermally stimulated transitions from one energy state to another. However, transitions between energy states can also take place optically. Optical processes can be important in the interpretation of thermoluminescence in three major ways - namely:

(i) *Optical Bleaching* of a radiation-induced signal. Here one exposes an irradiated sample to light before heating the sample; subsequent measurement of the TL will reveal the signal to be smaller than expected due to light-induced transfer of trapped charge out of the TL traps.

(ii) *Phototransferred TL* (PTTL). This is the TL induced by the optical transfer of charge from populated traps into empty traps. To observe PTTL one irradiates a sample at a temperature T_{irr}, after which the sample temperature is lowered to a new temperature $T_{ill} < T_{irr}$, at which point the sample is exposed to light. Alternatively, the irradiated sample may be preheated to a temperature T_{ph} in order to empty traps corresponding to lower temperature TL peaks, while leaving deeper traps still populated. The sample is then cooled back to T_{irr} at which point it is illuminated with light (i.e. $T_{ill} = T_{irr}$). In another example, a sample may have been irradiated in nature over very long periods for which the irradiation time $t_{irr} >> \tau$, the lifetime of the trapped charge in some of the shallower traps. In this instance, the traps corresponding to the lower temperature TL peaks are already empty at the end of the irradiation period and the preheat stage noted above can be omitted. In any of the above cases, TL peaks are observed in the lower temperature ranges when the sample is subsequently heated after illumination. The cause of these peaks is the optical transfer of charge from the deep, populated traps into the shallow, unpopulated traps - hence the term phototransferred TL.

(iii) *Light-Induced TL*: Here one exposes an *unirradiated* sample to light, and finds that a TL signal has been induced by the illumination.

Each of the above processes can have profound effects on the TL glow curve of a sample, and for some technologically or scientifically important materials (e.g. materials used in dosimetry or dating) these processes have been

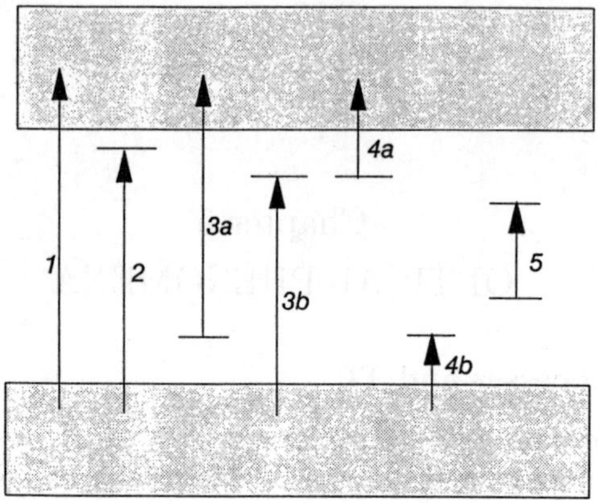

Figure 5.1 Expected optical transitions shown for a flat-band diagram. (1) Ionization (excitation across the band gap). (2) Exciton formation. (3) Defect ionization. (4) Trap ionization. (5) Internal transition.

studied in some detail. In the sections to follow we examine each of these effects separately. First, however, we make some general comments about optical processes in solids.

5.2 General Optical Processes in Solids

5.2.1 *Optical Transitions and Optical Trap Depths*

The intensity of light $I(\lambda, x)$ of wavelength λ, at depth x, in an insulator or semiconductor is described by Beer's law, namely:

$$I(\lambda, x) = I_o(\lambda) \exp\{-\alpha(\lambda) x\} \tag{5.1}$$

where $I_o(\lambda)$ is the incident intensity at the surface and $\alpha(\lambda)$ is the absorption coefficient. The value of α for a given wavelength depends upon the type of optical absorption process occurring. Some of the optical transitions that are important in insulators and semiconductors are illustrated in Fig. 5.1. Free carrier absorption and excitation to higher lying bands are not illustrated since these do not affect the TL properties, which is the focus of our consideration here. Transition (1) - i.e. band-to-band ionization - is not important for wide band gap insulators but is important for narrow band gap (semiconducting) materials. For example, ZnSe has a band gap at room temperature of ~ 2.71 eV and illumination with light of wavelength ≤ 455 nm creates free electrons and holes which subsequently become trapped at deep levels within the material. Subsequent heating of the sample induces TL[1].

Exciton formation (transition (2)) can lead to TL in wide band gap insulators, but the required optical transition energies are in the vacuum ultraviolet[2]. However, defect ionization (transitions (3a) and (3b)) can be very important in wide band gap materials. In unirradiated samples localized states below the Fermi level can act as donors, or states above the Fermi level can act as acceptors, during illumination, creating free electrons or holes, respectively. The free carriers so produced may then become trapped, yielding TL. An example is the ionization of the Bi anti-site deep donor in $Bi_{12}GeO_{20}$ when these samples are illuminated with 476 nm light[3]. Alternatively, radiation-induced trapped charge may be ionized by transitions of the type (4a) and/or (4b), thereby removing potential sources of TL (i.e. optical bleaching). Removal of TL by light is an important process in the TL dating of sediments; the method was first introduced by Wintle and Huntley[4] and involves the assumption that during deposition, the sediment grains (quartz and feldspars) are exposed to sunlight and lose their accumulated trapped charge due to solar bleaching. I.e. the TL signal is re-set to zero. Upon burial the grains are shielded from further light exposure and the surrounding natural radioactivity ensures that trapped charge slowly accumulates over the period of deposition (i.e. the "age" of the sediment). Upon retrieval and heating in the laboratory a TL signal is produced which is related to the natural dose absorbed since burial, and, hence, to the age of the sample.

If an irradiated sample is intentionally illuminated in the laboratory, optically stimulated luminescence (OSL) is often observed. Here charges optically released from radiation-induced traps recombine with oppositely charged carriers to produce a luminescence signal which is a measure of the absorbed radiation dose. This principle has now also been exploited successfully in the dating of sedimentary deposits, since the first suggestion of this application by Huntley and colleagues[5]. The advantage of using light to stimulate the luminescence is that one directly monitors the light-sensitive centers which will themselves have been bleached by the original sunlight exposure.

OSL may also be used in radiation dosimetry. Here synthetic materials which have been exposed to radiation are illuminated with light and the subsequent OSL signal is recorded. The OSL intensity is then related to the absorbed dose and a method of dosimetry is thus available[6],[7].

OSL is to be distinguished from photoluminescence which is the luminescence emitted following absorption of light by an internal transition, such as that depicted in transition (5)[8],[9]. Since these do not involve transport of charge from one defect state to another, such transitions do not effect the TL signal, but they can yield useful information about the available luminescent

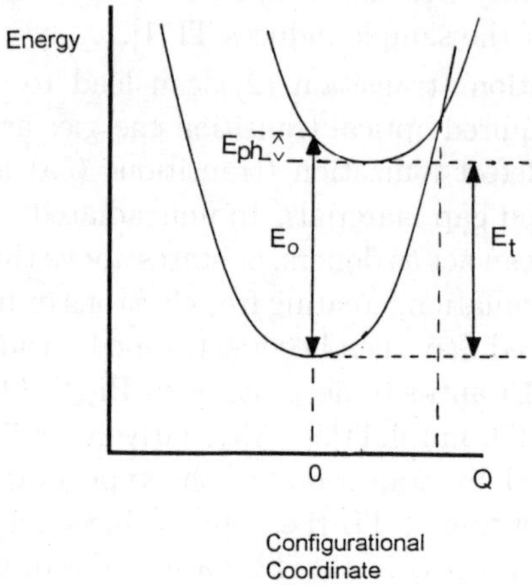

Figure 5.2 Configurational coordinate diagram showing the electron + elastic energy as a function of configurational coordinate Q. The deep level is optically ionized by absorption of energy E_o, which is larger than the trap depth by an amount E_{ph}.

sites; for example, the site symmetries of the rare earth emitters in rare-earth-doped alkaline earth fluorides[10].

Ionization of radiation-induced trapped charge (for example, during optical bleaching of TL, or during OSL production) requires photons of energy equal to, or greater than, the defect ionization energy (trap depth). Figure 5.2 indicates a configurational coordinate representation of the energy bands for a deep level, defect state exhibiting strong phonon coupling to the host lattice. The Franck-Condon principle dictates that optical ionization of an electron from the deep level to the conduction band occurs without adjustment of the lattice coordinate Q and requires absorption of an amount of energy E_o. The optical trap depth E_o can be substantially greater than the trap depth E_t by an amount $E_{ph} = S\hbar\omega_p$, where $\hbar\omega_p$ is the phonon energy and S is the Huang-Rhys factor. Estimates of the ratio E_o/E_t are due to Mott[11] and yield an approximation of

$$\frac{E_o}{E_t} \approx \frac{\epsilon_o}{\epsilon} \tag{5.2}$$

where ϵ_o and ϵ are the high frequency and static dielectric constants, respectively.

5.2.2 *Photoionization Cross-Section*

The absorption coefficient for a defect-to-band transition is given by

$$\alpha(\lambda) = N\sigma_o(\lambda) \tag{5.3}$$

where N is the defect concentration (m^{-3}) and $\sigma_o(\lambda)$ is the photoionization cross-section (m^2). For deep levels and parabolic bands the photoionization cross-section may be described by an expression of the form[12]:

$$\sigma_o = C\sqrt{E_o}\frac{(h\nu - E_o)^{3/2}}{h\nu(h\nu - \gamma E_o)^2} \tag{5.4}$$

where C is a constant and $h\nu$ is the incident photon energy. $\gamma = 1 - m^*/m_e$ is a constant dependent upon the effective mass m^* of the electron. For an effective mass equal to the free electron mass m_e, $\gamma = 0$. The photoionization cross-section can be estimated experimentally from the intensity of light $I(\lambda)$ required to yield the same photoconductivity, for different wavelengths[13]. Consider a state of equilibrium between the rate of emission of electrons from a trap (of trapped charge concentration n, under illumination with light of intensity $I(\lambda)$) and the rate of recapture. We have

$$I(\lambda)\sigma_o(\lambda)n = n_c(N - n)A \tag{5.5}$$

where N is the total concentration of traps, A is the capture probability, and n_c is the free electron concentration. Under conditions of constant photocurrent, n_c and n can be considered constant with respect to wavelength. Under these circumstances Eq. 5.5 may be re-written

$$\sigma_o(\lambda) = K/I(\lambda) \tag{5.6}$$

where $K = n_c(N/n - 1)A$ is a constant. The absorption coefficient (Eq. 5.3) due to defect-to-band ionization should be edge-like as a function of wavelength since the absorption line shape contains contributions from all relevant continuum states into which the electron is being excited[14]. The absorption band will not exhibit a resonance[12]. Following the transition, the charge on the defect will change by one electronic charge and significant lattice relaxation may occur. This is also true during the inverse process - namely, capture of a free carrier by the trap - and is the origin of the configurational coordinate diagram of Fig. 5.2, which implicitly assumes that the capture process proceeds directly into the ground state and not via excited states.

5.3 Optical Bleaching

5.3.1 *Kinetics*

Removal of an electron from a trap by optical absorption reduces the subsequent TL signal. Assuming the simplest model of one trap and one recombination center, the rate of detrapping may be expressed by the usual rate equations, thus

$$\frac{dn}{dt} = -nf + n_c (N - n) A \tag{5.7}$$

with

$$\frac{dn_c}{dt} = nf - n_c (N - n) A - n_c m A_m. \tag{5.8}$$

Here f is the optical excitation rate, given by

$$f(\lambda) = \phi(\lambda)\, \sigma_o(\lambda) \tag{5.9}$$

in which ϕ is the photon fluence (photons/m^2/s), and all the other terms in Eqs. 5.7-5.9 have been defined in previous chapters. For the simplest case of no retrapping ($nf \gg n_c(N-n)A$) we arrive at

$$n = n_0 \exp\{-ft\} \tag{5.10}$$

where n_0 is the original trapped charge level at time $t = 0$, and $1/f = \tau_d$ is the decay constant. For a single trap the subsequent TL signal S decays in a similar first-order process as a function of illumination time

$$S = S_o \exp\{-t/\tau_d\}, \tag{5.11}$$

equivalent to the isothermal decay of phosphorescence, given by Eq. 3.4. Eventually the TL signal due to this concentration of trapped charge must asymptotically decay to zero under constant illumination. This is true even if retrapping events become important at longer illumination times, although in this latter case the decay is no longer exponential. With retrapping, the second term in Eq. 5.7 cannot be ignored and to proceed one has to eliminate n_c by adopting the same procedure as used in the case of thermal detrapping, described in Ch. 2 (section 2.3.4). Here one makes the usual assumption of quasiequilibrium and, using the rate equations Eqs. 5.7 and 5.8, arrives at the optical equivalent of the General One Trap (GOT) equation, namely

$$\frac{dn}{dt} = -\frac{nfmA_m}{(N-n)A + mA_m}. \tag{5.12}$$

With $N \gg n$, $R = A/A_m$, $n = m$, and fast retrapping ($(N-n)A \gg mA_m$), Eq. 5.12 becomes

$$\frac{dn}{dt} = -\frac{n^2 f}{NR}. \tag{5.13}$$

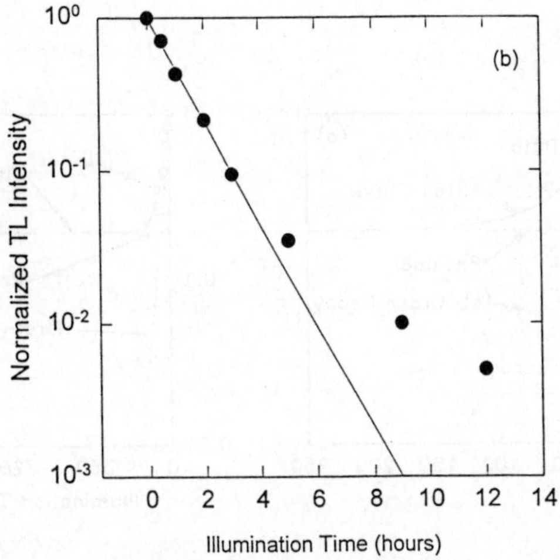

Figure 5.3 The decay of the gamma-ray-induced 450 K TL peak in Al_2O_3 as a function of illumination time, for an illumination wavelength of 540 nm. (After Walker et al.[15].)

Rearranging yields

$$\frac{dn}{n^2} = -\frac{f}{NR}dt \tag{5.14}$$

or

$$\frac{n_0}{n} = 1 + n_0\frac{ft}{NR}. \tag{5.15}$$

As $t \to \infty$, $n \to 0$. Thus, a TL signal due to this charge concentration will approach zero for long optical bleaching times. An example of a TL peak which is reduced by prolonged bleaching is shown in Fig. 5.3 for Al_2O_3. The initial first-order decay is obvious, as is the departure from first-order at longer times.

Sometimes, however, it may be observed that a TL signal is reduced by prolonged illumination, but does not go to zero. Experimentally, there may be several causes for this - e.g. an overlap of several TL peaks, some of which are emptied by the particular wavelength used, and some of which are not. However, notwithstanding experimental problems such as this, one may ask if it is intrinsically possible that a TL peak, due to a single trap level, may be reduced initially, but not to zero - no matter for how long the illumination is applied. An apparent example of this type of behavior is illustrated in Fig. 5.4 in which we see the decay of the 245 °C TL peak from a sample of natural (Arkansas) hydrothermal quartz following ^{60}Co irradiation and illumination with 460 nm light at room temperature. The data obtained can be described by the summation

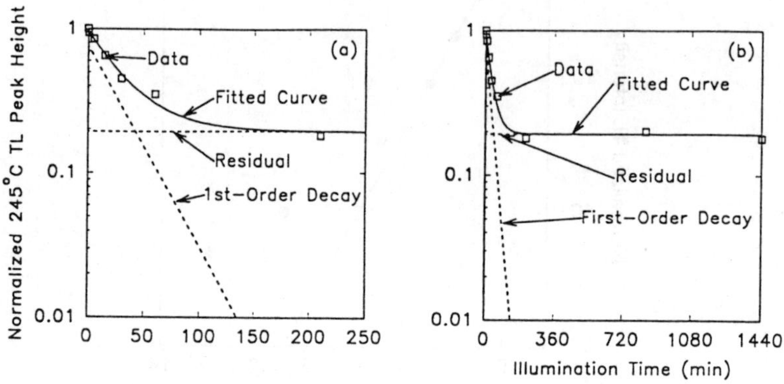

Figure 5.4 (a) The decay of the 245 ^oC TL peak of a quartz sample irradiated with ^{60}Co gamma rays and illuminated with 460 nm light at room temperature. (b) The same data as (a), but on a time scale up to 24 hours.
(After Morris and McKeever[16].)

of a first-order decay plus an unbleachable residual.

In order to explain these effects it is necessary to consider a model which is significantly more complex than the simple one-trap/one-center model described above. Chen et al.[18] introduced a model in which they allowed for retrapping into the electron trap, but also included the optical excitation of electrons from the recombination level into the conduction band during illumination. A similar mechanism is discussed by Shlukov et al.[19]. The model of Chen et al. is illustrated in Fig. 5.5(a). The significance of this latter transition is that the recombination center lies below the Fermi level, and thus the concentration of holes at this center actually *increases* during the illumination phase. As a result, it is impossible for the subsequent TL signal to decay to zero as a result of the illumination. Example simulated data obtained by Chen et al. using this model are shown in Fig. 5.6. An important prediction is that the final residual level is independent of the original absorbed dose, or the intensity of illumination. In fact, from Fig. 5.6 we see that the residual level at long illumination times is the same whether the traps are initially empty or initially full. Thus, with this model, the illumination is seen to generate a TL signal in a previously unirradiated specimen (a phenomenon to be discussed more fully in a later section). With respect to the results for quartz insufficient data presently exist to determine if the residual level illustrated in Fig. 5.4 is dose dependent or independent.

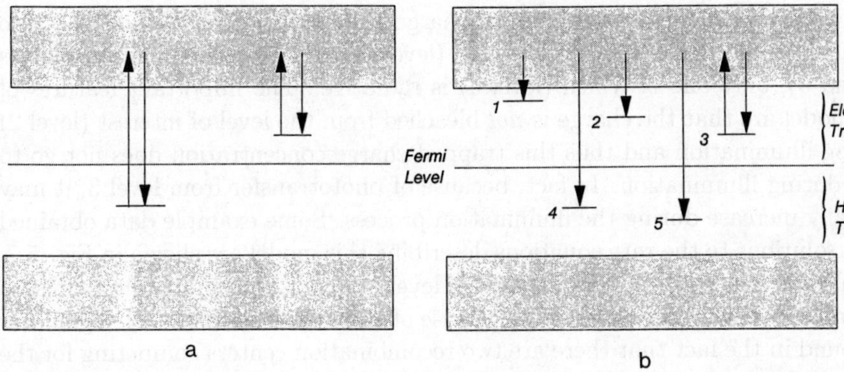

Figure 5.5 The models of (a) Chen et al., and (b) McKeever, to explain optical bleaching of TL in which a finite residual level is reached at long illumination times. (After Chen et al.[18], and McKeever[20].)

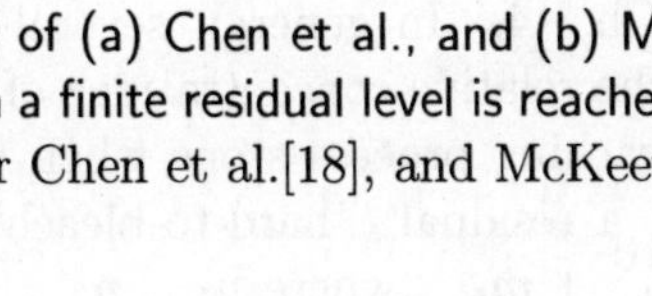

Figure 5.6 Example bleaching and excitation curves from the model of Chen et al.. The curves (a), (b) and (c) refer to three different illumination intensities. (After Chen et al.[18].)

An alternative model for the explanation of a non-zero residual level after bleaching is illustrated in Fig. 5.5(b). Here the TL signal of interest is due to the release of electrons from level 2, but the illumination does not bleach this level; instead it empties charge from a deep level - level 3. Also included in the model is a shallow trap (level 1) and two recombination centers (4 and 5), only one of which (level 4) is radiative. The important features of the model are that the charge is *not* bleached from the level of interest (level 2) during illumination and thus this trapped charge concentration does not go to zero during illumination. In fact, because of phototransfer from level 3, it may actually increase during the illumination process. Some example data obtained from solutions to the rate equations describing this model are shown in Fig. 5.7. In this particular example, the TL from level 2 decreases from its initial level to an unbleachable, residual level. The cause of the reduction in the TL signal can be found in the fact that there are two recombination centers competing for the charge released from level 3 during bleaching. Furthermore, during heating, the model of Fig. 5.5(b) represents a classic competition-during-heating situation, as described in detail in Ch. 4. In general, several possible behaviors are possible, dependent upon the relative concentrations of the trapped charges in the various levels. The bleaching process stops when the charge from level 3 is depleted. At this stage, a residual ("hard-to-bleach") component remains, such that $n_{1_\infty} + n_{2_\infty} = m_{4_\infty} + m_{5_\infty}$, where n_{1_∞}, n_{2_∞}, m_{4_∞}, and m_{5_∞} are the concentrations of electrons and holes in level 1, 2, 4 and 5 remaining when the concentration of electrons in level 3 has been depleted. If n_3 is large enough at the start of the bleaching procedure, m_{4_∞} may be ~ 0, and thus the final residual TL level will also be zero. However, if n_3 is such that $m_{4_\infty} \neq 0$, a non-zero residual TL level will be reached.

5.3.2 Wavelength Dependence

Not all TL peaks are bleached equally by exposure to light because of variations in the photoionization cross-section from trap to trap. Fig. 5.8 shows the TL from a natural sample of sedimentary quartz, along with the TL from a similar sample after prolonged bleaching by a solar simulator filtered to remove UV and blue light. One can observe, in addition to a small phototransferred TL peak at ~ 180 °C, that the high temperature TL has been reduced only in one small region. Subtraction of the bleached TL curve from the unbleached curve reveals that only the so-called "325 °C" peak was removed by the bleaching procedure. Other peaks remained unchanged. (The "325 °C" TL peak from quartz is so-called because, at a heating rate of ~10 °C/s, it appears at a temperature of ~325 °C. It retains this label even when other heating rates are used.) This behavior gives rise to the description of "easy-to-bleach" traps (viz. TL at ~325 °C) and "hard-to-bleach" traps (viz. TL at ~370 °C, and others). By using bleaching light of shorter wavelengths, however, even the deeper TL

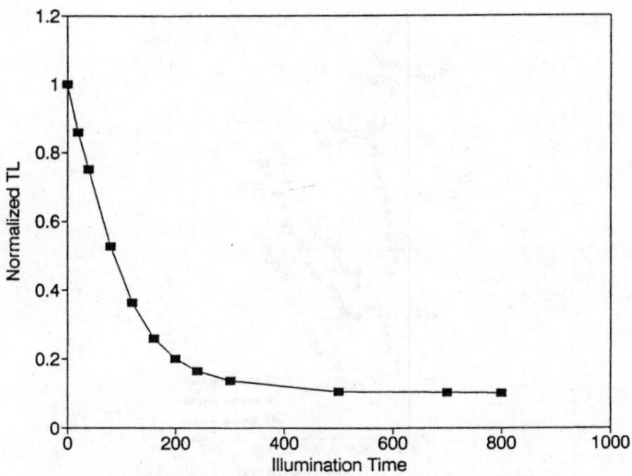

Figure 5.7 (a) Trapped charge concentrations (refer to Fig. 5.5(b)) versus bleaching time for the model of McKeever. (b) Predicted TL versus bleaching time, illustrating the non-zero residual TL level at long bleaching times.

(After McKeever[17].)

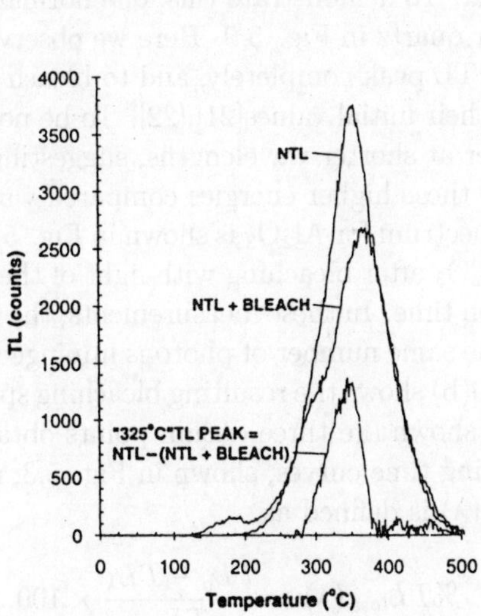

Figure 5.8 Glow curves of natural and bleached quartz measured at a heating rate of 10 K/s, showing the separated "325 ^{o}C" TL peak obtained by subtraction of the bleached curve from the unbleached curve.

(After Spooner[22].)

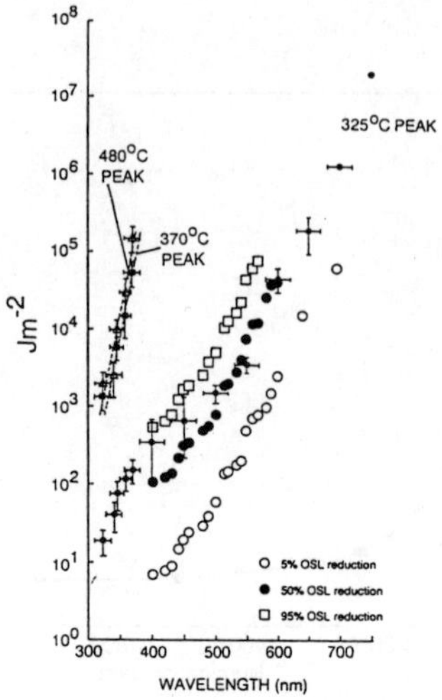

Figure 5.9 The bleaching spectrum for the TL from natural quartz, expressed as the energy required to reduce the TL to a certain level, as a function of energy. (After Spooner et al.[21].)

peaks can be bleached. To demonstrate this, one normally plots the "bleaching spectrum", shown for quartz in Fig. 5.9. Here we observe the energies required to bleach the 325 °C TL peak completely, and to bleach the 370 °C and 480 °C TL peaks to 80% of their initial values[21],[22]. To be noted is the fact that the slopes become steeper at shorter wavelengths, suggesting that additional traps are being emptied at these higher energies compared with the lower energies.

A bleaching spectrum for Al_2O_3 is shown in Fig. 5.10. Fig. 5.10(a) shows the TL peak from Al_2O_3 after bleaching with light of the wavelengths indicated for a fixed illumination time. In these measurements, the illumination power was adjusted such that the same number of photons impinged on the sample at each wavelength. Fig. 5.10(b) shows the resulting bleaching spectrum, expressed here as $\%TL_{lost}(\lambda)$. Also shown are three datum points obtained from the slope of the TL versus bleaching time curves, shown in Fig. 5.3, normalized to the data for 380 nm. $\%TL_{lost}(\lambda)$ is defined as

$$\%TL_{lost}(\lambda) = \frac{TL_o - TL_\lambda}{TL_o} \times 100, \tag{5.16}$$

where TL_o is the TL obtained immediately after irradiation, with no illumination, and TL_λ is the TL obtained after illumination for a fixed time at a wavelength λ.

One has to be careful when attempting to interpret data of the type shown in these figures as being a reflection of the photoionization cross-section

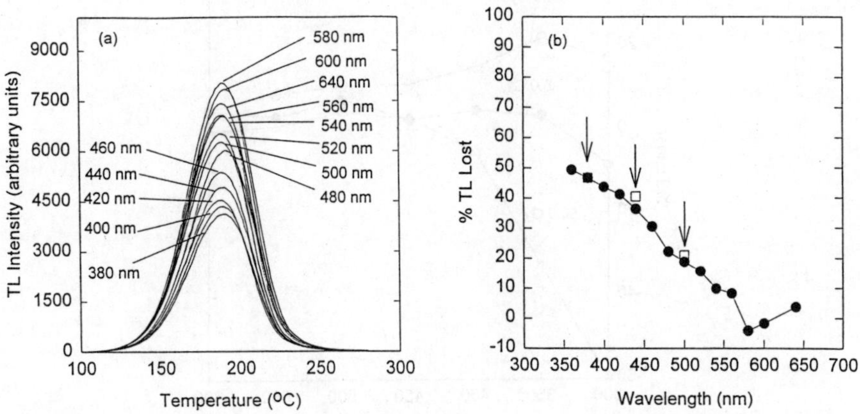

Figure 5.10 (a) TL from Al_2O_3 as a function of the wavelength of the bleaching light, for a fixed illumination time. (b) Bleaching spectrum, expressed as the percentage TL lost (see text), as a function of wavelength.
(After Walker et al.[15].)

$\sigma_o(\lambda)$. This is because simultaneously with bleaching of the charge out of the traps in question, one can have phototransfer of charge into the traps, from deeper traps. This can be expressed by writing $\%TL_{lost}(\lambda)$ as

$$\%TL_{lost}(\lambda) = F_f(\lambda) - F_t(\lambda) \tag{5.17}$$

where $F_f(\lambda)$ is a "fading", or "bleaching", function, directly proportional to $\sigma_o(\lambda)$, and $F_t(\lambda)$ is a "transfer" function. The latter is related to the wavelength dependence of the transfer of charge from deep traps into the traps under study. Clearly this function will depend upon the occupancy of the deep traps, which, in turn, will depend upon the past irradiation and thermal history of the specimen. In particular, for a large population of deep traps one can expect there to be a significant transfer of charge. Under such circumstances $F_t(\lambda)$ will dominate and $\%TL_{lost}(\lambda)$ will be < 0. Alternatively, for a small population of deep traps, $F_f(\lambda)$ should dominate and $\%TL_{lost}(\lambda)$ will be > 0. Such effects are illustrated in Fig. 5.11 for Al_2O_3.

5.3.3 *Thermal Assistance*

It may be that transitions from a defect ground state to a delocalized band do not take place directly, but instead occur via a localized excited state. Illumination of LiF containing F-centers at a wavelength of 250 nm is able to transfer electrons from the F-centers to stable electron trapping states. The transfer is

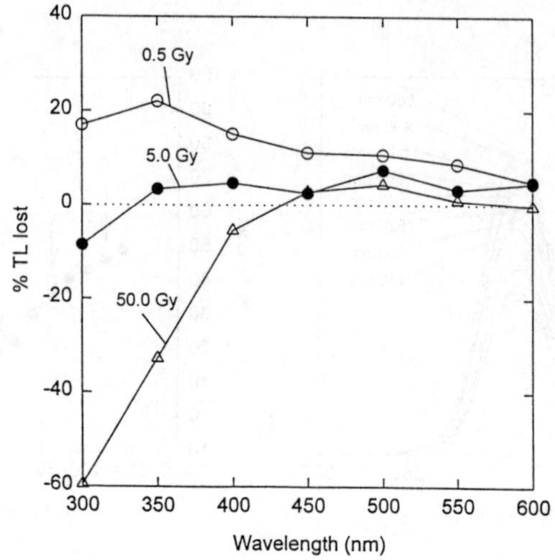

Figure 5.11 Percentage TL lost for the Al_2O_3 glow curve as a function of wavelength, for three different pre-doses (0.5 Gy, 5.0 Gy and 50.0 Gy). For the larger pre-doses the transfer of charge from the deep traps dominates and a growth in the TL signal (i.e. percentage TL lost < 0) is observed.
(After Walker et al.[15].)

observed to be temperature dependent, with an activation energy of 0.12 eV[23]. The F-center in LiF is believed to have an excited state 0.16 eV below the conduction band and the measured activation energy for phototransfer is suggested to correspond to thermal activation from this excited state into the conduction band. Similar mechanisms have been suggested by Hütt and colleagues[24],[25] who monitored the optically stimulated luminescence (OSL) which results from illumination of irradiated feldspar samples with infra-red light ($\sim$ 825-950 nm). The production of OSL in this manner was observed to have an activation energy of $\sim$0.10 eV. The TL traps emptied by this illumination are deep, however, and thus it was suggested by Hütt et al. that the infra-red illumination excited the trapped electrons from a localized ground state to a localized excited state, from where thermal excitation into the conduction band (activation energy 0.10 eV) leads to bleaching of the TL traps and production of OSL.

Poolton et al.[26]-[28] examined this proposal further for a variety of feldspars. Using the Bohr model for the hydrogen atom one can calculate the energy E_n and radius R_n of the nth excited state using

$$E_n = -E_h \frac{(m^*/m_e)}{(\epsilon_r n)^2} \tag{5.18}$$

and

$$R_n = \frac{R_h n \epsilon_r}{(m^*/m_e)} \tag{5.19}$$

where E_h and R_h are the ionization energy and Bohr radius of free hydrogen, respectively (13.6 eV and 0.53 Å), m^* is the electron effective mass, m_e is the free electron mass, and ϵ_r is the relative permittivity. Assuming a hydrogenic model for the defects responsible for IR-stimulated OSL, with $m^* = 0.757m_e$ for each feldspar type, the transition energies to the first excited state are calculated to be 1.441 eV, 1.422 eV and 1.225 eV for orthoclase, albite and anorthite, respectively. This compares very well with the measured IR absorption energies required for production of OSL in these materials - namely, 1.440 ± 0.003 eV, 1.422 ± 0.003 eV and 1.275 ± 0.004 eV, respectively. This gives broad support to the model proposed by Hütt and colleagues for thermal assistance of the OSL mechanism. However, Poolton et al.[27] demonstrate that the measured thermal activation energies for the three feldspar types are inconsistent with the Hütt et al. model. The thermal activation energies, measured using the IR energies noted above, are 0.105 eV (orthoclase), 0.065 eV (albite) and 0.03 eV (anorthite) - the opposite of what one might expect based on the measured (and calculated) excited state energies. These authors point out, however, that the radii of the first excited states, calculated using Eq. 5.19, increases as one goes from orthoclase to anorthite. Thus, Poolton et al.[27] propose that the measured thermal activation energy is the hopping energy from the excited state of the "donor" defect to an "acceptor" defect, which acts as a recombination center. This donor-acceptor recombination model predicts a zero value for the thermal activation energy if the overlap between the excited state wavefunction is high enough, and Poolton et al. suggest that this explains why the measured activation energy decreases as one goes from orthoclase to anorthite.

Spooner also observed that the OSL from quartz when illuminated with visible light is dependent upon temperature[22]. In this material, the measured thermal activation energy for OSL production depends smoothly upon the wavelength of the light used. This is also inconsistent with a Hütt-type mechanism for which one would expect that either the thermal activation energy would be independent of the wavelength, or it would vary discretely as higher excited states are populated. Because of this Spooner suggests a model based on the thermal excitation of charge within an array of ground state levels, from where optical excitation into the conduction band then occurs.

5.4 Phototransferred TL (PTTL)

5.4.1 General Observations

PTTL is the TL from shallow traps resulting from the transfer of charge from deeper traps. PTTL has been observed and studied for a wide variety of materials and has been suggested as a tool in radiation dosimetry and in dating of ceramic samples[29]-[38]. McKeever[2] reviewed the early (1960-1970) literature on PTTL effects in which it was used initially as a tool to monitor absorption spectra in materials with small absorption coefficients. The procedure was to irradiate the sample, cool and illuminate with light of a certain wavelength. By varying the wavelength and repeating the experiment one can build up an absorption spectrum revealing the absorption wavelengths corresponding to the radiation-induced donor centers. Early examples concentrated on the alkali halides, including work by Stoddard[39], Fieschi and colleagues[40],[41], Tournon and Bergé[42] and several others. However, other materials were also studied by these techniques, including diamond[43] and quartz[44]. A feature of these early studies was that often the PTTL excitation spectrum could be clearly identified with known optical absorption bands such that the sign of the charge carrier being transferred could be identified with some certainty, thereby identifying the sign of the charge carrier in the TL trap in question. An example is LiF where the main peak used in TL dosimetry ("peak 5") can be re-filled after heating by illuminating the sample with 250 nm light. Light of this wavelength is absorbed by F-centers and thus "peak 5" is regenerated by electrons being released from F-centers by the illumination and retrapped by the "peak 5" traps. This phenomenon gave rise to the ability to "re-read" the TL signal from LiF after the first heating[23],[45]-[47]. Other dosimetric materials in which PTTL has been examined include Al_2O_3[48],[49]. The interest in the PTTL properties of this material stem from its potential as an OSL dosimeter. Examinations of the PTTL properties lead to useful information regarding the optical energies likely to be most efficient in transferring charge between centers and help to identify the mechanisms involved in the optically stimulated and TL processes.

5.4.2 Models

The simplest model to describe phototransfer is that of one shallow trap into which the charge is transferred, one deep trap from which the charge is excited, and one recombination center, as already described in Fig. 4.10. If n_1 and n_2 are the concentrations of electrons in the shallow and deep traps, respectively, and m is the concentration of holes in the centers, an analysis of PTTL may start from the initial condition that $n_{10} = 0$, and $n_{20} = m_0$ at some point after the irradiation and immediately before illumination. If the illumination excites electrons from the deep traps at rate f then we may write the following rate equations during the illumination period:

$$\frac{dn_2}{dt} = -n_2 f + n_c (N_2 - n_2) A_2 \tag{5.20}$$

$$\frac{dn_1}{dt} = n_c (N_1 - n_1) A_1 \tag{5.21}$$

and

$$\frac{dm}{dt} = -n_c m A_m = \frac{m}{\tau} \tag{5.22}$$

where $\tau = (n_c A_m)^{-1}$ is a lifetime and all the other terms have their usual meanings. With the additional stipulations of quasiequilibrium and no re-trapping into the source traps ($n_2 f >> n_c (N_2 - n_2) A_2$) the solutions to the above equations are

$$n_2 (t) = n_{20} \exp\{-tf\}, \tag{5.23}$$

$$n_1 (t) = N_1 [1 - \exp\{-Bt\}] \tag{5.24}$$

and

$$m (t) = m_0 \exp\{-t/\tau\} \tag{5.25}$$

where $B = n_c A_1$. Note that B and τ are considered approximately constant only if the quasiequilibrium approximation ($dn_c/dt \approx 0$) is true. Thus, at the end of the illumination period, i.e. at $t = t^*$, there will be a certain concentration of charges in each of the traps and centers, as defined by Eqs. 5.23-5.25.

We now consider warming of the sample after the illumination has been switched off. This is a classic competition-during-heating situation and one which has already been fully analyzed in section 4.4. With the assumptions of quasiequilibrium and $n_1 (t^*) << N_2 - n_2 (t^*)$, we arrive at Eq. 4.61, re-written here as

$$S (t^*) = \frac{C m (t^*) n_1 (t^*)}{(N_2 - n_2 (t^*))} \tag{5.26}$$

or

$$S (t^*) = \frac{C \exp\{-t^*/\tau\} N_1 [1 - \exp\{-Bt^*\}]}{(N_2/n_{20} - \exp\{-t^* f\})}. \tag{5.27}$$

Under the restrictions described, Eq. 5.27 describes the variation in the PTTL peak due to the shallow trap as a function of the illumination time during PTTL stimulation. Note that as $t^* \to \infty$, $n_2 \to 0$, $m \to n_1$ and S will start from 0 at $t^* = 0$, and increase smoothly to a maximum, given by

$$S (\infty) = \frac{C n_{1\infty}^2}{N_2} \tag{5.28}$$

where $n_{1\infty} = m_\infty$ = the final concentrations at $t^* = \infty$. Fig. 5.12 illustrates solution to Eqs. 5.26 and 5.27, compared with a numerical solution to the

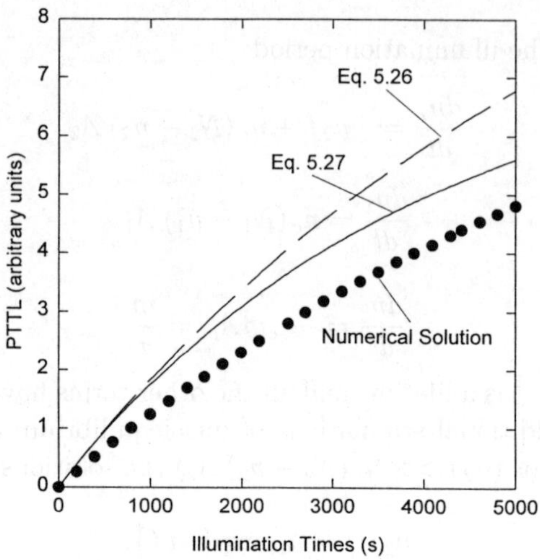

Figure 5.12 Simulated PTTL versus illumination times curves using the simplified expressions for PTTL (Eqs. 5.26 and 5.27) and calculated using numerical solutions to the rate equations, without assumptions. The chosen parameters are: $N_1 = 2\text{x}10^{17}$ m^{-3}; $N_2 = 3\text{x}10^{17}$ m^{-3}; $M = 6\text{x}10^{17}$ m^{-3}; $A_1 = A_2 = A_m = 10^{-15}$ m^{3}s^{-1}.

rate equations without the use of any simplifying assumptions. For the latter calculation, the Eqs. 5.20-5.22 were solved numerically for the illumination phase, and a similar set solved for the heating phase (i.e. by setting $f = 0$ and adding a thermal detrapping term for the shallow traps, i.e. $n_1 s \exp\{-E_t/kT\}$). This is compared with the solution obtained by using the numerical solutions from the illumination phase, but then adopting Eq. 5.26 to estimate the PTTL. These two approaches are compared with the solution obtained by using the initial starting values for the trapped charge populations and adopting Eq. 5.27 to calculate the PTTL. For each of these calculations the initial starting conditions were $n_{10} = 0$ and $n_{20} = m_0 = 10^{17}$ m^{-3}.

The analysis of PTTL based on the simple energy level scheme of Fig. 4.10 results in a PTTL versus time curve which grows monotonically from zero to a maximum level. However, one often finds experimental PTTL versus time curves which show an initial increase up to a maximum, and are then followed by a decrease at longer illumination times. One obvious interpretation of this behavior is that the illumination, in addition to filling the traps in question, is also emptying the traps via optical bleaching. Such processes have already been described for the main TL peak from Al_2O_3 in which transfer of charge from deep traps can occur simultaneously with optical bleaching out of the trap, dependent upon wavelength (see previous section). Sometimes, however,

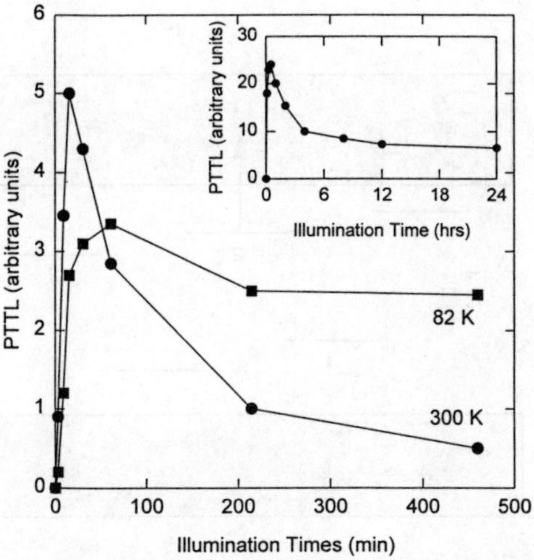

Figure 5.13 PTTL curves, at room temperature and at 82 K, for the 363 K peak in crystalline quartz. The inset shows the behavior of the 270 K peak illuminated with 325 nm at 100 K for 24 hours.
(After Alexander et al.[50].)

it is uncertain whether or not the illumination does in fact optically empty the trap in question. An example of this is the PTTL signal for some of the low temperature (<RT) peaks in crystalline quartz. Typical PTTL versus time curves for some peaks in this material are shown in Fig. 5.13. It was noted above in connection with Fig. 5.4 that McKeever and colleagues have discussed a model which explains how one can get partial bleaching of a TL signal down to a residual level, with no further bleaching of the signal being possible (see section 5.3). A similar model can be used to explain how a PTTL peak can exhibit a growth, and then a decay, *without* invoking the requirement of optical excitation out of the trap. Referring to Fig. 5.14, the model allows for the transfer of charge into the trap in question (in this case, transfer from level 2 to level 1) but does not include the optical emptying of charge from level 1. Nevertheless, an increase in the PTTL signal from level 1, followed by a decrease at longer illumination times is observed.

To explain this result we consider the non-radiative recombination center in Fig. 5.14. This center provides a competing, non-radiative recombination pathway and the conditions for charge neutrality are

$$n_c + n_1 + n_2 + n_3 = m_4 + m_5. \tag{5.29}$$

Here n_1, n_2 and n_3 are the concentrations of electrons in the three electrons traps, and m_4 and m_5 are the concentrations of holes in the radiative and non-

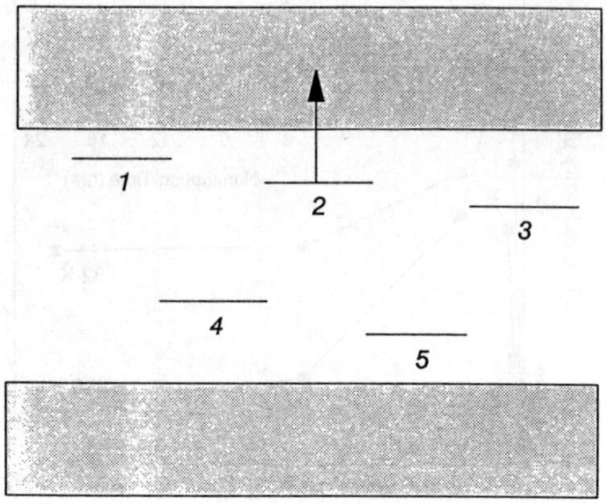

Figure 5.14 Model to explain PTTL, containing a shallow trap (level 1), an optically active trap (level 2), a deep, optically inactive trap (level 3), a radiative recombination center (level 4) and a nonradiative recombination center (level 5).

radiative recombination centers, respectively. The description of PTTL using this model now follows the principle described earlier in section 5.3, in which the reduction in the PTTL as the illumination proceeds is caused by the removal of holes, during the illumination period, from the radiative recombination sites. This removal occurs via recombination and is the source of the OSL signal observed during illumination. Although a similar removal also takes place with the simpler model described above, it is important to recognize that in that model n_1 is always less than (or equal to) m, because of the charge neutrality condition. Thus, if n_1 is increasing during illumination, there are always enough holes available such that the resulting PTTL signal simply follows n_1. However, in the more complex model of Fig. 5.14, n_1 may be less than, or greater than m_4, according to the new charge neutrality condition (Eq. 5.29). This means, that even though n_1 may be increasing, there may, for a given illumination time, no longer be enough holes in the radiative recombination site to accommodate this. Of course, the *total* number of available holes ($m_4 + m_5$) will always be greater than (or equal to) the number of electrons in the shallow traps. Thus, at the start of the illumination, the number of electrons in the shallow traps will be less than the number of holes in the radiative recombination center and the PTTL intensity will grow as n_1 grows. At longer times, however, n_1 may (depending upon the values of the various parameters) become greater then m_4. The PTTL intensity will now follow m_4 and will decrease as m_4 decreases. This reasoning follows that given earlier in Ch. 4 and can be summarized by Eq.

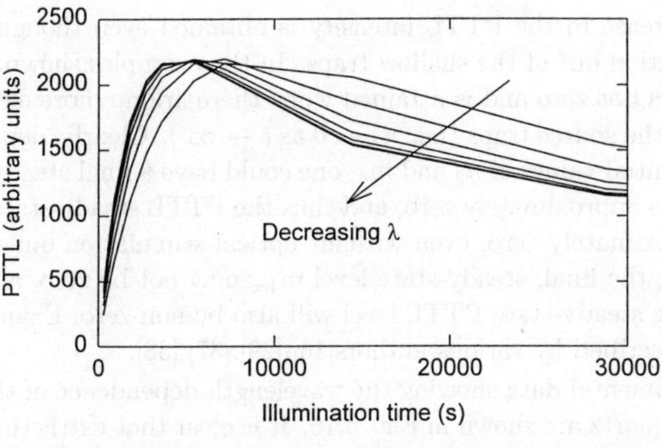

Figure 5.15 Numerical simulations of the PTTL versus illumination time using the model described in the text. The concentrations of the various centers used in this calculation are $N_1 = N_2 = N_3 = M_4 = M_5 = 10^{17}$ (all in m^{-3}). The trapping and recombination probabilities were $A_1 = 10^{-15}$, $A_2 = A_3 = 10^{-16}$, $A_4 = A_5 = 10^{-15}$, $A_{m4} = 10^{-13}$, and $A_{m5} = 10^{-14}$ (all in $m^3 s^{-1}$). The optical excitation rate from the donor trap (level 2) was 10^{-1} s^{-1}. The attempt-to-escape frequency for level 1 was 10^{12} s^{-1}, the trap depth was 0.90 eV, and the temperature was 200 K. (After Alexander et al.[50].)

4.36 which is re-written here, using the notation of this chapter, as

$$S_{PTTL} = \min(n_1, m_4). \tag{5.30}$$

However, as explained in Ch. 4, this simple, hand-waving argument is too simplistic when there is competition and multiple recombination/retrapping pathways for the electrons. In these circumstances the PTTL intensity will not always follow the minimum of n_1 or m_4. In fact, the only way to be sure where the charge is going, and what the resultant PTTL intensity will be for a given set of conditions and parameter values, is to solve the appropriate rate equations numerically. Fig. 5.15 shows the numerical solution to this more complex model for PTTL. The wavelength dependence was built into the model by using the relationship between excitation intensity f, the intensity of the illuminating light ϕ, and the photoionization cross-section σ from deep traps (Eq. 5.9) and the expression for the wavelength dependence of σ (Eq. 5.4).

The solution shown in Fig. 5.15 is typified by an increase in the PTTL intensity at short illumination times, and a decrease at longer times. The exact form of the curve is seen to vary with the wavelength of the light used to stimulate the effect. At shorter wavelengths both the growth and decay are more rapid than at longer wavelengths. The point to observe about the result is that a decrease in the PTTL intensity is obtained even though there is no optical excitation out of the shallow traps. In the examples shown, the steady-state PTTL is not zero and is attained when there are no more electrons to be excited from the source traps (i.e. $n_2 \rightarrow 0$ as $t \rightarrow \infty$). Clearly, depending upon the relative initial values of n_2 and m_4, one could have a final steady-state value of m_4 which is approximately zero, and thus the PTTL steady-state value must also be approximately zero, even without optical stimulation out of the traps. Alternatively, the final, steady-state level $m_{4\infty}$ may not be zero, and hence the corresponding steady-state PTTL level will also be non-zero. Examples of such curves are described by various authors[16],[29],[37],[38].

Experimental data showing the wavelength dependence of the PTTL response from quartz are shown in Fig. 5.16. It is clear that extracting the PTTL excitation spectrum from such data by simply taking the PTTL intensity at a fixed time can be equivocal in that the result will depend upon the time chosen to extract the PTTL intensity. At short illumination times the PTTL after short wavelength illumination is greater than that after long wavelength illumination. However, at longer illumination times, the reverse is true. Nevertheless, published examples of PTTL excitation spectra extracted in this way are common[29],[37],[38],[51].

Using the numerical solutions to the rate equations, as described above, Alexander et al.[50] show that the true excitation spectrum is obtained by taking the initial slope of the PTTL versus time curves. Fig. 5.17 shows the data for three PTTL peaks in quartz. Alexander et al. show that the data can be fitted well with the sum of three curves, each of which is described by Eq. 5.4 with $\gamma = 0.559$ for quartz.

5.5 Optically Stimulated Luminescence (OSL)

5.5.1 *Constant Excitation Intensity (cw OSL)*

OSL has become a popular method for determining equivalent doses for sedimentary deposits in dating applications and is growing in popularity in dosimetric applications. The method was first introduced for dating by Huntley et al.[5], and since then the technique been developed into an important dating procedure. For details of the progress in this arena the reader is referred to the proceedings of the Luminescence and ESR Dating conferences[52]. Applications in dosimetry have been less extensively reported, despite obvious advantages - e.g. no need to heat the sample, avoidance of thermal quenching, use of plastic dosimeters, etc. Occasional descriptions of OSL from a variety of dosimetric materials have been published, primarily describing the general OSL properties and potential use in dosimetry[6],[53]-[55]. In each of these studies the optical excitation is continuous wave (cw) and is either a high power arc lamp source,

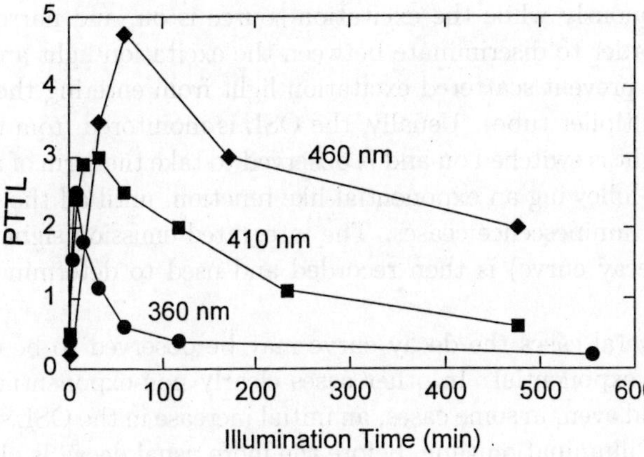

Figure 5.16 The 358 K PTTL peak of a quartz extraction from sand versus illumination time for three different illumination wavelengths. (After Alexander et al.[50].)

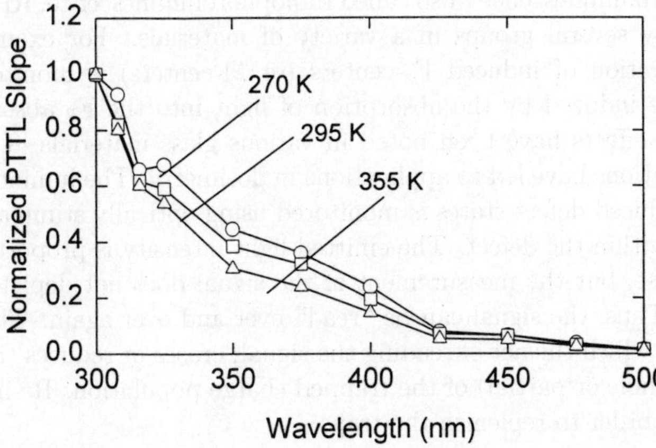

Figure 5.17 Excitation spectra for three PTTL peaks from natural Arkansas quartz. The similarity in the response for all three peaks indicates that the same source trap(s) is(are) being used for these PTTL traps over this wavelength range. (After Alexander et al.[50].)

along with a monochromator or filter system to select the excitation wavelength, or a laser operating at or near the desired wavelength. The luminescence is monitored continuously while the excitation source is on, and narrow band filters are used in order to discriminate between the excitation light and the emission light, and to prevent scattered excitation light from entering the detector (e.g. the photomultiplier tube). Usually, the OSL is monitored from the instant the excitation light is switched on and is observed to take the form of a luminescence decay curve, following an exponential-like function, until all the traps are emptied and the luminescence ceases. The integrated emission signal (i.e. the area under the decay curve) is then recorded and used to determine the absorbed dose.

In several cases the decay curve may be observed to be exactly, or approximately, exponential. In other cases clearly non-exponential decay curves are found, and even, in some cases, an initial increase in the OSL signal is seen as a function of illumination time, before the more usual decay is observed. These varying behaviors suggest a multiplicity of possible recombination mechanisms for the production of OSL, some of which are discussed in the next section.

It is worthwhile comparing OSL with another, related technique which works in a similar (but different) way and which is also used in dosimetry. We noted above that photoluminescence monitors transitions of type (5) in Fig. 5.1, whereas OSL monitors transitions of type (4a) and (4b). Radiation-induced photoluminescence (also called radiophotoluminescence, RPL) has been monitored by several groups in a variety of materials. For example, in LiF the concentration of induced F_2-centers (or M-centers) is monitored by the luminescence induced by the absorption of light into the F_2 absorption band [56]. Similar effects have been noted in various glass materials [8],[9],[57] and both observations have led to applications in dosimetry. The luminescence from radiation-induced defect states is monitored using optically stimulated internal transitions within the defect. The emitted light intensity is proportional to the absorbed dose, but the measurement of the signal does not deplete or remove the defect. Thus, the signal can be "read" over and over again. This contrasts with OSL in which the act of reading the signal, erases or reduces the signal, by depletion (whole or partial) of the trapped charge population. Re-irradiation is necessary in order to replenish the traps.

OSL Models

In general, the shape of an OSL decay curve does not lend itself to analysis using simple descriptions. The decay is usually non-exponential, typically exhibiting a long tail at long illumination times. Under some circumstances, and for some samples, the OSL may even display an initial slow increase after the illumination is applied, followed by the more usual decrease at longer times. Overall, the decay shape is dependent upon the sample, the absorbed dose, the illumination intensity and the temperature. An earlier analysis of the infra-red stimulated OSL decay from a collection of feldspar samples by Bailiff and Poolton[58] illustrated that, for microcline, sanidine and albite, the decay followed a $A(1+Bt)^{-P}$ law, with $P = 1$ and where A and B are constants and t is time. Furthermore, the constant A was observed to be exponentially related to temperature, suggesting a thermally activated process as previously proposed by Hütt and colleagues[24]. In contrast, for labradorite the OSL followed a $(1+Bt)^{-P}$ law and was independent of temperature, and $P = 2$. Later measurements on potassium feldspars by Bailiff and Barnet[59] gave values of P varying between 1 and 2. Poolton et al.[60] suggested that the infra-red stimulated OSL from feldspar is influenced by localized, donor-acceptor pair recombination. Evidence for this was obtained by observing the decay of the OSL signal after successively longer exposures of an irradiated sample to green light. For long green light exposures, the IR-stimulated OSL displayed a clear increase as a function of IR illumination time, up to a maximum, before a decay was observed. The longer the green light exposure, the longer the IR-illumination time required for the IR-induced luminescence to reach a maximum. This was interpreted as being caused by an increase in the mean donor-acceptor distances for lower concentrations of donors and acceptors. It should be stressed that donor-acceptor recombination is not predicted for OSL stimulated by green light for which delocalized transport via the conduction band is believed to be the case.

Smith and Rhodes[61] attempted to fit the decay of the green light stimulated OSL from quartz using a summation of three exponentials. One might expect a multi-component OSL decay curve if more than one trap is being emptied at the illumination wavelength being used - although all available data indicate that at 514 nm (the wavelength used by the above authors to stimulate the OSL) only one trap is being emptied, at least at short exposure times (e.g.[22],[50],[62]). Smith and Rhodes[61] also note an increase in the OSL decay rate as the temperature increases. The thermal dependence of the OSL from quartz was studied further by Spooner[22] who presented evidence for a thermally assisted process for OSL production.

A long tail is often observed in the OSL decay curve. This may be caused by either the influence of shallow traps - which localize charges released during illumination and slowly release them again at a rate determined by the trap depth and the sample temperature - or a contribution from traps which are emptying slowly at the excitation wavelength used in the experiment. The shallow traps may also yield an optically stimulated phosphorescence component to the OSL decay, and they give rise to a temperature dependent OSL component, since at higher temperatures the shallow traps become less effective at trapping

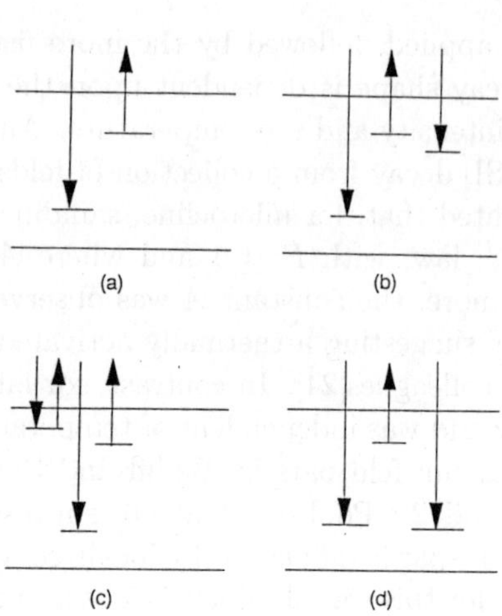

Figure 5.18 (a) Simplest model for OSL, involving one electron trap, one hole trap (which acts as a radiative recombination center) and transport through the conduction band. (b) Model containing an additional, deep competing trap. (c) Model containing a shallow, competing trap. (d) Model containing two recombination sites - one radiative and one non-radiative.

charge. These effects were shown in computer simulations by McKeever and colleagues[63],[64] and were demonstrated experimentally for Al_2O_3 by Markey et al.[7],[65].

Most models for OSL assume transport of the optically excited charge through the delocalized bands in order to reach the recombination site. Thus, the simplest model by which OSL can be produced is shown in Fig. 5.18(a). Here light stimulates trapped electrons, concentration n, into the conduction band at rate f, followed by recombination with trapped holes, concentration m, to produce OSL of intensity I_{OSL}. With the usual definitions, the rate equation describing the charge flow is

$$\frac{dn_c}{dt} = -\frac{dn}{dt} + \frac{dm}{dt} \tag{5.31}$$

which can be derived from the charge neutrality condition

$$n_c + n = m. \tag{5.32}$$

With the assumptions of quasiequilibrium ($dn_c/dt << dn/dt$, dm/dt and $n_c << n$, m) and negligible retrapping we have

$$I_{OSL} = -\frac{dm}{dt} = -\frac{dn}{dt} = nf \tag{5.33}$$

the solution of which is

$$I_{OSL} = n_0 f \exp\{-tf\} = I_0 \exp\{-t/\tau\}. \tag{5.34}$$

Here, n_0 is the initial concentration of trapped electrons at time $t = 0$, I_0 is the initial luminescence intensity at $t = 0$, and $\tau = 1/f$ is the decay constant. One can observe a straightforward relationship in which the initial intensity is directly proportional to the excitation rate and the decay of the OSL with time is a simple exponential. Recall also that the excitation rate is given by the product of the excitation intensity ϕ and the photoionization cross-section ($f = \phi\sigma$).

For two optically active traps (concentrations n_1 and n_2 and excitation rates f_1 and f_2) the same assumptions yield

$$\frac{dm}{dt} = -\frac{dn_1}{dt} - \frac{dn_2}{dt}. \tag{5.35}$$

Along with the charge neutrality condition of $n_1 + n_2 = m$ we have

$$n_1 = n_{10} \exp\{-tf_1\} \tag{5.36}$$

and

$$n_2 = n_{20} \exp\{-tf_2\}. \tag{5.37}$$

Thus

$$I_{OSL} = n_{10} f_1 \exp\{-tf_1\} + n_{20} f_2 \exp\{-tf_2\} = I_{1o} \exp\left\{-\frac{t}{\tau_2}\right\} + I_{20} \exp\left\{-\frac{t}{\tau_2}\right\} \tag{5.38}$$

and the OSL decay curve is the sum of two exponentials.

Alternatively, if the additional trap is optically (and thermally) inactive - i.e. it acts as a deep, competing trap only (Fig. 5.18(b)), the OSL intensity may be written

$$I_{OSL} = n_{10} f \exp\{-tf\} - \frac{dn_2}{dt} \tag{5.39}$$

where

$$\frac{dn_2}{dt} = n_c (N_2 - n_2) A_2, \tag{5.40}$$

adopting the standard definitions. With an additional assumption of $N_2 >> n_2$ then $n_c N_2 A_2 \approx$ constant, C. Therefore

$$I_{OSL} = n_{10} f \exp\{-tf\} - C \tag{5.41}$$

and the OSL is reduced in intensity by the extent of the retrapping into the deep traps. In the limit, as $t \to \infty$, $n_c \to 0$ and therefore $C \to 0$. Thus, C is in fact a very slowly varying function of time.

If the competing traps are shallow traps which are thermally metastable at the temperature of the OSL measurement (Fig. 5.18(c)), Eq. 5.40 becomes

$$\frac{dn_2}{dt} = n_c (N_2 - n_2) A_2 - n_2 p, \tag{5.42}$$

where p is the rate of thermal excitation out of the trap. Now we have

$$I_{OSL} = n_{10} f \exp\{-tf\} + n_2 p - n_c (N_2 - n_2) A_2. \tag{5.43}$$

The last two terms in Eq. 5.43 combine to produce a long-lived, temperature-dependent tail to the OSL decay. The form of this component will be an initial increase, followed by a decrease at longer times. Depending upon the relative size of this component compared with the first term, the overall OSL decay curve may exhibit an initial increase, followed by a decrease. The relative size of the two components also depends upon the excitation rate f such that at low values of f the temperature-dependent term may be significant.

For the case of two recombination centers (Fig. 5.18(d)), one of which (m_1) is radiative and the other (m_2) is non-radiative, we have $n = m_1 + m_2$, $n = n_0 \exp\{-tf\}$, and

$$I_{OSL} = n_0 f \exp\{-tf\} - \frac{dm_2}{dt}. \tag{5.44}$$

As with the case of the deep traps, the OSL intensity is reduced by the existence of a non-radiative pathway. Since

$$\frac{dm_1}{dt} = -n_c m_1 A_{m1} \tag{5.45}$$

and

$$\frac{dm_2}{dt} = -n_c m_2 A_{m2} \tag{5.46}$$

then, with quasiequilibrium ($dn_c/dt \approx 0$) we have

$$m_1 \approx m_{10} \exp\{-tn_c A_{m1}\} \tag{5.47}$$

and

$$m_2 \approx m_{20} \exp\{-tn_c A_{m2}\} \tag{5.48}$$

from which we have

$$\frac{m_1}{m_2} \approx \frac{m_{10}}{m_{20}} \exp\{-tn_c [A_{m1} - A_{m2}]\} \tag{5.49}$$

from which we see that the ratio m_1/m_2 is a time-dependent function. However, if $A_{m1} = A_{m2}$, then $m_1/m_2 =$ a constant, k. In these circumstances the charge neutrality condition is written

$$m_1 + \frac{m_1}{k} = n \tag{5.50}$$

or

$$m_1 = \frac{n}{1 + 1/k} = \frac{n}{K}. \tag{5.51}$$

From this

$$\frac{dm_1}{dt} = \frac{1}{K}\frac{dn}{dt} = -\frac{nf}{K} \tag{5.52}$$

and

$$I_{OSL} = \frac{1}{K} n_0 f \exp\{-tf\} = I_0' \exp\{-t/\tau\}. \tag{5.53}$$

In this way we see that the OSL decay curve may still be exponential in the case of two types of recombination center, but only if the recombination cross-sections are the same. Comparing Eq. 5.34 with Eq. 5.53 we observe that, as expected, the latter case describes a weaker OSL signal (by a factor $1/K$).

An alternative approach is to assume that retrapping into the optically active trap is allowed. For the simple case of one-trap and one-center we may write

$$I_{OSL} = \frac{dm}{dt} = -\frac{dn}{dt} = nf - n_c (N - n) A \tag{5.54}$$

and again the decay is no longer exponential. For the case of $N >> n$, $A/A_m = R$ and $R >> n/(N - n)$, a bimolecular (second-order) function results, thus

$$I_{OSL} = \frac{n^2 f}{NR} = \frac{dm}{dt} = -\frac{dn}{dt}. \tag{5.55}$$

Re-writing

$$\frac{dn}{n^2} = -\frac{f}{NR} dt \tag{5.56}$$

which yields, after integration,

$$I = I_0 \left(1 - \frac{n_0 f t}{NR}\right)^{-2} \tag{5.57}$$

where $I_0 = n_0^2 f/NR$. For the more general case, where $I = n^b f/NR$, the OSL decay curve is described by

$$\left(\frac{I}{I_0}\right)^{\frac{1-b}{b}} = \left(1 - \frac{n_0 f t}{NR}\right) \tag{5.58}$$

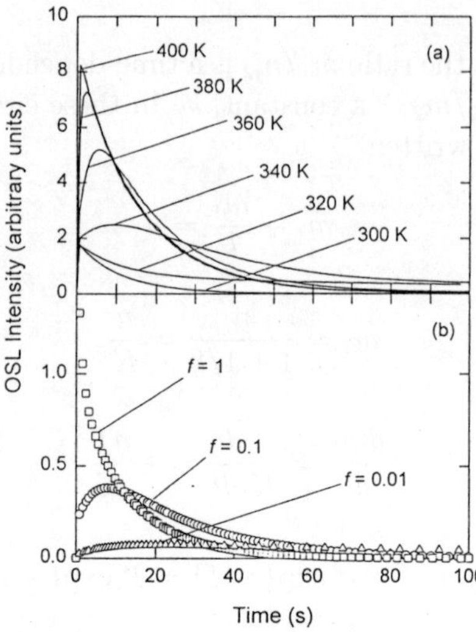

Figure 5.19 Theoretical OSL curve shapes obtained from a solution to the rate equations described by Fig. 5.14. (a) Variation as a function of temperature. (b) Variation as a function of excitation rate.

(After McKeever et al.[64].)

or

$$I = I_0 \left(1 - \frac{n_0 f t}{NR}\right)^{-\frac{b}{1-b}}. \tag{5.59}$$

It must be stressed that a real material will likely have several optically active traps, and a combination of deep competing traps, shallow competing traps, and both radiative and non-radiative centers. In consideration of this it should be recognized that a simple exponential shape for the OSL decay curve will only occur in real materials under certain special circumstances and sets of experimental conditions. The curve shape will, in general, be temperature and excitation rate dependent.

Examples of OSL decay curves calculated by solving numerically the rate equations for the model of Fig. 5.14 are shown in Fig. 5.19. At low temperatures, the shallow traps, deep traps and non-radiative recombination centers conspire to produce a weak, non-exponential OSL decay curve. As the temperature increases, so the shallow traps begin to play less of a role. The OSL intensity increases accordingly and a peak appears in the OSL curve at short times. At the highest temperatures the shallow traps play no role and the intensity is the highest. The deep traps and non-radiative centers still produce a non-exponential decay curve, however. Similar results to these are obtained experimentally, and examples are shown for quartz in Fig. 5.20. Although the

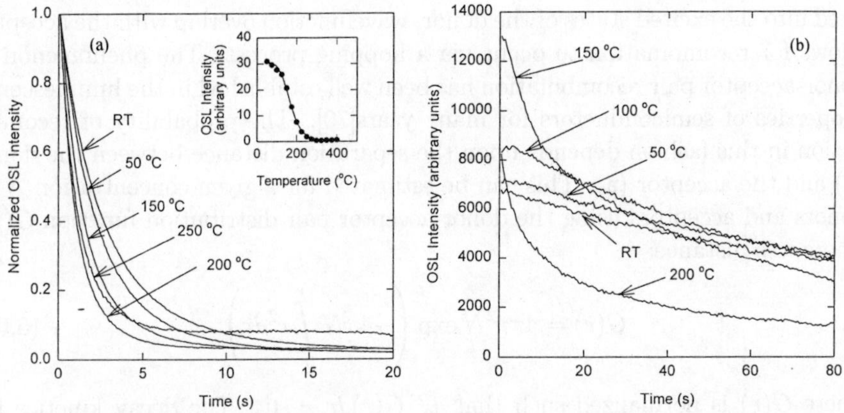

Figure 5.20 Experimental OSL data for quartz. (a) High excitation intensity OSL data as a function of temperature. Thermal quenching causes the overall emission to decrease with temperature, rather than the expected increase as predicted in Fig. 5.19. (b) At lower excitation powers, a small peak can be observed in the OSL curve at short times and for intermediate temperatures.
(After McKeever et al.[64].)

luminescence emission from quartz quenches as the temperature increases[66], the same basic behavior as predicted in Fig. 5.19 can be seen.

The preceding analyses presume that the released charge is transported via the delocalized bands to the recombination site. As such, the analyses will be valid whether or not the optical transition is direct to the conduction band, or is thermally assisted via an excited state[24],[25]. The latter process merely imposes a temperature dependence on the excitation rate, thus

$$f = f_o \exp\left\{\frac{W}{kT}\right\} \tag{5.60}$$

where W is the required thermal activation energy. Since the OSL decay constant $\tau = f^{-1}$, then the OSL decay rate will also be thermally activated according to

$$\tau = \tau_o \exp\left\{-\frac{W}{kT}\right\}. \tag{5.61}$$

This provides a means by which thermally assisted processes may be recognized experimentally.

However, Poolton et al.[26]-[28] suggest that IR-stimulated OSL from feldspars may not proceed via the conduction band, but instead may occur via a donor-acceptor recombination mechanism. These authors argue that once excited into the excited states of the donor, wavefunction overlap with the acceptor allows for recombination to occur via a hopping process. The phenomenon of donor-acceptor pair recombination has been well established in the luminescence properties of semiconductors for many years[70]. The probability of recombination in this fashion depends upon the separation distance between the donor (d) and the acceptor (a). This can be estimated for a given concentration N of donors and acceptors using the donor-acceptor pair distribution function $G(r)$, where r is distance:

$$G(r) = 4\pi r^2 N \exp\left(-4\pi N \int_0^r r^2 dr\right) \tag{5.62}$$

where $G(r)$ is normalized such that $\int_0^\infty G(r)dr = 1$. The decay kinetics for d-a recombination luminescence are complex and depend critically upon concentration, with faster decays occurring for higher concentrations. For high concentrations the decay proceeds as t^{-2} at long times, while it proceeds as t^{-1} for low concentrations. At short times the decay is much faster and cannot be expressed analytically[70].

5.5.2 Non-Constant Excitation Intensity

Linearly Increasing Intensity

In all of the above analysis, f was assumed to be constant. The measurement of OSL, however, does not require that the intensity of the illuminating light be maintained constant. A novel method for observing OSL curves was described by Bulur[67] who considered the case of a linearly increasing intensity, according to

$$f(t) = gt \tag{5.63}$$

where the constant g has units of $m^{-3}s^{-2}$. Thus, for the simple model (Fig. 5.18(a)) with quasiequilibrium and no retrapping, one has

$$I_{OSL} = -\frac{dn}{dt} = gtn \tag{5.64}$$

from which

$$n = n_0 \exp\left\{-\frac{gt^2}{2}\right\} \tag{5.65}$$

and

$$I_{OSL} = n_0 gt \exp\left\{-\frac{gt^2}{2}\right\}. \tag{5.66}$$

The OSL curve shape now starts from zero at $t = 0$ and exhibits a maximum at $t = t_m = \sqrt{1/g}$. The OSL maximum is $n_0\sqrt{g}\exp\{-1/2\}$. Fig. 5.21

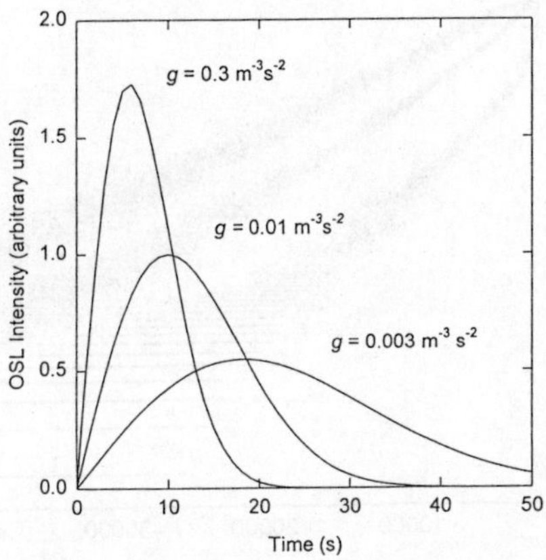

Figure 5.21 Calculated OSL curves for a linearly increasing excitation power, obtained from Eq. 5.66, for three values of g.

demonstrates several OSL curves predicted for a linearly increasing excitation intensity, obtained with different rates g, using Eq. 5.66.

For more general kinetics in which retrapping is allowed we have

$$I_{OSL} = -\frac{dn}{dt} = \frac{g}{n_0^{b-1}} t n^b \tag{5.67}$$

where b is a dimensionless, positive, real parameter ($b > 0$, $b \neq 1$) representing the kinetic order. The OSL intensity is now

$$I_{OSL} = n_0 g t \left[(b-1) \frac{g t^2}{2} + 1 \right]^{b/(1-b)} . \tag{5.68}$$

The OSL maximum appears at $t = t_m = \sqrt{2/g\,(b+1)}$, and the maximum value is $(2n_0/(b+1))\,(1/t_m)\,(2b/(b+1))^{b/(1-b)}$. For the specific case of second-order kinetics ($b = 2$) we have $t_m = \sqrt{2/3g}$ with a maximum of $3n_0/8t_m$.

Pulsed OSL

With pulsed OSL the excitation intensity is periodically pulsed, usually on a time scale (i.e. frequency and pulse width) appropriate to the time constants of the processes being observed. Sanderson and Clark[68] and Clark and Bailiff[69]used 10 ns laser pulses to observe fast OSL processes in feldspar samples. Example data from the latter authors are illustrated in Fig. 5.22 in

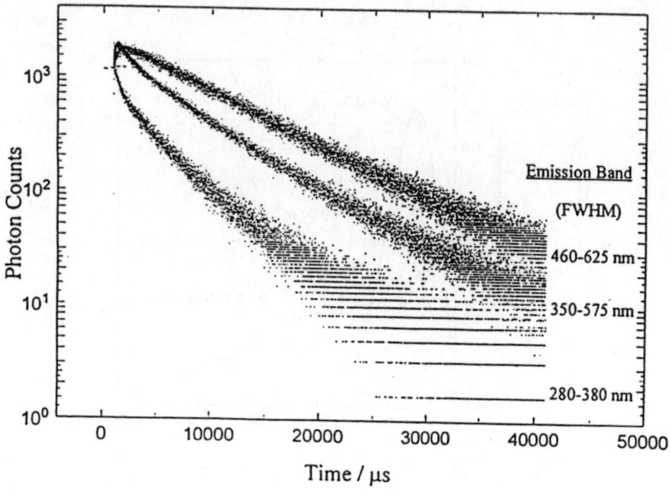

Figure 5.22 Pulsed OSL decay curves in three wavelength ranges from feldspar. The decays can be resolved into multiple exponential components with lifetimes varying from 2 ns to 300 μs.

(After Clark et al.[69].)

which the OSL decay after the cessation of the laser pulse is observed. Several separate processes could be resolved. Markey et al.[7],[65] examined the pulsed OSL from Al_2O_3:C. Since the emission center in this material is an F-center, with a lifetime of ~35 ms, the pulse width used by these authors was 100 ms. The OSL emission was followed both during and after the pulse and from the temperature dependence of the OSL decay rate the influence of shallow traps on the OSL process in this material was evaluated. Fig. 5.23 illustrates a pulsed OSL curve from Al_2O_3:C in response to a 100 ms pulse of light from an Ar-ion laser.

5.5.3 *Excitation Spectra*

An important parameter for the measurement of OSL is the wavelength for the excitation. The optimum (i.e. the most efficient) wavelength can be determined from an excitation spectrum. Fig. 5.24(a) shows the OSL excitation spectrum for quartz, shown on a log(OSL)-wavelength plot. Fig. 5.24(b) shows similar excitation spectra for feldspars. The resonance at about 1.4 eV in the latter spectrum is consistent with the model of Hütt and colleagues[24],[25] who proposed a thermally assisted transition into the conduction band from an excited state. The calculated thermal activation energies are also shown in the figure and the position of the resonance indicates that the excited state is approximately 1.4 eV above the ground state. Huntley and colleagues[62],[72] analyze

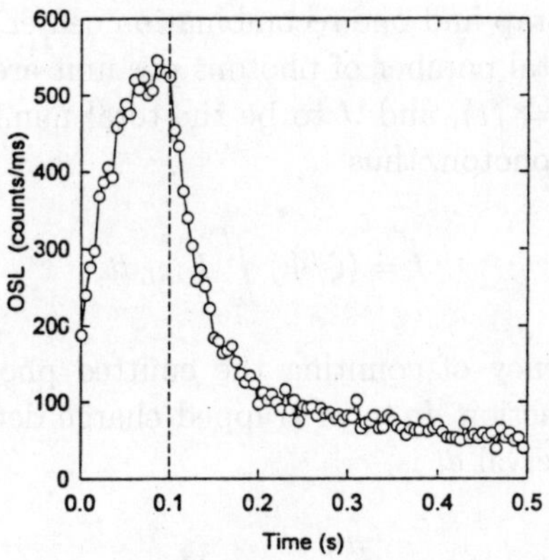

Figure 5.23 Pulsed OSL curve from Al_2O_3:C due to a 100 ms pulse of excitation from an Ar-ion laser at room temperature showing the build-up and decay of the OSL. The build-up has a lifetime of 35 ms. The decay of the OSL after the pulse is made up of a 35 ms component and a slower component (545 ms at room temperature). The lifetime of the slow component is temperature dependent and is due to trapping into shallow traps.

(After Markey et al.[7].)

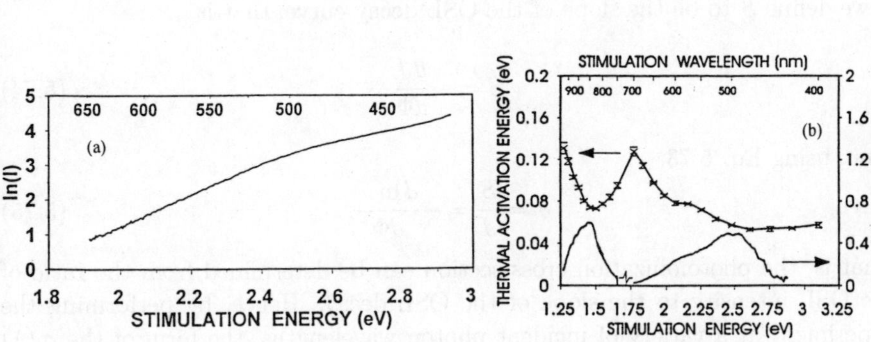

Figure 5.24 Excitation spectra for (a) quartz, and (b) amelia albite. (Also shown in the latter diagram is the thermal activation energy required for the production of OSL from amelia albite, as a function of the stimulation wavelength.)

(After Bøtter-Jensen et al.[71].)

OSL decay curves to determine the photoionization cross-section. Consider the simple model of one trap and one recombination center, with weak retrapping. Define Φ to be the total number of photons per unit area incident on the sample in time t (i.e. $\Phi = ft$), and J to be the total number of emitted photons counted per incident photon, thus

$$J = (\zeta/\Phi) \int_0^t I_{OSL} dt, \tag{5.69}$$

where ζ is the efficiency of counting the emitted photons. From Eq. 5.33 it follows that the fraction dn/n of trapped charge detrapped by an incident photon flux $d\Phi$ in interval dt is

$$\frac{dn}{n} = -\sigma d\Phi, \tag{5.70}$$

the solution of which is

$$n = n_0 \exp\{-\sigma\Phi\} \tag{5.71}$$

where n_0 is the initial trapped electron concentration. From the above definitions

$$J = -\zeta \frac{dn}{d\Phi} = n\sigma \tag{5.72}$$

and hence

$$J = \sigma n_0 \exp\{-\sigma\Phi\} = J_0 \exp\{-\sigma\Phi\}. \tag{5.73}$$

If we define S to be the slope of the OSL decay curve, that is

$$S = -\frac{dJ}{d\Phi} \tag{5.74}$$

then, using Eq. 5.73,

$$\sigma = \frac{S}{J} = -\frac{d \ln J}{d\Phi}. \tag{5.75}$$

That is, the photoionization cross-section can be determined from the ratio of the OSL intensity to the slope of the OSL decay. Hence, by performing the experiment at a variety of incident photon wavelengths, the form of the $\sigma(\lambda)$ curve can be determined. Huntley et al.[72] perform this analysis for quartz and obtain a curve similar to that shown in Fig. 5.24(a), with values of σ ranging from 10^{-24} to 10^{-21} m^2 over the incident photon energy range from 1.9 to 2.7 eV. Furthermore, for a single trap, this ratio should be constant throughout the decay curve, assuming that the simple exponential decay expected from Eq. 5.33 prevails.

5.6 Optical Generation

The optical transitions important in OSL are those indicated by arrows 4a and 4b in Fig. 5.1. However, if an unirradiated sample is exposed to light transitions of the type indicated by arrows 3a and 3b may occur. Some examples are the excitation of electrons due to the absorption of light by deep donors in sillenite crystals[3],[73],[74], the excitation of F-center electrons due to the absorption of F-band light in Al_2O_3 crystals[75], the excitation of F-center electrons by the absorption of F-band light in LiF crystals[2], and many others. In each of these instances the electron ground state is located below the Fermi level in the equilibrium configuration. Absorption of the appropriate wavelength of light excites the electrons into the conduction band from where they become trapped. When subsequently heated, the electrons become detrapped and recombination occurs. However, such a process does not necessarily yield TL. For example, absorption of light by the donor centers in bismuth germanium oxide (BGO), or bismuth silicon oxide (BSO) releases electrons, but no free holes are released. During heating, a thermally stimulated conductivity (TSC) signal is induced, but a TL signal is not observed[3]. Evidently, the electron hole recombination at the original donor site is non-radiative. In Al_2O_3, however, a similar process occurs in which F-centers are ionized when illuminated with UV light (205 nm) according to $F \rightarrow F^+ + e^-$. The electrons become trapped and, once released by thermal excitation (at approximately 450 K) they induce TSC and recombine with the F^+-centers forming excited F-centers. Relaxation of the excited F-centers yields light emission (TL) at 420 nm[75],[76]

As noted earlier, Chen et al.[18] describe a mechanism for the bleaching of a TL peak to a non-zero, equilibrium level based on a mechanisms such as that described above for Al_2O_3. The light is considered to both empty the electron traps by bleaching, and, at the same time, fill the electron traps by the excitation of free electrons from the recombination sites.

5.7 References

1. H.L. Cotal, B.G. Markey, S.W.S. McKeever, G. Cantwell and W.C. Harsch, *Physica B*, **185** (1993) 103.
2. S.W.S. McKeever, *Thermoluminescence of Solids*, Cambridge University Press, Cambridge (1985).
3. D. Bloom and S.W.S. McKeever, *J. Appl. Phys.* **77** (1995) 6511, 6521.
4. A.G. Wintle and D.J. Huntley, *Can. J. Earth Sci.* **17** (1980) 348.
5. D.J. Huntley, D.L. Godfrey-Smith and M.L.W. Thewalt, *Nature* **313** (1985) 105.
6. A.S. Pradhan, Bhuwan Chandra and R.C. Bhatt, *Radiat. Prot. Dosim.* **5** (1983) 159.

7. B.G. Markey, L.E. Colyott and S.W.S. McKeever, *Radiat. Meas.* **24** (1995) 457.
8. E. Piesch, B. Burgkhardt and M. Vilgis, *Radiat. Prot. Dosim.* **33** (1990) 215.
9. B. Lommler, E. Pitt, A Scharmann and S. Gudebakke, *Radiat. Prot. Dosim.* **47** (1993) 285.
10. J.L. Merz and P.S. Pershan, *Phys. Rev.* **162** (1967) 217, 235.
11. N.F. Mott and R.W. Gurney, *Electronic Processes in Ionic Crystals*, Clarendon Press, Oxford (1948).
12. K.W. Böer, *Survey of Semiconductor Physics*, Van Nostrand Reinhold, New York (1990).
13. H.G. Grimmeiss and L.-Å. Ledebo, *J. Appl. Phys.* **46** (1975) 2155.
14. A. M Stoneham, *Theory of Defects in Solids*, Clarendon Press, Oxford (1975).
15. F.D. Walker, L.E. Colyott, N. Agersnap Larsen and S.W.S. McKeever, *Radiat. Meas.* **26** (1996) 711.
16. M.F. Morris and S.W.S. McKeever, *Radiat. Meas.* **23** (1994) 323.
17. S.W.S. McKeever, *Radiat. Meas.* **23** (1994) 267.
18. R. Chen, W.F. Hornyak and V.K. Mathur, *J. Phys. D: Appl. Phys.* **23** (1990) 724.
19. A.I. Shlukov, S.A. Shakhovets, L.T. Voskovskaya and M.G. Lyashenko, *Nucl. Instrum. Meth. Mater. Res.* **73** (1993) 373.
20. S.W.S. McKeever, *Nucl. Tracks Radiat. Meas.* **18** (1991) 5.
21. N.A. Spooner, J.R. Prescott and J.T. Hutton, *Quat. Sci. Rev.* **7** (1988) 325.
22. N.A. Spooner, *Radiat. Meas.* **23** (1994) 593.
23. C.M. Sunta and S. Watanabe, *J. Phys. D: Appl. Phys.* **9** (1976) 1271.
24. G. Hütt, I. Jaek and J. Tchonka, *Quat. Sci. Rev.* **7** (1988) 381.
25. G. Hütt and I. Jaek, *Nucl. Tracks Radiat. Meas.* **21** (1993) 95.
26. N.R.J. Poolton, L. Bøtter-Jensen and G.A.T. Duller, *Radiat. Meas.* **24** (1995) 57.
27. N.R.J. Poolton, L. Bøtter-Jensen and O. Johnsen, *Radiat. Meas.* **24** (1995) 531.
28. N.R.J. Poolton, L. Bøtter-Jensen and O. Johnsen, *J. Phys. C: Condens. Matter* **7** (1995) 4751.
29. V.K. Jain, in *Thermoluminescence and Thermoluminescent Dosimetry, Vol II* , Y.S. Horowitz, editor (1984) 173.
30. Y.S. Horowitz, *Phys. Med. Biol.* **26** (1981) 766.
31. I.K. Bailiff, S.G.E. Bowman, S.K. Mobbs and M.J. Aitken, *J. Electrostat.* **3** (1977) 269.
32. S.G.E. Bowman, *Eur. PACT J.* **3** (1979) 381.

33. M.J. Aitken, *Phys. Rep.* **40C** (1978) 277.

34. G. Walner, E. Wild, H. Aref-Azar and P. Hille, *Radiat. Prot. Dosim.* **34** (1990) 69.

35. S.D. Miller, G.W.R. Endres, J.C. McDonald and K.L. Swinth, *Radiat. Prot. Dosim.* **25** (1988) 201.

36. P.A. Eschbach and S.D. Miller, *Radiat. Prot. Dosim.* **47** (1993) 289.

37. I. Milanovich-Reichhalter and N. Vana, *Radiat. Prot. Dosim.* **33** (1990) 211.

38. I. Milanovich-Reichhalter and N. Vana, *Nucl. Tracks and Radiat. Meas.* **18** (1991) 67.

39. A.E. Stoddard, *Phys. Rev.* **120** (1960) 114.

40. A. Bossachi, R. Fieschi and P. Scaramelli, *Phys. Rev. B*, **10** (1965) 5235.

41. R. Fieschi and P. Scaramelli, in *Thermoluminescence of Geological Materials*, D.J. McDougall, editor, Academic Press, London (1968) 291.

42. J. Tournon and P. Bergé, *Phys. Stat. Sol.* **5** (1964) 117.

43. J. Nahum and A. Halperin, *J. Phys. Chem. Sol.* **24** (1963) 823.

44. M. Schlesinger, *J. Phys. Chem. Sol.* **26** (1965) 1761.

45. W.G. Buckman and M.R. Payne, *Health Phys.* **31** (1976) 501.

46. V.K. Jain, *Phys. Stat. Sol. (a)*, **60** (1980) 531.

47. F. Sagastibelza and J.L. Alvarez-Rivas, *J. Phys. C: Sol. Stat. Phys.* **14** (1981) 1873.

48. L. Oster, D. Weiss and N. Kristianpoller, J. Phys. D: Appl. Phys. **27** (1994) 1732.

49. L.E. Colyott, M.S. Akselrod and S.W.S. McKeever, *Radiat. Prot. Dosim.* **65** (1996) 263.

50. C.S. Alexander, M.F. Morris and S.W.S. McKeever, *Radiat. Meas.* **27** (in press, 1997).

51. N. Kristianpoller, M. Abu-Rayya and R. Chen, *Radiat. Prot. Dosim.* **47** (1993) 37.

52. The proceedings which follow the development of OSL in dating are particularly those of the meetings at Cambridge (*Nucl. Tracks and Radiat. Meas.* **14** (1988)); Clermont-Ferrand (*Nucl. Tracks and Radiat. Meas.* **18** (1992)); Krems (*Radiat. Meas.* **23** (1994), and Canberra (*Radiat. Meas.* **27** (1997)).

53. P. Allen and S.W.S. McKeever, *Radiat. Prot. Dosim.* **33** (1990) 19.

54. H. Nanto, K. Murayama, T. Usada, S. Taniguchi and N. Takeuchi, *Radiat. Prot. Dosim.* **47** (1993) 281.

55. H. Nanto, T. Usada, K. Murayama, S. Nakamura, K. Inabe and N. Takeuchi, *Radiat. Prot. Dosim.* **47** (1993) 293.

56. S.D. Miller and G.W.R. Endres, *Radiat. Prot. Dosim.* **33** (1990) 59.

57. B. Burgkhardt, E. Piesch, M. Vilgis, T. Ishidoya and T. Ikegami, *Radiat. Prot. Dosim.* **34** (1990) 369.
58. I.K. Bailiff and N.R.J. Poolton, *Nucl. Tracks Radiat. Meas.* **18** (1991) 111.
59. I.K. Bailiff and S.M. Barnet, *Radiat. Meas.* **23** (1994) 541.
60. N.R.J. Poolton, L. Bøtter-Jensen, P.J.M. Ypma and O. Johnsen, *Radiat. Meas.* **23** (1994) 551.
61. B.W. Smith and E.J. Rhodes, *Radiat. Meas.* **23** (1994) 329.
62. C. Ditlefsen and D.J. Huntley, *Radiat. Meas.* **23** (1994) 675.
63. S.W.S. McKeever and M.F. Morris, *Radiat. Meas.* **23** (1994) 301.
64. S.W.S. McKeever, L. Bøtter-Jensen, N. Agersnap Larsen and G.A.T. Duller, *Radiat. Meas.* **27** (in press, 1997).
65. B.G. Markey, S.W.S. McKeever, M.S. Akselrod, L. Bøtter-Jensen, N. Agersnap Larsen and L.E. Colyott, *Radiat. Prot. Dosim.* **65** (1996) 185.
66. A.G. Wintle, Geophys. *J.R. Astr. Soc.* **41** (1975) 107.
67. E. Bulur, *Radiat. Meas.* **26** (1996) 701.
68. D.C.W. Sanderson and R.J. Clark, *Radiat. Meas.* **23** (1994) 633.
69. R.J. Clark, I.K. Bailiff and M.J. Toolley, *Radiat. Meas.* **27** (in press, 1997).
70. P.J. Dean, *Prog. Solid State Chem.* **8**, (1973) 1.
71. L. Bøtter-Jensen, G.A.T. Duller and N.R.J. Poolton, *Radiat. Meas.* **23** (1994) 613.
72. D.J. Huntley, M.A. Short and K. Dunphy, *Can. J. Phys.* **74** (1996) 81.
73. J.J. Martin, I. Földvári and C.A. Hunt, *J. Appl. Phys.* **70** (1991) 7554.
74. R. Oberschmid, *Phys. Stat. Sol. (a)* **89** (1985) 263.
75. G.P. Summers, *Radiat. Prot. Dosim.* **8** (1984) 69.
76. D.W. Cooke, H.E. Roberts and C. Alexander, *J. Appl. Phys.* **49** (1978) 3451.

Chapter 6
MATHEMATICAL CONSIDERATIONS

6.1 The Integrals in the Theory of Thermoluminescence and Other TSPs

As explained in the preceding chapters, in the theory of thermoluminescence, as long as a single activation energy E_t is involved, associated with a given frequency factor s, the expressions appearing involve the integrals

$$\int_{t_o}^{t} s \exp\{-E_t/kT\}dt. \tag{6.1}$$

This is the case in the discussed situations of first-order, second-order, general-order and mixed-order kinetics, and it occurs in the expressions developed for TL, TSC, TSEE, TSD, thermal desorption, DTA and TG. In many cases, the frequency factor is considered to be constant, in which case the expression to be evaluated is

$$s\int_{t_o}^{t} \exp\{-E_t/kT\}dt. \tag{6.2}$$

In the prevalent case when a linear heating function $T = T_0 + \beta t$ is utilized, the expressions in 6.1 and 6.2 change into

$$(s/\beta)\int_{T_0}^{T} \exp\{-E_t/k\theta\}d\theta, \tag{6.3}$$

which appeared in the expressions in previous chapters several times, e.g. Eqs. 1.4 and 3.61. As mentioned in Ch. 3, a hyperbolic heating scheme,

$$T = T_0/(1 - \beta' t) \tag{6.4}$$

has a significant theoretical advantage, since it changes the integral in expression 6.2 into

$$\alpha\int_{T_0}^{T} (1/\theta)^2 \exp\{-E_t/k\theta\}d\theta$$

where $\alpha = \beta'/T_0$, as shown in Eq. 3.66. This is an elementary integral since

$$\int_{T_0}^{T} (1/\theta^2) \exp\{-E_t/k\theta\} d\theta = (k/E_t) \left[\exp\{-E_t/kT\} - \exp\{-E_t/kT_0\}\right]. \quad (6.5)$$

It is interesting to note that this utilization of the hyperbolic heating function has been independently suggested by several researchers. These include Arnold and Sherwood[1], and Halperin et al.[2] who studied TL, Ehrlich[3], Redhead[4] and Carter[5] who worked on thermal desorption curves, and Müller and Teltow[6] who investigated TSC.

Since most workers in these different fields of thermally stimulated processes usually employ a linear heating function, the integral in expression 6.3 is to be dealt with. As shown below, it appears that the additional computational effort needed in the calculation of the integral with the linear heating scheme is a reasonable price to pay for avoiding the experimental difficulty in producing a hyperbolic heating function. Furthermore, since the original integral 6.1 includes s which may be temperature dependent, the inclusion of this temperature dependence in the integral should be included in the discussion. Different authors (see, for example, Chen[7]) showed that a common temperature behavior is $s = s''T^a$ where s'' is a constant and $-2 \leq a \leq 2$. This makes expression 6.1

$$s'' \int_{t_o}^{t} T^a \exp\{-E_t/kT\} dt. \quad (6.6)$$

It is obvious here that the use of a hyperbolic heating function is of no use. The expression for conventional constant heating rate is

$$(s''/\beta) \int_{T_0}^{T} \theta^a \exp\{-E_t/k\theta\} d\theta \quad (6.7)$$

which in the specific case of $a = -2$ yields the very simple situation in which the integral is elementary (this is similar to the case of $a = 0$ and hyperbolic heating). However, we would like to give a general treatment which can be used for any value of a within the specified range. We will start, however, the discussion with the frequently occurring case in which the frequency factor s is temperature independent, namely, $a = 0$.

The integral in expression 6.3 can obviously be written as

$$\int_{T_0}^{T} \exp\{-E_t/k\theta\} d\theta = F(T, E_t) - F(T_0, E_t) \quad (6.8)$$

where

$$F(T,E_t)=\int_0^T \exp\{-E_t/k\theta\}d\theta\ . \tag{6.9}$$

Our problem reduces, therefore, to the evaluation of the integral in Eq. 6.9 since the integral in Eq. 6.8 has been shown here to be the difference between two expressions of the same form, one with T and the other with T_0. In fact, as will be shown below, $F(T,E_t)$ is a very fast increasing function of T, and unless $T \cong T_0$, the approximation made by writing

$$\int_{T_0}^T \exp\{-E_t/k\theta\}d\theta \cong F(T,E) \tag{6.10}$$

is normally very good.

Repeated integration of Eq. 6.9 by parts, as shown by Grossweiner[8] and Haake[9], would yield

$$F(T,E_t)=T\exp\{-E_t/kT\}\sum_{n=1}(kT/E_t)^{n-1}(-1)^n n!. \tag{6.11}$$

The number of terms to be taken in the summation will be discussed below. Here it can be seen that due to the exponent, $F(T,E_t)$ is a very fast growing function of T, for a given E_t. Although we are going to continue discussing expression 6.11, it should be noted, as shown by Biegen and Czanderna[10], that one can write

$$F(T,E_t)=T\exp\{-E_t,T\}+(E_t/k)E_i(-E_t/kT) \tag{6.12}$$

where $E_i(-x)$ is the exponential integral defined by

$$-E_i(-x)=\int_x^\infty (e^{-u}/u)du \tag{6.13}$$

which can be written as the asymptotic series

$$-E_i(-x)=e^{-x}(1/x-1/x^2+2!/x^3-3!/x^4+...). \tag{6.14}$$

In fact, expression 6.11 is practically the same asymptotic series. The main feature of an asymptotic expansion such as Eq. 6.11 or Eq. 6.14 is that the terms go to infinity as $n\to\infty$. Thus, an asymptotic series is a divergent series. However, these series have properties that permit a very good approximation for the integral of interest. In fact, under specific circumstances explained below, even the first one or two terms may give a reasonably good approximation, which has been used in some approximate semi-analytical expressions for the thermally stimulated expressions. Gartia et al.[11] showed, however, that although taking only the first term in the approximation yields a relatively small change in the appearance of the peak, the consequences in terms of the evaluated activation energy, as found by the curve fitting method, may be substantial. Therefore, the use of more terms in the series is advisable. In this section, we shall mainly discuss the best approximation that can be achieved by these asymptotic series. Before going to that, a formal definition due to Poincarè (see Dingle[12]) is in order. Given a function $f(z)$, the series $\sum_{r=1}^{n} a_r/z^{r-1}$ is its asymptotic expansion if

$$\lim_{|z|\to\infty}\left\{f(z)-\sum_{r=1}^{n} a_r/z^{r-1}\right\}=0, \tag{6.15}$$

for every value of n.

We shall be more interested here in the somewhat more restricted, and easier-to-handle case of alternating sign asymptotic series as seen in expressions 6.11 and 6.14. As shown by Airey[13], when truncating such a series at a certain term n, the possible deviation from the desired function does not exceed the $(n+1)$th term. The absolute values of the terms in Eqs. 6.11, 6.14 and similar series decrease with n from $n = 1$ to a certain $n = N$, from which the terms start increasing in an unbounded manner. This can be seen by looking at the absolute value of the ratio between two subsequent terms,

$$|a_{n+1}/a_n| = [(kT/E_t)^{n+1}(n+1)!]/[(kT/E_t)^n n!], \tag{6.16}$$

which yields

$$|a_{n+1}/a_n| = (kT/E_t)(n+1). \tag{6.17}$$

The range of interest in the study of thermoluminescence and other thermally stimulated processes is normally $E_t/kT \geq 10$. In fact, it is more characteristic that E_t/kT is 20 or more, in which case taking just two terms in Eq. 6.11 results in a possible error which does not exceed 1.5%. This utilization of two terms has been reported by some researchers in the developing of semi-analytical expressions for thermally stimulated processes. However, we would like to study the asymptotic series more deeply and find the best accuracy that one can achieve using it. Inspecting Eq. 6.17, one can immediately see that the absolute values of the terms decrease from $n = 1$ to the term for which $(kT/E_t)n \simeq 1$, namely, $n \simeq E_t/kT$, which, as just stated, is 10 or more. Hence, it would be a good strategy to continue the series down to the smallest term in absolute value, i.e., the one with $N = [E_t/kT]$. The possible error in truncating the series in this way would be ([14])

$$|R_N| = |a_{N+1}| = (kT/E_t)^{N+1}(N+1)!. \tag{6.18}$$

The next step, as explained by Chen[14], is to add $\frac{1}{2}a_{N+1}$, to the series up to a_N (see also Olver[15]). For $N \simeq E_t/kT$, a_N and a_{N+1} differ only slightly from one another, but have opposite signs. Thus, the partial sums S_{N-1} and S_{N+1} have about the same value, differing from S_N by $\sim a_N$. Since the possible errors in S_{N-1}, S_N and S_{N+1} do not exceed a_{N+1} by much, adding $\frac{1}{2}a_{N+1}$ to S_N (or adding $\frac{1}{2}a_N$ to S_{N-1}) would reduce the possible error by, at least, a factor of 2.

A much more elaborate discussion on truncated asymptotic series was given by Dingle[12] who showed that improved results can be achieved by multiplying the last term taken, namely, the smallest in absolute value, by a "converging factor". This was given by a power series, but as shown by Chen[16], the sums of these series attain values of 0.49-0.51 for the present case of alternating sign asymptotic series for $E_t/kT \geq 10$. Thus, adding $\frac{1}{2}a_{N+1}$ to S_N proves to be much better than estimated before, yielding a possible error of

$$|R_N| \cong \frac{1}{100}\,|a_{N+1}| = \frac{1}{100}\,(kT/E_t)^{N+1}\,(N+1). \tag{6.19}$$

By the use of the Stirling formula for $(N+1)!$, Chen[14] showed that while N is related to the smallest term, the absolute value is,

$$|a_{N+1}| \cong \sqrt{2\pi E_t/kT}\,\exp\{-E_t/kT\}. \tag{6.20}$$

By using Eq. 6.19, we can see that[17]

$$|R_N| \cong \frac{1}{100}\sqrt{2\pi E_t/kT}\,\exp\{-E_t/kT\} \tag{6.21}$$

which gives the possible accuracy without actually calculating the series. It is more interesting, however, to evaluate the *relative* possible error, namely, the ratio of the possible error to the sum of the series. This is not significantly different from the ratio of the possible error to the first term, namely,

$$|R_N/a_1| \cong \frac{1}{100}\sqrt{2\pi (E_t/kT)^3}\,\exp\{-E_t/kT\}\,. \tag{6.22}$$

For $E_t/kT = 10$, this gives about 3.6×10^{-5}, and it becomes smaller very fast as E_t/kT grows larger. Thus, for $E_t/kT = 15$, $|R_N/a_1| \cong 4.5 \times 10^{-7}$[17].

A similar treatment was given by Chen[18] to the more general integral in Eq. 6.7. The integration in parts yields here

$$\int_0^T \theta^a \exp\{-E_t/k\theta\}d\theta = \left[T^{a+1}/(a+1)\right]\exp\{-E_t/kT\}$$

$$-\left[E_t/\left(k(a+1)\right)\right]\int_o^T \theta^{a-1}\exp\{-E_t/k\theta\}d\theta \tag{6.23}$$

which can be utilized as a recurrence equation for $a > 1$ or $a < 0$; $a \neq -1$. This may help us to reduce the problem with any given a to a one with $0 \leq a \leq 1$, unless a is an integer less than -2. The expression for the integral, which is denoted by $F(T, E_t, a)$, is now given by

$$F(T, E_t, a) = (kT^{a+2}) \exp\{-E_t/kT\} \times$$

$$\left[1 + \frac{1}{\Gamma(a+2)} \sum_{n=2} (kT/E_t)^{n-1}(-1)^{n-1}\Gamma(a+n-1)\right] \tag{6.24}$$

where $\Gamma(y)$ is the gamma function of y, namely, the extension of the factorial to non-integer values of the argument, which reduces to the factorial for positive integer numbers k, $\Gamma(k+1) = k!$. The series is finite and exact when a is a negative integer, less than -2, as shown by Petty[19]; Eq. 6.24 now becomes

$$F(T, E_t, a) = (k/E_t)^{|a|-1}(|a|-2)! \exp\{-E_t/kT\} \sum_{n=0}^{|a|-2} (-E_t/kT)^n/n!. \tag{6.25}$$

Expression 6.24 may look easier for application if one remembers that

$$\Gamma(a+n+1)/\Gamma(a+2) = (a+2)(a+3)...(a+n). \tag{6.26}$$

Using the same strategy of going down to the smallest term in absolute value, and adding one half of the next term, one gets the relative possible error as[16]

$$|R_n/a_1| \cong \frac{1}{100}\sqrt{2\pi}(E_t/kT)^{a+3/2} \exp\{-E_t/kT\}/\Gamma(a+2). \tag{6.27}$$

As pointed out, in cases where E_t/kT is large enough, the first two terms in Eq. 6.11 give a reasonably good approximation to $F(T, E_t)$. This will be (see, for example, Sviszt[20])

$$F(T, E_t) \cong (kT^2/E_t)(1 - 2kT/E_t) \exp\{-E_t/kT\}. \tag{6.28}$$

A similar treatment can be found in the work by Balarin[21].

It has been shown by Gorbachev[22] that a somewhat better two-term approximation can be given by

$$F(T, E_t) \cong \left[kT^2/(E_t + 2kT)\right] \exp\{-E_t/kT\}. \tag{6.29}$$

A further improvement with the same number of terms has been given by Balarin[23] who suggested the use of

$$F(T, E_t) \cong (kT^2/E_t) \exp\{-E_t/kT\}/\sqrt{1 + 4kT/E_t}. \tag{6.30}$$

In a later work Gorbachev[24] has extended the approximation in Eq. 6.29 to include the case of a temperature-dependent frequency factor, and gave the expression

$$F(T,E_t,a) \cong \left\{kT^{a+2}/\left[E_t+(a+2)kT\right]\right\}\exp\{-E_t/kT\}\,. \qquad (6.31)$$

Simpler approximations based on the first one or two terms in the asymptotic series were utilized by a number of researchers. Keating[25] used the approximation based on the first two terms and related to the temperature dependent pre-exponential factor

$$\int_{T_0}^{T}\theta^b\exp\{-E_t/k\theta\}\,d\theta \cong \frac{kT^{b+2}}{E_t}(1-\Delta)\exp\{-E_t/kT\} \qquad (6.32)$$

where $\Delta=(b+2)kT/E_t$, and derived from it the methods discussed in section 3.4. An even simpler single term approximation has been utilized by Podgoršak[26] and utilized later by a number of researchers. For the case of a temperature independent pre-exponential factor, they used

$$\int_{T_0}^{T}\exp\{-E_t/k\theta\}\,d\theta \cong \frac{kT^2}{E_t}\exp\{-E_t/kT\}\,. \qquad (6.33)$$

The possible error here may be as high as 10%, and since the integral appears in the exponent, the resulting error may be rather significant.

Another version based on the asymptotic series was given by Jenkins[27]. His approximation is

$$F(T,E_t) \cong Te^{-x}\left[1-A(x)/B(x)\right] \qquad (6.34)$$

where $x=E_t/kT$ and where

$$A(x)=\sum_{n=0}^{4}a_nx^n\ ;\ B(x)=\sum_{n=0}^{4}b_nx^n\,. \qquad (6.35)$$

the coefficients a_n and b_n are:

$a_o=0.26777$; $a_1=8.63476$; $a_2=18.05902$; $a_3=8.57332$; $a_4=1$,

and

$b_0=3.95850$; $b_1=21.09965$; $b_2=25.63296$; $b_3=9.57332$; $b_4=1$.

A different approach of using the continued fraction for evaluating the integral has been employed by a number of authors. Paterson[28] who studied TSC curves, has used the expression

$$\varphi(x)=\int_{x}^{\infty}\exp\{-u\}\,u^{-2}du \qquad (6.36)$$

which is the same as $F(T, E_t)$ with $x = E_t/kT$. The continued fraction expression he wrote was

$$\varphi(x) \cong \exp\{-1\}x^{-1}\{\frac{1}{x+2}\text{-}\frac{2}{x+4}\text{-}\frac{6}{x+6}\text{-}....\text{-}\frac{n(n+1)}{x+2(n+1)}\text{-}....\}. \tag{6.37}$$

Once again the accuracy in utilizing this expression depends on the number of terms taken. Paterson[28] went up to three terms, in which case the approximation is

$$\varphi(x) \approx \frac{\exp\{-x\}}{x}\frac{x^2+10x+18}{x^3+2x^2+36x+24}. \tag{6.38}$$

Senum and Yang[29] used four terms and found

$$\varphi(x) \approx \frac{\exp\{-x\}}{x}\frac{x^3+18x^2+88x+96}{x^4+20x^3+120x^2+240x+120}. \tag{6.39}$$

Some other investigators gave similar expressions of the ratio between two polynomials with different numbers of terms. These include Squire[30] who wrote

$$F(T, E_t) \approx T\exp\{-x\}(x+3.0396)/(x^2+5.0364x+4.1916) \tag{6.40}$$

and Zsakó who suggested the empirical formula

$$F(T, E_t) \approx (E_t/k)\frac{\exp\{-x\}}{(x-d)(x+2)} \tag{6.41}$$

where $d = 16/(x^2 - 4x + 84)$.

Another approach was used by Krystek[32] who used the Taylor expansion for the exponential integral (see also Abramowitz and Stegun[33])

$$E_i(-x) = c + \ln|x| + \sum_{i=0}^{\infty}(-x)^i/(i \times i!) \tag{6.42}$$

with $c = 0.5772....$, being the Euler constant. Although this is a convergent series rather than a divergent asymptotic series, it is useful mainly for small values of x and, therefore, is not very efficient for our case. For $x = 10$, for example, the first term in the series is $(-10)^1/(1 \times 1!) = -10$. The 10th term is $(-10)^{10}/(10 \times 10!) = 276$. One should, therefore, go to much higher values of i before convergence is reached, and thus, the computational effort needed for a given accuracy is much larger with this series than with the asymptotic one. The situation gets even worse in the present case for larger values of x, as opposed to the asymptotic series for which the accuracy improves significantly for higher values of $x = E_t/kT$, and for the same number of terms utilized.

Additional ways of evaluating $F(T, E_t)$ or $E_i(-x)$ through which $F(T, E_t)$ can be found as well as $F(T, E_t, a)$ are given in the literature. Some of these are Šesták[34], Norwisz and Hajduk[35], Anderson et al.[36], Christodoulides[37], Gartia et al.[38] and Singh et al.[39].

6.2 Numerical Solutions to the Differential Equations

As we have seen in previous chapters different sets of simultaneous differential equations, related to different physical circumstances, can be written to describe the traffic of charge carriers between trapping states and recombination centers. An example set was given by Eqs. 2.17, 2.19 and 2.20, in this case for the transitions occurring during the heating between one trapping state, one kind of recombination center and the conduction band. An example set of equations for transitions taking place during the excitation phase can be seen in Eqs. 4.64-4.68. Both sets of simultaneous equations are solved numerically as described below, however, there is some difference in the numerical effort required in the two cases. In the processes taking place during the excitation phase, the coefficients are assumed to be constant and the numerical solution is normally straightforward. When the equations simulating the heating stage are solved, however, the numerical solution is complicated by the very fast variation of the coefficient $\exp\{-E_t/kT\}$ with temperature and, therefore, with time. However, modern methods for numerically solving differential equations, mainly different versions of the Runge-Kutta algorithm, are robust enough to yield numerical solutions with excellent accuracy.

6.2.1 Partial Separation of Variables

Before discussing the various methods presently used, we would like to briefly discuss the work by Shenker and Chen[40] in which, prior to the utilization of the numerical method, a partial separation of the variables is performed. In this work, the set of three simultaneous differential equations governing the heating phase in a relatively simple scheme of one trapping state, one kind of recombination center, and transitions of the thermally released electrons through the conduction band. The set of equations, re-iterated here from previous chapters, is

$$-dm/dt = A_m m n_c, \tag{6.43}$$

$$-dn/dt = sn\exp\{-E_t/kT\} - A_n(N-n)n_c, \tag{6.44}$$

$$dn_c/dt = dm/dt - dn/dt, \tag{6.45}$$

where the terms retain their definitions from earlier chapters. In order to solve the equations, one has to know (or choose in a simulation) the set of parameters A_m, A_n, s, E_t and N as well as the initial values of the concentrations m_0, n_0 and n_{c0}. Also, the heating function $T = T(t)$ is to be determined; this may be $T = constant$ for phosphorescence, $T = T_0 + \beta t$ for the prevalent linear heating function or, in fact, any (usually monotonically increasing) heating function. As pointed out earlier, the TL intensity is given by $I(T) = -\alpha(dm/dt)$ where α is a constant; the TSC intensity is expected to be associated with $n_c(T)$ resulting from the calculation by $\sigma = e\mu n_c$ where σ is the conductivity, μ the mobility and e, the absolute value of the electron charge. In addition to the difficulty associated with the the rapidly varying function $\exp\{-E_t/kT\}$, another difficulty encountered in the numerical solution of the equations in their original form occurs due to the fact that in most cases $|dm/dt| \approx |dn/dt| >> |dn_c/dt|$. Thus, a small computational error in Eqs. 6.43 and 6.44 would cause a large error in Eq. 6.45, namely, in the evaluation of dn_c/dt and the solution tends to blow up after a short time. Although modern numerical methods can solve this by providing good enough accuracy, the method by Shenker and Chen[40], which also manages to overcome this difficulty, deserves a brief description.

The solution suggested by Shenker and Chen involved the definition of a new variable

$$x(t) = \int_{t_o}^{t} n_c(\tau)d\tau \tag{6.46}$$

where t is the time, with an initial value t_o. For any function of time y, one denotes dy/dt by $\dot{y}$ and dy/dx by y' and one gets

$$\dot{y} = y'(dx/dt) = y'n_c. \tag{6.47}$$

From Eqs. 6.43, 6.44 and 6.47, one obtains a set of three differential equations with the independent variable x and the three dependent variables m, n and t, thus

$$m' = -A_m m \tag{6.48}$$

$$n' = -s(n/n_c)\exp\{-E_t/kT\} + A_n(N-n) \tag{6.49}$$

$$t' = 1/n_c. \tag{6.50}$$

In the usual case of a linear heating function, T is given by $T = T_0 + \beta t$; for any other case we assume that $T = T(t)$ is some known function. n_c is computed separately such as to cause minimal error with the aid of Eq. 6.45 (for details see reference[40]). The main point is that Eq. 6.48 is independent of Eqs. 6.49 and 6.50, and can therefore be solved separately and directly

$$m = m_0 \exp\{-A_m x\}. \tag{6.51}$$

Equations 6.49 and 6.50 constitute a set of only two coupled linear differential equations which are to be solved numerically. Shenker and Chen[40] also extended the method of solution to the case of several trapping states and several recombination centers by using the same new integration variable x given in Eq. 6.46 and the same improved Runge-Kutta method previously described by Barkley[42].

6.2.2 The Runge-Kutta Methods

Different methods for the numerical solution of sets of simultaneous differential equations have been developed. Kelly et al.[41] were the first who utilized the Runge-Kutta method for the solution of the TL-TSC simultaneous differential equations. In the 1980s these calculations were performed usually on large, mainframe computers, and the tool often utilized was the sixth-order predictor-corrector Runge-Kutta method (see e.g. [43] and [44]). In fact, while simulating a situation with a continuum of trapping levels with a given distribution, even 100 simultaneous equations were solved in this manner[45]. With the advance of stronger personal computers, it turned out that results of such numerical solutions could be achieved with PCs in very reasonable computing time. It also evolved that fifth-order Runge-Kutta methods can give results with practically any desirable accuracy (see e.g. Chen and Hag-Yahya[46]). For the sake of demonstration, we will describe briefly the steps in the fifth-order Runge-Kutta method as given by Zonneveld[47], although fourth-order and even third-order methods have proven to be adequate in many situations. Such methods have been very successfully utilized with multiple simultaneous equations, dealing with the excitation, relaxation and heating of a system with several unknown functions (see, for example, Ch. 4), while using 386-, 486- or Pentium-class desktop computers.

Let us assume a set of n simultaneous equations,

$$\frac{dy_j}{dt} = f_j\left(t, y_1(t), ...yn(t)\right).\ j = 1, ..., n \tag{6.52}$$

Note that these equations include a number of parameters, some of which may be dependent on the independent variable t. In our case, the independent variable is the time, and normally, the most important time dependent parameter is $\exp\{-E_t/kT\}$ where $T = T(t)$, which thus changes rather strongly from one point to the next during the heating stage.

The initial time t_o is given and is denoted by k_{oo}. Also are given $y_j(t_o)$ and are denoted by k_{jo}, $j = 1, ..., n$. It is obvious that by the use of Eqs. 6.52, $dy_i/dt\mid_{t=t_o}$, $j = 1, ..., n$ are also known. Let us denote the integration step by h. The value of h can be varied at each step of the integration, and is determined by employing the calculated possible error at each step. If at the ith step, the error is larger than a prescribed value, the step h is reduced to, say, half of its previous value. In other versions each step is performed twice, once by a full step h and, again, by two steps of $0.5h$. Only if the difference between the two results is smaller than a predetermined "small" accuracy parameter is the step accepted. Otherwise the stepsize is halved again, and the process is repeated until the required accuracy is reached. This method of varying the stepsize is very useful, particularly when the strong temperature-dependent function is one of the parameters (i.e. during heating).

The following magnitudes are now calculated in the $(i+1)$th step

$$k_{oj} = hf_j(t, y_1, ..., y_n) \tag{6.53}$$

$$k_{1j} = hf_j\left(t + \frac{2}{9}k_{oo}, y_1 + \frac{2}{9}k_{o1}, ..., y_n + \frac{2}{9}k_{on}\right) \tag{6.54}$$

$$k_{2j} = hf_j\left(t + (k_{oo} + 3k_{1o})/12, y_1 + (k_{o1} + 3k_{11})/12, ..., y_n + (k_{on} + 3k_{1n})/12\right) \tag{6.55}$$

$$k_{3j} = hf_j\left(t + (k_{oo} + 3k_{20})/8, y_1 + (k_{o1} + 3k_{21})/8, ..., y_n + (k_{on} + 3k_{2n})/8\right) \tag{6.56}$$

$$\begin{aligned} k_{4j} = hf_j(t &+ (53k_{oo} - 135k_{1o} + 126k_{2o} + 56k_{3o})/125, \\ y_1 &+ (53k_{o1} - 135k_{11} + 126k_{21} + 56k_{31}\}/125, ..., \\ y_N &+ (53k_{on} - 135k_{1n} + 126k_{2n} + 56k_{3n})/125) \end{aligned} \tag{6.57}$$

$$\begin{aligned} k_{5j} = hf_j(t &+ (-63k_{oo} + 189k_{1o} - 36k_{2o} - 112k_{3o} + 50k_{4o})/28, \\ y_1 &+ (-63k_{o1} + 189k_{11} - 36k_{21} - 112k_{31} + 50k_{41})/28, ..., \\ y_n &+ (-63k_{on} + 189k_{1n} - 36k_{2n} - 112k_{3n} + 50k_{4n})/28) \end{aligned} \tag{6.58}$$

$$\begin{aligned} k_{6j} = hf_j(t &+ (133k_{oo} - 378k_{1o} + 276k_{2o} + 112k_{3o} + 25k_{4o})/168, \\ y_1 &+ (133k_{o1} - 378k_{11} + 276k_{21} + 112k_{31} + 25k_{41})/168, ..., \\ y_N &+ (133k_{on} - 378k_{1n} + 276k_{2n} + 112k_{3n} + 25k_{4n})/168). \end{aligned} \tag{6.59}$$

The value of the function in the $(i+1)$th step is now given by

$$y_j^{i+1} = y_j^i + (35k_{oj} + 162k_{2j} + 125k_{4j} + 14k_{5j})/336, \tag{6.60}$$

and the possible error is given by

$$\Delta y_j = (21k_{oj} - 162k_{2j} + 224k_{3j} - 125k_{4j} + 42k_{6j})/14. \tag{6.61}$$

Details and proofs are given by Zonneveld[47].

The usual way to determine whether the Runge-Kutta calculations are sufficiently accurate is to recompute the value at the end of each interval with the stepsize cut in half. If the resultant change is of negligible magnitude, the value is accepted; if not, the value is recomputed with the stepsize halved again, and so on until acceptable accuracy is achieved. Press and Teukolsky[48] described the family of adaptive stepsize Runge-Kutta methods. They state that a good ordinary differential equations (ODE) integrator should exert some adaptive control over its own progress, making frequent changes in its size h. Usually the purpose of this adaptive stepsize control is to achieve some predetermined accuracy in the solution with minimum computational effort. Many small steps should be made in "treacherous terrain", while a few great strides should speed through smooth uninteresting "countryside". The resulting gains in efficiency may be by a factor as large as 100 or more. With the fourth-order Runge-Kutta with which Press and Teukolsky are dealing, the most straight forward technique for adaptive stepsize control is step doubling, as previously proposed by Gear[50]. One takes each step twice, once as a full step and then, independently, as two half steps. Press and Teukolsky[48] show that the overhead cost in this procedure is just a factor of 1.375.

Going further with this analysis, let us denote the exact solution for an advance from x to $x+2h$ by $y(x+2h)$, and the two approximate solutions by y_1(one step $2h$) and y_2(two steps, each of size h). These authors[48] show that if one starts with a fourth-order method, the true solution and the two numerical approximations are related by

$$y(x+2h) = y_1 + (2h)^5\phi + O(h^6) + ... \tag{6.62}$$

$$y(x+2h) = y_2 + 2(h^5)\phi + O(h^6) + ... \tag{6.63}$$

where, to order h^5, the value ϕ remains constant over the step. ϕ is a number which has the order of magnitude $y^{(5)}(x)/5!$. The expression 6.62 involves $(2h)^5$ because the stepsize is $2h$, whereas the expression 6.63 involves $2(h^5)$ because the error on each step is $h^5\phi$. The difference between the two numerical estimates is a convenient indicator of the truncation error

$$\Delta = y_2 - y_1. \tag{6.64}$$

One should try to keep this difference to a desired degree of accuracy, neither too large nor too small, by adjusting h. Press and Teukolsky[48] give a "numerical recipe" based on these concepts as well as on vast numerical experience, which proved to be very reliable in practice.

Among other schemes proposed to minimize the error in a Runge-Kutta computation with minimum effort is the Runge-Kutta-Fehlberg method. Here, instead of computing the y-value for stepsizes h and $h/2$, one determines the values for the same stepsize, but computed with different orders of errors. Thus, one compares values computed using fourth- and fifth-order Runge-Kutta calculations. Each computation makes use of the same k's, minimizing the number of function calculations required. The local error term is $O(h^6)$.

In the same way that the Runge-Kutta-Fehlberg method is a fifth-order method with order 4 embedded, so Nørsett and Thomsen[49] proposed two methods which are, respectively, order 3 with order 2 embedded (NT1), and order 4 with order 3 embedded (NT2). The local error E_n is determined from the difference of the values computed for the m and $m-1$ orders

$$y_{n+1} = y_n + h_n \sum_{j=1}^{m} b_i f(t_n + c_i h_n, Y_i) \tag{6.65}$$

and

$$^{*}y_{n+1} = y_n + h_n \sum_{j=1}^{m-1} b_i f(t_n + c_i h_n, Y_i) \tag{6.66}$$

where

$$Y_i = y_n + h_n \sum_{j=1}^{i} a_{ij} f(t_n + c_i h_n, Y_j). \tag{6.67}$$

The values for the constants a_{ij}, b_i, c_i and d_i are given by the so-called Bütcher Scheme, described by Nørsett and Thomsen[49]. The local error in the two methods is $O(h^3)$. The methods proceed by defining a local error tolerance and modifying the stepsize so that the estimated local error is less than a certain fraction f_{safety} (typically 0.6-0.9) of the local error tolerance. The local error tolerance E_T is

$$E_T = E_A + \|y_n\| \, E_R \tag{6.68}$$

where E_A and E_T are the defined absolute and relative local errors, respectively, and the factor f_{mod} by which the stepsize is modified is

$$f_{mod} = f_{safety} \left(E_T / \|E_n\| \right)^{1/m}. \tag{6.69}$$

The next stepsize h_{n+1} is then calculated from $h_{n+1} = f_{mod} h_n$.

6.2.3 Utilization of a Monte Carlo Method for Evaluating TL and TSC Curves

Kulkarni[51] has quoted Kelly et al.[41] saying that the stepsize in the Runge-Kutta method cannot exceed the lifetime of electrons in the conduction band, and therefore, getting exact solutions using the Runge-Kutta method is impractical if the lifetime of the electrons in the conduction band is very small. Kulkarni has offered an alternative, namely a Monte Carlo method which deals quite successfully with the computation of TL and TSC peaks under these conditions. The essence of the Monte Carlo method can be described by considering the simple one-trap/one-center model. The method proceeds by generating a random number R to represent the probability that an electron will be released from the trap, thus

$$R = 1 - \exp\left\{ -n \int_{t_i}^{t} p(T(\tau)) d\tau \right\} \tag{6.70}$$

which is the cumulative distribution function of the time t, after some initial time t_i, at which the first of n electrons escapes from the trap, with escape probability $p(T(t)) = s\exp\{-E/kT(t)\}$. In the above equation, τ is a dummy variable representing time. Similarly, a random number R' is used to represent the probability that an electron will be removed from the conduction band at time t', after initial time t_i, according to

$$R' = 1 - \exp\left\{ -n_c \left[A(N-n) + A_m m \right] (t_i - t') \right\}. \tag{6.71}$$

Removal of an electron from the conduction band can occur via either retrapping, or via recombination. Equations 6.70 and 6.71 are solved to determine the values of t and t' and thereby establish whether the first event after t_i is a detrapping event, or a trapping/recombination event. If the former, the conduction band population is incremented by one; if the latter, the band population is decreased by one. If an electron is removed from the band, one then has to determine if the event was a retrapping event (in which case the trapped electron population is incremented by one) or a recombination event (in which case the electron is removed from the system). This is done using a third random number R'' and comparing this with the probability of retrapping, given by the ratio $A(N-n)/[A(N-n)+A_m m]$. If $R'' > A(N-n)/[A(N-n)+A_m m]$, a recombination event is assumed; if $R'' < A(N-n)/[A(N-n)+A_m m]$ a retrapping event is assumed.

After adjusting the concentrations according to the above calculations, the time t_i is incremented to either t or t' (whichever is smaller) and the whole process is repeated. All through the calculation, temperature is evaluated using $T = T_0 + \beta t$. The TL intensity is estimated at each temperature by evaluating how many recombination events took place in each temperature interval, assuming that each event produces one photon.

Typically, the whole calculation is performed many times in order to determine the statistical error. Each cycle tracks several thousand electrons (the more the better), but it is easier (i.e. quicker) to perform (say) 10 cycles tracking 1000 electrons in each cycle, than it is to track 1 cycle containing 10,000 electrons. Adding the results of each cycle produces the valid result. A total of about 500,000 electrons is required for good statistics, and the error is then estimated using standard statistical techniques.

In ranges of the parameters where both the Runge-Kutta and the Monte Carlo methods could be employed Kulkarni[51] showed that the results of the two methods were quite close to each other, except that the Runge-Kutta results are smoother. However, in the Monte Carlo method, the calculation time remains practically the same when the lifetime of free electrons is very short, whereas with the Runge-Kutta method it may become prohibitively long. Kulkarni[51] also describes how the above method can be extended to models more complex than the simple two-level model.

6.2.4 Using Approximate Integro-Differential Equations for the description of Complex TL Curves

In a series of papers, Kantorovich et al.[52]-[55] presented a way of transforming the set of simultaneous differential equations describing the complex situation of several kinds of recombination centers and several trapping states by integro-differential equations. In the first paper[52], the situation of one kind of recombination center and a series of trapping states was studied, and an approximation was given by a single integro-differential equation. This case will be described here in some detail. Let us re-write the basic set of equations governing the excitation and heating stages, as given by Kantorovich et al.[52], in a slightly different notation, more similar to our notation of Chapters 2 and 4.

$$dn_i/dt = A_{ni}(N_i - n_i) - s_i n_i \exp\{-E_{ti}/kT\} \tag{6.72}$$

$$dm/dt = A_v m_v (M - m) - A_m m n_c \tag{6.73}$$

$$dn_c/dt = \sum_{i=1}^{p} A_{ni}(N_i - n_i) - A_m m n_c + X \tag{6.74}$$

$$dm_v/dt = X - A_v m_v (M - m) \tag{6.75}$$

where n_c and m_v are the concentrations of free electrons and holes, respectively; n_i is the concentration of electrons in the ith trap and N_i is the total concentration of the ith trap. A_{ni} is the trapping probability into the ith trap; m is the concentration of holes in the center and M is the total concentration of the relevant centers. A_v is the probability of the free holes being trapped in M, and A_m is the recombination probability of free electrons and trapped holes. E_{ti} is the activation energy (trap depth) of electrons in the ith trap and s_i is its frequency factor. X is the rate of production of free electrons and holes in the conduction and valence bands, respectively, during the excitation by the irradiation; k is the Boltzmann constant, t is time and T is the temperature. Kantorovich et al.[52],[53] made the "usual" quasiequilibrium simplifying assumptions mentioned in Ch. 2, concerning the small relative changes in the concentration of the free electrons and holes,

$$|dn_c/dt| << |dn_i/dt| \, ; \, i = 1, ..., p \tag{6.76}$$

$$|dm_v/dt| << |dm/dt| \, . \tag{6.77}$$

They also assumed that all trapping states are far from saturation, namely,

$$n_i << N_i, \, i = 1, ..., p \tag{6.78}$$

$$m << M, \tag{6.79}$$

and defined a total trapping factor $\mathcal{D}$

$$\mathcal{D} = \sum_{i=1}^{p} A_{ni} N_i . \tag{6.80}$$

For dealing with the heating stage they took $X = 0$. They further defined $\mathcal{D}_i = A_{ni}N_i/\mathcal{D}$ and, assuming a dominant retrapping of electrons, they took the first terms in the Taylor expansion of the relevant functions, and found

$$\frac{dm}{dt} = -\frac{A_m m}{\mathcal{D}} \sum_{j=1}^{p} w_j n_j \tag{6.81}$$

$$\frac{dn_i}{dt} = \mathcal{D}_i(1 - \frac{A_m m}{\mathcal{D}}) \sum_{j=1}^{p} w_j n_j - w_i n_i \tag{6.82}$$

where $w_i = s_i \exp\{-E_{ti}/kT\}$.

These expressions were further developed into a single integro-differential equation in $z(t) = \ln(m_0/m)$ where m_0 is the initial concentration of trapped holes. Defining now

$$F(z) = z + (A_m m_0/\mathcal{D})e^{-z} \tag{6.83}$$

and

$$\varphi_i(t) = \int_o^t w_i(\tau)d\tau, \tag{6.84}$$

they reached the single integro-differential equation

$$\dot{z} = \sum_{j=1}^{p} w_j \left(\frac{A_m}{\mathcal{D}} e^{-\varphi_j} N_j^o + \mathcal{D}_j F(z) - \mathcal{D}_j e^{-\varphi_j} \int_o^t e^{\varphi_j} w_j F(z) d\tau \right) \tag{6.85}$$

where $N_j^o = n_j^o + \mathcal{D}_j m_0$. Kantorovich and Fogel[53] further showed that the TL intensity is given by

$$I \propto -\dot{m} = m_0 \dot{z} \exp\{-z\}. \tag{6.86}$$

The idea of having a single equation replacing a set of simultaneous differential equations is very appealing. However, quite a number of simplifying assumptions have been introduced which limit the generality of the approach. Moreover, once a set of parameters is given, Eq. 6.85 should be solved numerically. In order to check the validity of their method, Kantorovich et al.[53] solved a specific case with one kind of recombination center and three traps, with the present method and with the Runge-Kutta method, and found very good agreement with the first peak but rather noticeable deviations in the higher temperature peaks.

In later work, Kantorovich et al.[54] further developed the method to take care of a series of trapping states and a series of recombination centers. As a result, they got two simultaneous integro-differential equations. The numerical solution was compared again with the numerical results reached by the Runge-Kutta method. Deviations were found here too due to the simplifying assumptions made. The method was even utilized for certain cases of continuous distributions of traps and centers.

6.2.5 Applications of Analog Computers to the Analysis of Thermally Stimulated Processes

The use of digital computers for the study and analysis of thermally stimulated phenomena has been prevalent in the last quarter century, as it has been in practically all the sciences. In the early seventies, there have been a few attempts to utilize analog computers in the study of TL, which will be only very briefly mentioned here as a historical remark. Razdan et al.[56],[57] used an analog computer for simulating TL curves. First- and second-order peaks were produced, and the possible dependence of the pre-exponential factor on temperature was taken into account. Böhm and Scharmann[58] used an analog computer to simulate peaks under first- and second-order conditions, and showed that the TL peak precedes its TSC counterpart, as proven generally by Chen[59] (see Ch. 9). As could be expected, they found that both the peak width methods and the various heating rates methods would give good results for the TL peaks (for which they had been developed in the first place), and much poorer results for the TSC peaks.

6.3 Peak Fitting and Deconvolution Methods for Analyzing Complex Glow Curves

The above methods concentrate on procedures to simulate mathematically both simple and complex glow curves. However, they are impractical as tools in the analysis of real glow curves since they require microscopic information about the trapping levels and recombination centers that is rarely known beforehand. Rather, the usual challenge is to estimate values for these parameters using the experimentally recorded glow curve. In Ch. 3 we discussed various methods of analysis of experimental glow curves, including peak fitting methods, but we reserved a description of the mathematical procedures until this chapter. Here we discuss mathematical aspects of the techniques of peak fitting and deconvolution.

6.3.1 Peak Fitting

As described in Ch. 3, the technique of peak fitting proceeds by establishing (or guessing!) the number of individual glow peaks present in the glow curve and thereby representing the glow curve thus

$$I(T) = \sum_{i=1}^{\nu} \alpha_i f_i(T) \tag{6.87}$$

where α_i is a scaling constant, $f_i(T)$ is the chosen mathematical function for the description of the individual glow peaks (e.g. the Randall-Wilkins equation, the Garlick-Gibson equation, the General-Order equation, etc.), and ν is the assumed number of individual peaks present in the complete glow curve. Thus, fitting parameters for each peak are α_i, s_i, E_{ti} and possibly the kinetic order b_i. Hence, a total of 4ν fitting parameters is possible and it is clear that one is operating in multidimensional parameter space. The basic procedure is to guess at values for each parameter, evaluate $I(T)$ using Eq. 6.87, and to compare with the experimental glow curve. The parameters are then changed until a minimum error is obtained between the evaluated TL signal and the experimental signal. The most popular minimization routine used in these applications is the Marquardt method of nonlinear least squares fitting.

Consider a proposed fitting function

$$y = f(x, \mathbf{a}) \tag{6.88}$$

where $\mathbf{a} = (a_1, a_2, a_3, ..., a_m)$ is a vector of m adjustable parameters. Our aim is to arrive at the best fit of this function to the n data points $(x_1, y_1, ...x_n, y_n)$. To reach this goal, we determine to minimize the quantity

$$S = \sum_{i=1}^{n} \left(\frac{f(x_i, \mathbf{a}) - y_i}{w_i} \right)^2 \tag{6.89}$$

where w_i is a weighting parameter used to weight the data set to the most critical part, although most often w_i is set to unity for all i. S is more usually known as the "chi-squared" value and, with the weights set to unity, we wish to determine those values of $\mathbf{a}$ for which

$$\frac{\partial S}{\partial a_j} = \Phi(\mathbf{a}) = \sum_{i=1}^{n} [f(x_i, \mathbf{a}) - y_i] \left[\frac{\partial f(x_i, \mathbf{a})}{\partial a_j} \right] = 0, \text{ for } j = 1....m. \tag{6.90}$$

Since the TL equations in which we are interested are non-linear functions of the fitting parameters (i.e. f is a non-linear function of $\mathbf{a}$) the contours of constant S are ellipsoids. The minimum of S lies in a valley, of which there may be several (local minima). The aim is to find those values of $\mathbf{a}$ which correspond to the deepest valley (the global minimum). One may move initially in the direction of a minimum by considering a small increment $\delta\mathbf{a}$ which is chosen by evaluating the gradient of S at given values of $\mathbf{a}$ and moving in the direction of steepest descent. $\delta\mathbf{a}$ is chosen such that the step is always downhill. Although for non-linear equations this steepest descent method does not always point in the direction of the minimum, the method is a good one to choose initially when far from the minimum[60]. A better sense of which direction to move toward minimization is obtained by taking account of the curvature of S . One has the Taylor series expansion

$$\Phi_j(\mathbf{a} + \delta a) = \Phi_j(\mathbf{a}) + \sum_{k=1}^{m} \frac{\partial \Phi_j}{\partial a_k} a_k, \text{ for } j = 1....m. \tag{6.91}$$

for which the second-order and higher-order terms have been ignored. This is valid if the chosen function and parameters represent a good approximation to the data set and, therefore, is good close to the minimum. This leads to a set of new parameter values calculated using steps of the form

$$\delta a_j = \left(\frac{\partial \Phi_j(\mathbf{a})}{\partial a_k}\right)^{-1} \Phi_j(\mathbf{a}) = \mathbf{H}^{-1}\Phi_j(\mathbf{a}), \text{ for } j = 1....m. \quad (6.92)$$

Marquardt[61] used a much earlier suggestion by Levenberg[62] to combine the above two approaches in which the steepest descent method is used when far from the minimum and the curvature method (known as the Gauss-Newton method[60]) is used when close to the minimum. Marquardt introduces a parameter λ and defines a new matrix $\mathbf{H}'$ such that

$$H'_{jk} = H_{jk} \text{ for } j \neq k \quad (6.93)$$

and

$$H'_{jk} = H_{jk}(1+\lambda) \text{ for } j = k. \quad (6.94)$$

λ is a dimensionless parameter which is adjusted to alter the size and direction of each step. Thus, when λ is very large ($>> 1$) the diagonal terms in the matrix dominate the calculation of the direction and stepsize and we are effectively using the steepest descent method to determine the step. Alternatively, when λ is small ($<< 1$) we are using the Gauss-Newton method. The procedure suggested by Marquardt to decide upon which is the appropriate method (i.e. to decide upon the appropriate value for λ) is:

(a) Choose initial values for the parameters $\mathbf{a}$.

(b) Calculate $S(\mathbf{a})$.

(c) Choose an initial guess for λ (e.g. 0.001).

(d) Evaluate $\delta\mathbf{a}$ and $S(\mathbf{a}+\delta\mathbf{a})$.

(e) If $S(\mathbf{a}+\delta\mathbf{a}) > S(\mathbf{a})$, increase λ by a factor of 10, and repeat (d).

(f) If $S(\mathbf{a}+\delta\mathbf{a}) < S(\mathbf{a})$, decrease λ by a factor of 10, replace $\mathbf{a}$ by $\mathbf{a}+\delta\mathbf{a}$, and repeat (d).

This iterative process is continued until $S(\mathbf{a})$ decreases by some predetermined small amount. This condition should be defined to be commensurate with the statistical accuracy of the initial data. Alternatively one might simply say that once the change in $S(\mathbf{a})$ is less than (say) 10^{-3} of its last value, the iteration should stop; or, for example, one might say that whenever this condition occurs (say) 3 times in a row, then convergence has occurred.

It is to be noted that, in particular when the problem has a large number m of adjustable parameters, the procedure of inverting the $m \times m$ matrix $\mathbf{H}$ may be time consuming. Moreover, a common problem occurring with the Newton method in all its forms is that at the final stages, the iterations produce oscillations between two points rather than convergence into the minimum. A group of procedures termed quasi-Newton methods has been developed for solving these two problems; for details see e.g. Avriel[63]. It should be remembered, however, that all these methods - steepest descent, Newton and quasi-Newton - can guarantee only a local minimum and therefore, the global minimum is reached only if the first iteration starts rather close to this minimum. Methods have also been developed for *increasing the probability* of approaching the global minimum even when the initial guess of the set of parameters is rather far from the final optimum. These include the *simulated annealing*[64],[65], *tabu search*[65],[66] and a group of methods which has grown in popularity, namely the *genetic algorithms*[66]-[69].

Whatever criterion is chosen to define convergence the "goodness" of fit needs to be determined. Since the standard deviation of each datum point is generally not known, and the weighting factors are normally unity, one gets a chi-squared value which is intensity dependent, and S is not close to zero for a good fit. Comparison of the goodness of fit for a variety of glow peaks with different intensities is thus difficult. One possible solution is to normalize all peaks (to a maximum of unity) before fitting, but with multiple peak glow curves this is often difficult, and in circumstances where one wants to retain the absolute intensity information this procedure is not feasible. As a result, chi-squared is often not a useful parameter to determine the goodness of the fit. A popular criterion is to use the Figure of Merit (FOM)[70] defined as

$$FOM = \sum_{j_{initial}}^{j_{final}} \frac{100\,|f(x_i, \mathbf{a}^*) - y_i|}{\Delta} \tag{6.95}$$

where $\mathbf{a}^*$ is the final set of fitted parameters, and Δ is the integral under the curve over the region of interest, from $j_{initial}$ to j_{final}. Since the FOM is normalized by the integral under the curve, the goodness of fit may be compared from peak to peak, and from glow curve to glow curve, while retaining the absolute intensity information. FOM values of a few percent generally indicate acceptable fits.

Most methods of peak fitting described in the literature use basically the same fitting routine as described here, some with modifications to allow for rapid analysis for applications such as radiation dosimetry. Several of these are described by Horowitz and Yossian[71].

6.3.2 Deconvolution

A logical extension of the curve fitting procedures described above is to fully deconvolute the glow curve. In Eq. 6.87, α_i is an amplitude dependent upon the trap energy E_{ti}, and frequency factor s_i. Likewise f_i is the glow curve shape, also dependent upon E_{ti} and s_i. That is $\alpha_i = A_i(E_{ti}, s_i)$ and $f_i(T) = F_i(E_{ti}, s_i, T)$. Thus, for many traps, for which the separation in E_t is dE_t and the separation in s is ds, we may rewrite Eq. 6.87 as

$$I(T) = \int_{E_1}^{E_2} \int_{s_1}^{s_2} A(E_t, s)\, F(E_t, s, T)\, dE_t ds \tag{6.96}$$

where we sum all the glow curves corresponding ranges of E_t (from E_1 to E_2) and s (from s_1 to s_2). Note that E_t and s are not completely independent variables since the combination of E_t and s for a given peak is constrained to yield a TL peak in the range experimentally observed. Thus, as E_t increases, so s must increase so that the TL peak remains at the same temperature. Eq. 6.96 has the form of the Fredholm integral equation in which a function $F(E_t, s, T)$ is convolved with a distribution function $A(E_t, s)$ to produce the observed data $I(T)$. The problem now is to use a known form of the function $F(E_t, s, T)$ (for example, the Randall-Wilkins equation) to extract the unknown two-dimensional distribution function $A(E_t, s)$ from the data set. A difficulty, however, is that while we seek a two-dimensional function $A(E_t, s)$, the data set itself $I(T)$ is only one-dimensional. Several approaches can be made to this problem.

The simplest approach is to assume that the frequency factor s is a (known) constant for all traps. Equation 6.96 then simplifies to

$$I(T) = \int_{E_1}^{E_2} A(E_t)\, F(E_t, T)\, dE_t \tag{6.97}$$

where the variable s is removed from the computation and becomes a fixed constant contained within $F(E_t, T)$. One needs to assume a realistic value for s (say, $10^{10} - 10^{12}$ s^{-1}) and the problem is reduced to finding that distribution in E_t which is necessary to produce the required glow curve shape for the chosen value of s. The one-dimensional distribution $A(E_t)$ can be regarded as a density of states function for the traps within the band gap, over the range E_1 to E_2. One then proceeds by treating the experimental glow curve as the sum of a large number of "glow curvelets", each with its own amplitude A, its own energy E_t, and with a separation ΔE_t between curvelets in the value of the individual E_t's. Clearly, the larger the number of curvelets used, the finer the resolution in energy for the density of states function. One then fits the sum of the glow curvelets to the data set using a non-linear least squares fitting routine, as described in the previous section. The best fit yields the unknown distribution function $A(E_t)$, in steps of ΔE_t.

Examples of this procedure are shown in Figs. 6.1 and 6.2. In Fig. 6.1 is shown a simulated glow curve (a) with an activation energy of 1.253 eV and a frequency factor of 10^{12} s^{-1}. In curve (b) is shown the deconvolution of the glow

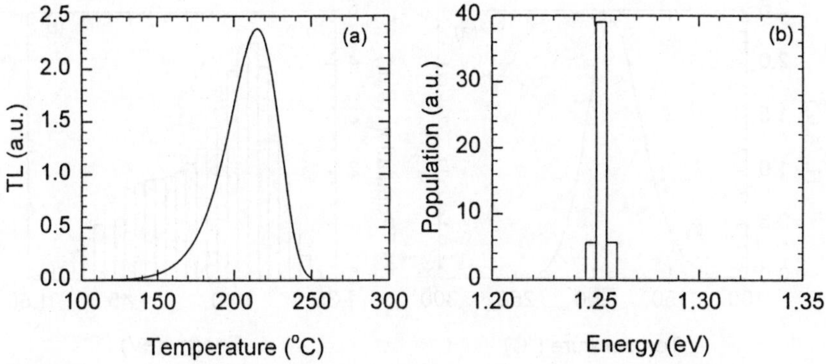

Figure 6.1 (a) A simulated glow curve with $E_t = 1.253$ eV and $s = 10^{12}$ s^{-1}. (b) The deconvolution of the glow peak in (a), using $s = 10^{12}$ s^{-1}. The heating rate is 2.0 °C/s.

peak, using $s = 10^{12}$ s^{-1}. Since in this case $A(E_t)$ is a delta function one gets a very narrow density of states distribution from the deconvolution analysis. An example of a deconvolution of a real glow curve is shown in Fig. 6.2 for the glow curve from beta-irradiated Al_2O_3:C. A value for s of 10^{14} s^{-1} is assumed for this analysis. Although the E_t values subsequently obtained cannot accurately be treated as "trap depths", the analysis does indicate that this glow peak is best described as arising from a distribution of trapping states between 1.35 eV and 1.50 eV. This highlights that the procedure is particularly useful for determining the density of states for a trap or set of traps with a distribution of activation energies, as one might expect, for example, for an amorphous or glassy material. Restraint must be exercised in the interpretation of E_t, however. It is incorrect to interpret the obtained E_t values as trap depths, since an accurate determination of the trap depth requires a knowledge of the actual s factor. As we have seen, the latter is only guessed at in this simple analysis. Furthermore, if one has a distribution in E_t, one can also reasonably expect a distribution in s values, and not a single-valued s as we have assumed here.

A better approach to account for the likely variation in s is to repeat the above calculations many times, each time with a different s. If this was done enough times, with a sufficiently small increment in s on each occasion, one could produce a density of states function distributed in s also. The difficulty here is that one still assumes the same s value for each peak in the glow curve. A more realistic approach would be to allow E_t and s to vary independently.

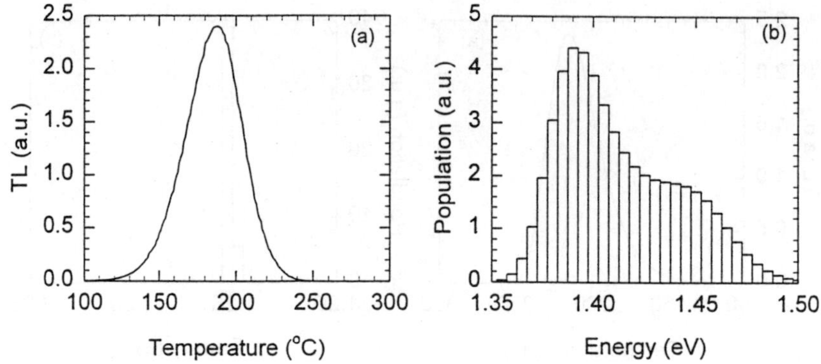

Figure 6.2 (a) Glow curve from beta-irradiated (6.4 rad) Al_2O_3:C, and (b) its deconvolution assuming a constant s value of 10^{14} s^{-1}, and using a heating rate of 2.0 °C/s.

However, now one requires a constraint to be placed upon the plausible range of allowed s values for a given E_t value, and vice-versa. To see the difficulty more clearly, imagine a single TL peak arising from a distribution of E_t values over an unknown range E_1 to E_2. As low values of E_t are tested, the s values have to be reduced commensurately in order to produce a TL peak in the same temperature range as is observed experimentally. Smaller s values produce wider TL peaks, and soon one finds that the range of s is constrained by the fact that wider and wider glow peaks will not fit within the experimental glow peak. Hence, a lower limit to s (and E_t) results. However, the same is not true for higher values of E_t (and s). TL peaks with high s values are narrow, and hence one can imagine large numbers of narrow peaks being summed together to produce the desired experimental glow curve shape. Hence, an upper limit to E_t (and s) is less clearly defined.

What is required is an additional constraint to define more clearly *both* the upper and lower limits. This can be provided by experiments in which the heating rate is varied. We have seen in previous chapters how changing the heating rate β causes the TL peak to shift to higher or lower temperatures, as β is increased or decreased, respectively. We discussed in Ch. 3 how the observed shift enables the E_t value to be extracted. That is, the experimentally observed shift in the temperature of the peak maximum T_m defines the trap depth, and, consequently, the frequency factor. Although many pairs of E_t and s can produce a TL peak at the same T_m, only a particular pair of E_t and s

values will produce the observed shift in T_m as a function of β. A single TL peak caused by a single-valued activation energy will be characterized by a unique E_t and s in this way, and this formed the basis of the Heating Rate methods of analysis, described in Ch. 3. However, a TL peak caused by a distribution in E_t values will produce a range of E_t, s pairs.

In view of the above, a possible procedure for the deconvolution of a complex glow curve would be to record several glow curves at a variety of heating rates. Ideally, one would record many glow curves for many heating rates and subsequently fit the summation of a matrix of computed glow curves to the $I(T, \beta)$ surface. Eq. 6.96 would now become

$$I(T,\beta) = \int_{E_1}^{E_2} \int_{s_1}^{s_2} A\left(E_t, s\right) F\left(E_t, s, T, \beta\right) dE_t ds \tag{6.98}$$

where we have overcome the objection of the dimensionality of the data set. Unfortunately, examples of such procedures are presently not available.

6.4 References

1. W. Arnold and H. Sherwood, *J. Phys. Chem.* **63** (1959) 2.
2. A. Halperin, M. Leibovitz and M. Schlesinger, *Rev. Sci. Inst.* **33** (1962) 1168.
3. G. Ehrlich, *J. Appl. Phys.* **32** (1961) 4.
4. P.A. Redhead, *Vacuum* **12** (1962) 203.
5. G. Carter, *Vacuum* **12** (1962) 245.
6. P. Müller and J. Teltow, *Phys. Stat. Sol. (a)* **12** (1972) 471.
7. R. Chen, *J. Appl. Phys.* **40** (1969) 570.
8. L.I. Grossweiner, *J. Appl. Phys.* **24** (1953) 1306.
9. C.H. Haake, *J. Opt. Soc. Am.* **47** (1957) 649.
10. J.R. Biegen and A.W. Czanderna, *J. Therm. Anal.* **4** (1972) 39.
11. R.K. Gartia, S.D. Singh, E.D. Singh, N.C. Deb and P.S. Mazumdar, *Phys. Stat. Sol. (a)* **150** (1995) 749.
12. R.B. Dingle, *Proc. Roy. Soc.* **A244** (1958) 456, 476.
13. J.R. Airey, *Phil Mag.* **24** (1937) 521.
14. R. Chen, *J. Comput. Phys.*, **4** (1969) 415.
15. F.W.J. Olver, *Asymptotic and Special Functions*, Academic Press (1974).
16. R. Chen, *J, Comput. Phys.*, **8** (1971) 156.
17. R. Chen, *J. Therm. Anal.* **6** (1974) 585.
18. R. Chen, *J. Comput. Phys.* **6** (1970) 314.
19. H.R. Petty, E.T. Arakawa and J.K. Baird, *J. Therm. Anal.* **11** (1977) 417.

20. P. Sviszt, *Acta Academiae Scientiarium Hungaricae* **80** (1975) 167.
21. M. Balarin, *Phys. Stat. Sol. (a)* **54** (1979) K137.
22. V.M. Gorbachev, *J. Therm. Anal.* **8** (1975) 349.
23. M. Balarin, *J. Therm. Anal.* **12** (1977) 169.
24. V.M. Gorbachev, *J. Therm. Anal.* **10** (1976) 447.
25. P.N. Keating, *Proc. Phys. Soc.* **78** (1961) 1408.
26. E.B. Podgoršak, P.R. Moran and J.R. Cameron, *Proc. 3rd Conf. Lumin. Dosim.* Risø Report **249** (1971) 1.
27. T.R. Jenkins, *J. Comput. Phys.* **29** (1978) 302.
28. W.L. Paterson, *J. Comput. Phys.* **7** (1971) 187.
29. G.I. Senum and R.T. Yang, *J. Therm. Anal.* **11** (1977) 445.
30. W. Squire, *J. Comput. Phys.* **6** (1970) 152.
31. J. Zsakó, J. *Therm. Anal.* **8** (1975) 593.
32. M. Krystek, *Phys. Stat. Sol. (a)* **40** (1977) K65.
33. M. Abramowitz and I.A. Stegun, *Handbook of Mathematical Functions*, Dover Pub. Inc., New-York (1988).
34. J. Šesták, *Thermochim. Acta* **3** (1971).
35. J. Norwisz and N. Hajduk, *J. Therm. Anal.* **13** (1978) 223.
36. H. Anderson, D. Haberland and E. Witte, *J. Therm. Anal.* **17** (1979) 409.
37. C. Christodoulides, *J. Comput. Phys.* **79** (1988) 469.
38. R.K. Gartia, T.S.C. Singh and P.S. Mazumdar, *J. Therm. Anal.* **37** (1991) 203.
39. S.D. Singh, R.K. Gartia and P.S. Mazumdar, *Phys. Stat. Sol. (a)* **146** (1994) 825.
40. D. Shenker and R. Chen, *J. Comput. Phys.* **10** (1972) 272.
41. P. Kelly, M.J. Laubitz and P. Bräunlich, *Phys. Rev. B* **1** (1971) 1960.
42. Y.R. Barkley, *A Runge-Kutta for all Seasons, The University of Wisconsin, Madison, AD-642910 (1966).*
43. R. Chen, S.W.S. McKeever and S.A. Durrani, *Phys. Rev. B1*, **24** (1981) 4931.
44. S.W.S. McKeever, J.F. Rhodes, V.K. Mathur, R. Chen, M.D. Brown and R.K. Bull, *Phys. Rev. B15*, **32** (1985) 3835.
45. W.F. Hornyak, R. Chen and A.D. Franklin, *J. Lumin.* **46** (1990) 251.
46. R. Chen and A. Hag-Yahya, *Radiat. Prot. Dosim.* **65** (1996) 17.
47. J.A. Zonneveld, *Automatic Numerical Integration*, 2nd edition, Matematisch Centrum, Amsterdam (1970).
48. W.H. Press and S.A. Teukolsky, *Computers in Physics* **6** (1992) 188.
49. S.P. Nørsett and P.G. Thomsen, *BIT* **26** (1986) 100.

50. C.W. Gear, *Numerical Initial Value Problems in Ordinary Differential Equations,* Prentice-Hall, Englewood Cliffs, (1971).

51. R.N. Kulkarni, *Radiat. Prot. Dosim.* **51** (1994) 95.

52. L.N. Kantorovich, G.M. Fogel and V.I. Gotlib, *J. Appl. Phys.* **21** (1988) 1008.

53. L.N. Kantorovich, and G.M. Fogel, *J. Appl. Phys. D: Appl. Phys.* **22** (1989) 817.

54. L.N. Kantorovich, G.M. Fogel and V.I. Gotlib, *J. Phys. D: Appl. Phys.* **23** (1990) 1219.

55. L.N. Kantorovich, A.I. Livshicz and G.M. Fogel, *J. Phys.: Condens. Matter* **5** (1993) 7503.

56. K.N. Razdan, W.D. Brennan and L.I. Grossweiner, *J. Appl. Phys.* **41** (1970) 832.

57. K.N. Razdan, W.G. Wiatrowski and W.D. Brennan, *J. Appl. Phys.* **44** (1973) 5483.

58. M. Böhm and A. Scharmann, *Phys. Stat. Sol.* (a) **5** (1971) 563.

59. R. Chen, *J. Appl. Phys.* **42** (1971) 5899.

60. W. Schreiner, M. Kramer, S. Krischner and Y. Langsam, *PC Tech Journal* (1985) 72.

61. D.W. Marquardt, *J. Soc. Indust. Appl. Math.* **11** (1963) 431.

62. K. Levenberg, *Quart. Appl. Math.* **2** (1944) 164.

63. M. Avriel, *Nonlinear Programming: Analysis and Methods*, Prentice-Hall Inc. Engelwood Cliffs, (1976).

64. P.J.M. van Laarhoven and E.A. Aarts, *Simulated Annealing: Theory and Applications*, Reidel Pub. Co., Dordrecht (1987).

65. R. Battiti and G. Tecchioli, *Comput. Math.* **28** (1944) 1.

66. F. Glover, *Discr. Appl. Math.* **49** (1994) 231.

67. J.R. Koza, *Genetic Programming, On the Programming of Computers by Means of Natural Selection*, The MIT Press, Cambridge (1992).

68. P. Sutton and S. Boyden, *Am. J. Phys.* **62** (1994) 549.

69. K.A. Dowsland, *J. Oper. Res. Soc.* **47** (1996) 550.

70. S.K. Misra and N.W. Eddy, *Nucl. Instrum. Meth.* **166** (1979) 537.

71. Y.S. Horowitz and D. Yossian, *Radiat. Prot. Dosim.* **60** (1995) 1.

Chapter 7
APPLICATIONS OF THERMOLUMINESCENCE

7.1 Radiation Dosimetry

One of the earliest suggested applications of TL was the detection and measurement of absorbed radiation. As materials came into notice which possessed high sensitivity - that is detectable luminescence emission for small absorbed doses of radiation - so emerged the potential use of TL in dosimetry. Earliest among those to suggest this application specifically were Watanabe[1], Daniels and colleagues[2],[3], and, later, Cameron and colleagues[4]. The development advanced to the extent that the first book on the topic was published in 1968[5].

Dosimetry applications can be most conveniently divided into several general categories, including the detection of the absorbed dose to people and to the environment. The basic premise in the use of TL in dosimetry is that the intensity of the TL can be related, in a straightforward way, to the absorbed radiation dose. In an ideal system, the TL signal is directly proportional to the absorbed dose over a wide range of doses. As a result, the dose response can be calibrated, and unknown doses determined by simply measuring the TL intensity resulting from the exposure and comparing with the calibration signal. In fact, although complications are abundant (see Ch. 4), this simple concept can essentially be considered to have been achieved in reality. The sensitivity and wide dynamic range of TL in this application are unrivaled by other established dosimetric techniques. The method can be used routinely to measure doses as small as those delivered by the natural environment over a period of several days, or doses as high as those delivered inside nuclear reactors - a dose difference of a factor of approximately 10^{11}. In what follows we describe briefly the range of uses of TL in radiation dosimetry and, in doing so, we introduce the term thermoluminescence dosimeter, or TLD.

7.1.1 Dosimetry Applications

Personnel Dosimetry

The primary objective of personnel dosimetry is to monitor the radiation dose delivered to personnel during routine occupational exposure. Examples include workers at nuclear reactors, hospital radiotherapy technicians, or naval personnel on nuclear powered vessels. In addition to routine monitoring, the field of personnel dosimetry also includes the determination of absorbed doses caused by accidental exposure to radiation. The overall desired aim is to limit an individual's total exposure to below the maximum permissible dose values recommended by the regulatory agencies. Broadly, personnel dosimetry covers: (i) *Extremity Dosimetry* - determination of the maximum dose equivalent to any part of the hands, arms or feet. (ii) *Whole Body Dosimetry* - determination of the dose equivalent absorbed at a depth of 400-1000 mg cm^{-2} in human tissue below the surface of the body, and/or the dose equivalent in critical organs. The primary interest is in penetrating radiation (gamma rays, x-rays (>15 keV) and neutrons). (iii) *Tissue Dosimetry* (i.e. skin dose) - determination of the dose equivalent absorbed at a depth of 5-10 mg cm^{-2}. Here one is interested in non-penetrating radiation (beta particles, and <15 keV x-rays).

In all such applications the primary requirement of a TLD material is that of tissue equivalency. That is, one seeks a TLD material that absorbs the same dose of radiation in a given position in a given radiation field as human tissue would absorb if placed at the same position in the same field. Additional desirable properties include an independence on the energy of the radiation (more specifically, the same energy dependence as displayed by human tissue making the material *tissue equivalent* at all energies), and a high sensitivity. The dose equivalent range of interest is from $\sim 10^{-5}$ Sv to 10^{-1} Sv, with a required uncertainty of better than $\pm 10 - 20\%$. (*Note*: using SI units, dose equivalent in tissue, H (in Sv) = absorbed dose, D (in Gy) $\times$ quality factor, Q[6]; 1 Gy = 1 J/kg.)

Environmental Dosimetry

The continuous monitoring of radiation released to the environment has become a major concern for industrialized nations and TLD usage is an important factor in this type of activity. Since exposure levels are low (with equivalent doses of, typically, 10^{-5} Gy) long exposure times are required and, thus, long-term stability of the TLD becomes vitally important, along with extreme sensitivity. Gamma and neutron emitters are the main radiation sources of interest, and potential sources include the escape of gaseous radionuclides from nuclear power stations during day-to-day operations, low-level waste disposal, contamination from nuclear fuels reprocessing, nuclear power station accidents, and others. Throughout the United States and Europe TLD systems are in place near nuclear installations for the purpose of monitoring radiation levels, above the natural background, which can be linked to the operation of these facilities. In addition, since the advent of manned space flight there has been a growing interest in dosimetry in space where the primary radiation contribution is from galactic cosmic rays. One obvious cause for this interest is the increased exposure to harmful radiations that people (astronauts) can expect on long-term, manned space flights. Additional concern lies in the prolonged exposure of electronic components where the decreasing scale of the active regions in these devices (a few square microns in cross-section) and the shift from analog to digital circuitry means that device failure due to discrete radiation events (i.e. interactions with single particles, known as Single Event Upsets) is becoming a serious concern. The main component of galactic cosmic rays is energetic protons (1 - 1000 MeV), and heavy charged particles (HCPs) from the solar wind. TLDs have been used on several recent manned space flights for the purpose of determining the dose from such penetrating radiation[7].

Medical Dosimetry

The small size and high sensitivity of TLD materials has long been exploited in clinical studies since the TLD samples can be inserted into appropriate openings in the human body before exposing the patient to ionizing radiations during diagnosis and/or therapy. The exposed TLDs are then retrieved and analyzed. In this way physicians are able to determine actual doses delivered to critical internal organs during these procedures and from such information are able to prescribe necessary additional treatments. Such utility is hardly possible with any other form of radiation dosimeter. The two areas of use for clinical radiation exposure of humans are diagnostic radiology (e.g. x-ray exposure in mammography, dentistry and general health screening) and radiotherapy (primarily cancer therapy of various types). Radiation types include x-rays (as low as 10 keV), gamma-rays (from ^{137}Cs or ^{60}Co), beta particles (up to 40 MeV), protons and other heavy charged particles, and neutrons. Doses equivalent range from 10^{-5} to 1 Sv in radiology, and up to 10 Sv in radiotherapy. Dose estimation accuracy of better than $\pm 3\%$ is aimed for since accuracy outside this range can adversely affect the outcome of the treatment.

Clearly, the foremost requirement of a TLD material to be used in this fashion is tissue equivalency. High sensitivity is desirable so that TLD sizes can be kept as small as possible for *in vivo* measurements. Non-toxicity is also a required property. Furthermore, since the doses used can be quite high, a linear dose response over a wide dose range is advantageous.

Retrospective Dosimetry

Worldwide, interest is growing in the development of new and improved methods for retrospectively assessing the radiation dose in accidentally contaminated areas. For a number of years TL techniques have been used with some success in this application. Examples include Chernobyl (Byelorussia and Ukraine) and the Techa River (Russia), along with similar efforts at Hiroshima and Nagasaki (Japan) and the Nevada Bomb Test Site (US). The thrust has been to use commonly available materials, located at the actual sites, which may be used as local dosimeters. Example TL materials which are proving to be useful in this context are pottery, tile, porcelain, and other ceramic objects. From these materials quartz and feldspar can be extracted and the absorbed dose to these components can be determined using various TL dosimetry techniques. The most popular techniques used thus far are those developed for dating applications (see a later section in this chapter) and include the Pre-Dose Effect and so-called quartz-inclusion method (e.g.[8],[9]). Optical readout techniques (OSL; Ch. 5) have also been used in this application[10].

High Dose

Materials testing, food sterilization and other similar applications require the use of very high (10^2 to 10^6 Gy) doses of irradiation and TLDs have also found application in these high dose regimes. Many materials suffer from the onset of severe super(supra)linearity at these dose levels and therefore the number of materials which can be used in these applications is somewhat limited. Nevertheless, LiF, Al_2O_3 and $CaSO_4$ and others have been found to be useful to some degree. A challenge to TLD materials development in this arena is to fabricate (or find) materials with a high saturation dose, which defines the upper limit of measurable dose in this context[11]. Apart from monitoring the direct TL response to dose at such high dose levels, however, some researchers have advocated monitoring the decrease in the TL sensitivity (so-called "radiation damage") to provide assessment of the dose. The mechanisms responsible for the sensitivity decrease, however, are unknown in most materials.

7.1.2 General Properties of TLDs

The general properties of a wide variety of the main-stream TLD materials used in dosimetry have been reviewed in a book by McKeever et al.[12]. Below we summarize each of the necessary dosimetry properties.

Dose Response

Obvious, desirable properties of TLD materials include a linear response to absorbed dose and an adequate sensitivity. The dose response of most TL materials is far from linear, however, displaying a variety of non-linear effects depending upon the dose range and the irradiation type (see Ch. 4). However, at sufficiently low doses most popular TLD materials display an approximately linear response, along with a sufficient sensitivity, to enable their use in several of the fields discussed above.

The dose response function (also known as the supralinearity index) $f(D)$ is defined as the functional dependence of the intensity of the measured TL signal upon the absorbed dose. Using Eq. 4.10 we define the normalized dose response function, as modified by Chen and McKeever[13]

$$f(D) = \frac{(S(D) - S_0)/D}{(S(D_l) - S_0)/D_l} \tag{7.1}$$

where S_0 is the intercept on the TL axis of the extrapolation of the linear region of the dose response curve, and $S(D)$ and $S(D_l)$ are the TL intensity values at doses D and D_l. The dose D_l is a dose at which the dose response is linear. For most TLD materials $S_0 = 0$ and thus Eq. 7.1 takes the more usual form

$$f(D) = \left[\frac{S(D)/D}{S(D_l)/D_l}\right]. \tag{7.2}$$

Thus, an ideal dosimeter would satisfy $f(D) = 1$ over a wide dose range, say from $D = 0$ Gy to several MGy. Unfortunately, $f(D) = 1$ is found only over a narrow dose range, up to a few Gy, in many TLD materials. Supralinearity, defined as $f(D) > 1$, is commonly observed, while sublinearity ($f(D) < 1$) is most often observed during the approach to saturation. Non-linearity is not necessarily a serious problem in dosimetry so long as the non-linear characteristic is single valued and a reliable calibration curve can be generated. Often, however, $f(D)$ is found to be a *functional*, rather than just a *function* - that is, the actual value of $f(D)$ at any given dose is governed by the dose and thermal history of the sample, as well as a wide range of material and external factors - including the type and energy of the radiation used, and the detailed procedure used to heat the sample during glow curve measurement. Of particular importance is the observation that $f(D)$ depends upon the LET (linear energy transfer) or ionization density of the radiation field. For example, $f(D)$ values, obtained at about 200 Gy, for LiF TLD-100 have been observed as high as $\sim$3.5 for ^{60}Co irradiation, but only $\sim$1.5 for 20 kVp x-ray irradiation[14]-[16], and $\sim$1.0 for alpha particle irradiation[17],[18]. The dose at which the onset of supralinearity becomes evident is also LET-dependent[18] and spectral changes have also been reported as a function of LET[19].

Important material factors include the nature of the host material itself. For example, Fig. 7.1 shows the growth of TL with dose for quartz (110 °C TL peak), CaF_2:Mn (the main dosimetry peak) and LiF:Mg,Ti (TL peak 5). The dose response for the 110 °C peak from quartz is highly superlinear ($f(D) > 2$) while for the TL from CaF_2:Mn the response is almost linear ($f(D) = 1$). Peak 5 from LiF:Mg,Ti, however, displays a linear-supralinear-sublinear growth. Furthermore, Chen et al.[20] have demonstrated that in SiO_2 the superlinear

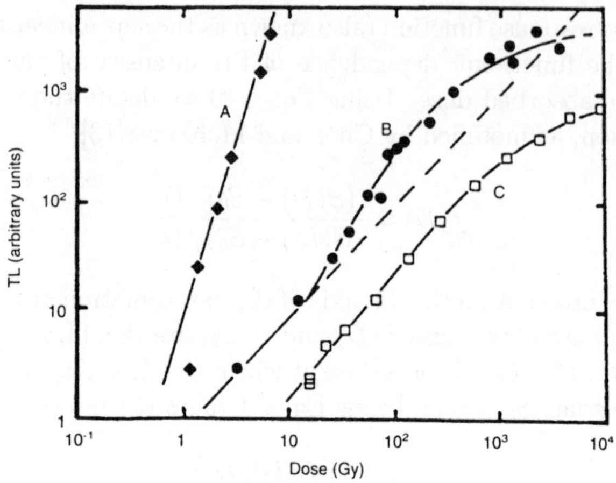

Figure 7.1 Example dose response curves for three materials used in TL dosimetry. A: the 110 °C peak in quartz; B: peak 5 from TLD-100 (LiF:Mg,Ti); C: the main dosimetry peak in CaF_2:Mn.

(After McKeever et al.[12].)

properties can be entirely removed by appropriate annealing treatments. In LiF:Mg,Ti, decreasing the Mg content enhances supralinearity[18], while increasing the OH impurity content decreases supralinearity[21],[22]. All these factors indicate a dependence of $f(D)$ on the nature of the host material and its defect structure, including the impurity content. The various theories forwarded to explain the shapes of the dose response curves shown in Fig. 7.1 have been discussed in depth in Ch. 4.

We might also mention an additional property required of a TLD material, namely that its dose response should be independent of the *rate* of irradiation. Measurements at extremely high dose rates using pulsed x-ray sources have revealed this to be the case for several common TLD materials. For example, LiF has been shown to have a dose rate independent response from 5 Gy/s to 1.5×10^8 Gy/s[23]-[25]. Fortunately, dose rate dependence does not seem to be a problem for most materials.

Aside from the super(supra)linear properties of TLD materials, the dose at which saturation sets in is an important property in high dose applications. Most materials display saturation at doses near 10^2 - 10^3 Gy. Many high dose applications require saturation limits of 10^5 - 10^6 Gy, however. Exercises at materials engineering for high dose applications have achieved dose limits of 10^4 - 10^5 Gy[11],[26].

Sensitivity

A formal definition of the sensitivity of a particular TLD material is the TL signal strength per unit absorbed dose. It is difficult, however, to define such a parameter in an absolute sense since sensitivity also depends upon the TL readout system used in the measurement, along with the optical filters, the heating rate and the method of measurement of the TL signal (usually the glow curve area between two chosen temperatures, or the height of a particular TL peak). To overcome uncertainties associated with the absolute measurement of sensitivity, one normally defines a *relative sensitivity* by comparing the TL signal from the material of interest with the TL signal from LiF TLD-100. Thus, if TLD-100 is assigned a sensitivity of 1, the relative sensitivity $R(D)$ of the TLD material is defined

$$R(D) = \frac{S(D)_{material}}{S(D)_{\mathrm{TLD-100}}} \tag{7.3}$$

where $S(D)_{material}$ and $S(D)_{\mathrm{TLD-100}}$ are the TL signals, at dose D, of the material and TLD-100, respectively. Since each material has its own $S(D)$ function, with differing degrees of nonlinearity compared to LiF TLD-100, $R(D)$ itself is dose dependent. Since the dose response of a given material is also dependent upon dose type, $R(D)$ is also LET dependent. Furthermore, some materials display a strong dependence of the sensitivity upon heating rate. For example, Al_2O_3:C undergoes strong thermal quenching of the F-center luminescence efficiency at high temperatures. We have seen how TL peaks move to higher temperatures as the heating rate increases (Ch. 2) and so we find that the luminescence efficiency of Al_2O_3:C decreases as the heating rate increases[27], and thus, $R(D)$ can also be heating rate dependent. TLD-100 does not show this behavior. Similarly, $R(D)$ depends on the filter selection during TL measurement because of the different emission wavelengths from the different materials. Hence, $R(D)$ is a somewhat loosely defined term.

Many of the materials used in TL dosimetry have poor sensitivity in the "as grown" state and have to be sensitized by various annealing and/or pre-exposure treatments. A popular sensitization procedure for TL peak 5 in LiF TLD-100 is to pre-irradiate with a dose of $\sim 10^3$ Gy and to subsequently heat the sample at 300 °C for one hour[5],[12]. This is believed to fill deep traps which may act as competitors to peak 5. In this way, sensitization can be seen to be intimately related to the properties of supralinearity in this material (Ch. 4). As one would expect from the suggested competition models for TL, a sensitized sample does not show the same degree of supralinearity as an unsensitized sample[21],[28].

The lowest level of detection (or minimum measurable dose) is a parameter related to sensitivity and is defined as two (some researchers quote three) standard deviations of the signal from an unexposed dosimeter. This definition applies, however, not just to the material but includes the complete TL system (i.e. material, sample (size and shape), badge construction, filters, readout instrumentation, algorithm and analysis routines). Material-dependent factors include regeneration of signals following either HCP irradiation[29], or neutron irradiation (for example by the production of tritium in LiF via the reaction ^{6}Li(n,α)^{3}H[30]), and the possibility of a "self-dose" caused by internal, natural radioactivity within the sample[31].

Energy Response

Tissue equivalency is an important consideration for those materials to be used in personnel or medical dosimetry. For photon radiation the important material factor is the so-called effective atomic number, Z_{eff}, which dictates the amount of energy absorbed by the material in a given radiation field. For human tissue $Z_{eff} = 7.4$, and thus any material with an effective atomic number near this value will mimic human tissue and is known as "tissue equivalent". The higher the value of Z_{eff}, and the lower the energy of incident radiation photons, the larger the TL response to a given dose. Thus, non-tissue equivalent materials tend to exhibit an "over response" at lower photon energies. Normally, one defines the energy response of a TLD material with reference to the response at a gamma photon energy of 1.25 MeV. Thus, CaF_2 ($Z_{eff} = 16.3$) has an over response of 3 at an energy of 10^{-1} MeV, while LiF ($Z_{eff} = 8.14$) has almost no over response at the same energy and is approximately tissue equivalent. TLD materials should exhibit high stability of the signal over exposure and analysis time scales, and preferably should be insensitive to light exposure, chemical reaction and humidity. The most popular material for dosimetry applications is LiF doped with Mg and Ti (TLD-100). This tissue equivalent material has found application in all personnel and medical applications. CaF_2 (doped with Mn, Dy or Tm impurities) is a popular environmental dosimeter, and some dosimetry services still use natural fluorite, even for personnel dosimetry. Other popular materials include $CaSO_4$ (Mn, Dy or Tm doped) and $Li_2B_4O_7$ doped with Mn or Cu. Developments include new, ultrasensitive forms of LiF (LiF:Mg,Cu,P) and sapphire (Al_2O_3:C)[12].

The energy response is the variation of the detected TL output, for a fixed dose, as a function of the energy of the absorbed radiation. This variation stems from the dependence of the material's absorption coefficient on radiation energy and for photon irradiation this is defined in terms of the mass energy absorption coefficient of the material, namely μ_{en}/ρ . The photon energy response $S_E(E)$

is defined

$$S_E(E) = \frac{(\mu_{en}/\rho)_m}{(\mu_{en}/\rho)_{ref}} \tag{7.4}$$

where the subscripts m and ref refer to the TLD material and a reference material, respectively. Air is normally used as the reference material.

Energy loss can take place by pair production, Compton scattering or the photoelectric effect, and which process dominates depends not just upon the energy of the incident photon but also upon Z_{eff}. The photoelectric component of the mass absorption coefficient varies approximately as Z_{eff}^3; the Compton component is proportional to Z_{eff}/M (where M is the molar mass of the TLD material); and the pair production component varies as Z_{eff}^2. Which process dominates depends also upon the energy. Photoelectric interactions dominate at low photon energies but the Compton effect becomes dominant as the energy increases. Pair production is only important at high energies ($E >$ several MeV). The exact energy ranges for which these effects predominate are governed by Z_{eff}. For materials with a low effective atomic, the Compton effect is the dominant energy loss mechanism for energies between ~25 keV to ~10 MeV. The practical use of photon energy response is aided by the definition of the *Relative Energy Response* (RER), defined with respect to 1.25 MeV photons from a ^{60}Co source, thus

$$(RER)_E = \frac{S_E(E)}{S_E(1.25 \text{ MeV } ^{60}\text{Co})}. \tag{7.5}$$

Since the most desirable response is flat, it is clear that low Z_{eff} materials are preferred. The photon energy response of "nearly tissue-equivalent" LiF:Mg,Ti shows an over-response for energies 20 keV $< E <$ 100 keV, with a decrease in response for $E <$ 20 keV due to the increased LET of the radiation at these low energies (e.g.[12],[32]). However, LiF:Mg,Cu,P, which has the same Z_{eff}, shows an "anomalous" energy response compared to LiF:Mg,Ti due to a microdosimetric ionization energy effect in which the response becomes saturated within microscopic volumes along the tracks of the secondary electrons.[33],[34] and results in a minimum in the photoionization density at the point where Compton scattering replaces the photoelectric effect as the main mode of energy transfer.

The response of a TLD material to β particles is more complex. Charged particles lose their energy in small increments, undergoing many collision and radiative interactions. The key parameter is the stopping power $dE/dx\,|_{E,Z_{eff}}$, where x is the distance into the material. The important material property is again the effective atomic number with both the radiative and the collision stopping power depending upon Z_{eff}. Since the stopping power is energy-dependent, one finds, for a given material, a variety of electron penetration depths depending upon the β-particle energy. The absorbed dose and, therefore, the measured TL signal will depend upon energy when the thickness of the TLD material is greater than the range of the β particles. In general, the energy dependence for β particles shows an increasing TL response with β-particle energy, leveling off when the β-particle range exceeds the material thickness. For this reason, β-particle TLDs are manufactured from thin materials of thicknesses 5 - 10 mg cm^{-2}. Examples of TLD dosimeters manufactured in this way include LiF[35] and Al_2O_3:C[36], both of which show a response to beta particles which is essentially energy independent down to 237 keV (from ^{204}Tl sources). The degree of over-, or under-, response for lower energies depends upon the filters used and the thickness of the samples.

For heavy charged particles, the energy is deposited in densely ionized tracks, as noted in Ch. 4. The early saturation of these local volumes gives rise not just to a linear dose response, but also to a lower TL sensitivity compared to β or high energy photon irradiation. Thus, it is found that as the LET of the HCP increases, so the TL efficiency (i.e. sensitivity) decreases. Track structure theory has been used to relate the HCP response (relative to gamma irradiation) to the gamma dose response function, $f(D)$[37] .

Since neutrons are neutral particles, the TL response to these particles stems solely from the interaction of the material with secondary charged particles. For thermal neutrons, TLD materials are manufactured with isotopes of high thermal neutron reaction cross-section. Thus, $Li_2B_4O_7$ and LiF are manufactured enriched in ^{6}Li and (for lithium borate) ^{10}B. The fast neutron TL response arises from the production of recoil protons and heavy charged particles (fission fragments). These particles have a low TL sensitivity and as a result, the response of TLD materials to fast neutrons is weaker than that for thermal neutrons. Neutron sources usually contain a significant gamma-ray component, and hence there is a need for TLD materials which discriminate between these two radiation types in mixed gamma-neutron fields. Materials which exhibit high TL sensitivity to gamma photons, but have a relatively low saturation dose would appear to be excellent neutron discriminators. This is because, within the tenants of track structure theory, the low saturation dose leads to a greatly reduced TL sensitivity to heavy charged particles and this in turn means a low neutron sensitivity.

7.1.3 Some Example TLD Materials

Some brief examples of TLD materials and their properties are noted below. More detailed information can be obtained from specialist texts such as Cameron et al.[5], Horowitz[38], and McKeever et al.[12], and the proceedings of the triannual Solid State Dosimetry Conference (e.g. [39]). Some representative glow curves of popular TLD materials are shown in Fig. 7.2.

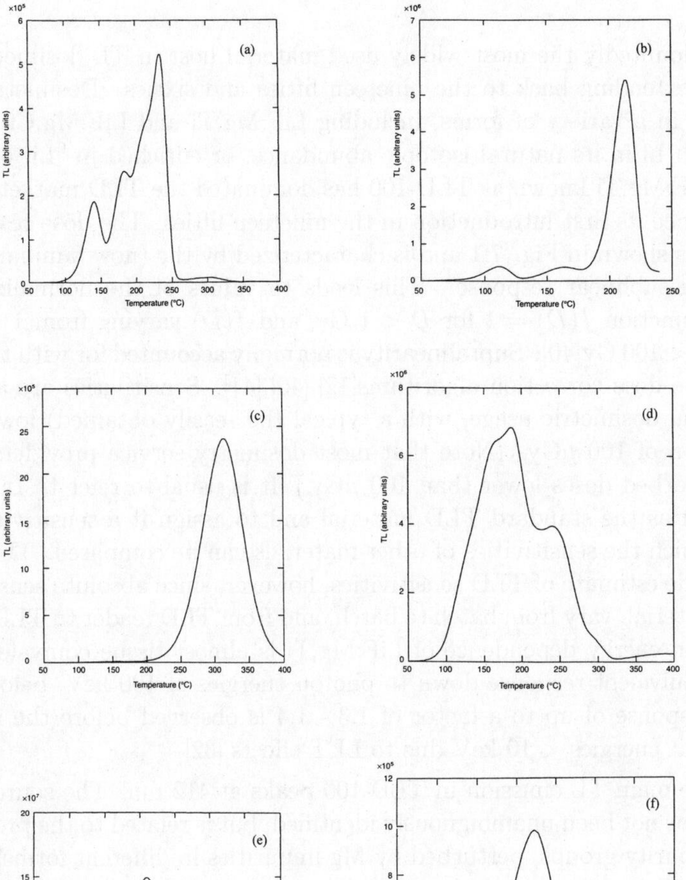

Figure 7.2 Representative glow curves of some of the main TLD materials. (a) LiF:Mg,Ti; (b) LiF:Mg,Cu,P; (c) CaF_2:Mn; (d) CaF_2:Dy; (f) Al_2O_3:C; (g) $CaSO_4$:Dy.

(After McKeever et al.[12].)

Lithium Fluoride

LiF is undoubtedly the most widely used material host in TL dosimetry, with a history extending back to the nineteen fifties and sixties. Dosimetric-grade LiF exists in a variety of forms, including LiF:Mg,Ti and LiF:Mg,Cu,P. Each comes with Li in its natural isotopic abundance, or enriched in ^{6}Li, or ^{7}Li. A form of LiF:Mg,Ti known as TLD-100 has dominated the TLD market for four decades since its first introduction in the nineteen fifties. The dose response of TLD-100 is shown in Fig. 7.1 and is characterized by the (now familiar) linear-supralinear-sublinear response. This leads to values of the normalized dose response function $f(D) = 1$ for $D < 1$ Gy, and $f(D)$ varying from 1 to ~ 2.5 for $1 < D < 100$ Gy[40]. Supralinearity is normally accounted for with the use of appropriate dose correction algorithms[12],[40],[41]. Sensitivities are adequate for personal dosimetric usage, with a typical (i.e. easily obtained) lowest level of detection of 100 μGy. (Note that most dosimetry service providers do not report absorbed doses lower than 100 μGy.) It is usual to refer to LiF:Mg,Ti (TLD-100) as the standard TLD material and to assign it a sensitivity of 1.0 against which the sensitivities of other materials can be compared. This allows only a crude estimate of TLD sensitivities, however, since absolute sensitivities, for any material, vary from batch to batch, and from TLD reader to TLD reader. The photon energy dependence of LiF:Mg,Ti is almost tissue equivalent, with a tissue equivalent response down to photon energies of 100 keV, below which an over-response of up to a factor of 1.3 - 1.4 is observed before the response decreases at energies $<$ 10 keV due to LET effects.[32]

The main TL emission in TLD-100 peaks at 412 nm. The source of the emission has not been unambiguously identified, but is related to the presence of Ti-OH impurity groups, perturbed by Mg impurities in different forms[42],[43]. LiF:Mg,Cu,P is characterized by an emission centered at 410 nm which is evidently related to P[44].

Perhaps the greatest disadvantage of LiF:Mg,Ti is its propensity to change its sensitivity as a result of usage. This property requires that the sensitivity be reset before the material can be used again in dosimetry. The stimuli for these sensitivity changes are the irradiation itself, and the heating of the sample during TL readout. Irradiation of the material results in the filling of deep traps. Empty deep traps act as competitors to the TL process such that as the deep centers fill, as a result of previous exposure, the competition effect of these defects is lessened. The result is an enhancement of the sensitivity and, at the same time, the removal of the supralinear growth of the TL with absorbed dose[12],[38],[45]-[21]. This can be understood from a discussion of the role played by competing traps in the dose response of TL materials, as discussed in detail in Ch. 4. The effect of heating is to alter the defect equilibria within the material such that the concentrations of the various defect types are altered. The main defect species giving rise to TL peak 5 in this material is a trimer cluster of three Mg^+_{Li}-V^-_{Li} monomers (or dipoles). The concentration of the trimers is maximized by pre-irradiation annealing of the material at a temperature between 300 and 400 °C, for about 1 hour, followed by an anneal at 80 °C for 24 hours. The 300-400 °C anneal serves to disperse the Mg impurities into dipole form, while the lower temperature anneal enables the clustering of the dipoles into trimers. It is clear then that the defect equilibrium existing at room temperature prior to, and during, irradiation is a metastable equilibrium state, and the slow return to a lower free energy state causes annealed LiF:Mg,Ti to have a limited "shelf-life" before re-annealing is required. The return to a lower free energy state occurs via the diffusion of impurities and the formation of non-luminescent, Mg precipitates. If left for long enough, sufficient precipitation can occur and a "fading" of the TL signal becomes evident (i.e. a loss of sensitivity; see Ch. 10 for additional remarks concerning this type of "fading").

Heating the material during TL readout, however, stimulates the dispersion of the trimers into dipoles and, as a result, the defect equilibrium after readout is not the same as that before readout - hence the need for the two-stage annealing procedure mentioned in the previous paragraph. Note that the higher temperature of 400 °C is usually selected for the anneal since this meets the dual purpose of resetting the defect equilibrium, and emptying the deep traps which would otherwise remain filled after TL readout.

Other annealing protocols have been suggested for this material, including post-irradiation anneals, so-called "reader anneals" in which the sample is heated only in the TL readout process, and, indeed, no annealing at all (viable for low dose usage). These and other suggested treatments for LiF:Mg,Ti are summarized in detail by Horowitz[38] and McKeever et al.[12].

The dose response of LiF:Mg,Cu,P is linear-sublinear, rather than linear-supralinear-sublinear (as for LiF:Mg,Ti). The lack of supralinearity is a particular advantage for radiotherapy applications in which the absorbed dose is typically of the order of 10 Gy. Linearity extends from $\sim 3 \times 10^{-6}$ Gy to $\sim$ 10 Gy, where sublinearity starts[47]. This difference in dose response has been explained on the basis of track interaction and track structure theory[48] (see also Ch. 4). The primary advantage of this material in TLD usage is its high sensitivity making it very attractive for several personal and environmental dosimetry applications. Its sensitivity with respect to TLD-100 is from 23 to 35 times greater, depending upon the selected detection wavelength and TL reader[47]. The energy dependence of LiF:Mg,Cu,P is considerably different from LiF:Mg,Ti showing a significantly reduced response at lower photon energies. As noted above, this is explained[33],[34] as a microdosimetric effect arising from a saturation of the TL in microscopic volumes along the tracks of the secondary electrons.

The biggest difficulty with using LiF:Mg,Cu,P in dosimetry is related to its response to thermal treatments. Both the sensitivity and the glow curve shape are dependent upon the maximum temperature used in TL readout, and the pre-irradiation annealing parameters. These considerations have led to a recommendation of a pre-irradiation anneal of 240 °C for 10 min., and a limitation of 240 °C during TL readout. A difficulty with this protocol, however, is the build up of a residual signal which can hinder the application of this material when measuring low doses[49]. To overcome this some authors have increased the readout temperature to 270 °C, but only for a short duration[50].

Calcium Fluoride

Dosimetric grade calcium fluoride comes in several forms, namely CaF_2 doped with Mn, Dy or Tm, or natural fluorite. The major impurities in fluorite important for TL dosimetry are Dy, Tm and Ce. The simplest glow curve from these materials is that of CaF_2:Mn (see Fig. 7.2) which exhibits a single glow peak (actually made up of a series of overlapping peaks; possibly indicating a distribution of similar trapping states). The peak is only slightly supralinear with dose (Fig. 7.1) although the exact dose response depends upon the physical form of the material (thin chips, thick chips, powders) due to radiation-induced coloration and self-absorption of the TL[51]. Because of the high Z_{eff} for this material (16.3) all CaF_2 TLDs exhibit a significant overresponse at photon energies < 80 keV, limiting their usefulness in personal dosimetry. Sensitivities, relative to TLD-100, are variously reported to vary from ~1.7 to ~6.7, depending upon the measurement conditions[12]. The glow curve is quite stable with re-use (i.e. in response to prior irradiation and heating) and residual signals can be kept to acceptable levels by a pre-irradiation anneal at 400 °C for 10 min.

This material's biggest drawbacks are the background which builds up with time, possibly caused by potassium impurities within the growth starting material, and a surprisingly high degree of fading for a glow peak which appears at such a high temperature. The cause of the fading appears to be partially due to tunnelling (i.e. anomalous fading; see Ch. 10).

CaF_2:Dy and CaF_2:Tm each have more complex glow curves than the Mn-doped version of this material. The dose response for both exhibits supralinearity and while the sensitivity of CaF_2:Tm is comparable to that of CaF_2:Mn, the sensitivity of CaF_2:Dy is ~30 times greater than TLD-100. Since both materials feature TL emission at quite low glow curve temperatures, a high thermal fading rate is both expected and observed. A post-irradiation annealing treatment (for example, 115 °C for 10 min. for CaF_2:Dy) to remove the unstable parts of the glow curve is often adopted when using these materials in dosimetry. Emission from the Mn-doped material is at 495 nm, due to the Mn^{2+} d-electron first excited $^4T_{1g}(^4G)$ state to the $^6A_{1g}(^6S)$ ground state. Dy-doped material emits at 480 nm (due to $^4F_{9/2}$-$^6H_{15/2}$ transitions of the Dy^{3+} 4f-electrons) and 575 nm (due to $^4F_{9/2}$-$^6H_{13/2}$ transitions). In the Tm-doped material the TL emissions are at 360 nm, 450 nm, 465 nm and 650 nm (due to $^1D_2$-$^3H_6$, $^1D_2$-$^3F_4$, $^1G_4$-$^3H_6$ and $^1G_4$-$^3F_4$ transitions, respectively).

Aluminum Oxide

Aluminum oxide comes in two main forms for dosimetry. An earlier suggested form was Al_2O_3:Mg,Y[26], and, more recently, highly sensitive, anion-deficient Al_2O_3:C has been produced[27]. Other forms (i.e. with other dopants) have been suggested for TLD applications[12], with varying degrees of success. The dose response of Al_2O_3:C is linear up to approximately 1 Gy, after which it becomes very slightly supralinear until its saturation at $\sim$ 30 Gy. In contrast, the dose response of Al_2O_3:Mg,Y is linear all the way up to 10 kGy, making it a candidate for high dose dosimetry. The biggest difference between the two materials, however, lies in the sensitivity. The sensitivity of Al_2O_3:C has been variously reported to be up to 30-50 times greater than that of TLD-100, enabling this material to detect doses as low as several μGy under optimum conditions. By contrast, the sensitivity of Al_2O_3:Mg,Y allows a lowest level of detection of only ~1 mGy. The high sensitivity of Al_2O_3:C is a result of the growth conditions for this material in which a large concentration of oxygen vacancy centers is produced - hence the term, "anion deficient". The F- and F^+- centers which result, yield large numbers of radiative recombination sites and, hence, a high TL sensitivity. Thus, the TL emission (at 420 nm) is characteristic of the 3P-1S transition of F-centers. In Al_2O_3:Mg,Y, however, the number of such centers is much reduced, with a consequent loss of sensitivity. Al_2O_3:Mg,Y exhibits TL emission at 320 nm due to the 1B-1A transition in F^+-centers. Occasionally, the 695 nm emission characteristic of the ^2E-$^4A_{2g}$ 'R-line' transition in Cr^{3+} impurities may also be observed.

Another major difference lies with the glow curve peak temperature - the main dosimetric peak for Al_2O_3:C appears at ~200 °C (at ~10 °C/s; Fig. 7.2) whereas Al_2O_3:Mg,Y shows a much weaker peak at 200 °C and a large one at 300 °C. A difficulty with the F-center emission from Al_2O_3:C is that it is strongly thermally quenched at around the temperature of the TL peak. As a result, not all of the recombinations result in luminescence and this, in turn, gives rise to a strong heating rate dependence of the sensitivity. As described in earlier chapters, TL peaks shift to higher temperature when the heating rate is increased. At higher temperatures, however, the degree of thermal quenching is enhanced, with the result that less light (rather than more light, as expected from kinetic arguments; Ch. 2 and Ch. 3) is obtained as the heating rate is increased[52]. This is one of the properties of this material which has held back its popular adoption in TLD applications.

The second property of this material to prevent its widespread use as a mainstream TLD material has already been discussed in Ch. 5 - namely, its light-induced fading properties[53]. The main dosimetric signal from this material is optically sensitive and displays both optically induced fading, and phototransfer into the main dosimetric traps from deeper traps. This makes thorough shielding from light essential when operating this material as a dosimeter. It should be noted, however, that the light sensitivity of this material is exploited by using the material as a very sensitive OSL dosimeter (see Ch. 5). The optical fading properties of Al_2O_3:Mg,Y have yet to be fully investigated.

Apart from its high sensitivity, an advantage of Al_2O_3:C is the fact that thermal treatments are unnecessary between dose measurements when being used for assessment of low doses. The material is usually given a one-time anneal at 900 °C for 1 hour before application in the field. After that, it can be re-used at will, as long as high doses are not being measured. After high doses, or after frequent application of low doses, a residual trapped charge can build up in deep traps and this can affect the sensitivity of the material to both radiation (through competition effects) and to light[54]. To re-set the material, therefore, a high temperature anneal is performed. For Al_2O_3:Mg,Y the recommended thermal treatment is a pre-irradiation anneal at 600 °C for 1 hour, followed by 100 °C for another hour[12].

The effective atomic number of Al_2O_3 is 10.2, causing an over-response (with respect to human tissue) for photon energies less than 100 keV, reaching $\sim$2.8 at $\sim$40 keV compared with 1.25 MeV[55]. As is to be expected, the photon energy response of Al_2O_3:Mg,Y displays similar features[56]. The fading of the TL from Al_2O_3:C is very small, but fading from Al_2O_3:Mg,Y is unexpectedly high, considering the temperature of the main glow peak[12].

Calcium Sulphate

The main emissions from the Dy-doped version of this material are at 475 nm (due to $^4F_{9/2}$-$^6H_{15/2}$ transitions of the 4f- electrons), and 575 nm (due to $^4F_{9/2}$-$^6H_{13/2}$ transitions). Longer wavelength emissions are observed at 660 nm ($^4F_{9/2}$-$^6H_{11/2}$) and 750 nm ($^4F_{9/2}$-$^6H_{9/2}$). The Tm-doped material exhibits main TL emissions at 360 nm, 450 nm, and 465 nm (due to $^1D_2$-$^3H_6$, $^1D_2$-$^3F_4$, and $^1G_4$-$^3H_6$ transitions, respectively). The dose response of the main TL peak (220 °C) from $CaSO_4$:Dy is linear up to a dose of 10 Gy, after which it is supralinear up to its saturation dose of 10^4 Gy. High doses can be followed (up to 10^5 Gy) using the high temperature TL peak (450 °C)[57]. Unusually, the degree of supralinearity is dependent upon the grain size in powdered $CaSO_4$, possibly related to self-absorption effects, as with CaF_2:Mn. The sensitivity is approximately 10 compared with TLD-100, although values of 40-60 have been reported[38],[58]. The sensitivity is dependent upon the concentration of Dy (or Tm) and optimum values lead to minimum measurable doses of 2 μGy for $CaSO_4$:Dy[59].

With $Z_{eff} = 15.5$, $CaSO_4$ displays a significant over response for photon energies between 25-200 keV.[59]. Recommended pre-irradiation annealing treatments are similar to TLD-100 and consist of 400 °C for 1 hour[60] although improved sensitivity and reduced residual signals can be obtained by annealing at temperatures up to 700 °C[38]. The fading characteristics can be improved with a post-irradiation anneal at 120 °C for 10 minutes[59]. Under such circumstances, fading of just 5% in 6 months may be expected. As with Al_2O_3, however, $CaSO_4$ is sensitive to light, and optically induced fading of 50% can be observed after just 3 hours of exposure to sunlight[61].

Other Materials

A host of other materials have been suggested at one time or another for use in TL dosimetry, some of which appear to show excellent properties. However none have been able to displace LiF:Mg,Ti (TLD-100) from its established position. Among the other materials are included $Li_2B_4O_7$:Mn, $Li_2B_4O_7$:Cu, MgB_4O_7:Dy, MgB_4O_7:Tm, BeO and MgO, along with other forms (dopants) of the materials mentioned above. We should also include natural forms of CaF_2 (fluorite) and NaCl (halite), as well as many other minerals. Particularly we should mention quartz and various feldspars used in TL dating - to be discussed in the sections to follow. We finish this section with a summary of the salient TL properties of the main TLD materials in Table 7.1.

7.2 TL Dating of Archaeological and Geological Samples

7.2.1 *Introduction*

Another important application of thermoluminescence is its use in the dating of archaeological and geological samples. In principle, this application is very closely related to that of dosimetry measurements discussed in the previous section, but the details are very different and, in most cases, the difficulties in this kind of investigation are much larger.

As explained by Aitken[63], one of the fathers of this technique, if a sample of ancient pottery, brick, tile, or from a terra-cotta figure is heated rapidly to, say, 500 °C, there is a weak but measurable emission of light.

Phosphor	Commercial Names (1)	Main TL Peak (°C) (3)	Main Emission (nm)	Z_{eff}	Max. Sensitivity	Linear Dose Range (Gy)	Saturation Dose (Gy)	Energy Response (30 keV/^{60}Co)
LiF:Mg,Ti (2)	TLD-100	~235	415	8.2	1	5×10^{-5} - 1	$\sim10^{3}$	1.3
LiF:Mg,Cu,P (2)	GR-200	~210	410	8.2	~35	3×10^{-6} - 10	~10	anomalous (4)
CaF_2:Mn	TLD-400	~310	495	16.3	~7	10^{-5} - 10	$\sim2\times10^{2}$	~13
CaF_2:Dy	TLD-200	~185	480,575	16.3	~30	10^{-5} - 10	$\sim5\times10^{2}$	~13
CaF_2:Tm	TLD-300	~170	360,450	16.3	~3	1 - 10	~10	~13
Al_2O_3:C	TLD-500	~210	420	10.2	~30	10^{-6} - 10^{2}	~50	~2.8
Al_2O_3:Mg,Y	D-2/D-3	~280	320	10.2	~1	10^{-3} - 10^{4}	$\sim10^{4}$	~2.8
$CaSO_4$:Dy	TLD-900	~225 (1 °C/s)	480,575	15.5	~50	10^{-6} - 10	$\sim10^{5}$	10-12
$CaSO_4$:Tm	-	~225 (1 °C/s)	360,450	15.5	~60	10^{-6} - 30	$\sim10^{2}$	10-12
$Li_2B_4O_7$:Mn	TLD-800	~190	600	7.4	~0.4	10^{-4} - 3	$\sim10^{4}$	~0.9
$Li_2B_4O_7$:Cu (2)	-	-	368	7.4	~2	10^{-4} - 10^{3}	$\sim10^{3}$	~0.8
MgB_4O_7:Dy	-	~230 (1 °C/s)	480,575	8.4	~7	10^{-5} - 1	~10	~1.5
BeO	-	~335 (1 °C/s)	330	7.1	~1	10^{-4} - 0.5	$\sim10^{2}$	~1.4
MgO	-	~200	750	10.0	-	10^{-4} - 10^{4}	$\sim10^{4}$	-

Table 7.1: Summary of the properties of several TLD materials. *Notes*: (1) Popular names, as assigned by various manufacturers and commercial suppliers. (2) ^{6}Li- and ^{7}Li-enriched versions of these materials also available. (3) At 10 °C/s, unless noted. (4) Anomalous response; see text. (Data extracted from McKeever et al. [12], Tawil [41] and McKeever [62].)

In a second heating of the same sample, the emission consists only of black-body radiation (incandescence). The extra light emitted in the first heating is TL that comes from minerals in the sample and is due to their prolonged exposure to the weak flux of natural nuclear radiation emitted by radioactive impurities in the pottery and the surrounding burial soil. These are potassium, thorium and uranium at concentrations of a few ppm. These radioactive materials have long half-lives of 10^9 years or more, and therefore, the radiation flux is constant and the amount of TL is proportional to the time that has elapsed since the pottery was fired by ancient man. The firing drains the relevant minerals of TL acquired during geological times, thus setting the "clock" to zero. In this way, by measuring both the sensitivity of the sample to acquiring TL by exposure to a calibrated radioactive source, and the radioactive content of the clay and surrounding soil, it is possible to calculate the age, at least in principle.

Whereas in TL dosimetry one can choose the best dosimeter available for a given radiation field, out of many candidates, in TL dating one can use only the naturally occurring materials in the sample under study. Moreover, the TL dosimeter can be studied and calibrated before its use whereas in dating, the calibration can be performed only after the exposure and signal read-out. As compared to TL dosimeters, samples used for dating are usually less sensitive to the excitation radiation. Also, non-linear dependence on the dose, such as superlinearity or saturation may decrease the accuracy of the age determination. In addition, some of the accumulated signal may be lost due to normal, thermal fading as well as anomalous fading (described in Ch. 10). An effect which may or may not be related is the dependence on dose rate, which means that there may be a difference between applying the same dose on a given sample over a period of hundreds or thousands of years in antiquity, or within a few minutes at a high dose-rate in the laboratory. Also, effects of sensitization of the sample (i.e. sensitization of the luminescent minerals, e.g. quartz) due to previous heating and/or irradiation may affect the correct comparison of the natural and laboratory doses.

If all these effects are small or can be compensated for in some way, the relevant age can be determined. The basic age relation is

$$age = \frac{TL\ acquired\ during\ antiquity}{TL\ per\ unit\ radiation\ dose \times annual\ dose\ from\ surrounding} \,. \tag{7.6}$$

Farrington Daniels who was the first to develop the use of TL for dosimetry, also suggested its utilization for geological and archaeological age determination (see Daniels et al.[64]). TL from ancient pottery was first reported by Grögler et al.[65] and by Kennedy and Knopff[66]. During the 1960s, thermoluminescence was developed for archaeological dating at Oxford (see e.g. Aitken et al.[67], [68]) as well as in other centers by Ichikawa[69], Mazes and Zimmerman[70], Ralph and Han[71] and Mejdahl[72].

Göksu et al.[73] extended TL dating to materials other than baked clay. This extension, in which burnt flint was used, potentially took the method well beyond the age range of radio-carbon dating. The same is true for a later extension to burnt stone and stalagmitic calcite. Studies in the dating of volcanic lava were reported by Wintle[74] and by Guérin and Valladas[75], but turned out to be rather difficult in practice. Vast amounts of work on other geological samples has been done over the last 30 years, starting with Morozov[76] who studied the TL from various types of sediment.

Windblown and some waterlain sediments can also be dated, and in these cases the TL clock appears to have been set (or nearly set) to zero by the bleaching effect of sunlight, as first suggested by Wintle and Huntley[77],[78]. It is to be noted that the upper date limit in these applications is within the last million years.

TL has also been utilized for dating the fall of meteorites to Earth. For details see, for example, papers by Sears and Mills[79], McKeever[80] and Benoit et al.[81].

It should be mentioned that an alternative way of detecting trapped carriers is ESR, which will be discussed in chapters 8 and 9, and OSL, which was discussed in Ch. 5. A review on the use of ESR for dating was given by Hennig and Grün[82], and a review of OSL dating is provided by Wintle[83].

An essential feature of TL necessary for its utilization for dating is the stability of the portion of the glow curve being used. The question may be asked how this stable region is recognized and confirmed. This is done by the *plateau test* (see, for example, Aitken[63]), in which the natural glow curve, namely, the TL observed from a sample which has not received any irradiation in the laboratory, is compared with the "artificial" glow curve observed as a result of irradiation by a radioactive source in the laboratory. As shown in Fig. 7.3, in the curve labeled N+β, there is substantial TL below 200 °C, due to the laboratory irradiation. The dashed line represents the ratio of the two glow curves. This ratio is seen to rise from zero below 200 °C to a plateau which begins around 300 °C. The reason is that the trapping states associated with the glow above 300 °C are deep enough and therefore the thermal leakage from the traps is negligible during antiquity. The stability corresponding to a given position in the glow curve increases very quickly with temperature. As stated by Aitken[63], whereas 200 °C corresponds to a lifetime of a few years, 300 °C corresponds to a lifetime of a few thousand years, and 400 °C to 10^7 years. Thus, a constant ratio between the natural and the N+β glow curves gives a strong indication that in this plateau region, there has been negligible leakage of electrons from the associated traps over the period of time that has elapsed since all the traps were emptied in the firing by ancient man.

The value of the ratio in the plateau region enables the evaluation of

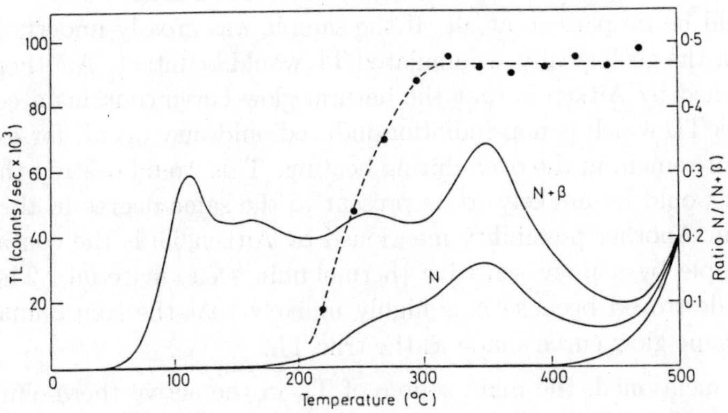

Figure 7.3 Curve N is the "natural" glow-curve and N+β the "natural+artificial" glow-curve obtained from another portion of the same sample to which a laboratory dose of β radiation has been administered. The dashed line represents the ratio of the two curves.

(After Aitken[63].)

the natural dose. The plateau level of 0.47 in Fig. 7.3 indicates that the dose equivalent is 0.47/(1 − 0.47), i.e. 0.88 times the artificial dose used which in the present case was 10 Gy[63]. Thus, in Fig. 7.3, the thermoluminescence due to the artificial dose of 10 Gy is 12% larger than the natural TL, which is equivalent to a natural dose of 8.8 Gy. This is not necessarily the accurate value for the paleodose that the sample received during antiquity due to the possible non-linear effects discussed in Ch. 4. Ignoring this possibility, the basic age equation may be written as

$$age = \frac{paleodose}{dose\ rate}. \tag{7.7}$$

Alternative names to the paleodose are *accrued dose, accumulated dose, archaeodose* and *total dose.* The term *equivalent dose* (ED) is reserved for the value found with no correction for superlinearity (if it exists). As discussed by Aitken[63], checking the occurrence of the plateau is essential and, due to some possible complicating factors, it is not sufficient just to go to the high temperature region of ~400 °C where the thermal lifetime is very long. The possible complicating factors include anomalous fading, which has been mentioned already. Also, if the sample was poorly fired by ancient man, the TL clock would not have been properly set to zero. Thus, the deeper traps would still contain electrons accumulated during geological times. This would cause the ratio to increase above the plateau value at higher temperatures, and it is possible that there would be no plateau at all. If the sample was grossly underfired, or not fired at all, the geologically accumulated TL would be intact. Another possibility mentioned by Aitken is that the natural glow curve contains a component of spurious TL which is non-radiation-induced, and may occur, for example, if traces of air remain in the oven during heating. This would destroy the plateau because it would be unlikely to be present to the same degree in the artificial glow curve. Another possibility mentioned by Aitken[63] is the contamination of the sample by a more sensitive thermoluminescent material. The plateau would be destroyed because it is highly unlikely that the contaminant would have the same glow curve shape as the true TL.

As mentioned, the main source of TL in the active thermoluminescent minerals used for dating archaeological samples are thorium, uranium and potassium. The radiation from rubidium and cosmic rays adds a small percentage to the total measured TL. ^{40}K emits β radiation which has a range of up to a few mm in pottery as well as γ-radiation which penetrates to a distance of about 0.3 m. The TL contribution from β particles arises from radioisotopes within the sample, except for the surface layer which may be influenced by the soil radioactivity. On the other hand, the TL contribution from γ-radiation is almost entirely from the soil. The TL is proportional to the amount of ionization and therefore, it is proportional to the energy absorbed from the radiation. The TL produced for a given amount of absorbed energy (i.e. in Gray) is the same for β and γ radiations. From thorium and uranium there are α particles as well. This kind of radiation is heavily ionizing radiation with ranges in pottery of 0.01-0.05 mm. Therefore, the TL per gray is significantly smaller for α particles than for β and γ radiation, being usually between 5% and 50% of that of β and γ depending on the material. The reason is that the ionization density produced is so large that the TL traps lying in the central core of the tracks of the particles get saturated and as far as TL is concerned, a large proportion of the electrons thus produced are lost as compared to β and γ radiations. Cosmic rays contribute about 3% of the total radiation dose[63].

The relative effects of the different kinds of radiation can cause TL dates to appear to be systematically too recent, as first pointed out by Fremlin and Srirath[84]. Special sub-methods have been devised to overcome these complications. Fleming[85],[86] developed the quartz inclusion technique in which large grains are extracted, and after etching away the outer surface to remove the α-irradiated component, the TL measurements are made on the residual cores of the grains so that in calculating the age, only the β and γ dose rates are considered. Zimmerman [87],[88] developed the fine-grain technique in which grains of 1-8 μm are separated for the TL measurement. These are small enough so that practically no attenuation of the α dose occurs. It is therefore appropriate here to calculate the age using the full annual dose rate including the α contribution.

Whereas quartz is low in radioactivity, there are other minerals such as zircon, apatite and potassium feldspars, in which the internal radioactivity is high. In large grains of such minerals the TL is due mainly to the internal dose within the grain which, as explained by Aitken[63], has important advantages for dating.

7.2.2 Evaluation of the Annual Dose

A number of methods are used to evaluate the annual dose imparted to a sample. One of these methods, which is of special interest in the context of the present book, is thermoluminescence itself. One can use TLD phosphors such as calcium fluoride, calcium sulphate, or aluminum oxide which are so sensitive that an exposure of such samples for a few weeks to the rather weak natural radioactivity causing TL in archaeological material, induces a measurable level of TL. In an arrangement developed by Bailiff[89], about 1 gm of powdered pottery is put in a perspex container. In the bottom of this container, there is a thin window of plastic which allows β particles to emerge, but not α particles. A TL dosimeter is placed underneath the window; this consists of a copper tray containing the TL phosphor set in silicone resin. The TL accumulated in several weeks of storage is measured using a standard TL oven where the tray is placed directly on the heater plate. In this case, it is obvious that the dose measured by the phosphor is less than that inside the pottery because β particles reach the phosphor from one side only and because of some absorption in the window. Using standard samples with known contents of potassium, thorium and uranium, the ratio between the two can be evaluated. For the specific arrangement described here, this ratio was found to be 3.55. During the storage, the unit is placed in a lead container to shield it from cosmic rays and external γ-rays (e.g. from the bricks of the building).

The dose rate inside a sample is affected by its moisture content because of absorption of part of the dose by water. However, it is convenient[63] to measure the β dose rate with the sample dry and then to make a correction to allow for the estimated average moisture content during burial. Whereas the range of β particles in pottery is only a few millimeters, γ-rays have a range of up to 0.3 m. Thus, the γ dose rate experienced by the quartz grains in a pottery fragment of a few centimeters originates almost entirely from the surrounding soil. The measurement of the γ dose rate is easily performed by burying the TL phosphor in the archaeological site. The phosphor is inserted into a capsule which is buried on-site into a 0.3 m deep hole and left there for several months. It is then taken to the laboratory and the accumulated dose is evaluated.

It is not easy to carry out the evaluation of the α dose rate by TL dosimetry, due to the short range of the α particles. A technique of *thick-source* α counting is normally used. A powdered layer of sample, thick compared to the 50 μm range of the most energetic α particles from thorium and uranium, is placed on top of a scintillation screen which is viewed from beneath by a photomultiplier. Each α particle striking the screen produces a scintillation which, in turn, produces a burst of photoelectrons from the PMT photocathode. After amplification, this becomes an electrical pulse which is registered on a suitable counting device. The scintillation screen is made by sprinkling ZnS powder onto sticky tape. In ZnS, pulses produced by α particles are much larger than those produced by β and γ radiations, which enables the rejection of the latter. In fact, the β and γ dose rates can also be derived from the α count-rate, provided that the ratio of thorium/uranium is known[63].

7.2.3 Authenticity Testing of Art Objects

Whereas many factors limit the accuracy of absolute dating of archaeological objects, TL has been very successful in authenticity testing. In this, one has to decide between an age of less than a hundred years (in the case of a fake) and over (say) 500 years for a genuine antique object. Uncertainties in the annual dose which may be typically $\pm 20\%$ and the exact burial circumstances are thus unimportant. A rather simple and inexpensive test may determine in a reliable way the authenticity of the object under test. The sample is obtained by drilling a small hole which yields about 50 mg of powder, sufficient to prepare a few fine-grain samples and leave a residue for evaluating the annual dose. Radioactivity measurements are made using α counting for the thorium and uranium and flame photometry for the potassium.

There have been claims that forgers of ceramic art objects used artificial γ irradiation in order to give the object an apparently authentic TL. There are, however, some more sophisticated methods to discover this situation. One of these is the subtraction technique which utilizes the combination of two methods, namely the fine-grain and the quartz inclusion techniques [63]. The total dose for the fine-grain method can be written

$$P_{FG} = age \times (D_{\alpha}^{'} + D_{\beta} + D_{\gamma} + D_{c}) \tag{7.8}$$

where P_{FG} is the paleodose evaluated for the fine grains; D_{β}, D_{γ} and D_c are the annual dose rates of β, γ and cosmic rays, respectively and $D_{\alpha}^{'}$ is the *effective* α dose rate. For the quartz inclusions, grains of around 0.1 mm diameter are selected before etching, and the attenuation of the β dosage requires a small correction of about 10%. Thus, the dose equation is

$$P_I = age \times (0.9D_{\beta} + D_{\gamma} + D_c)\ . \tag{7.9}$$

After subtraction, one gets

$$age = \frac{P_{FG} - P_I}{D'_\alpha + 0.1 D_\beta} \,. \tag{7.10}$$

This result is independent of the external dose from γ radiation and cosmic rays, and, of course, the effect of an external, artificial γ irradiation should cancel out in this procedure. An alternative method of using zircon ($ZrSiO_4$) grains for this purpose has also been used. The details can be found in the work by Zimmerman et al.[90].

7.2.4 Pre-Dose Dating

The possible effect of sensitization, which is the increase of the sensitivity of a sample following a cycle of irradiation and heating, has been discussed in Ch. 4. For the presently discussed application of dating, we shall briefly consider the sensitization of the TL peak at ~110 °C in quartz which is used for an independent dating technique called the pre-dose dating.

Fleming[91] proposed the pre-dose method, based on the pre-dose effect (sensitization) previously described and explained by Zimmerman[92] for the 110 °C peak in quartz. The 110 °C peak has a half-life at room temperature of about 1-2 hours and, therefore, is not observed in the natural TL. The peak is, however, sensitive to β dose and thus, exposure to a small dose of, say, 0.01 Gy is normally sufficient to induce a measurable TL. The change in *response* to this "test-dose" is the tool for measuring the archaeological dose. The main feature discovered by Fleming and by Zimmerman was that in order for the contents of the "memory" to the archaeological dose to be revealed, the sample has to be heated to ~500 °C. This thermal activation causes a strong increase in the subsequent response to the test dose. A typical example, taken from Aitken[63], is shown in Fig. 7.4. Assuming linearity of the sensitization effect with the dose, which is usually the case up to relatively large doses, the increase is proportional to the total dose received since the pottery was fired, namely, to the paleodose. The sequence of measurements is[63]:

1. Give a test dose and measure the 110 °C TL response, S_o.
2. Heat to the activation temperature, say, 500 °C.
3. Give a test dose and measure the 110 °C TL response, S_N.
4. Give a laboratory calibrating dose, β, of several Gray. Repeat step 3, and measure the quenched sensitivity S'_N which may be lower than S_N.
5. Repeat step 2.
6. Give a test-dose and measure $S_{N+\beta}$.

The reason why the *quenched* sensitivity S'_N is somewhat smaller than S_N was explained in Ch. 4; the difference between S_N and S'_N is usually rather small. Sometimes, a *quenching correction* is performed, as explained

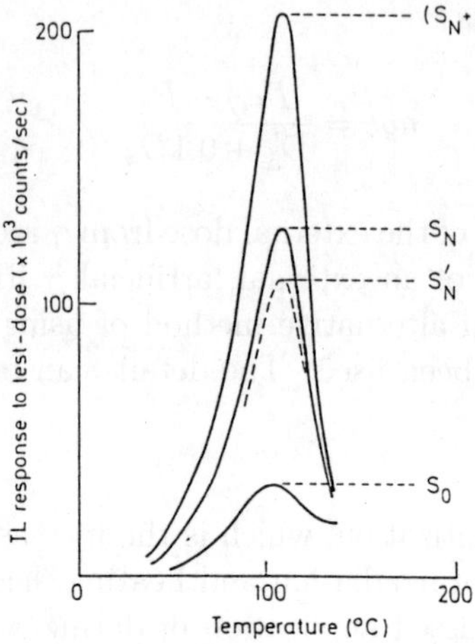

Figure 7.4 Magnitudes measured for pre-dose dating. The glow curves, all for the same portion, show the response to a β dose of 0.01 Gy at various stages; S_o before any heating, S_N after heating to 500 °C, S'_N after giving a laboratory dose of 1 Gy and heating to 150 °C, and $S_{N+\beta}$ after a second heating to 500 °C. (After Aitken[63].)

by Aitken[63]. Since $S_N - S_o$ is proportional to the paleodose N, and the further increase between steps 3 and 6 is proportional to the calibrating dose, we get, assuming that the quenching correction is small

$$N = \frac{S_N - S_o}{(S_{N+\beta}) - S_N}\beta. \tag{7.11}$$

Chen[93] suggested a method by which the natural dose N can be evaluated in cases where saturation effects occur, and tested it using laboratory doses in Norwegian α-quartz. The simplest assumption we can make about a saturating system is asserting that its functional form is exponential, namely,

$$S - S_o = (S_\infty - S_o)\left[1 - \exp(-D/B)\right] \tag{7.12}$$

where S_∞ is the maximum sensitivity attainable for very high values of the dose D, where B is a constant with dose units. Once the sensitivities (corrected for quenching, if required) are known as a function of the dose, one can use a numerical method to evaluate S_∞ and B and to extrapolate the curve to $S - S_o$, thus finding the equivalent dose N. Chen[93] also suggested an alternative graphical method, the details can be found in the book by Chen and Kirsh[94]. In view of the model for sensitization given by Zimmerman[92], Chen[93] made further distinction between the reservoir and center saturation (see Ch. 4), and discussed the possibility of extracting the archaeological dose in each of these cases. It is to be noted that although pre-dose dating is a rather powerful, and certainly an independent method for dating, even when extended into the ranges of approach to saturation, its use is limited to the last 1000-2000 years.

The pre-dose method was successfully used for the evaluation of the dose absorbed in roof tiles at Hiroshima[96] and of the fall-out dose associated with the Nevada nuclear bomb testing in the bricks of houses as reported by Bailiff and Haskell[95]. Pre-dose dating has also been used for authenticity testing in porcelain[97] where the direct TL is masked by spurious light or, in some cases, is affected by anomalous fading.

7.2.5 TL Dating of Geological Specimens

Thermoluminescence has also been used by several investigators to obtain the date of deposition of sediments[63]. Exposure to sunlight causes bleaching to near zero of previously acquired TL, and when the sunlight cannot reach a certain layer due to the later deposition of additional layers, TL starts accumulating in a similar way to that in pottery after firing. A sediment which is often dated is loess, which consists of grain sizes of 2-60 μm. In dry, windy conditions, these grains may have been carried up to several hundred kilometers from their source. During this transportation, previously acquired TL may indeed be totally bleached. In addition to loess, other sediments have been dated by TL. These include sand dunes, ocean sediments, sediments in lakes filled by melting glaciers and dust incorporated into the ice of a glacier. The first works on this subject were by Shelkoplyas[98] and the subject got an important boost by the work by Wintle and Huntley[77],[78].

Although sediments originate from different sources, their glow curves are rather similar. Usually, the TL from feldspars in sediments is stronger than that of quartz. The former is more easily bleachable and, therefore, its zeroing is better. On the other hand, in feldspars there is anomalous fading. It is often advisable to separate quartz from feldspars, and to perform the dating separately.

The most recent trends in TL dating can be found in the proceedings of the triennial Solid State Dosimetry Conferences and the triennial International Conference on Luminescence and EPR Dating.

7.2.6 Optical Dating and Related Effects

In Ch. 5 we described the phenomenon of optically stimulated luminescence (OSL), and its utilization for dating of sedimentary deposits as well as other geological and archaeological specimens has been mentioned. As noted, this application was first suggested by Huntley et al.[99]. These authors also mentioned the main advantages of using light rather than heat to stimulate the luminescence, namely, that one directly monitors the light-sensitive centers. In this section, we will mention rather briefly some examples of the use of optical dating. As described in a short review paper by Aitken[100], for routine dating of quartz, the signal is usually stimulated by green light from an argon ion laser, from filtered xenon or halogen lamps or from light-emitting diodes. The dominant emission wavelength in quartz is in the UV. The limitation in the measurable age range in quartz is due to saturation rather than to instability. For different types of feldspars, there is a strong stimulation of the signal in the near infra-red (IR), thus there is a distinction in the literature between OSL meaning stimulation by visible light, and IRSL when the stimulation is by infra-red. Here, IR-emitting diodes are often used for stimulation. In these cases, emission in the whole range of visible and UV light is available for detection. IRSL has an advantage over TL and OSL since it is more easily bleachable by daylight.

The 325 °C TL peak, often used for TL dating in quartz, was shown to have a lifetime of $\sim$30 Ma (3×10^7 years) while held at 20 °C. The OSL signal is apparently related to the same trapping level and recombination center, and Stokes[101] reported an age of $\sim$130 ka for sediments from the lowest lake at Bir Tafawi in Egypt, an age consistent with other, non-luminescent methods. It is obvious here that the age measured is not affected by the very long lifetime decay. Huntley et al.[102] evaluated the age of the Woakwine dune in Southern Australia as being 120-125 ka; this is associated with the highest sea level of the last interglacial period.

It is to be noted that shallow traps can contribute to OSL. Since these have much shorter lifetimes, their interference should be avoided. This can be done by pre-heating to, say, 220 °C for 5 minutes. The validity of the preheat can be established by the *preheat plateau test*[100].

Concerning IRSL, Hütt et al.[103] showed that K-feldspars extracted from sediments are very effectively stimulated by infra-red in the range of 750-1000 nm. The signal from feldspars is much brighter than from quartz which enabled the development of the *single aliquot dating* method (see Duller[104]). Single aliquot dating has been found to be useful for poorly bleached samples.

As pointed out, "optical dating" in its two versions, OSL and IRSL, has become popular since the mid-1980s, and although TL is still the leading candidate for many routine applications (e.g. for archaeological authenticity tests), research into the optical methods is definitely more active, in particular with regard to geological samples. As an example of a typical application, Berger and Huntley[105] studied glass samples from ash deposits, and determined an age of 7.7 ka for Mazama ash using IRSL. This age agreed with previous TL results. Similarly, Wood[106] evaluated quartz OSL dates for deposits which may be related to tsunami or a severe storm. The ages for different depths varied between 130 ka and 1 Ma. Perkins and Rhodes[107] reported the age of fluvial sediments as determined by optical dating. OSL and IRSL were used on the quartz components, and ages commensurate with other dating techniques were determined. Duller[108] and Lang[109] demonstrated the possibility of using IRSL for the dating of poorly bleached sediments. IRSL was also used by Clarke[110] who evaluated ages between 6000 and 8000 years for aeolian sand and alluvial fan deposits from the Mojave desert. Fuller et al.[111] described the partial bleach methodology as applied to IRSL of an alluvial sediment from the Danube. Krbetschek et al.[112] discussed some OSL and IRSL properties of natural feldspars as related to optical dating. The bleaching of IRSL and OSL from feldspars was reported by Spooner[113]. McKeever et al.[114] studied the OSL sensitivity changes during single aliquot procedures, using computer simulations, and Bailey et al.[115] reported the partial bleaching and the decay form of quartz OSL. Richardson[116] investigated the residual OSL signals in a modern sample of intertidal estuarine sediment, and found an overestimate of the age of an old sample of this kind, both by TL and IRSL. Anomalous fading of IRSL in archaeological ceramics was reported by Shepherd[117]. Optical dating of different sediments has been reported by Porat et al.[118], Lang and Wagner[119]and Feathers[120].

7.3 References

1. S. Watanabe, *Phys. Rev.* **83** (1951) 785
2. F. Daniels and D.F. Saunders, *Science* **111** (1950) 461.
3. F. Daniels, C.A. Boyd and D.F. Saunders, *Science* **117** (1953) 343.
4. J.R. Cameron and G.N. Kenney, *Radiat. Res.* **19** (1963) 199.
5. J.R. Cameron, N. Suntharalingam and G.N. Kenney, *Thermoluminescence Dosimetry,* University of Wisconsin Press, Madison (1968).
6. *Radiation Quantities and Units,* Report 19; *Dose Equivalent,* supplement to Report 19. International Commission on Radiation Units and Measurements, Washington DC (1973).
7. N. Vana, W. Schöner, M. Fugger and Y. Akatov, *Radiat. Prot. Dosim.* **66** (1996) 145.
8. E.H. Haskell, *Radiat. Prot. Dosim.* **47** (1993) 514.
9. I.K. Bailiff, *Radiat. Meas.* **24** (1995) 507.
10. L. Bøtter-Jensen, H. Jungner and N.R.J. Poolton, *Radiat. Meas.* **24** (1995) 525.
11. A.C. Lewandowski and V.K. Mathur, *Radiat. Prot. Dosim.* **66** (1996) 213.
12. S.W.S. McKeever, M. Moscovitch and P.D. Townsend, *Thermoluminescence Dosimetry Materials: Properties and Uses,* Nuclear Technology Publishing,

Ashford (1995).

13. R. Chen and S.W.S. McKeever, *Radiat. Meas.* **23** (1994) 667.
14. Y.S. Horowitz and M. Moscovitch, *Nucl. Instrum. Meth.* **A243** (1986) 207.
15. Y.S. Horowitz and B. Ben Shachar, *Radiat. Prot. Dosim.* **23** (1987) 401.
16. Y.S. Horowitz and S.W.S. McKeever, *Radiat. Phys. Chem.* **36** (1990) 35.
17. M. Moscovitch and Y.S. Horowitz, *J. Phys. D: Appl. Phys.* **21** (1988) 804.
18. E.F. Mische and S.W.S. McKeever, *Radiat. Prot. Dosim.* **29** (1989) 159.
19. P.D. Townsend, K. Ahmed, P.J. Chandler, S.W.S. McKeever and H.J. Whitlow, *Radiat. Eff.* **72** (1983) 245.
20. R. Chen, X.H. Yang and S.W.S. McKeever, *J. Phys. D: Appl. Phys.* **21** (1988) 1452.
21. T.G. Stoebe and S. Watanabe, *Phys. Stat. Sol. A,* **29** (1975) 11.
22. T.G. Stoebe and L.A. DeWerd, *J. Appl. Phys.* **57** (1985) 2217.
23. C.J. Karzmark, J. White and J.F. Fowler, *Phys. Med. Biol.* **9** (1964) 273.
24. E. Tochilin and N. Goldstein, *Health Phys.* **12** (1966) 1705.
25. N. Goldstein, *Health Phys.* **22** (1972) 90.
26. M. Osvay and T. Biró, *Nucl. Instrum. Meth.* **175** (1980) 60.
27. M.S. Akselrod, V.S. Kortov and E.A. Gorelova, *Radiat. Protect. Dosim.* **47** (1993) 159.
28. S.W.S. McKeever, *J. Appl. Phys.* **68** (1990) 73.
29. Ch. Petridou, C. Christodoulides and S. Charalambous, *Nucl. Instrum. Meth.* **150** (1978) 247.
30. E. Piesch, B. Burgkhardt and A.M. Sayed, *Nucl. Instrum. Meth.* **157** (1978) 179.
31. K.S.V. Nambi, *Nucl. Instrum. Meth.* **175** (1980) 71.
32. E. Tochilin, N. Goldstein and T.J. Layman, *Proc. 2nd Int. Conf. on Luminescence Dosimetry,* CONF-680920 US Atomic Energy Commission, Washington D.C., (1969) 424.
33. P. Olko, P. Bilski, E. Ryba and T. Niewiadomski, *Radiat. Prot. Dosim.* **47** (1993) 31.
34. P. Olko, *Radiat. Prot. Dosim.* **65** (1996) 151.
35. P. Bilksi, M. Budzanowski, P. Olko and P. Christensen, *Radiat. Prot. Dosim.* **66** (1996) 101.
36. M.S. Akselrod, S.W.S. McKeever, M. Moscovitch, D. Emfietzoglou, J.S. Durham and C.G. Soares *Radiat. Prot. Dosim.* **66** (1996) 105.
37. J. Kalef-Ezra and Y.S. Horowitz, *Int. J. Radiat. Isot.* **33** (1982) 1085.
38. Y.S. Horowitz, editor, *Thermoluminescence and Thermoluminescent Dosimetry, Vols. I, II and III*, CRC Press, Boca Raton (1984).

39. Á. Petö and Gy. Uchrin, editors, *Proceedings of the 11th Solid State Dosimetry Conference*, Budapest, *Radiat. Protect. Dosim.* **65** & **66** (1996).
40. M. Moscovitch, P.L. Taylor, J.E. Rodgers and D.A. McRae, *Radiat. Prot. Dosim.* **47** (1993) 505.
41. R.A. Tawil, in *Applications of New Technology: External Dosimetry*, J. Higginbotham, editor, Medical Physics Publishing, Madison (1996) 123.
42. L. Delgado and A. Delgado, *J. Appl. Phys.* **55** (1984) 515.
43. T.M. Piters, W.H. Meulemans and A.J.J. Bos, *Rev. Sci. Instrum.* **64** (1993) 109.
44. S.W.S. McKeever, *J. Phys. D: Appl. Phys.* **24** (1991) 988.
45. N. Suntharalingam and J.R. Cameron, *Phys. Med. Biol.* **14** (1969) 397.
46. J. Zimmerman, *J. Phys. C: Sol. St. Phys.* **4** (1971) 3277.
47. D.-K. Wu, F.-Y. Sun and H.-C. Dai, *Health Phys.* 46 (1984) 1063.
48. Y.S. Horowitz, *Radiat. Prot. Dosim.* **47** (1993) 135.
49. S. Wang, Y. Wang, G. Cai, Sh. Wang, and Z. Zha, *Radiat. Prot. Dosim.* **47** (1993) 223.
50. L. Oster, Y.S. Horowitz and A. Horowitz, *Radiat. Prot. Dosim.* **49** (1993) 407.
51. S.G. Gorbics and F.H. Attix, *Health Phys.* **25** (1973) 499.
52. G. Kitis, J. Papadopoulos, S. Charalambous and J.W.N. Tuyn, *Radiat. Prot. Dosim.* **55** (1994) 183.
53. F.D. Walker, L.E. Colyott, N. Agersnap Larsen and S.W.S. McKeever, *Radiat. Meas.* **26** (1996) 711.
54. M.S. Akselrod and E.A. Gorelova, *Nucl. Tracks Radiat. Meas.* **21** (1993) 143.
55. M.S. Akselrod, V.S. Kortov, D.J. Kravetsky and V.I. Gotlib, *Radiat. Prot. Dosim.* **33** (1990) 119.
56. M. Ranogajec-Komor and M. Osvay, *Radiat. Prot. Dosim.* **17** (1990) 379.
57. F. Hasan, G. Kitis and S. Charalambous, *Nucl. Instrum. Meth. B* **9** (1985) 218.
58. M. Prokić and L. Bøtter-Jensen, *Radiat. Prot. Dosim.* **47** (1993) 195.
59. M.G. Guelev, I.T. Mischev, B. Burgkhardt and E. Piesch, *Radiat. Prot. Dosim.* **51** (1994) 35.
60. C. Furetta and J.W.N. Tuyn, *Appl. Radiat. Isot.* **37** (1986) 935.
61. A.S. Pradhan, *Radiat. Prot. Dosim.* **47** (1993) 151.
62. S.W.S. McKeever, *Thermoluminescence of Solids*, Cambridge University Press, Cambridge (1985).
63. M.J. Aitken, *Thermoluminescence Dating*, Academic Press (1985).
64. F. Daniels, C.A. Boyd and D.F. Saunders, *Science* **117** (1953) 343.
65. N. Grögler, F.G. Houtermans and H. Stauffer, *Helv. Phys. Acta* **33** (1960) 595.

66. G.C. Kennedy and L. Knopff, *Archaeology* **13** (1960) 147.
67. M.J. Aitken, M.S. Tite and J. Reid, *Nature* **202** (1964) 1032.
68. M.J. Aitken, D.W. Zimmerman and S.J. Fleming, *Nature* **219** (1968) 442.
69. Y. Ichikawa, *Bull. Inst. Chem. Res. Kyoto Univ.* **43** (1965)1.
70. R.B. Mazes and D.W. Zimmerman, *Science* **152** (1966) 347.
71. E.K. Ralph and M.C. Han, *Nature* **210** (1966) 245.
72. V. Mejdahl, *Archaeometry* **11** (1969) 99.
73. H.Y. Göksu, J.H. Fremlin, H.T. Irwin and R. Fryxell, *Science* **183** (1974) 651.
74. A.G. Wintle, *Nature* **245** (1973) 143.
75. G. Guérin and G. Valladas, *Nature* **286** (1980) 697.
76. G.V. Morozov, *Proc. 8th Int. Quaternary Assoc. Congress,* US Geol. Survey Library, Washington D.C. (1968) 167.
77. A.G. Wintle and D.J. Huntley, *Nature* **279** (1979) 710.
78. A.G. Wintle and D.J. Huntley, *Canad. J. Earth Sci.* **17** (1980) 348.
79. D.W.G. Sears and A.A. Mills, *Nature* **242** (1973) 25.
80. S.W.S. McKeever, *PACT J.* **9** (1983) 322.
81. P.H. Benoit, A.S.T. Jull, S.W.S. McKeever and D.W.G. Sears, *Meteoritics* **28** (1993) 196.
82. G.J. Hennig and R. Grün, *Quater. Sci. Rev.* **2** (1984) 157.
83. A.G. Wintle, *Radiat. Prot. Dosim.* **47** (1993) 627.
84. J.H. Fremlin and S. Srirath, *Archaeometry* **7** (1964) 58.
85. S.J. Fleming, *Archaeometry* **9** (1966) 170.
86. S.J. Fleming, *Archaeometry* **12** (1970) 133.
87. D.W. Zimmerman, *Archaeometry* **10** (1967) 26.
88. D.W. Zimmerman, *Archaeometry* **13** (1971) 29.
89. I.K. Bailiff, *Europ. PACT J.* **6** (1982) 72.
90. D.W. Zimmerman, M.P. Yuhas and P. Meyers, *Archaeometry* **16** (1974) 19.
91. S.J. Fleming, *Archaeometry* **15** (1973) 13.
92. J. Zimmerman, *J. Phys. C: Sol. St. Phys.* **4** (1971) 3277.
93. R. Chen, *Europ. Pact J.* **3** (1978) 49.
94. R. Chen and Y. Kirsh, *Analysis of Thermally Stimulated Processes,* Pergamon Press, Oxford (1981).
95. I.K. Bailiff and E.H. Haskell, *Radiat. Prot. Dosim.* **6** (1984) 245.
96. Y. Ichikawa and T. Nagatomo, *Proc. US-Japan Workshop for Reassessment of Atomic Bomb Dosimetry in Hiroshima and Nagasaki,* (1983) 104.
97. D. Stoneham, *Europ. PACT J.* **9** (1983) 227.
98. V.N. Shelkoplyas, *Chronology of the Glacial Age*, Geographic Society of the

USSR, Pleistocene Commission, Leningrad (1971) 155.

99. D.J. Huntley, D.L. Godfrey-Smith and M.L.W. Thewalt, *Nature* **313** (1985) 105.
100. M.J. Aitken, *Quater. Geochron.* **13** (1994) 503.
101. S. Stokes, *The Middle Paleolithic in the Egyptian Sahara During the Last Interglacial*, F. Wendorf, A.E. Close and S.C. Schilder, editors, Southern Methodist University Press (1993).
102. D.J. Huntley, J.T. Hutton and J.R. Prescott, *Quater. Sci. Rev.* **12** (1993) 1.
103. G. Hütt, I. Jaek and J. Tchonka, *Quater. Sci. Rev.* **7** (1988) 381.
104. G.A.T. Duller, *Nucl. Tracks Radiat. Meas.* **18** (1991) 371.
105. G.W. Berger and D.J. Huntley, *Quater. Geochron.* **13** (1994) 509.
106. P.B. Wood, *Quater. Geochron.* **13** (1994) 513.
107. N.K. Perkins and E.J. Rhodes, *Quater. Geochron.* **13** (1994) 513.
108. G.A.T. Duller, *Quater. Geochron.* **13** (1994) 521.
109. A. Lang, *Quater. Geochron.* **13** (1994) 521.
110. M.L. Clarke, *Quater. Geochron.* **13** (1994) 525.
111. I.C. Fuller, A.G. Wintle and G.A.T. Duller, *Quater. Geochron.* **13** (1994) 539.
112. M.R. Krbetschek, U. Rieser and W. Stolz, *Radiat. Prot. Dosim.* **66** (1996) 407.
113. N.A. Spooner, *Radiat. Meas.* **27** (in press, 1997).
114. S.W.S. McKeever, L. Bøtter-Jensen, N. Agersnap Larsen and V. Mejdahl, *Radiat. Meas.* **27** (in press, 1997).
115. R.M. Bailey, B.W. Smith and E.J. Rhodes, *Radiat. Meas.* **27** (in press, 1997).
116. C.A. Richardson, *Quater. Geochron.* (in press, 1997).
117. L. Shepherd, *Quater. Geochron.* (in press, 1997).
118. N. Porat, R. Amit, E. Zilberman and Y. Enzel, *Quater. Geochron.* (in press, 1997).
119. A. Lang and G.A. Wagner, *Quater. Geochron.* (in press, 1997).
120. J.K. Feathers, *Quater. Geochron.* (in press, 1997).

Chapter 8
RELATED PHENOMENA

Several physical and chemical effects are similar in certain ways to thermoluminescence. In some cases, like thermally stimulated conductivity (TSC) and thermally stimulated electron emission (TSEE), the effects are inherently related to TL and, therefore, in many cases there is a point in measuring them simultaneously with TL, as will be shown in Ch. 9. In other phenomena, such as thermal desorption of gases from the surface of a solid or in differential thermal analysis (DTA) and differential thermogravimetry (DTG), there is a close analogy between TL and the effect in question, but not necessarily a direct relation. In all these phenomena, a certain property of the material under examination is measured as a function of temperature. Of course, not all properties measured as a function of temperature qualify to be included in this group of phenomena. In most cases, an initial excitation of the sample is required prior to the heating, for example, an irradiation of the sample. This, however, can be replaced by some other physical initiation of the process in some of the other phenomena. For example, even in some rare cases of TL, applied pressure can induce the effect.

In some of these analogous phenomena, the similarity between the different effects is not so obvious, and the main association is in the form of the mathematical expressions representing the effect. Quite often this is the reason people working on a sub-group of these phenomena do not know about advances in the methods of analysis of the measurements made in other subgroups. Chen and Kirsh[1] made the point that ideas should be exchanged between people working on different thermally stimulated phenomena - even if, at first sight, they look orthogonal to each other. In this respect, it should be mentioned that some methods of analysis, e.g. the initial rise and the various heating rates methods were developed independently for the analysis of TL, DTA and thermal desorption.

A somewhat similar phenomenon which is *not* included can be mentioned as an opposite example; this is black-body radiation of a heated sample. This effect is merely the incandescence of the material due to its temperature and is not related to any initial manipulation of the sample. Furthermore, all the effects related or analogous to TL appear in the form of peaks whereas the black-body-radiation grows with temperature in an unbounded manner.

Concerning the analogies between TL and the other thermally stimulated phenomena, both those physically related and those which are not, like all analogies their applicability is limited. For example, as will be shown later in this chapter, the shapes of TL curves and their associated TSC and TSEE curves are not identical due to additional physical factors which are inherent in the latter. It can be stated that the fact that additional information can be extracted from simultaneous measurements (e.g. TL-TSC) is hinged upon these basic differences. Thus, the use of methods of analysis developed for TL may turn out to be inaccurate for the study of, say, TSC curves, and the early investigators who attempted such analyses may have reached inaccurate results.

All in all, our main task in this chapter is to describe these various effects and to illustrate the similarities and dissimilarities between them. In this way we hope to be able to convey to the reader the main considerations to be made when choosing methods of analysis for dealing with different thermally stimulated processes. A discussion concerning the additional information that can be reached from the simultaneous measurements of TL with these other phenomena is postponed until the next chapter.

8.1 Thermally Stimulated Conductivity (TSC)

8.1.1 General Considerations

In a number of thermally stimulated processes current is measured as a function of temperature following an excitation of the sample, usually an insulator or a semiconductor, by an appropriate means. This sub-group of thermally stimulated phenomena includes thermally stimulated conductivity (TSC) in which the measured conductivity as a result of the initial excitation is associated with the transport of charge carriers, electrons or holes, thermally released from trapping states in which they had been trapped during the excitation. This kind of conductivity is the subject of the present section. Other kinds of electrical conductivity measurements will be discussed in later sections. These include currents due to the motion of ions - ionic thermoconductivity (ITC), or thermally stimulated depolarization currents (TSD, TSDC), both associated with the depolarization of a state of polarization in dielectric samples - as well as thermally stimulated electron emission which has to do with the current measured when electrons from the surface are emitted from the sample into vacuum, and then captured by a collector plate.

Returning to TSC, it can be excited in the same way as TL by many kinds of nuclear radiation - α particles, β particles, and γ rays, as well as by neutrons and heavy particles. Also, x-rays, UV light and visible light can cause the appearance of TSC signals. The first account of this phenomenon seems to be a short letter by Herman and Hofstadter[2]. They found that if a willemite sample is irradiated by UV light at low temperature and then heated in the dark with an electric field applied across two electrodes on the sample, the dark-current shows a large increase at about -60 °C. Herman and Hofstadter[2] suggested that this dark current effect was very probably connected with the emission of "trapped light" (TL) which had previously been reported by Johnson[3] in the same samples.

We shall describe here, very briefly, the experimental set-up and procedure for monitoring the thermally stimulated conductivity. In its simplest form, this comprises of the application of two contacts to the sample such as silver paint, evaporated gold thin films or other contact materials. The sample is then put in series with a dc voltage supply, and an electrometer, and the current is continuously recorded during the heating. The temperature is measured, usually by a thermocouple, and the current is recorded as a function of temperature (or time). This can be done simultaneously with the measurements of the TL glow curve, so that comparison between the two phenomena can be made; this subject of simultaneous measurements will be elaborated upon in the next chapter. A difficulty may arise when the irradiation-induced TSC curve occurs on top of a high conductivity background. Agarwal and Fritzsche[220] developed a method to measure a small TSC signal on top of a strongly temperature dependent conductivity which is not due to the initial excitation. This has been done by using two identical samples, one irradiated and the other unirradiated, in a bridge configuration.

It has been pointed out by Henisch[5] that the conclusions to be drawn from measurements performed with two contacts are somewhat limited. The main point is that whereas one is mainly interested in the bulk conductivity of a sample, one always measures the system conductance, a magnitude which includes the contact effects. The seemingly obvious solution, namely, to reduce the contact effects to negligible magnitudes, is not very feasible since usually, one does not have a good control over the relevant contact parameters. As argued by Henisch, the only way to properly measure conductivity is by the four probe method. In this way contact and volume properties can be separated, but even this can be done only when the two are indeed separable, and when the material is homogeneous in the bulk, and has a desirable geometrical shape. In the case of samples in the form of arbitrarily shaped thin flat plates (or films), the conventional four probe measurement cannot be performed. One should then turn to the method suggested by van der Pauw[6]. In this method, one applies four small contacts on four arbitrarily chosen points A,B,C,D on the circumference of the flat sample. Provided that the sample is singly connected (no isolated cavities), the equation

$$\exp\{-\pi R_{AB,CD} d/\rho\} + \exp\{-\pi R_{BC,DA} d/\rho\} = 1 \tag{8.1}$$

should hold, where ρ is the resistivity of the sample and d is its thickness. The resistance $R_{AB,CD}$ is defined as the potential difference $V_D - V_C$ between the contacts D and C per unit current passed through the contacts A and B; $R_{BC,DA}$ is defined similarly. Thus, the experiment comprises of two voltage measurements when a given current is applied to the contacts. One should then solve numerically the transcendental equation 8.1 to get the value of the resistivity ρ, from which the conductivity $\sigma = 1/\rho$ is readily found. There may be some complication in doing this repeatedly at small intervals when the temperature progresses in a TSC measurement, but it is still recommended for future research in view of the mentioned difficulties in the traditional method. A variant of the van der Pauw method, particularly applicable to high resistivity materials has been described by Hemenger[7]. The van der Pauw method has also been used, by Kuriyama et al.[8], for the measurement of TSC related to Cu antisite vacancy complex defects in Cu-diffused semi-insulating GaAs. Ohmic electrodes were produced by evaporating the Au-Ge-Ni alloy. Using this sensitive method, these researchers discovered two new traps, T_1 with energy of 0.25 eV and T_2 with 0.52 eV. The latter is associated with a Cu_{Ga}-V_{As} complex. More work on TSC in thin layers of GaAs was reported by Lau et al.[9] who also identified the electron trap with an activation energy of 0.52 eV in annealed buffer layers. In addition, a continuum of states was also detected, and was attributed to the interface states at the interface of arsenic precipitates and bulk GaAs.

It is to be noted that even the van der Pauw method may be inaccurate when substantial non-equilibrium effects such as injection and extraction take place. As pointed out by Henisch[5], contact and volume properties are sometimes hard to differentiate, for example in the presence of minority carriers; therefore, if this is the case, even the four probe measurement is of limited help. Popescu and Henisch[10],[11] dealt with the current continuity at an interface between the metal (the contact) and the semi-insulator (the sample), which is governed by the "current composition ratio". This ratio is defined as the fraction of the total current carried by minority carriers. If this ratio changes discontinuously at an interface, the transport relationships - current densities, continuity equations and Poisson's equations - would directly imply a departure from the equilibrium carrier concentrations which extend from the interface to some distance in both materials. The consequence is non-ohmic behavior due to diffusion currents resulting from concentration gradients. Henisch[5] also showed that the absence of non-linearity between current and contact voltage, usually considered a criterion for "good" contact, does not necessarily mean that the contacts have no significant role in the process. For true ohmic contacts it is required that not only is the $I - V$ plot linear, but also that the specific contact resistance is zero, i.e. $R_c = (\partial I/\partial V)^{-1}_{V=0} = 0$. The relation of these considerations to TSC experiments is that when a measurement is performed at a constant voltage (which is usually the case), the injection effects will be unimportant although not altogether negligible as compared to the total current at low current densities, and may substantially increase at high currents. As a result, one can expect a temperature dependent effect which may interfere in a complicated manner with the thermally stimulated current. This would make the extraction of the bulk free carriers concentration $n_c(T)$ in which we are interested, quite difficult. The case of a constant current (rather than a constant voltage) would be of some advantage in this sense. However, even at constant current, the injection ratio is expected to be temperature dependent during a TSC run. Therefore, the conditions which prevail at one temperature cannot be directly compared with those occurring at another.

Henisch[5] suggested the following comparative measurements that should indicate (though not conclusively prove) that contact and interface effects, problems of current continuity, and space charge effects play only a minor role. These are:

I. Measurements on specimens having different distances from contact to contact, with the same contact material. A "good" result is when the resistance depends linearly on this distance.

II. Measurements of specimens with the same thickness and different kinds of contacts; a positive result is when the current is the same with different contact materials.

III. Measurements with different contact voltages. If the shape of the TSC curve is independent of the voltage, the mentioned possible complications are probably insignificant.

Another method for evaluating the conductivity of semiconductors which bypasses the contact complications has been suggested and utilized by various investigators. These include Henisch and Zucker[12], Weingarten and Rothberg[13] and Bryant and Gunn[14]. The method is the contactless measurement of semiconductor conductivity by radio-frequency free carrier power absorption. To the best of our knowledge, no account of TSC measurements of this sort has been given in the literature so far. A later version, appropriate for measuring very low conductivities, has been suggested by Chatain et al.[15]. These authors revise a previous method by Ogawa[16]. In this method, a Fourier transform of the frequency depending torque in Ogawa's torsion pendulum is evaluated. This reduces by a factor of ~100 the relatively long times needed for measuring the very low conductivities of insulating materials, thus making these measurements feasible. Yet another contactless method for the determination of conductivity of semi-insulators has been described by Müller[17]. Using a boxcar integrator for measuring the current-time signals, the proposed method is applied to investigate the properties of semi-insulating GaAs and AgBr at room temperature. This may possibly be attempted in the future in spite of the additional complication of having to heat the sample in a controlled manner within the compartment used for the contactless procedure. To conclude this subsection, taking into account the points made by Henisch[5], one should examine the TSC experimental results with some caution, as far as the measured current is used for determining the behavior of the conductivity $\sigma(T)$.

8.1.2 Theory and Experimental Evidence

In this subsection we will assume that the contact effects already mentioned, are controllable, and therefore do not make a large contribution to the measured quantities, so that we are able to get a good evaluation of the excited conductivity $\sigma(T)$ from our measurements. Most of the theories discussed here assume that homogeneous conditions prevail throughout the specimen, namely, that there are no longitudinal concentration gradients, not only in equilibrium, but also during the release of non-equilibrium carriers from traps. Precisely, this can only be true in an infinite sample since, as pointed out, in real cases we have to cope with continuity relationships. These may depend on the kind of contacts utilized, but their occurrence is inevitable even with "good" contacts. The measurements suggested in the previous subsection, namely, testing different specimen thicknesses, various contacts and different contact voltages, can give us an indication on how disturbing are the non-equilibrium effects. As an example, we can mention an elaborate study of optimizing the material to be used for the electrodes on corona charged polypropylene, which has been performed by Baba and Ikezaki[18]. These authors discuss the thermal effect of the vacuum deposition of metallic electrodes, which is different for different contact materials, and show that among four materials utilized - Bi, Au, Ag and Al - the best is Bi.

Once we are satisfied that the conductivity measurement gives the bulk conductivity $\sigma(T)$ quite accurately, we can use the relation with the concentration of free carriers $n_c(T)$,

$$\sigma = e\mu n_c \tag{8.2}$$

where e is the electronic charge and μ the mobility. This is precise only as a local relation, and in order for it to be correct on the macroscopic level, one has to assume homogeneity of the bulk material, absence of space charge and absence of minority carriers. If one further assumes that μ is temperature independent, one gets a $\sigma(T)$ curve which has the same shape as $n_c(T)$. (Some cases in which μ is temperature dependent will be discussed below; this dependence is, however, rather weak.) This assumption has been made by practically all the investigators of TSC. Concerning the evaluation of μ, it should be noted that the usual Hall mobility measurement requires the application of contacts on the sample. To avoid the possible contact problems in this respect, Molnar and Kennedy[19] suggested the use of microwave measurements of magneto-conductivity which serves as a contactless mobility measurement. By this, they evaluated the mobilities of S and Se implanted GaAs, and got results that compared favorably with previously known Hall mobilities.

Several of the early investigators of TSC assumed a constant lifetime τ of the thermally released charge carriers in the conduction band. These include Haering and Adams[20], Keating[21], Nicholas and Woods[22], Bube et al.[23], Devaux and Schott[24], Cowell and Woods[25], Haine and Carley-Read[26], Buehler [27] and Garofano and Morelli[28]. In fact, this assumption was also used in later works, such as Kang et al.[29], Datta et al.[30]and Suryanarayana et al.[31]. In the notation of Ch. 2, this would mean that m in Eq. 2.20 is practically constant; τ in this case is $1/(mA_{mn})$. This may indeed be nearly correct if m varies only slightly along a TSC peak; the peak shape may well be associated with the exhaustion of carriers from traps (rather then opposite sign carriers in centers). Making this assumption, one gets TSC curves which look exactly the same way as the corresponding TL ones. Most of the authors mentioned here measured TSC peaks that were of first order and applied the various methods of analysis, originally developed for the study of TL single peaks.

Haering and Adams[20], while assuming a constant τ, investigated two extreme cases, slow retrapping and fast retrapping. Slow retrapping is characterized by $(N_i - n_i)Sv \ll \tau^{-1}$, where N_i is the concentration of trapping states, n_i is the concentration of electrons in traps, S is the cross-section for trapping an electron, and v is the thermal velocity of electrons in the conduction band. Under these conditions, and assuming a linear heating function $T = T_0 + \beta t$, one gets

$$\sigma(T) = N_c Sve\mu n_0 \exp\{-E_t/kT\} \exp\left\{-(N_c Sv/\beta)\int_{T_0}^{T} \exp\{-E_t/k\theta\}\,d\theta\right\} \tag{8.3}$$

where N_c is the density of thermally available states in the conduction band. This is a first-order expression similar to Eq. 1.4 where the product $N_c Sv$ (with dimensions of s^{-1}) replaces the frequency factor in the usual TL equation. An application of the Haering and Adams[20] analysis for slow retrapping was made by Takizawa et al.[32] who studied a composite TSC peak in visible-light-excited $CdInGaS_4$. These authors found an envelope peak at ~120 K, which they deconvoluted into four components of the Eq. 8.3 type, and extracted the relevant trapping parameters.

The other case is fast retrapping in which $(N_i - n_i)Sv \gg \tau^{-1}$. The conductivity curve is given here as

$$\sigma(T) = \left(\frac{N_c}{N_i}\right) \mu e n_0 \exp\left\{-\frac{E_t}{kT}\right\} \exp\left\{-\left(\frac{N_c}{N_i \beta \tau}\right) \int_{T_0}^{T} \exp\left\{-\frac{E_t}{k\theta}\right\} d\theta\right\}. \quad (8.4)$$

Although Eq. 8.4 results from an entirely different physical assumption than the one leading to Eq. 8.3, the outcome is very similar, namely, a first-order curve where now $N_c/(N_i\tau)$ (with units of s^{-1}) now replaces the frequency factor s. Both N_cSv in the former case and $N_c/(N_i\tau)$ in the latter are temperature dependent and, therefore, in a precise treatment, should appear within the integral. This dependence is expected, however, to behave in many cases like T^a similarly. The mathematical evaluation of the integral with T^a is discussed in Ch. 6 (see Eq. 6.7). Buehler[27] discussed this case with $a = 2$.

A rather thorough investigation of first-order and second-order TSC peaks, assuming a constant τ, was conducted by Nicholas and Woods[22] who also evaluated the cross-sections for retrapping from the effective frequency factor, related to several peaks in CdS. Rabie and Rumin[33] studied TSC in Zn-compensated Si. Although first-order peaks were found with the value of τ being constant along the heating process, τ was found to depend strongly on the concentration of negatively charged Zn ions. Thus, they prepared a series of samples with different amounts of the impurity, that turned out to have the same activation energy E_t, but different values of τ. Assuming fast retrapping, the maximum condition derived from Eq. 8.4 is

$$\exp\{E_t/kT_m\} = N_c k T_m^2/(N_i \beta \tau E_t) \quad (8.5)$$

which is very similar to Eq. 1.5; the maximum temperature T_m would depend, however, on τ. Equation 8.5 can now be written as

$$1/T_m = (k/E_t) \ln\left(T_m^2/\tau\right) + A \quad (8.6)$$

where $A = (k/E_t)\ln(N_c k/N_i \beta E_t)$. In a way somewhat similar to the various heating rates method (see Ch. 3), one can plot $\ln(T_m^2/\tau)$ vs. $(1/T_m)$ and get a straight line, the slope of which yields the value of E_t. The different values of τ were separately determined from photo-current measurements. It is to be noted that since τ does not appear as a parameter in Eq. 8.3, the effect is not expected to be seen in the slow retrapping case. The concentration dependence effect of Zn compensated Si indicates, therefore, that fast retrapping prevails here.

All the above analyses assume quasiequilibrium (QE). Lewandowski and colleagues[34],[35] discuss a generalized expression for TSC for the slow retrapping case in which QE is not assumed. The expression for the TSC *current* obtained in this case is

$$I(T) = \frac{AFe\mu n_0}{\beta} \exp\left\{-\frac{1}{\beta}\int \frac{dT}{\tau(T)}\right\} \int_{T_0}^{T} \exp\left\{\frac{1}{\beta}\int \frac{d\theta}{\tau(\theta)}\right\} s(\theta) \exp\left\{-\frac{E_t}{k\theta}\right\}$$

$$\times \exp\left\{-\frac{1}{\beta}\int_{T_0}^{\theta} s(\Omega) \exp\left\{-\frac{E_t}{k\Omega}\right\} d\Omega\right\} d\theta. \tag{8.7}$$

in which the temperature dependencies of τ and s are included, but μ is assumed to be constant. Here A is the electrode area and F is the magnitude of the applied field. θ and Ω are dummy variables representing temperature. This expression is to be compared with the equivalent expression for TL in Eq. 2.116.

As already noted, the assumption that the lifetime τ is constant during the heating of the sample is an approximation which is valid only in particular cases. Therefore, trying to predict the shape of a TSC curve from a set of given trapping parameters cannot be done using this assumption, and of course, it is far from being obvious that methods developed for the analysis of thermoluminescence glow peaks are applicable for TSC peaks. Most generally, the evolution of the current with increasing temperature should be associated, in the relatively simple case of one trapping state and one kind of center with $n_c(T)$ which can be extracted by solving (usually numerically) the set of simultaneous differential equations 2.17, 2.19 and 2.20. Of course, if more trapping states and recombination centers are involved, more differential equations should be added to the set. For different sets of parameters in the equations, one can expect many different TSC peak shapes. Let us concentrate here on two cases, given as examples to many different outcomes of a single TSC peak.

Saunders[36] seems to have been the first who questioned the constancy of τ. He suggested that since $\tau = 1/(A_{mn}m)$, τ should increase with temperature as m decreases. For the second-order expression 2.30, he obtained the expression for conductivity

$$\sigma(T) = e\mu\tau I = \left(\frac{s_1}{s}\right) \mu e N_c \exp\left\{-\frac{E_t}{kT}\right\} \left[1 + \frac{s_1}{\beta}\int_{T_0}^{T} \exp\left\{-\frac{E_t}{k\theta}\right\} d\theta\right]^{-1} \tag{8.8}$$

where s is the frequency factor, $s_1 = n_0 s/N$, and n_0, N, and N_c are the same as defined above. The right hand side includes

$$\left[1 + (s_1/\beta)\int_{T_0}^{T} \exp\left\{-E_t/k\theta\right\} d\theta\right]^{-1} \tag{8.9}$$

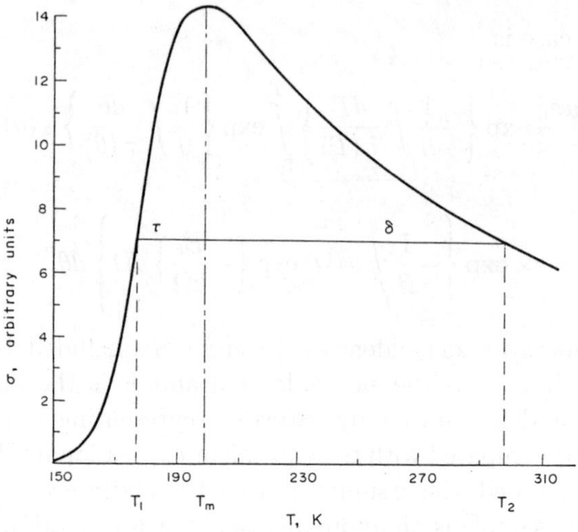

Figure 8.1 An $n_c(T)$ peak calculated by the use of Eq.8.8 for the parameters E_t=0.4 eV, s_1=10^{10} s^{-1} and β=0.5 Ks^{-1}.
(After Chen[37].)

rather than the same expression to the power of -2 as in second-order kinetics in TL. This indicates that the fall-off at high temperature is very slow. Chen[37] further investigated this expression and showed that in this kind of peak, the characteristic value of the shape factor $\mu_g = \delta/\omega$ (see section 3.4) is about 0.8. Figure 8.1 depicts a peak calculated by Chen from Eq. 8.8. Gartia et al.[38] further studied this expression and showed that μ_g varied between 0.77 and 0.91 for different values of $U_m = E_t/kT_m$. Singh et al.[39] went one step further by examining the shape of the TSC peaks relative to those of TL with kinetic orders of 1.5, 2 and 2.5, and suggested new versions of the peak-shape methods and the various heating rate method for evaluating the activation energy from the TSC peaks. Filina and Margulis[40] extended the theoretical investigation of TSC under similar conditions by solving in detail the kinetic equations for two, coupled trapping levels. An expression was derived for the conductivity which reflects the mutual effect of these levels, regardless of their difference in energy, for the case of fast, repeated trapping of carriers.

Some caution is advised while deriving conclusions from such rather unusual TSC peak shapes. Experimental results by Wynands and Cocivera[41] of TSC in CdSe show a broad peak at $\sim$ 160 K with $\mu_g \approx 0.60$. Although this is lower than the above-mentioned values, at first sight it may look as an example of much higher than the conventional values of the shape factor (0.42-0.52) in the kinetics of first or second order, and looks as an indication to an effect similar to that shown in Fig. 8.1. However, these authors[41], used a

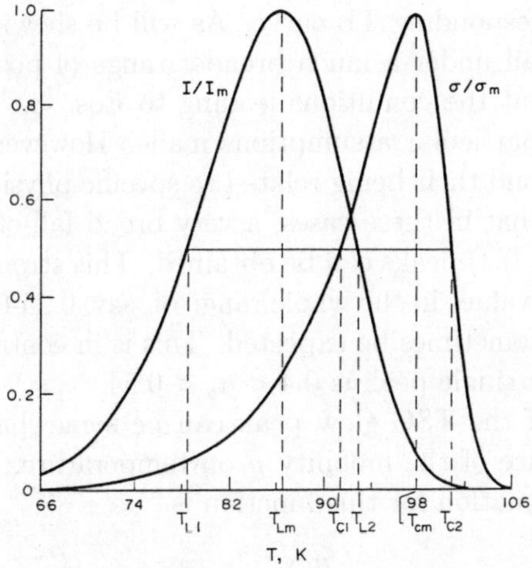

Figure 8.2 Normalized curves of TL and $\sigma(T)$ as calculated by the use of Eq. 8.10. The parameters chosen are E_t=0.1 eV, s=10^5 s^{-1}, A_{mn}=10^{-11} m^3 s^{-1} and α=10^{-3}.
(After Chen[43].)

deconvolution method, based on repeated partial heating, and showed that in this case, the apparent peak is just an envelope of the sum of five peaks, each having a symmetry factor characteristic of simple, TL-like, first-order peaks.

A similar treatment has been given by Kelly and Bräunlich[42], and by Chen[43] to the TSC corresponding to a first-order peak. The expression for σ is

$$\sigma(T) = \left(\frac{se\mu}{A_{mn}}\right) \exp\left\{-\frac{E_t}{kT}\right\} \left[1 + \alpha \exp\left\{\frac{s}{\beta}\int_{T_0}^{T} \exp\left\{-\frac{E_t}{k\theta}\right\} d\theta\right\}\right]^{-1}, \quad (8.10)$$

where $\alpha = c/n_0$ and where c is the concentration of charge carriers in "thermally disconnected" traps. This means that in the range of occurrence of the peak, one has $n + c = m$. For very high values of α, Eq. 8.10 reduces to a simple first-order expression, similar to Eq. 8.3 or Eq. 8.4, since the unity can be neglected with respect to the expression containing α. On the other hand, when $\alpha < 1$, the shape of the peak is expected to be entirely different. This is characterized by very low values of the shape factor $\mu_g = \delta/\omega$. As shown by Chen[43], for $\alpha = 10^{-4}$, a value of $\mu_g \approx 0.27$ was found. Fig. 8.2 depicts an example of a peak of $\sigma(T)$ calculated by Eq. 8.10, along with its corresponding TL curve. The TSC peak is seen to occur at a higher temperature than its TL counterpart; this happens also[37] in the second-order peak expressed by

Eq. 8.8 and its corresponding TL curve. As will be shown in Ch. 9, this shift is expected to prevail under a much broader range of possible circumstances. It is to be noted that the conditions leading to Eqs. 8.8 and 8.10 are rather limited due to the particular assumptions made. However, the importance of these equations beyond their being related to specific physical situations, lies in the demonstration that in these cases, a very broad fall-off ($\mu_g \approx 0.8$) or very narrow fall-off ($\mu_g < 0.3$) peaks can be obtained. This strongly indicates that in the general case, μ_g values in the whole range of, say, 0.25-0.9 and perhaps even broader range, can sometimes be expected. This is in contrast to TL where the expected range for a single peak is $0.4 < \mu_g < 0.54$.

The shape of the TSC glow peak can be somewhat dependent on the functional dependence of the mobility μ on temperature. Most authors agree that a good approximation for this function is

$$\mu(T) = \mu_c T^b \tag{8.11}$$

where μ_c is a constant, the dimensions of which depend on the value of b. Lax[44] mentioned values of b between -1.5 and -2.3. Siegel et al.[45] also reported the dependence of the mobility on temperature, and improved the results they got for the activation energy from the initial-rise method (see section 3.3) by plotting $\ln(n/T^{3/2})$ vs. $1/T$. This proved the $T^{3/2}$ dependence in their GaAs samples. Tomaszewicz[46], following Bube[47], suggested that the parameter b depends on the carrier scattering mechanism. In the case of covalent semiconductors, for scattering by acoustic phonons, $b = -3/2$, and for scattering by charged impurities, $b = 3/2$. Anyway, it should be kept in mind that temperature dependence contributes only a relatively small effect since the main temperature dependence of conductivity is due to the Boltzmann function $\exp\{-E_t/kT\}$, on which the concentration of charge carriers depends. In fact, it should be mentioned (see e.g. Kivits and Hagebeuk[48] and Tomaszewicz[46]) that the free carrier concentration itself may be temperature dependent (in addition to the exponential function), due to the possible dependence of the frequency factor. This dependence is also expected to be a power of the temperature, T^a with $-2 \leq a \leq 2$, which is also not very significant as compared to the Boltzmann function. Shaw and Hoghaddam[49], developed a method for evaluating trapping parameters from TSC peaks, based on the same assumption concerning the dependence of the frequency factor on temperature, and on the performance of a series of measurements at different heating rates, in a manner somewhat similar to that described in detail in section 3.5, with regard to TL peaks.

Further work along the same lines has been reported by Elkomoss et al.[50], in a study of TSC curves in CdTe crystals. These authors consider a dependence of $S \propto T^\alpha$ where S is the cross-section for free electron trapping, and $-1 \leq \alpha \leq -4$. Taking into consideration the temperature dependence of the concentration of conduction *trapping states* and that of the thermal velocity, this yields the dependence of the effective frequency factor on temperature of $\propto T^a$ with a between -2 and +1. The authors[50] consider the possibilities of slow and fast retrapping. They claim to have separated fourteen overlapping peaks, and evaluate the kinetic parameters for each of the components.

Rinkevičius and Kavaliauskienė[51] considered a more elaborate model, taking into account the simultaneous majority and minority carriers in the sample. The authors wrote five simultaneous differential equations for a situation with three types of levels in the forbidden gap: 1 - traps for majority carriers (electrons in the model), 2 - traps for minority charge carriers (holes), and 3 - recombination centers. These equations are rather similar, in principle, to sets of simultaneous equations governing the TL process:

$$dn/dt = \gamma_n N_c \exp\{-E_{t1}/kT\} - \gamma_n(N_t - n_t)n - \gamma_{nr} p_r n \tag{8.12}$$

$$dn_t/dt = -\gamma_n N_c n_t \exp\{-E_{t1}/kT\} + \gamma_n(N_t - n_t)n \tag{8.13}$$

$$dp_t/dt = -\gamma_p P_v p_t \exp\{-E_{t2}/kT\} - \gamma_p(P_t - p_t)p \tag{8.14}$$

$$dp/dt = \gamma_p P_v p_t \exp\{-E_{t2}/kT\} - \gamma_p(P_t - p_t)p - \gamma_{pr}(N_t - p_t)p \tag{8.15}$$

and the neutrality condition

$$n + n_t = p + p_t + p_r. \tag{8.16}$$

Here, T is the temperature and t is the time, normally related by a linear heating function $T = T_0 + \beta t$, where β is the heating rate. n and n_t are, respectively, the concentrations of free and trapped electrons, and p_v and p are the concentrations of free and trapped holes. E_{t1} and E_{t2} are, respectively, the depth of the trapped electrons and holes. γ_n, γ_p, γ_{nr} and γ_{pr} are the capture coefficients of electrons and holes by traps and recombination centers, respectively. By making the traditional quasiequilibrium assumptions, $n \ll n_t$, $p \ll p_t$, $dn/dt \ll dn_t/dt$ and $dp/dt \ll dp_t/dt$, they found the simplified expressions

$$n = \frac{\gamma_n N_c n_t \exp\{-E_{t1}/kT\}}{\gamma_n(N_t - n_t) + \gamma_{nr} p_r} \tag{8.17}$$

$$p = \frac{\gamma_p P_v p_t \exp\{-E_{t2}/kT\}}{\gamma_p(P_t - p_t) + \gamma_{pr}(N_r - p_r)} \tag{8.18}$$

and the neutrality condition simplified to

$$p_r = n_t - p_t. \tag{8.19}$$

The authors[51] further make the assumption of weak retrapping of both electrons and holes, and end up with the following expressions for trapped electrons and holes

$$n_t = n_{to} \exp\left\{ -\beta^{-1} \int_{T_0}^{T} \gamma_n N_c \exp\{-E_{t1}/k\theta\}\, d\theta \right\} \tag{8.20}$$

and

$$p_t = p_{to} \exp\left\{ -\beta^{-1} \int_{T_0}^{T} \gamma_p P_v \exp\{-E_{t2}/k\theta\}\, d\theta \right\}. \tag{8.21}$$

n_t and p_t are seen to be monotonically decreasing functions of temperature. The rate of their decrease is determined by the trap depths, E_{t1} and E_{t2}, and capture coefficients, γ_n and γ_p. If $E_{t1} < E_{t2}$ but $\gamma_p > \gamma_n$, it is possible that, with increasing temperature, n_t diminishes faster than p_t and the electron lifetime $\tau_n = 1/[\gamma_{nr}(N_t - n_t)]$ will rise. After that, the fast decrease of p_t will cause a reduction of τ_n and, therefore, of n. This means that when hole traps start emptying very fast, the holes from the valence band are captured by the recombination center, thus reducing the electron lifetime τ_n. If reduction of τ_n takes place in the vicinity of the TSC peak, its shape may be distorted or it may even split into a double peak. The authors[51] term this effect as "thermal quenching of thermally stimulated conductivity" (TQTSC). In a later work, Rinkevičius and Kavaliauskienė[52] study the influence of potential fluctuations on TSC. They take into consideration the cases of slow and fast retrapping, and demonstrate that the TSC peak broadens due to these fluctuations.

Rinkevičius and Kavaliauskienė[51] simulate numerically the described effect, and also analyze previous results of TSC in GaAs, given by Kiliulis et al.[53]. They explain two peaks, occurring at 184 and 204 K, both found by repeated cooling and heating cycles to be attributed to the same trap of 0.34 eV, as being the result of such splitting. Further work on GaAs has been given by Fang and Look[54]. They concentrate on the most prominent trap in undoped GaAs, T_5, with an activation energy of 0.27-0.31 eV. The authors suggest that the thermal quenching of TSC in the region where T_5 is active, namely, ~150 K, can be explained by the formation of a current-limiting, high-field domain that is nearly stationary. This domain can form when a high electric field instability in a particular region cause field enhanced trapping of the TSC electrons in that region, thus causing even higher local fields and further trapping. The high-field, high-resistivity domain thus formed can limit the current, as indeed is observed. The T_5 trap appears to be As_{Ga}-related defect, possibly $As_{Ga}-V_{Ga}$. This defect can trap an electron during the illumination at 83 K, and then release it into the conduction band at a higher temperature, which is the normal TSC process. However, at high electric field, it may be possible for the electron to recapture immediately at the As_{Ga} part of the defect, and cause a metastable transition of the As_{Ga}. Above 140 K, the As_{Ga} returns to the normal state, and the electron can finally be emitted into the conduction band. This model is consistent with the thermal quenching being triggered by an emission process with an activation energy close to that of T_5, as is found from an analysis of the thermal quenching at different heating rates as reported by Fang and Look[54]. Look et al.[55] reported a TSC peak at 100 K in semi-insulating GaN. A main trap was identified, with a cross-section $\sigma \simeq 3 \times 10^{-26}$ m^{-2} and $N\mu\tau \simeq 3 \times 10^{16}$ $m^{-1}V^{-1}$, where N is the trap concentration, μ the mobility and τ the free-carrier lifetime. This trap is much deeper than the typical shallow donor in conducting GaN, but shallower than any of the centers previously reported by DLTS measurements.

Another interesting experimental study on TSC was reported by Bloom and McKeever[56],[57]. They examind the trapping levels in Ga doped $Bi_{12}GeO_{20}$ (BGO) by measuring TSC as well as TL. They showed that excitation with photons of energy greater than the band gap induces both TSC and TL. Examination of the TSC and TL signals as a function of excitation wavelength allows the distinction between electron and hole trapping states for which trapping parameters have been determined. In particular, the analysis of TSC shows that the main peaks below 160 K in undoped BGO arise from two trapping centers, each characterized by a distribution of activation energies centered at ~0.24 and ~0.29 eV, with distribution widths of ~0.065 eV. In the Ga-doped sample, the large peak seen below 180 K arises from a trapping center at ~0.29 eV, with a distribution width of ~0.085 eV.

It is to be noted that other kinds of thermally stimulated conductivity have been reported in the literature, namely, effects which are not directly related to the release of trapped carriers, or otherwise different from conventional TSC. Although these are also normally observed as peaks of measured current as a function of temperature, they usually are not directly related to TL in the same materials (in many cases, TL does not appear in these samples), and in most cases, their shape is not similar to that of TL glow curves. Excluding thermally stimulated depolarization, which we discuss in depth in a following section, these will be mentioned only very briefly here. They include, for example, TSC in amorphous silicon, a-Si:H, as described by Misra et al.[58] and by Fritsche and Ibaraki[60]. Due to the amorphous nature of the material, a continuous density of states occurs in the gap, which makes the analysis rather different from the discrete levels case. Ibaraki and Fritsche[59] explain their results in terms of the multiple trapping theory. The TSC shows pronounced structure as a function of temperature, and decreases rapidly by several orders of magnitude as defects are created by light exposure. Fritsche and Ibaraki[60] measure by TSC the quantity mobility-lifetime-density of states ($\mu\tau g$) as a function of temperature, which, in turn, is related to an energy in the gap. They show that only when this product is large enough, the density of states $g(E)$ can be determined from the TSC and photoconductivity measurements. More work on the multiple trapping model as used for TSC was given by Schrader and Kryszewski[61]. They carried out some detailed computer simulations for systems with exponential trap distributions. The resulting drift peaks show a linear dependence of the logarithm of current density on temperature in the initial rise region. The maximum temperature of the drift peak reveals dependencies on the heating rate, the applied field strength and the sample thickness. The main model parameters such as the mobility-lifetime product of mobile carriers, attempt-to-escape frequency of trapped carriers, and the mean depth of the exponential trap distribution, can be evaluated from the TSC drift peaks.

Zhou and Elliott[62] strongly challenge the validity of the interpretation of TSC in a-Si:H. They show that TSC is probably controlled by recombination rather than by thermal emission of trapped electrons. Therefore, the relationship between the TSC and the density of states is much more complicated than previously suggested.

In a series of papers, Fleetwood et al.[63]-[65] described the TSC seen in charged insulating films experiencing *arbitrary* time-dependent thermal environments and high electric fields. The method includes the effect of field-induced lowering of the trap energy scale. Trap distributions for irradiated metal-SiO_2-Si capacitors were accurately determined from TSC measurements spanning a factor of 50 in heating rate, providing an improved estimate of the *trapped-hole* energies in SiO_2. More quantitatively, the authors define a function $n(\phi, t)$ of the energy ϕ and time t which has the units of concentration per unit area per unit energy. They assume that the charge leaves the traps by an activated process that can be described by the generalized activation differential equation with negligible retrapping

$$\frac{\partial}{\partial t}n(\phi, t) = -n(\phi, t)F(t)\left[-\frac{\phi - R(t)}{A(t)}\right]. \tag{8.22}$$

The activation function is $A(t) = kT(t)/q$ where k is Boltzmann's constant, q is the electric charge, and $T(t)$ is the time-dependent temperature. In the presence of field-induced barrier lowering, $R(t) = \beta\sqrt{E(t)}$, where $E(t)$ is the time-dependent electric field. β depends on the particular circumstances, for example, for the Poole-Frenkel barrier in SiO_2, $\beta = \beta_{PF} = (q/\pi\varepsilon)^{1/2} = 3.8\times10^{-5}$ $V(V/m)^{-1/2}$. $F(t)$ is given by $F(t) = a\,[T(t)]^2$ where $a = 2\sqrt{3}(2\pi)^{3/2}gm^8 \times \sigma_t k^2/h^3$, g being the multiplicity of the state, m^* is the effective mass of the carriers in the dielectric, σ_t is the capture cross-section of the dielectric traps, and h is Planck's constant. Using $m^* \approx m_e$, $g \approx 1$ and $\sigma_t \approx 10^{-17}$ m^2 they find the value of $a \sim 3\times10^8$ K^{-2}s^{-1}. The authors go one step further by integrating

Eq. 8.22 and find the *emission front* $\phi_m(t)$. They conclude that the results in SiO_2 indicate the barrier lowering mechanism for holes is of the same form as Schottky emission, and that the attempt-to-escape frequency has a value of $\sim 10^{14}$ Hz, suggesting that the emission process may be associated with optical phonons.

The relation of space charge effects to TSC measurements has been studied by Mizutani et al.[66] while working on TSC in polyvinylidene fluoride (PVDF). In this material, they found that polarization arising from trapped charges vanished after the first heating cycle, and a recovery was observed only when charge injection from the electrodes occurred, or if some form of irradiation could create new free charges inside the sample. In a later work, Faria et al.[67], working on β-PVDF found that the thermally released space charges are re-trapped at the polymer surface traps. Thermally stimulated current peaks could therefore be observed in repeated heating-cooling cycles, provided that an electric field was applied to facilitate the return of the charges into the bulk.

An anomalous TSC phenomenon in CaS has been reported by Green et al.[68]. They studied a range of samples with different concentrations of various dopants such as Ce, Al, P and Cu and found a previously known peak at ~308 K, *independent* of the nature and concentration of the dopant. Evidence is presented which indicates that surface states are responsible for the currents rather than the release of charge carriers in the bulk; the source of the charge carriers was uncertain. Simultaneous TSC and TL measurements (see Ch. 9) on CaS:Ce,Cu indicate that the TL peak at 300 K and the TSC peak at 308 K are mutually independent.

An interesting phenomenon of the effect of pressure-induced plastic deformation on TSC has been reported by Bulanyĭ[69] et al. They applied pressure on ZnS:Cu,Mn samples and found some new peaks in the range of 180-230 K. These new peaks suggest that the trapping levels associated with defects get progressively more abundant with an increase in the extent of plastic deformation, which converts the ZnS crystal lattice into a cubic structure with a single orientation. A further deformation causes the number of stacking faults to fall off substantially, leading to a TSC curve with eight peaks. The low temperature peaks at 80-160 K are present in all samples, and are independent of the degree of plastic deformation, and the excitation intensity. These appear to be associated with surface states. The results indicate that dislocations moving through the crystal interact with various centers, and cause them to change; certain types are destroyed while others are generated.

A different kind of effect has been reported by Pai[70], Slowik[71] and Chen[72] who studied TSC in N-vinylcarbazole (PVK). These authors concluded that in this material, the charge carrier transport is by hopping. In this process, a carrier hops from a localized state to another without ever moving in an extended state. Thus, in spite of the similarity in the appearance of the current vs. temperature curves, the analysis in the hopping transport systems using the existing methods may not be justified. While in the usual band transport systems, the TSC is interpreted as being due to the increased *concentration* of free carriers at high temperatures, in the hopping transport system, it can be considered as being due to the increased *mobility* of all the carriers.

Barland et al.[73] investigated the TSC properties of chemical vapor deposition (CVD) Si_3N_4 and found that, in fact, it was a thermally stimulated ionic current (TSIC). The mobile ion was found to be potassium which was responsible for a peak at ~100 °C. The mobility of the potassium has been determined to be $\mu_o = 4.4 \times 10^{-13}$ m^2/Vs in the range of 25-150 °C. The activation energy for detrapping K^+ ions has been determined to be ~1.0 eV.

8.1.3 TSC Dosimetry

We shall finish this section with a brief mention of the applications of TSC, which are somewhat similar to those of TL discussed in Ch. 7. It is obvious that, at least in principle, in the same way as TL can be used as a tool for measuring the amount of absorbed dose, TSC can be utilized in a similar manner. However, the applications of TSC in dosimetry and dating so far are very far behind the advanced stage which TL dosimetry and TL dating have reached. We are aware of only one publication on TSC dating of an archaeological sample, by Hwang and Fremlin[74]. More work on TSC dosimetry can be found in the literature. Mitchell and DeNure[75] suggested that the use of TSC samples has several attractive features as dosimeters. They are small, simple and inexpensive, and they produce a direct electrical output which allows the read-out equipment also to be simple and inexpensive. Even more important is the fact that these dosimeters are useful in relatively high dose ranges of $>10^3$ Gy, above the range of some of the commercial TLD systems. Much of the work done on TSC dosimetry was performed in the late sixties and during the seventies. The groups of materials used were crystalline samples, organic materials and glasses.

In the crystalline materials, one of the first samples used was diamond. Bose and Henisch[76] studied the potential use of insulating diamonds with gold contacts as TSC dosimeters in the UV and visible regions. Measurements were performed on neutron-irradiated as well as unirradiated samples. Bowlt[77] studied TSC in type Ia insulating diamond with graphite electrodes. Excitation was, again, by UV light; a strong peak was found at ~540 K with weaker ones at lower temperatures. Vermeulen and Halperin[78] investigated the dose dependence of a TSC peak at ~520 K in UV irradiated semiconductor diamonds. The main result has been that the dependence is nearly linear at low doses and tends to saturation at higher doses. Some investigators studied the potential use of sapphire (Al_2O_3) for TSC dosimetry. Moran et al.[79],[80] found that the TSC peak height response to x-rays in sapphire, with evaporated aluminum contacts, is linear with the exposure between 10 mR and 100 R, and is useful up to 10^4 R. The samples are reusable with no additional treatment for exposures below 10^3 R. These samples are exposure-rate independent from 0.12 mR/s to 3.2 R/s; they are also stable and reasonably tissue equivalent as a function of energy. Lörincz[81] also studied TSC in sapphire, but used ac measurements which enabled the evaluation of higher doses. Lassabatere et al.[82] studied the dependence of TSC intensity on the dose of UV excitation (~5 eV light) in metal-oxide-semiconductor (MOS) devices of the Si-SiO_2-Au structure. Gomez and Sinencio[83] investigated TSC in sulphur single crystals excited by x-rays. They measured the growth curves and found a sublinear dependence on the dose with total saturation at 10^3 Gy. von Bardeleben et al.[84] studied the TSC response to visible light of $AgGaS_2$ with evaporated gold electrodes, and found a sublinear dependence on the dose. Gates et al.[85] investigated the TSC properties of a number of inorganic materials with the potential utilization as dosimeters in mind. The excitation of all these samples was by γ-irradiation from a ^{60}Co source. In MgF_2 they found a peak at 88 °C, a rather low temperature due to which it can be used for dosimetric purposes only if the readout is performed immediately after the irradiation. In CaF_2:Dy, which is applicable as a TLD only up to 10 Gy, a monotonous increase in the peak height was observed up to 10^4 Gy. In ThO_2, the dependence of a TSC peak at ~100 °C on the γ dose was found to be nearly linear (in fact, slightly superlinear) up to 5×10^4 Gy. Gates et al.[85] considered Y_2O_3:ThO_2 as the most promising TSC dosimeter material. The peak utilized appears at 295 °C, which makes it very stable at room temperature. The dose dependence between 10^3 and 5×10^4 Gy is sublinear with a slope of ~0.4 on a log-log scale. Harper and Thomas[86] studied some other inorganic materials. These include BeO, B_2O_3, MgF_3, Al_2O_3, SiO_2 and $Na_2B_4O_7$. The best one appeared to be MgO which they reported to be a suitable UV light dosimeter.

Some organic materials, mainly polymers, have been tested for their TSC properties. Ramsey and Joesoef[87] made measurements of neutron dose by using TSC in polyethylene. The dose dependence was found to be sublinear, and similar results were reported in x-ray irradiated polyethylene. The authors conclude that these samples are suitable for measuring neutron doses between 0.1 and 10 Gy in the energy range of 1-14 MeV. In x-rays, one can go up to as high as 200 Gy with this material. Polyethylene was also used by Perlman and Unger[88] for studies of the TSC response to 6 keV electrons. Lapke et al.[89] studied TSC in polyethylene irradiated by γ-rays at relatively low temperatures. The dose dependence of two peaks, at 80 °C and 110 °C, was found to be linear in the range of 30-300 Gy. Perlman and Unger[88] and Grinter and Bowlt[90] studied the TSC properties of teflon (PTFE-polytetrafluoroethylene), excited by 100 keV x-rays. The dose dependence was found to be linear between 0.01

and 10 Gy. Further results on PTFE have been reported by Conway et al.[91] who irradiated their samples using a ^{60}Co γ-source, a ^{90}Sr β-source and a ^{137}Cs γ-source. The dose dependence curve was nearly linear up to 1.0 Gy, with tendency to saturation between 1 and 10 Gy. In some of the samples, a permanent charge occurred which changed the TSC properties irreversibly. The remedy to this was a radiation and annealing procedure prior to the subsequent dose response measurement. Conway et al.[91] also studied TSC in polyflouoroethylene propylene (FEP) and polymethylepentene (TPX). The latter was found to be a tissue equivalent material. The results of *integrated* TSC as a function of dose was found to be linear between 10^{-3} and 10 Gy, which suggested a promising material for TSC dosimetry.

Some TSC dosimetry work was performed on amorphous glasses. Street and Yoffe[92] measured TSC in evaporated thin films of amorphous chalcogenides, As_4S_4, As_2S_3, As_2Se_3 and Se. They tested different electrodes and found that aluminum was most satisfactory for both As_2Se_3 and Se. Excitation was by visible light and the dependence of the peak height on the dose was basically linear. Mitchell and DeNure[75], worked on TSC radiation dosimeters made from borosilicate sealing glass, exposing it to ^{60}Co γ-rays or 175 keV x-rays. The dosimeters consisted of metal-glass-metal sandwich structures with electrodes bonded to the glass with glass-to-metal seals. The dose dependence curve was found to be practically linear between 10 and 5×10^3 Gy, and tended to saturation at higher doses. Bowlt and Wagget[93] studied various glasses and plastics as potential TSC dosimeters, and found spectrosil (OH - synthetic fused silica) to be appropriate for dosimetry of 90 keV x-rays. A TSC peak occurring at $\sim$170 °C was rather stable when the specimens were maintained at room temperature and exposed to normal room light for as long as one week. Both the peak height and the total charge collected between 125 °C and 200 °C depended linearly on the excitation dose.

8.2 Thermally Stimulated Electron Emission (TSEE)

8.2.1 General

Another thermally stimulated process which is directly related to the release of trapped electrons is *thermally stimulated electron emission* (TSEE). Actually, TSEE is one of a more general group of effects called *exoelectron emission* (EEE) which is the emission of slow electrons (few eV) out of the surface of solids following high energy radiation such as α, β, and γ-rays, x-rays and, in some cases, following mechanical treatment (see e.g. Scharmann[94]). EEE is also termed the *Kramer effect* after the investigator who substantially advanced its understanding[95]. In this case, if a measurement is performed at a constant temperature, the emission decays with time after the initial excitation.

A number of papers had been published between 1897 and 1902, which marked the beginning of EEE research. H. Curie[96] and Rutherford[97] observed a decaying "radioactivity" of materials previously exposed to radioactive materials, which may at least partly be explained by EEE. Russel[98] and Villard[99] observed that zinc and bismuth foils, mechanically treated or exposed to x-rays, could blacken photographic plates, which can possibly be attributed to exoelectron emission. McLennan[100] excited sulfates and chlorates by cathode rays and recorded the delayed electron emission. He also discovered that the charge involved was negative and that the particles involved were of low energy. In the 1930s, an abnormally high background count was observed in the newly invented Geiger-Müller counter, which was due to exoelectron emission (see Becker[101],[102]). This background decreased on aging of the counters, but an increase was observed after gas discharge or mechanical treatment of the electrodes.

As mentioned already, Kramer was the first to systematically investigate EEE. He studied the electron emission from a number of ionic crystals and metal surfaces after mechanical treatment or exposure to heat or light, and found emission maxima at different characteristic temperatures during heating[103], thus forming the basis to *thermally stimulated electron emission* (TSEE) research. Bohun[104] further discussed the similarities between TL and TSEE, and Nassenstein[105] extended a kinetic model of TL to explain TSEE.

The experimental procedure for recording a TSEE curve is rather simple, and will be very briefly described here. A sample, usually an insulator, is connected to an electrode, and is placed in a vacuum tube; another electrode is also present in the tube at a certain distance from the sample, the latter being positive as compared to the sample. A given potential difference is maintained between the two electrodes. Various types of the tube have been developed for specific purposes such as certain temperature ranges, or high or low TSEE rates, some examples of which were given by Becker[101],[102]. The sample is first excited in one of the ways mentioned in connection with TL, namely, UV light, x-rays, γ rays, etc. During this time, electrons may be trapped in trapping levels. The sample is then heated and when the electrons are thermally stimulated from the traps, they can be released from the sample and be accelerated to the other electrode. Current as a function of temperature (or time) can thus be recorded. The current increases with temperature in a certain range, but due to the limited number of electrons in a trap and due to their recombination from the conduction band into centers the electrons become exhausted, which results in a reduction in the intensity at higher temperatures, and the observation of a TSEE peak. The shapes and other properties of the peaks reveal information on the trapping states. Experimentally speaking, the special feature of TSEE is that single particles can be detected. Thus, open point counters, Geiger-

Müller counters and multipliers can be used instead of the collecting electrodes, in conjunction with nuclear counting electronics. This is quite similar to the photon counting technique which is rather prevalent in the measurements of TL. Due to the very low energy of exoelectrons, a window type counter cannot be utilized for the detection of TSEE unless the electrons are accelerated[101]. This can be done, for example, if the electrons are accelerated to, say, 6 keV and a very thin window is being used. It seems preferable, however, to employ a windowless Geiger-Müller counter. Various readout instruments have been used. These include high amplification proportional counters with a point counter as described by Fintelman[106], as well as structures in which the counting wire is a loop as described by Attix[107] or a straight line as discussed by Moreno et al.[108].

The use of a Geiger-Müller counter is advantageous when low levels of the effect are to be monitored. This is so since in this case, single particles can be detected which cannot be done, for example, in TSC. On the other hand, proportional counters, the dead-time of which is much smaller ($\sim$0.1 μs as compared to $\sim$200 μs in the G-M counter), are recommended for the measurement of intense TSEE up to saturation levels, as shown by Moreno et al.[108].

8.2.2 Theory

Experimental results of simultaneous measurements of TL and TSEE have shown close relationships between the two phenomena in many cases (see Ch. 9). The fact that both effects are related to the same process are obvious, for example in x-ray irradiated NaCl samples as shown in Fig. 9.8, taken from Bohun[109]. There are, however, experimental results in other materials in which TL peaks occur without TSEE counterparts or, vice versa. As explained already with respect to TSC, TL unaccompanied by TSEE may occur when the charge carrier, once thermally stimulated, is not brought into the conduction band but rather to an excited state within the forbidden gap, from which recombination is allowed. Another possibility, specific to the present case is that the carriers responsible for TL are holes, thermally stimulated into the valence band; these cannot, of course, be freed from the sample into the vacuum and therefore, cannot contribute to TSEE. However, some authors described a more complicated situation, in which, although the released primary charge carriers are holes, electrons are released from the sample as TSEE peaks due to the Auger effect. On the other hand, TSEE may occur without TL if the transitions involved are radiationless or if the light emission is in a spectral range undetectable by the detecting device being used. Another complication regarding the correlation between TL and TSEE is that whereas the former is basically a bulk effect, TSEE is quite often associated with surface levels which may not be related to bulk states. TSEE may be influenced by adsorption on

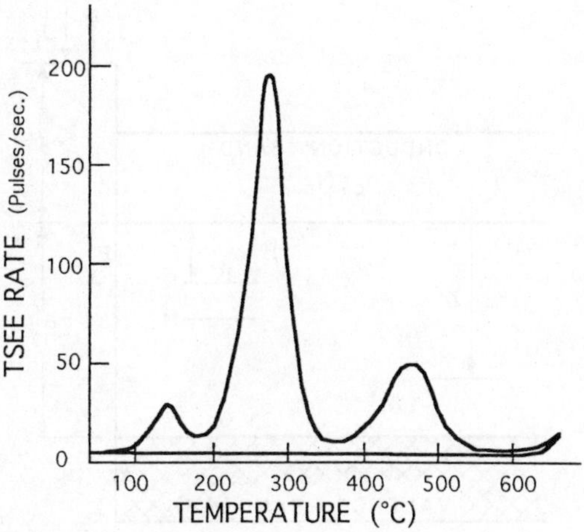

Figure 8.3 Exoelectron glow curve of lunar material from Apollo 12. 10 mg of γ -irradiated material was used.
(After Becker et al.[110].)

the surface or by polishing of the surface, consistent with the surface nature of the effect.

An unusual, impressive TSEE glow curve is shown in Fig. 8.3, taken from Becker et al.[110]. The curve is unusual in two respects. First, three separate distinct single peaks are seen, at 142, 268 and 465 °C. The second point is related to the rather exotic sample, which is γ-excited lunar material brought back by the crew of Apollo 12.

Since the general appearance of a TSEE peak is similar to that of TL, and since the understanding of the phenomenon is that electrons may be ejected from a sample after having been thermally stimulated from a trap in a similar way as in TL, several investigators applied the simple analysis of Randall and Wilkins[111] to the TSEE peaks (for a summary, see Becker[102]). Assuming a single trapping level, no retrapping and a linear heating function, a simple peak would be obtained and the activation energy and frequency factor could be chosen so as to best fit with the experimental curve. This approach, in which one assumes that the rate of release of electrons from traps is proportional to the Maxwell function $\exp\{-E_t/kT\}$, was termed by Bohun[112] and Holzapfel[113] the "Maxwell tail" theory.

The Randall-Wilkins approach is, certainly, an oversimplification since, as explained already in the TL case, the situation is usually much more complicated even if only one trapping level and one kind of recombination center exist. Moreover, in the present case of TSEE, even if an electron has been

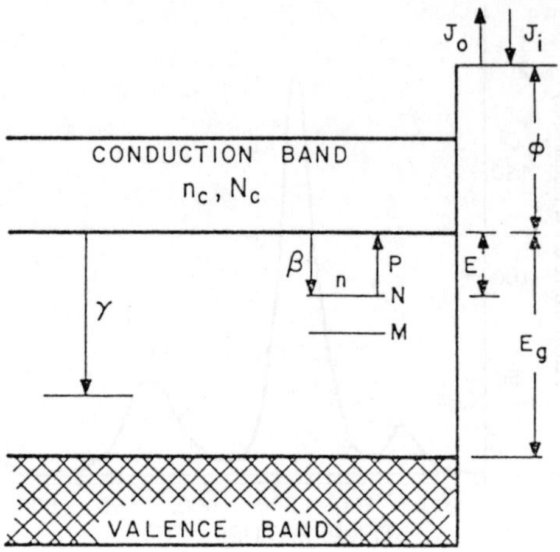

Figure 8.4 Energy level diagram, including the work function ϕ, taking part in the TSEE process.

(After Kelly[115].)

raised into the conduction band, it needs further energy - the work function ϕ - to get released into the vacuum. This has first been discussed by Pimley and Francis[114] and further by Holzapfel[113] and Kelly[115]. A simple schematic energy level model, including the work function ϕ , given by Kelly[115] is depicted in Fig. 8.4. It is to be noted that the model described by Kelly includes J_i and J_o, the current densities into and out of the surface, respectively. Kelly assumes that $J_i = fJ_o$ where $0 \le f \le 1$, f being a constant, and the net current is then $J = (1 - f)J_o$. In this case one gets

$$J(T) = n_c(T)\,[kT/(2\pi m^*)]^{1/2} \exp\{-\phi/kT\} \tag{8.23}$$

where $J(T)$ is the net current density out of the sample, m^* is the effective mass of the conduction electron, $n_c(T)$ is the momentary concentration of electrons in the conduction band, resulting from all the traps and centers involved (including, of course, possible retrapping) and ϕ is the work function. In the simplified situation in which $n_c(T)$ behaves like a first-order peak, it is simple to show that the TSEE curve is given by

$$J(T) \propto \exp\{-E_t/kT\} \exp\left\{-(s/\beta)\int_{T_0}^{T} \exp\{-E_t/k\theta\}\,d\theta\right\} \exp\{-\phi/kT\}. \tag{8.24}$$

By the procedure of equating the first derivative to zero, Holzapfel[113] has found the maximum condition to be

$$(s/\beta)\exp\{-E_t/kT_m\} = (E_t+\phi)/\left(kT_m^2\right) \tag{8.25}$$

which is the same as 1.5 in Ch. 1, except that $E_t + \phi$ replaces E_t in the right hand side (but not in the integral in Eq. 8.24!). The $T^{1/2}$ term in Eq. 8.23 has been ignored in Eq. 8.25; its possible influence has been considered along with the possible temperature dependence of the frequency factor in Ch. 3, while dealing with Eq. 3.92. The case discussed in Eqs. 8.23 and 8.25 is similar to that given by Wintle[116] to explain anomalous fading of TL. Here, $-\phi$ replaces the energy W associated with the dependence of the recombination probability on temperature[116]. The various heating rates method discussed in section 3.5 should yield the value of E_t, whereas the initial rise method should give $E_t + \phi$. Thus, ϕ can be evaluated as the difference between the two results.

If one wishes to study the more general case, one has to consider $n_c(T)$ as found from the solution of the three simultaneous differential equations 2.17, 2.19 and 2.20, and insert it into Eq. 8.23. In general, the multiplication of $n_c(T)$ by the increasing function $T^{1/2}\exp\{-\phi/kT\}$ should cause the TSEE maximum to appear at a higher temperature than the corresponding TSC peak (see Ch. 9).

As noted, an important feature of TSEE as compared to TSC is that, evidently, only electrons can be released from the sample whereas holes cannot. Thus, one can usually distinguish between electron and hole TSC peaks by performing a simultaneous TSC-TSEE measurements; normally, only the electron-peaks would show up in the latter. An opposite effect of the release of electrons from the surface of the material as a result of the primary thermal release of holes from traps was described by a number of investigators. The first account of this effect was given by Tolpigo et al.[117]. This was followed by a number of groups who considered the occurrence of this *Auger effect* in different materials. As explained by Holzapfel[118] and Kawanishi[119], the trapped hole may be thermally released into the valence band and subsequently recombine with an electron in a center. Instead of releasing the energy as a photon (which is the case of TL), the energy may be transferred to a trapped electron which can thus be released from the trap and then, from the sample. Bohun[109] considered this possibility while discussing TSEE in several ionic crystals. In a later work, Kortov et al.[120] base the explanation of their experimental results in α-Al_2O_3:Mg single crystals on the Auger process. Kutomi and Tomita[121] interpret the occurrence of one of the two main TSEE peaks in $Li_2B_4O_7$:Cu (at ~165 °C) on the Auger effect, whereas the two other peaks are ascribed to the Maxwell-tail emission of electrons. It should be noted, however, that according to Holzapfel[118], this case of Auger TSEE seems to be the exception rather than the rule since high concentrations of the filled traps are required, which is not generally the case following the application of "regular" doses to the sample.

An important advantage of TSEE as compared to TSC is that no contacts are needed for the measurements, which eliminates some of the problems mentioned above concerning electrical contacts. On the other hand, since TSEE peaks are sometimes related to surface levels rather than to those of the bulk it cannot always serve for the study of the bulk of the material. In contrast, it can sometimes be used as a sensitive tool for the investigation of the surface states.

Out of the many hundreds of papers discussing TSEE properties, we shall only mention a selected few published relatively recently. Biedrzycki[122] reported TSEE in triglycene sulfate (TGS) crystals. This author associates the observed TSEE peaks with the macroscopic polarizations being positive or negative following their being polarized by a dc field. Golek et al.[123] studied TSEE in the high-T_c superconductor $GdBa_2Cu_3O_{7-x}$, excited by an electron beam, in the temperature range of 90-350 K. Two peaks were observed at ~160 and ~220 K. The latter appears to be connected with the ordering of oxygen vacancies in the planes containing the Cu-O chains. The occurrence of the 160 K peak may be associated with the recombination of thermally released holes with electrons trapped at F-centers, which is another example of the Auger effect.

Watanabe et al.[124] investigated TSEE from $BaTiO_3$ ceramics. They found a difference between negative temperature coefficient (NTC) of resistivity and positive temperature coefficient resistivity (PTC) ceramics. In the latter, excited by x-rays, several peaks were detected between 40 and 250 °C with a main peak below 100 °C. No TSEE peak was found in the same region in the NTC ceramics. The difference between the two kinds of samples was attributed to the surface potentials. Watanabe et al.[125] studied TSEE from x-ray irradiated hydrogenated amorphous carbon (a-C:H) films on stainless steel substrates. Whereas the TSEE peaks from 80 nm films originate mainly from the oxide of the substrate, with 280 nm films, TSEE originates from the film itself. The authors suggest that the effect can thus be utilized for the characterization of the surface thin films.

Momose et al.[126] reported the TSEE glow curve from x-ray excited Ar-plasma-treated Si. In the temperature range of 298-623 K, two broad emission peaks appeared at ~373 and ~473 K, and an increase of intensity above 573 K. Further analysis showed that four types of electron states are involved in the SiO_2/Si interfacial region, resulting in four components, two in each of the observed broad peaks. It was found that the trap depth for each level decreased with increasing oxygen contents at the surface, but the electron affinity was almost independent of the oxygen contents. Huzimura and Nakahira[127] investigated the energy distribution of TSEE from low energy electron bombarded

MgO. The excitation was by electrons in the range of 75 eV to 3 keV, and the resulting main peak was at ~150 °C. In this case, it was found that the energy spectrum of exoelectrons was very sensitive to the charge distribution and gas adsorption on the surface of the MgO sample.

Sakurai and Momose[128] presented a model of TSEE which included one kind of electron trap characterized by the activation energy E_t and the frequency factor s, emptied by a first-order process into the conduction band. From there, the electrons leave the solid by thermionic emission, characterized by the electron affinity ϕ. Since the authors excluded retrapping and recombination, the equations can be solved analytically with a linear heating function. They applied their method to the analysis of TSEE from glass deposited on metal by argon plasma treatment. Rudlof and Glaefke[130] criticized this method of analysis due to its simplifying assumptions. They recommend the use of many of the methods mentioned in Ch. 3 and elsewhere, such as the fractional glow technique, the various heating rates method, the initial rise method, etc., and tried to elucidate the full picture from the combination of all the available information. In a later work, Sakurai[129] developed his previous method by including recombination and retrapping in the TSEE equations and extracting the relevant parameters by a best fit method.

Similarly to the cases of TL, and to some extent TSC, the presently discussed TSEE can also be utilized for dosimetry as long as the intensity of the measured peaks is a monotonically increasing function of the applied dose. A review paper on this subject has been written by Spurný and Käämbre[131]. In a summary of previous results they mention the main materials that can be used for this application - namely, BeO and Al_2O_3, - but they also note that LiF and $CaSO_4$ samples have been successfully employed. While comparing TSEE and TL dosimetry, they mention the high sensitivity of the former (10^{-8}-100 Gy range), the easy reusability, the ease of use for dosimetry of α and β radiation, the unlimited storage time (for peaks at high enough temperature), but they also note the rather limited range of linearity. The material used mostly for TSEE dosimetry is BeO (see e.g. Becker[101]), produced with different kinds of dopants, including Si and Li.

8.3 Thermally Stimulated Polarization and Depolarization Currents

8.3.1 General

The last two sections described two kinds of thermally stimulated current following an appropriate excitation (normally irradiation) in two different kinds of experimental situations, namely, TSC and TSEE. Both these phenomena are intimately related to thermoluminescence (TL) in the sense that an essential step in these processes is the filling of trapping states during the excitation, and their thermal release into the delocalized bands during the heating phase. Also, these phenomena are characterized by the occurrence of peaks in the curve of current as a function of temperature. As briefly mentioned in Ch. 1, there are other thermally stimulated processes in which electric current is measured as a function of temperature. In most cases, these are associated with electrical polarization and depolarization of the samples. Such dielectric relaxation processes occur in a variety of materials, crystalline as well as amorphous insulators and semiconductors, normally without a prior irradiation. In certain cases, the depolarization currents may occur along with a TSC contribution, which may make the interpretation of the experimental results more complicated. On the other hand, the polarization and depolarization currents may provide useful information concerning the dielectric properties and the internal structure of the sample. An interesting point to be noted is that although the physical reason for these currents is different from that causing TSC, the mathematical appearance of the differential equations governing the process is rather similar to those of the thermally stimulated processes described so far and, in fact, the analogy is closest to the simple form of TL.

Although most of the investigators who studied these phenomena utilized similar experimental procedures, sometimes with some variations in the details, different terminology has been used. We shall follow the common terminology (see e.g. Müller[132]) and refer to the technique in which a polarized dielectric is depolarized during the heating of the sample as *Thermally Stimulated Depolarization* (TSD). The terms *Ionic Thermocurrents* or *Ionic Thermoconductivity* (ITC) will also be sometimes used in TSD cases where the current is due to the motion of ionic charge carriers. The less common procedure in which an unpolarized sample is polarized during heating under an external field is called *Thermally Stimulated Polarization Currents* (TSPC). The term *Thermally Stimulated Dielectric Relaxation* (TSDR) will be used as a general name for both TSPC and TSD. The term thermally stimulated conductivity (TSC) is reserved for radiation-induced current peaks which result from the thermal release of trapped charge carriers, as discussed above. It should be noted, however, that some authors refer to TSDR phenomena as TSC, which may sometimes be confusing.

The application of a static electric field to a dielectric may cause its polarization. The polarization may be due to several microscopic mechanisms. These are:

1. The displacement of electrons within the atoms or ions, as well as the small displacement of ions within molecules or unit cells. This is called *induced dipole polarization.*

2. The alignment of permanent dipoles, also termed *dipolar polarization.* An important work by van Turnhout[133] discussed the non-isothermal dielec-

tric relaxation phenomena of polymers or other dielectrics bearing a persistent charge, namely *electrets*. Carrying opposite charges on the two sides, electrets are the analogues of magnets.

3. Migration of ions or ionic vacancies over macroscopic distances, which may cause the accumulation of ionic space charge near the electrodes. This is known as *ionic space charge polarization.*

4. Drift of electrons or holes towards the electrodes and their trapping (known as *space charge polariztion*).

5. The injection of excess charges (electrons or holes) from the electrodes into the sample.

In thermally stimulated depolarization (TSD), a polarized sample is heated from some "low" temperature, T_0, usually in the absence of an external field, while its electrodes are short-circuited through a sensitive current detectors. The current which appears in the external circuit due to depolarization processes is recorded as a function of time or temperature. Each depolarization process becomes detectable when its relaxation time is short enough by producing a depolarization current which initially increases with temperature, then reaches a maximum and drops off to zero as the equilibrium distribution is reached. The occurrence of simple first-order kinetics, which are a special case in TL, is the usual state in TSD. In an ideal situation, a separate current peak is monitored for each relaxation process. An example is shown in Fig. 8.5. If partial overlapping between two peaks occurs, separation of the high temperature peak can be achieved by the usual *thermal cleaning*. Referring to Figs. 8.5 and 8.6 taken from Kitts et al.[134], let us indicate the maximum temperatures of the two peaks by T_1 and T_2 ($T_1 < T_2$). Heating the sample to an appropriate intermediate temperature T', ($T_1 < T' < T_2$) and subsequently cooling it, erases the low temperature component. Then, the sample is heated again to exhibit the second peak which is now expected to be "clean". In order to obtain the first peak separately, the sample is polarized at a temperature T_p, such that $T_1 < T_p < T_2$ for a period of time $t \approx \tau_1(T_p) \ll \tau_2(T_p)$ where τ_1 and τ_2 are the lifetimes of the two components at the temperature T_p. Under these circumstances, the lower temperature polarization process approaches saturation, while the peak associated with the second process appears relatively weak, as seen in Fig. 8.6.

Early experiments of thermally stimulated depolarization were reported with no specification of the heating program, since the main purpose was to measure the total released charge. These include, for example, works by Frei and Groetzinger[135], Gross[136], Gubkin and Matsonashvili[137], and Murphy[138]. Since the mid-sixties, several investigators have used a linear heating function and recorded current vs. temperature, namely, the TSD curve. Driver and Wright[139] used cooling with a forward bias and heating with a reverse bias,

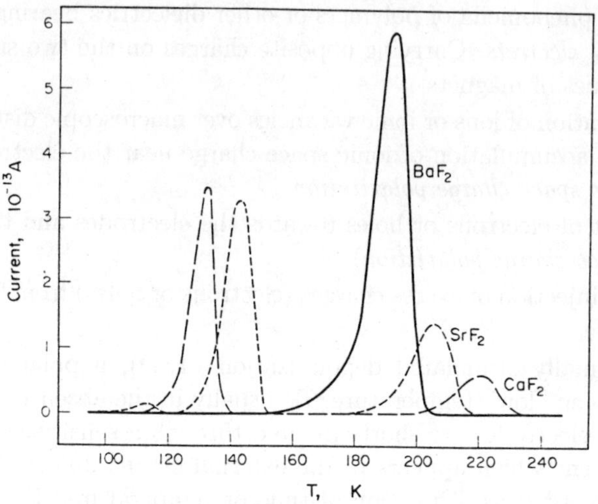

Figure 8.5 TSD curves for alkaline-earth fluorides doped with Gd^{3+}. In SrF_2 and CaF_2, the separate peaks indicate different processes of dipole relaxation. The main peak in BaF_2 includes a smaller component at 170 K.
(After Kitts et al.[134].)

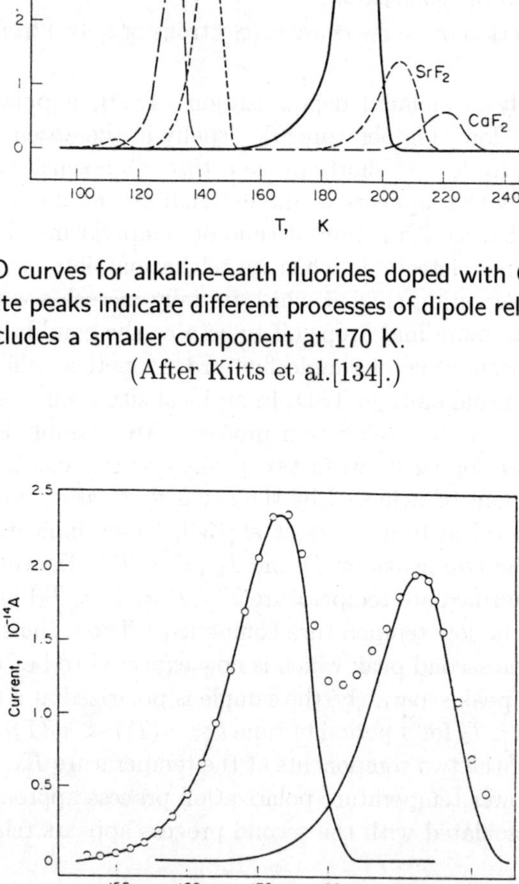

Figure 8.6 Resolution of the TSD peak of BaF_2:Gd^{3+} by polarization at 160 K. The area of the 193 K peak is only 2% of its saturation area, resulting from polarization near 190 K.
(After Kitts et al.[134].)

in order to study the thermal release of trapped space charge in CdS. Bucci et al.[140],[141] studied TSD in ionic crystals due to dipole orientation, which they termed ionic thermocurrents, ITC. Zolotaryov[144] et al. applied TSD to the study of electron and hole trapping. Since then, TSD techniques have been applied, by many investigators, to the study of various dielectrics, ionic and divalent crystals , semiconductors and polymers.

The polarization of the sample is usually achieved by applying an external field to a capacitor in which the sample is the dielectric material. Sometimes the sample is insulated from the electrodes by thin insulating foils, for example Teflon, in order to prevent the injection of charge carriers; this arrangement is called *insulating* or *blocking electrodes*. The advantages of alternatively using an air gap between one of the electrodes and the sample was discussed by van Turnhout[133]. Some investigators accomplished polarization by irradiating one side of the sample with UV light, x-rays, etc., either instead of the application of the electrical field or in addition to it. These include Alzetta et al.[145], Fields and Moran[146], Moran and Podgoršak[147] and Böhm and Scharmann[148].

8.3.2 Depolarization Currents due to Dipole Reorientation (Ionic Thermocurrents: ITC)

As mentioned above, the detection of dipole relaxation by TSD currents was introduced by Bucci et al.[140],[141], who termed their method Ionic Thermocurrents (ITC). Since then, the method has been used for many kinds of samples which contain dipolar defects that can be oriented by an external electric field. In ionic crystals, such dipoles may be formed when host ions are substituted by impurity ions of different valency. Charge compensation is usually attained by lattice vacancies, interstitial host ions, or impurity ions of different types. If the compensating charge is located near the impurity ion, a dipolar complex is formed[142],[143]. In polymers, ITC peaks appear as a result of the disorientation of the polar side-groups. For example, PMMA (polyethyl methacrylate) has a polar ester side-group -$COOCH_3$ with a dipole moment of 1.6 Debye, which can rotate *with* and *without* the main chain segments -C-CH_2-. When the polymer is heated to its softening temperature, the dipoles are disoriented by the motion of the main chain segments as described by van Turnhout[149].

An ITC measurement consists typically of the following stages:

1. The sample is polarized to saturation in a static electric field E_p, at a temperature T_p.

2. The specimen is cooled, with the field still applied, to a temperature T_0, where the motion of the dipole is hindered, namely, where the lifetime $\tau(T_0)$ is very long, at least of the order of several hours. Typically, T_0 is around 80 K or less.

3. At T_0, the external field is removed, and a sensitive current detector is directly connected between the electrodes. Because of the low temperature, the

dipoles are frozen in their ordered position, and the sample remains polarized even in the absence of the field.

4. The solid is warmed up, usually at a constant heating rate β (convenient rates are between 1 and 10 Kmin^{-1}), while the current through the detector is recorded as a function of temperature.

8.3.3 The Basic Equations of ITC

For the sake of simplicity let us assume a homogeneous dielectric containing one type of non-interacting dipoles. Let us denote by μ the electric dipole moment of a single dipole, and by N_d the concentration per unit volume of the dipoles. A dipole may reorient by a thermally activated jump of one of the ions which form the complex. Usually, such a jump can be characterized by a relaxation time τ, which is a function of the temperature,

$$\tau(T) = \tau_o \exp\{E/kT\}\,. \tag{8.26}$$

τ^{-1} is the probability for a single jump per unit time; the pre-exponential factor τ_o is assumed to be independent of temperature. E is the activation energy for the rotation of a dipole, T is the absolute temperature and k is Boltzmann's constant. As explained by van Turnhout[149], in the absence of an external electric field, the dipoles are oriented randomly. If an external field E_p is applied to the crystal at a temperature T_p, polarization of the dipoles is set up. The saturation polarization P_o is given by the Langevin function which can be approximated at high enough T_p ($kT_p \gg \mu E_p$) by

$$P_o = \mu^2 E_p N_d \alpha / kT_p. \tag{8.27}$$

α here is a dimensionless factor which depends on the structure of the crystals. For example, for free rotating dipoles, $\alpha = 1/3$. It should be noted that at a low temperature, a very long time may be needed for the polarization to saturate, since the dipoles are "frozen" in their positions. At high temperatures, the saturation of the dipoles polarization may be reached in a few seconds; this is the case in most ionic crystals at room temperature.

The model by Bucci et al.[141] for an ITC peak was developed for a perfect insulator containing one kind of dipole in a rather dilute concentration, so that mutual interactions between dipoles are negligible. Under these conditions, the rate of the dipoles reorientation is proportional to the concentration of dipoles still aligned. The depolarization kinetics are therefore represented by

$$-dP/dt = P/\tau(T) = \frac{P}{\tau_o}\exp\{-E/kT\}\,, \tag{8.28}$$

where P is the instantaneous polarization. The current density due to the depolarization is given by $j(T) = -dP/dt$. As opposed to TL, here Eq. 8.28

is *intrinsically* of first-order. The solution of the equation will therefore be a characteristic first-order peak. Taking the value of P_o from Eq. 8.27, and assuming a linear heating function $T = T_0 + \beta t$, we get

$$j(T) = \frac{N_d \mu^2 \alpha E_p}{kT_p \tau_o} \exp\{-E/kT\} \exp\left\{-(1/\beta\tau_o) \int_{T_0}^{T} \exp\{-E/k\theta\}\, d\theta.\right\} \qquad (8.29)$$

Due to the exact analogy with first-order TL peaks, the condition for the maximum can be written immediately as

$$kT_m^2 = \beta E \tau_o \exp\{E/kT_m\}. \qquad (8.30)$$

Of course, this can be found by equating the derivative of Eq. 8.29 to zero. The temperature of the maximum, T_m, is seen to be independent of T_p and E_p.

The concentration of the dipoles N_d can be evaluated from Eq. 8.27 in which P_o can be found from the area under the ITC peak,

$$P_o = \int_{T_0}^{\infty} j(\theta) d\theta = \frac{N_d \mu^2 E_p \alpha}{kT_p}. \qquad (8.31)$$

According to Eq. 8.29, the current intensity at any temperature is proportional to the polarizing field E_p. This can be sometimes used to distinguish between real ITC and *spurious* current peaks due to contact effects, which are not proportional to E_p, as explained by Bucci et al.[141].

In addition to the pure and doped alkali halides mentioned already (see also Figueroa et al.[150], ITC was measured in alkaline-earth fluorides (e.g. see Matthews and Crawford[151]) as well as in other ionic and non-ionic crystals (see, for example, Kunze and Müller[152] and Kessler[153]). ITC measurements have also been used in the study of dipoles relaxation in various polymers. For example, see Pfister and Abkowitz[154] and Tanaka et al.[155].

8.3.4 *ITC with Distributions of Activation Energies and Pre-Exponential Factors*

The model by Bucci et al.[140],[141] described above assumed discrete relaxation processes, each having a well defined single activation energy and a single pre-exponential factor. This is sometimes called a *single Debye relaxation.* As mentioned already, most of the investigators who studied ITC due to dipole relaxation in ionic crystals found first-order peaks which agreed with the model. We should mention, however, some exceptions. Shelley and Miller[156] studied ITC in CaF_2:Na^+ and found that for samples with relatively high concentrations of Na^+ ($\geq 0.02\%$), the main peak at 165 K did not look like a first-order peak. The peak was significantly broader than expected, and broadened further as the

Na^+ concentration increased. The authors attributed this peak to the relaxation of the dipolar complex Na^+-V_F^- in CaF_2:Na^+ where a Na^+ ion substitutes a Ca^{++} ion and a nearby vacancy of an F^- ion. Shelley and Miller[156] showed that the deviation of their experimental results from Bucci's model can be ascribed to a dipole-dipole interaction and leads to a relaxation process with a distribution of activation energies. In fact, the extension of ITC theory to this case had previously been described by Gross[157],[158].

Similar studies have highlighted concentration-dependent effects in the lanthanide-doped, alkaline-earth fluorides. In these materials dipoles are formed due to the substitution of divalent Ca^{2+} with trivalent lanthanide ions (M^{3+}), and the consequent formation of charge compensating F_{int}^- interstitial ions in nearest-neighbor positions. Consequently, these materials give rise to strong ITC peaks due to the reorientation of the M^{3+}-F_{int}^- dipoles. As the concentration of M^{3+} ions increases the ITC peaks become broadened, and these effects were attributed to electrostatic and elastic interactions within the crystals[159]-[162]. The interactions include electrostatic dipole-monopole and dipole-dipole effects, as well as strong elastic interactions due to dipole clustering. Other authors[163] have observed that the addition of divalent ions to CaF_2 doped with M^{3+} results in the reduction of the interaction effects and leads to a narrowing of the broadened ITC peak, as well as a reduction in the height of the peak, for a fixed M^{3+} concentration. Jassemnejad and McKeever[163] studied this effect in CaF_2 doped with Mn^{2+} and Ce^{3+} and explained it in terms of the electrostatic interaction energy for a dipole of moment $\vec{\mu}_1$ in an electric field $\vec{F}$, namely $U_i = -\vec{\mu}_1 \cdot \vec{F}$, where the electric field is the sum of monopole, dipole and higher-order terms ($\vec{F} = \vec{F}_1 + \vec{F}_2 + \vec{F}_3 + ...$). For dipole-dipole interactions the field is

$$\vec{F}_2 = \frac{[3\,(\vec{\mu}_1 \cdot \vec{r}_{12})\, r_{12} - r_{12}^2\, \vec{\mu}_2]}{4\pi\varepsilon r_{12}^5} \tag{8.32}$$

where $\vec{r}_{12}$ is the vector connecting dipoles of moments $\vec{\mu}_1$ and $\vec{\mu}_2$, and ε is permittivity. Jassemnejad and McKeever proposed a reduction in the $Ce^{3+}-F_{\text{int}}$ distance due to the addition of Mn^{2+}, thereby reducing the dipole moment and decreasing both the interaction strength and the size of the ITC peak (the former in accordance with Eq. 8.32 and the latter in accordance with Eq. 3.27).

Broad ITC peaks which are attributed to a continuous distribution of activation energies or pre-exponential factors are also common in polymers. Examples were give by van Turnhout[133],[149], Vanderschueren[164],[165],[166] and by Fischler and Röhl[167]. Van Turnhout[133],[149] showed that in the case of a distributed polarization, the shape of the peak and the maximum temperature depend strongly on the polarization temperature T_p, as well as on the storage time of the polarized sample prior to its heating. The depolarization current in the case of a continuous distribution of activation energies is derived from Eq. 8.29, and is given by

$$j(T) = P_o s \int_0^\infty F(E) \exp\left\{-E/kT - \frac{s}{\beta}\int_{T_0}^{T} \exp\{-E/k\theta\}\, d\theta\right\} dE. \tag{8.33}$$

Laredo et al.[168] gave an example of the shape of the ITC peak in the case of a Gaussian distribution of the activation energies. Here $F(E)$ is the distribution function for the activation energies.

In the case of a continuous distribution of pre-exponential factors, the current is given by

$$j(T) = P_o \exp\{-E/kT\} \int_0^\infty s\Phi(s) \exp\left\{-\frac{s}{\beta}\int_{T_0}^{T} \exp\{-E/k\theta\}\, d\theta\right\} ds, \tag{8.34}$$

where $s = 1/\tau_o$. $\Phi(s)$ is the distribution function for the pre-exponential factors. In an important book on this subject, Jonscher[169] also considered broader-than-Debye shaped peaks as a manifestation of distributions of relaxation times, which would normally correspond to distributions of activation energies. Studies by Halpern[170],[171] showed that in cases with a common activation energy E_o and a distribution of frequency factors, the effective activation energy found by Bucci's analysis is usually less than E_o. In systems with the frequently observed stretched exponential isothermal relaxation, of the form $\exp\left\{-(t/\tau)^\beta\right\}$ where τ is thermally activated with an activation energy E_o and $\beta < 1$, the activation energy deduced from TSDC experiments would usually be βE_o, thus less than E_o. When the distribution is that of activation energies with a common pre-exponential factor, the pre-exponential factor determined by Bucci's analysis differs from its true value. Halpern[171] suggested some simple criteria for establishing whether the parameters derived from a given set of measurements do, in fact, correspond to those of the system. Leal Ferreira et al.[172] further developed this approach and showed an approximate universality which applies to TSDC curves when properly re-scaled. The universality was seen to be very accurate as far as variation of the relevant activation energy and the heating rate are concerned, and was somewhat poorer when the frequency-factor or the stretched-exponential parameter were varied.

The above-mentioned technique of thermal cleaning, denoted here as *fractional depolarization*, is the simplest way to deal with such a distribution-related ITC signal. Solunov and Ponevsky[173] and Zieliński et al.[174] pointed out that in the case of a single activation energy and a distribution in s, the

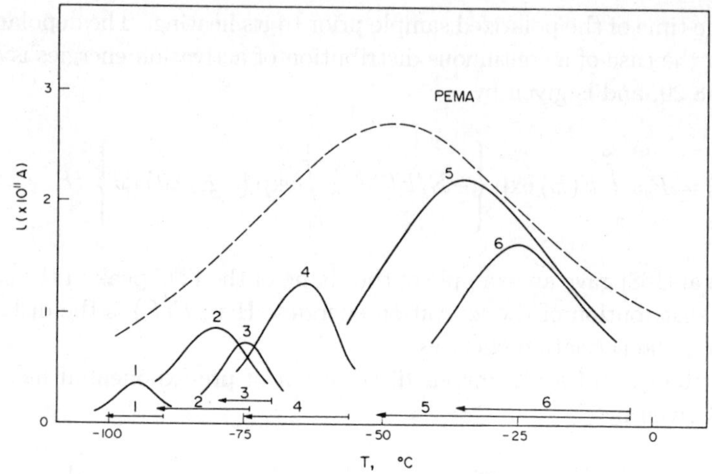

Figure 8.7 Application of the fractional polarization to the β peak of Polyethyl methacrylate (PEMA). The dashed curve corresponds to the saturated envelope peak, and the individual components are shown.
(After Vanderschueren[165].)

initial slopes on a semi-log scale are parallel, whereas a distribution in the activation energies results in different slopes.

Figure 8.7 presents a broad ITC peak, measured in polyethyl methacrylate (PEMA). The peak is shown to be deconvoluted to, at least, six components using the fractional depolarization, namely, heating and cooling the sample in subsequent cycles, thus covering different temperature ranges, and observing the separated peaks. Van Turnhout[133] utilized an alternative method of fitting the experimental results to Eqs. 8.33 or 8.34, with various empirical distribution functions, so as to find the best fit.

8.3.5 Further Models of ITC-TSD

Other models for dipole relaxation have also been suggested in which neither a Debye relaxation nor a distributed Debye relaxation is assumed for dipole relaxation. As shown by Hedvig[175], some dipole-TSD peaks in polymers follow the equation

$$j(T) = -dP/dt = -P/\tau(T) \tag{8.35}$$

which is the same as the left-hand equation in expression 8.28. However, the characteristic time τ is now given by

$$\tau(T) = \tau_o \exp\{-a(T - T_g)/(b + T - T_g)\} \tag{8.36}$$

for $T > T_g$ where T_g is the glass-rubber transition temperature. Of course, the variables t and T in these two equations are connected by the heating function $T(t)$. Equations 8.35 and 8.36 describe the major relaxation in polymers, which occurs when they transform from the glassy to the rubbery state, as described by van Turnhout[149]. This glass-rubber transition at T_g involves the configurational rearrangement of parts of the main long chains, which require some space to move. Another model for TSD in non-Debye dielectrics was developed by Jonscher[176]. In this model, the depolarization current is given by

$$j(T) = aP_o\omega_p(T)\left[\int_0^t \omega_p\left[\theta(t)\right]dt\right]^{-n} \tag{8.37}$$

for $t < 1/\omega_p$, where $\omega_p = 1/\tau$ is the reciprocal relaxation time and n is a positive number smaller than unity which is only weakly temperature dependent.

Jonscher[177] suggested a new approach to thermally stimulated depolarization. The motivation for the new approach was stimulated by experimental results, such as those by Careem and Jonscher[178], which showed clearly that stearic acid films with relatively very flat isothermal responses gave very pronounced TSD peaks. This does not agree with the Debye relaxation model. Jonscher[177] presented an alternative approach based on the notion that the temporal response under a variable temperature follows a different law to the isothermal response under variable time. In this way, it is proposed to decouple the thermally stimulated and the isothermal responses, which, in the conventional analysis, are coupled through the exponential decay

$$i(t) \propto \exp\{-t/\tau(T)\} \tag{8.38}$$

where $\tau(T)$ depends on temperature as in Eq. 8.26. Jonscher[177] points out that Eq. 8.38 represents the time dependence of polarization under a step-function field change in the isothermal case. It is, therefore, confined to this particular Debye form of analysis, and cannot tackle the 'universal' form of dielectric response defined in terms of the response function $f(t)$ which defines the discharge current under step-function field change by an amount ΔE,

$$i(T) = \varepsilon_o \Delta E f(T). \tag{8.39}$$

Jonscher *postulates* the existence of an analogous relation, applicable to the step-function increase of temperature by ΔT at a constant field which may be zero

$$i_T(t) = \dot{\varepsilon}(T)\Delta T\Theta(t) \tag{8.40}$$

where $\dot{\varepsilon} = d\varepsilon(T)/dT$ is the derivative of the static permittivity with respect to temperature, and $\Theta(t)$ is the thermal response function which is to the temperature variation what $f(t)$ is to field variation. The significance of Eq. 8.40

is that $i_T(t)$ depends on the strength of the temperature dependence of ε, in the absence of which there is no signal, as well as on the magnitude of the temperature change.

The response to an arbitrarily time-varying temperature $T(t)$ is given by the deconvolution integral[169]

$$i(T) = \dot{\varepsilon}(T) \int_{o}^{\infty} \Theta(t')\dot{T}(t-t')dt' \tag{8.41}$$

where $\dot{T}(t)$ is the heating rate and t' is the dummy variable. A step function temperature rise of amplitude ΔT gives $\dot{T}(t) = \Delta T \delta(t)$ where $\delta(t)$ is the delta function. The integral in Eq. 8.41 becomes exactly Eq. 8.40, which may be written as

$$i(t) = \Delta\varepsilon\Theta(t) \tag{8.42}$$

where $\Delta\varepsilon = \dot{\varepsilon}(T)\Delta T$ is the increment of ε in the temperature interval ΔT. This suggests a simple way of determining the function $\Theta(t)$ experimentally by measuring the time-domain response of the dielectric material to a step-function of temperature. Jonscher[177] makes a rather plausible guess that $\Theta(t)$ decays exponentially like

$$\Theta(t) = C \exp\{-t/\tau_T\}, \tag{8.43}$$

(where C is a constant) and further postulates that τ_T should follow the Arrhenius form of temperature dependence with an activation energy W_T,

$$\tau_T(T) = \tau_{T_\infty} \exp\{W_T/kT\} \tag{8.44}$$

where τ_{T_∞} is a suitable constant. It is not clear whether even in the limit of Debye behavior, the relaxation times $\tau(T)$ and $\tau_T(T)$ are the same quantity. In more general cases, there is no evident connection between $\tau_T(T)$ determined from TSD experiments and any isothermal relaxation times $\tau(T)$. In the view of Jonscher[177], this constitutes a strong advantage because the identity of the two relaxation times has never been properly established and, in some cases, has definitely been disproved, e.g. by Careem and Jonscher[178]. Jonscher[177] finally derives the relation for the polarization at constant field and with step-variable temperature

$$dP_T(t)/dt = -P_T(t)/\tau_T(T). \tag{8.45}$$

This is exactly equivalent to Eq. 8.28. This way, one may reproduce the entire analysis of TSD with a more physically realistic concept of the relaxation time $\tau_T(T)$, which does not determine the nature of the isothermal dependence.

Kato et al.[179] studied the thermally stimulated depolarization current under a temperature gradient (TG-TSD). They examined theoretically the effect of the temperature gradient on the initial rise region and found a good agreement between their experimental results on a cross-linked polyethylene (XLPE) film and the theoretical prediction.

Triantis et al.[180],[181] investigated theoretically the asymmetry of the ITC peak, and developed an approximate formula for evaluating the activation energy,

$$E = \frac{3.94kT_L T_M}{T_M - T_L} \tag{8.46}$$

where T_M is the temperature of the maximum and T_L is the temperature ($T_L < T_M$) for which $I(T_L) = I(T_M)\exp\{-3\}$. Of course, Eq. 8.46 is of the same class as the *shape methods* discussed in Ch. 4.

Vila and Jiménez de Castro[182] presented the equations governing TSD processes in a suspension of ellipsoidal semiconducting particles in an insulating matrix, which exhibits Maxwell-Wagner type polarization. A first-order TSD process is obtained when all particles are identical and arranged with one of their principal axes parallel to the electric field. For a random distribution of particle shapes, sizes and orientations, the TSD peak is broader. A somewhat similar approach was taken by Poźniak[183] who gave a simplified theory of TSD of a parallel plate condenser filled with heterogeneous solid consisting of two dielectrics. It is assumed that the particles, with different shapes and dimensions, of one dielectric are sparsely distributed in the other dielectric. It is also assumed that the average field in the solid is the same as the external electric field.

8.3.6 TSD-ITC Experimental Results

Out of the vast literature on TSD experimental results, we will mention only a few selected examples. We will start with TSD in inorganic materials. Vila et al.[184] investigated the TSD currents in neutron-irradiated Al_2O_3.e They found two peaks due to dipolar relaxation processes, at 242 and 273 K. Their results on the fluence dependence and thermal stability indicate that the second peak may be related to F-type centers or their corresponding interstitials, induced by the irradiation. Oliveira et al.[185] studied TSD currents in Cu^+-OCN^- coupled pairs in KCl doped with CuCl and KOCN. The TSD peak at ~54 K is best fitted when the OCN^- perturbed Cu^+ is considered to be a double relaxation type system. This conclusion is supported by the results of UV optical absorption. In a later work, Oliveira and Siu-Li[186] described the off-center effect of Cu^+-doped KCl polycrystalline films as studied by optical absorption and TSD current. The TSDC spectrum shows a weak and broad dipole relaxation band at 49K. The fitting analysis using three different relaxing dipole models reveal a continuous distribution of relaxation time and activation energy, and yields a mean activation energy of 147 meV. The results show strong dipole relaxation changes, which are different from those observed in single-crystal samples.

Pissis and Daoukaki-Diamanti[187] report the results of TSD currents in Na-13x zeolite ($[Na_{86}(AlO_2)_{86}(SiO_2)_{106}]\cdot nH_2O$) in the temperature range of 77-300 K. The main focus of this research was on a quantitative characterization of the effect of water on the motion of ions, and the reduction of the mobility of water molecules in the confined pore geometry. Five distinct relaxation processes were identified. Three of them were attributed to motions of ions, namely, in the order of increasing temperature, local jumps of ions in the supercages, a Maxwell-Wagner effect, and dc conductivity. The remaining two processes, present in the samples only above a certain critical water content, were ascribed to the reorientation of water molecules in the supercages and small β-cages. Papathanssiou et al.[188] investigated TSD phenomena in LiF:Ca^{2+} crystals. Results on as-grown samples indicate a space charge and/or an interfacial polarization phenomenon. Thermal perturbation is employed to release free impurity-cation vacancy dipoles. The subsequent bands are critically examined and one of them is attributed to jumps of the bound cation vacancy via nearest-neighbor paths.

A number of investigators have studied the TSD currents in glassy materials. Algería et al.[189] gave a theoretical account, taking into consideration a distribution of relaxation times. Such a distribution was assumed to arise from differences in free volume within the glass. The authors introduced a phenomenological parameter for describing the interaction among the different zones. They assumed a temperature independent symmetric distribution in the inverse of the free volume expansion coefficient. As a result, they found an asymmetric temperature dependent distribution of relaxation times which agreed favorably with their experimental results in PVC samples. Katyal et al.[190] reported the results of TSD in $(As_2S_3)_{1-x}(PbS)_x$ glasses with $x = 0.1, 0.3$ and 0.5. The authors associated their results in these amorphous compositions to the Bucci and Fieschi[140] model for dipole orientation and determined the relevant activation energies and frequency factors. Mikla et al.[191] investigated the TSD properties of Sb_xSe_{1-x} amorphous glassy semiconductors. The trap levels were studied by a combination of conventional TSD current spectroscopy and time-of-flight measurements. The addition of Sb atoms to the pure amorphous Se introduces a new set of shallow electron traps at 0.22, 0.34 and 0.45 eV, in addition to those existing in pure Se.

Another group of materials in which TSD measurements have been performed extensively is organic materials, in particular different kinds of polymers. Again, only a small number of examples will be given. Faubert[192] studied the TSD of polybutadiene liquid crystal side-chain polymer. Two TSD peaks have been found, at ~15 °C and ~42 °C. The low temperature peak is associated with the modification of the polarization induced by the molecular motion of T_g, the glass transition temperature. The high temperature peak characterizes the reorientation of the mesogenic part. Similar work on the molecular motions of side-chain crystalline polysiloxanes was performed by Mano et al.[193]. The upper glass transition relaxation seems to correspond to reorientation motion in the liquid crystalline phase of the longitudinal and transverse components of the dipole moment of the mesogenic side-group. Markiewicz et al.[194] investigated TL and TSD in low density polyethylene samples. They concluded that TSD currents measured between -190 and 30 °C originate in dipole relaxation processes, i.e. the space charge transport component is negligible. Charge injection commenced at ∼-10 °C at a field strength of 9000 kV/m applied via evaporated gold electrodes, and increased rapidly with increasing temperature. However, the charge injection did not generate a sufficient concentration of trapped space-charge to yield measurable TSD current, since nearly all the injected charge was transported across the sample to the opposite electrode without being trapped. It follows that such transport occurs only through the amorphous regions of the samples, since traps which can immobilize electrons above -10 °C are known to exist in the crystalline regions. Pissis and Apekis[195] reported TSD in polymers (poly(propylene glycols)), block co-polymers (polyurethanes) and simply continued liquids (propylene glycol). A sharp dipolar peak was always found in the TSD thermogram at the glass transition temperature. The complex TSD curves were experimentally decomposed into single responses by the "thermal sampling technique" which is the same as the thermal cleaning method used in TL. The spectra of activation energies and entropy factors were determined. Aldana et al.[196] studied the relaxation parameters from TSD spectra of polymers. Their method consists of finding N elementary curves that best fit the experimental curve. Computer-generated curves were analyzed, for monoenergetic peak and for a Gaussian energy distribution. Good results were reached even when random noise was added to simulate experimental errors. Complex TSD curves from epoxy resin were also separated into their components. Finally, the TSD peak corresponding to the glass transition temperature in bisphenol-A polycarbonate was also analyzed. Faruque[197] investigated the physical and morphological structure of materials, using fractional depolarization - a technique which again uses the idea of subsequent heatings and coolings over relatively small temperature ranges. Molecular parameters, such as relaxation times, the activation energy, the migration enthalpy and entropy, and the degree of disorder in polymers were thus calculated.

8.3.7 Depolarization Currents Due to Ionic Space Charge

In addition to current peaks involving the rotation of dipoles ITC peaks may also appear due to the migration of ionic charge carriers. In this kind of effect, an external polarizing field is applied at a temperature T_p, where the mobility of ionic interstitials or vacancies is considerable. A non-uniform distribution of ionic charge-carriers is created in the sample, and space charge is accumulated near the electrodes. The resulting polarization is termed *ionic space charge polarization.* At a low temperature T_0, the space-charge distribution is frozen in. Reheating the sample in the absence of the external field gives rise to a displacement current peak in the external circuit while the ionic defects return to a uniform distribution.

According to Bucci et al.[141], ITC peaks due to the release of ionic space charge have the following adverse properties:

1. The temperature of the maximum, T_m, is not well defined.
2. The area under the peak is not a linear function of the applied electric field, in particular at low fields.
3. The shape of the peak does not allow the measurement of the relevant activation energy.

Kunze and Müller[198] and Podgoršak and Moran[199] showed that the space-charge ITC peaks are nearly proportional to the polarizing field, and that T_m depends strongly on the impurity concentration in the sample. Although there is an agreement that the analysis of ionic space-charge ITC is rather a complex issue, simple models for its study have been developed. Kunze and Müller[198] investigated both types of ITC in ionic crystals, and used the terms DO-ITC (DO for Dipole Orientation) and CD-ITC (CD for Charge Displacement). Their model for CD-ITC assumes that the non-uniformity of the charge distribution can be represented by a charge density Q on the electrodes. In this case, the electric field in any point in the sample is

$$E_{el}(t) = Q(t)/\varepsilon\varepsilon_o, \tag{8.47}$$

and the current density during warming is given by

$$j(t) = -dQ/dt. \tag{8.48}$$

$j(t)$ is related to the conductivity $\sigma(t)$ by

$$j(t) = \sigma(t)E_{el}. \tag{8.49}$$

In cases where the contribution of one type of defect to the conductivity is dominant, the ionic conductivity over limited temperature ranges can be expressed approximately by

$$\sigma(T) = \sigma_o \exp\{-E/kT\}. \tag{8.50}$$

In the range of relatively low ionic conductivity, below 600 K in most ionic crystals, the conductivity is dominated by the presence of impurity ions. Assuming that the association between charge carriers and impurity ions is nearly complete, E is given by $E = E_m + E_a/2$, where E_a is the activation energy of the charge carriers. Substituting Eqs. 8.47 and 8.50 into 8.48 yields

$$j(t) = -\frac{dQ}{dt} = \frac{\sigma_o}{\varepsilon\varepsilon_o} Q \exp\{-E/kT\}. \tag{8.51}$$

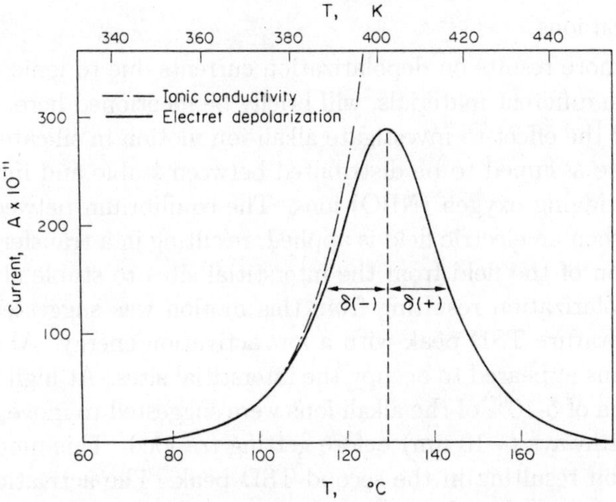

Figure 8.8 An ionic space-charge TSD peak in CaF_2 (solid line) and the ionic conductivity curve of the same crystal (dashed line).
(After Podgoršak and Moran)[199].)

Integrating Eq. 8.51, assuming a constant heating rate, $dT/dt = \beta$, one gets the depolarization current

$$j(T) = \frac{\sigma_o}{\varepsilon\varepsilon_o} Q_o \exp\{-E/kT\} \exp\left\{-\frac{\sigma_o}{\varepsilon\varepsilon_o\beta}\int_{T_0}^{T} \exp\{-E/k\theta\}\, d\theta\right\}, \qquad (8.52)$$

where Q_o is the charge density on the electrodes at temperature T_0.

Equation 8.52 is, of course, entirely analogous to that of a first-order TL peak. It was found, however, that in many cases the situation is different. Podgoršak and Moran[199] found a nearly symmetrical CD-ITC peak in high purity CaF_2 and suggested second-order kinetics to explain it. In fact, they distinguished between the ionic space-charge peak and the increasing ionic conductivity as shown in Fig. 8.8. Kristianpoller and Kirsh[200] found in SrF_2:Tb, TSD peaks of ionic space-charge origin which did not fit first-order kinetics, but rather fit general-order kinetics with $b = 1.5$. They pointed out that assuming homogeneity of the field throughout the sample during the heating may be wrong. Solutions for more realistic charge distributions were evaluated by numerical methods by van Turnhout[133],[149]. Also, Tanaka et al.[201] reported the high temperature ITC in CaF_2 and CaF_2:Eu, and attributed the results to the motion of mobile interstitial fluorine ions in CaF_2. The decrease of the newly found peak at 316 K was explained by assuming that mobile fluorine vacancies are compensated by fluorine ions introduced into CaF_2 as charge compensators for trivalent Eu ions.

Some more results on depolarization currents due to ionic space charge, as measured in different materials, will briefly be mentioned here. Agarwal and Day[202] used the effect to investigate alkali-ion motion in silicate glasses. The alkali ions were assumed to be distributed between stable and interstitial sites around non-bridging oxygen (NBO) ions. The equilibrium between these sites is disturbed when an electric field is applied, resulting in a transfer of alkali ions in the direction of the field from the interstitial sites to stable states, or vice-versa. The polarization resulting from this motion was suggested to produce the low-temperature TSD peak with a low activation energy. About 1 to 20% of the alkali ions appeared to occupy the interstitial sites. At high temperature, a small fraction of 5-10% of the alkali ions were suggested to move along limited conduction pathways ($\sim$10 nm) before getting trapped. This motion produces the polarization resulting in the second TSD peak. The activation energy for alkali motion along the pathways was found to be higher than that of the first TSD peak, but lower than that for dc conduction. At still higher temperatures, the alkali ions become capable of long-range traslational motion which was believed to be responsible for the dc conductivity.

Meuldijk and den Hartog[203] studied ITC in $Sr_{1-x}La_xF_{2+x}$ solid solutions, with LaF_3 concentrations between 0.01 and 42 mol%. They showed that the enhancement of the conductivity and the shift of the space-charge relaxation band to lower temperature with increasing rare-earth concentration can be explained with the same mechanism. Charge transport is supported by two kinds of interstitial jumps, free interstitial jumps occurring far from a rare-earth ion and energetically more favorable dipole-like jumps of interstitial fluoride ions in the neighborhood of a rare-earth ion. It was shown that the charge transport in rare earth doped fluorite type ionic crystals is supported by a percolation mechanism.

Ouwerkerk et al.[204] investigated TSD in solid solutions of $Ba_{1-x}La_xF_{2+x}$ and $Ba_{1-x}U_xF_{2+2x}$. From 10 to 400 K, the TSD scans showed four dipolar reorientation peaks to be present in the former and seven in the latter. Each scan also showed the relaxation peak due to macroscopic space charge. For all the $Ba_{1-x}La_xF_{2+x}$ and $Ba_{1-x}U_xF_{2+2x}$ solid solutions studied, a large depolarization peak was found, positioned at higher temperatures than the dipole- or cluster-reorientation peaks. The position of this peak was strongly dependent on the dopant concentration. For $Ba_{1-x}U_xF_{2+2x}$ solid solutions, the peak position changed gradually with x from $\sim$370 K at $x = 0$ down to $\sim$150 K at $x = 0.17$. Ionic conductivity measurements on the same solid solutions showed the activation enthalpy for fluoride ion interstitials migration to have a similar dependence on x. The large high-temperature TSD peak was therefore ascribed to the migration of unbound fluoride ion interstitials. The maximum temperature T_m of this peak was suggested to be related to the migration energy of fluoride ion interstitials.

Shindo[205] studied TSD due to polarization by space charge migration over macroscopic distances in polyethylene terephtalate (PET). Since most of the space-charge is supposed to stay near the surface, Shindo removed a thin layer from the surface, and indeed, found a substantially reduced TSD current. In another experiment, a stack of three sheets of polymer film, with or without metal foils between them, was poled to become a thermoelectret. After cooling, it was separated into three sheets. It was found that the TSD current of the outer two sheets were larger than the middle one. The results suggest that most of the space charge was due to ions.

8.3.8 Depolarization Currents Due to Electronic Charge Carriers

The motion of electrons and holes in a sample can also contribute to the polarization and depolarization of dielectrics. Applying an electric field to an electron-conducting material may cause the injection of electronic charge carriers into the sample followed by drift of both injected and intrinsic charge carriers. If the sample is cooled when the external field is on, some charge carriers may be trapped in trapping levels. Once the field is removed, the electrodes short-circuited and the sample heated, TSD currents can be measured. The experimental procedure is basically the same as in the measurements of dipole reorientation, with the addition that, here, the polarization can be caused also by electron bombardment or corona charging using a knife edge at high potential, as described by Murphy and Costa-Riberio[138], by Creswell and Perlman[206] and by Sessler and West[207]. A number of models have been developed for the analysis of this phenomenon, which appears to be more complicated than that due to ionic thermocurrents. Each of these models is based upon some specific assumptions about the trapping and retrapping processes, the band structure near the contacts, etc. Most of the existing models assume that once the electron (or hole) is thermally released from the trap, it can move freely in the conduction (valence) band until its recombination or retrapping. This band model is used as an approximation even for highly disordered amorphous solids. Only in the case of very strong retrapping, can the charge carriers be considered as hopping from one site to the next across potential barriers, in a similar manner to that of ionic charge carriers described above.

Perhaps the simplest model for this kind of TSD currents is that by Müller[208]. His model deals with the experimental situation in which the sample is insulated from the electrodes by thin foils of Mylar or Teflon, to prevent injection. The electric field in the sample is assumed to be homogeneous and not influenced by the boundary zones. The dark conductivity of the sample is described by $\sigma = \sigma_o \exp\{-H/kT\}$ where H is the activation enthalpy. The TSD is assumed here to be caused by small displacements of equilibrium charge carriers, and not by the redistribution of excess charges localized near the surface during the polarization process.

As mentioned above for the case of TL, the use of a hyperbolic heating function facilitates the derived expressions to some extent; Müller[208] used the same idea by taking a heating function for which

$$-d(1/T)/dt = (1/T^2)(dT/dt) = a = constant. \tag{8.53}$$

The equation governing the polarization P under these circumstances is

$$-dP(t)/dt = j(t) = \sigma(t)E(t) = \sigma(t)\frac{fP(t)}{\varepsilon_o\varepsilon_2} \tag{8.54}$$

where j denotes the current density in the sample, E is the electric field and σ is the conductivity. f is given by

$$f = \frac{2\varepsilon_2 D_1}{\varepsilon_1 d_2 + 2\varepsilon_2 d_1} \tag{8.55}$$

where ε_1 and d_1 are the relative permittivity and the thickness of the foils, respectively, and ε_2 and d_2 are those of the sample. Integration of Eq. 8.54 yields

$$P(T) = P_o \exp\left\{-\frac{f\sigma_o k}{aH\varepsilon_o\varepsilon_2}\left[\exp\{-H/kT\} - \exp\{-H/kT_0\}\right]\right\}. \tag{8.56}$$

For T significantly larger than T_0, $\exp\{-H/kT_0\}$ can be neglected, being much smaller than $\exp\{-H/kT\}$. Inserting into Eq. 8.54 results in

$$j(T) = \frac{f\sigma_o P_o}{\varepsilon_o\varepsilon_2}\exp\left\{-\frac{H}{kT} - \frac{f\sigma_o k}{aH\varepsilon_o\varepsilon_2}\exp\{-H/kT\}\right\}. \tag{8.57}$$

This represents an asymmetrical peak, similar to a first-order TL peak; the expression for the more conventional linear heating function is somewhat more complicated, but also yields a similarly asymmetrical peak. The maximum condition can be derived by equating the derivative of Eq. 8.57 to zero, which yields

$$f\sigma_o k = aH\varepsilon_o \exp\{-H/kT_m\} \tag{8.58}$$

where T_m is the maximum of the TSD peak. The activation enthalpy H can be determined from

$$H = \frac{j_m k e}{aP_o} \tag{8.59}$$

where j_m is the maximum current density, P_o is the initial polarization and e is the basis of the natural logarithm. Müller showed that when the assumptions of his model are valid, the thermally stimulated polarization (TSP) current measured when the crystal is heated under an external field using a hyperbolic heating rate, is given by exactly the same equation 8.57. This *linear theory* of TSD holds in cases where the mobility of the charge carriers is very small and, accordingly, the concentration of free carriers in the bulk by far exceeds the concentration of those accumulated near the surface. A comparison between TSD and TSP currents should be a test for the applicability of the assumptions. Another possible test is to check the simple relation which exists in this model between the dark conductivity $\sigma(T)$ and the position and maximum intensity of the TSD peak.

Müller[209] measured TSD currents in amorphous As_2Se_3. He found a peak at 270 K with an activation energy in full agreement with the present model. Thurzo et al.[210] studied TSD currents in amorphous As_2Se_3 between aluminum electrodes, and found a prominent peak at 300 K with an activation energy smaller than that of the dark conductivity and suggested that the interpretation of the results is not as simple as stated by Müller. They ascribed their peak to the relaxation of the charge accumulated at the metal-semiconductor interface, a situation which is excluded in Müller's model. The different arrangement of the electrodes and the influence of impurities on the TSD current can explain the discrepancies between the two results. In a later paper, Müller[211] showed that contact effects can indeed strongly influence the TSD and TSP peaks in electronic conducting materials.

Creswell and Perlman[206] analyzed the depolarization of corona charged foils, assuming a uniform distribution of free and trapped charge over the irradiated volume, using the following equations:

$$dj = \rho_f v(x)dx/d = J(x)dx/d \tag{8.60}$$

where $\rho_f dx$ is the free electron charge per unit area in an element dx, $v(x)$ is the velocity of the charge layer, d is the thickness of the sample, $J(x)$ is the local current due to motion of electrons in the space charge field, and j is the contribution to the thermally measured current. The expression for Ohm's law in this notation is

$$J(x) = \rho_f \mu E(x) \tag{8.61}$$

where μ is the electron mobility and $E(x)$ the space charge field. Poisson's equation can be written as

$$dE/dx = \rho/\varepsilon \tag{8.62}$$

where ρ is the total (free plus trapped) charge density and ε is the dielectric constant. Similarly to the approximations discussed with relation to "regular" TSC (see Eqs. 8.3 and 8.4 above), Creswell and Perlman[206] distinguished

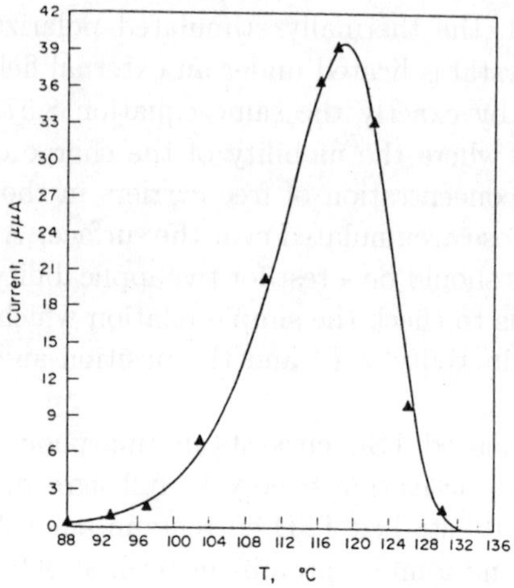

Figure 8.9 A "Thermal Current" of corona charged Mylar. The triangles represent the experimental points and the solid line is the theoretical curve with $E = 2.2$ eV. (After Creswell and Perlman[206].)

between slow and fast retrapping. The former yielded TSD current given by

$$j = \frac{\mu e^2 \delta^2 n_0^2}{2\varepsilon d} \frac{\tau}{\tau_o} \exp\left\{ -\frac{E}{kT} - \frac{2}{\beta \tau_o} \int_{T_0}^{T} \exp\left\{ -\frac{E}{k\theta} \right\} d\theta \right\} \tag{8.63}$$

whereas the latter resulted in

$$j = \frac{\mu e^2 \delta^2 n_0^2}{2\varepsilon d} \frac{N_c}{N_t} \exp\left\{ -\frac{E}{kT} - \frac{2N_c}{\beta N_t \tau} \int_{T_0}^{T} \exp\left\{ -\frac{E}{k\theta} \right\} d\theta \right\} \tag{8.64}$$

where δ is the penetration depth of the irradiation, e is the electronic charge, τ is the free electron lifetime, N_c is the effective concentration of conduction states, N_t is the density of traps, n_0 is the initial density of electrons in traps, and $\tau_o \exp\{E/kT\}$ is the average lifetime of an electron in the trap. It is immediately seen that these two expressions represent characteristic first-order, TL-like curves. The fit of the last equation to experimental results is shown in Fig. 8.9.

Simmons and co-workers[212]-[214] studied various aspects of dielectric relaxation, and investigated theoretically and experimentally the current in thin insulating films when heated from a low temperature T_0 under a constant voltage V_i, applied with blocking electrodes, after being cooled to T_0 under a different voltage V_d. They attributed the observed current peaks to the motion of electrons. Their theory assumes *Schottky Barriers* at the sample-electrode interface, and a cathodic depletion region, the width of which changes when the field is applied. The traps are assumed to be donors and their depth is lowered in an electric field due to the Poole-Frenkel effect. Experimental results with thin films of Al-CeF_3-Al were successfully interpreted in terms of the theory. Mar and Simmons[215] investigated TSD due to electrons in MOS structures and showed that the J-T curves, associated with bulk and surface generation of carriers, exhibit characteristics which allow the two processes to be distinguished from one another.

Zhdan et al.[216]-[218] used a slightly different technique for studying TSD of semiconductors which they termed *Thermally Stimulated Capacitor Discharge* (TSCD). They use thin film samples evaporated on insulating layers. Two metal contacts were utilized, one of which was attached directly to the sample and the other to the other side of the insulator. The insulator acts as a barrier between the contact and the sample. This Metal-Insulator-Semiconductor-Metal (MISM) arrangement makes a capacitor, one plate of which is on the sample. Zhdan et al.[216] have shown that in the absence of retrapping and assuming uniformly distributed traps of one type, the current in the external circuit in the course of a TSCD measurement is given by

$$i = \frac{eLSn_0}{\tau_o} \exp\left\{-\frac{E}{kT} - \frac{1}{\beta\tau_o}\int_{T_0}^{T} \exp\{-E/k\theta\}\, d\theta\right\} \tag{8.65}$$

where e is the electron charge, E is the activation energy of the trap, L and S are, respectively, the length and cross-section of the semiconducting film. n_0 is the concentration of filled traps at T_0; τ_o^{-1} is given by vS_tN_c, where v is the average thermal velocity of an electron, S_t is the capture cross-section of the trap, and N_c is the effective density of states in the conduction band. It should be noted that Eq. 8.65 is the same as those for first-order TL or TSC peaks except that the electron mobility does not take any part. Katzir and Halperin[219] and Agarwal[220] questioned whether the TSCD curve can be interpreted merely in terms of thermally released charge carriers. Agarwal[220] showed that depolarization of space charge due to the variation of the sample conductivity with temperature may have a dominant contribution. Yang[221] reported on short-circuit and open-circuit TSCD in polypropylene (PP) film samples. The two peaks, observed at 70 °C and 100 °C, were found to be caused by detrapping of charges from the surface and the bulk respectively, while the origin of a peak at 150 °C was not clear. The short-circuit TSCD showed the existence of a positive current peak between two negative peaks after irradiating the sample with a large dose. This current reversal phenomenon verifies the fact that the charge layer in electron-beam-charged PP broadens towards the rear electrode as the injected charge density is increased.

8.3.9 Thermally Stimulated Polarization Currents

A very closely related phenomenon to the TSD is the that of thermally stimulated polarization current (TSPC) which, in a sense, can be considered as the inverse effect and is associated with the same relaxation processes. The experimental procedure is practically the same as in TSC measurements except for the absence of the initial excitation. Another difference is that whereas TSC is measurable in many semiconductors and insulators, TSPC is limited to those materials which are easily polarizable. In the experiment, the sample is first cooled to some "low" temperature T_0 without any applied electric field. At T_0, an external electric field is applied to the sample which is then heated under the external field, while the current through the circuit is recorded vs. temperature. Trapped charge carriers or dipolar defects which become mobile at elevated temperatures, give rise to current peaks which appear as the sample becomes polarized. Like in the other thermally stimulated processes, the current increases with temperature, reaches a maximum and then drops to zero as the polarization is saturated. TSPC has been less extensively studied than TSD currents. One of the main reasons for this is experimental, namely, the signal to noise ratio is better when the sample is heated in the absence of an external field. On the other hand, TSPC measurements can reveal additional information which could not be obtained from TSD measurements alone.

TSPC due to dipole relaxation

Pfister and Abkowitz[154] studied TSPC of polyvinylidene-fluoride films which were first cooled to 163 K, and attributed a current peak at 223 K to the reorientation of dipoles in a process associated with the glass transition. McKeever and Hughes[222],[223] observed TSPC in alkali halides. They developed the following equation for a TSPC peak due to a single Debye relaxation

$$j(T) = -\frac{N\mu^2 E_p \alpha}{kT\tau_o} \exp\left\{-\frac{E}{kT} - \frac{1}{\beta\tau_o}\int_{T_0}^{T} \exp\left\{-\frac{E}{k\theta}\right\} d\theta\right\} \qquad (8.66)$$

where all the parameters have the same meaning as in Eq. 8.29. Equation 8.66 is similar to that of a first-order ITC peak, except for the variable T appearing in the denominator of the pre-exponential coefficient (which is the Langevin function, Eq. 8.27), rather than the constant temperature T_p. Manifacier et al.[224], included the temperature dependence of P_o, after which the TSPC

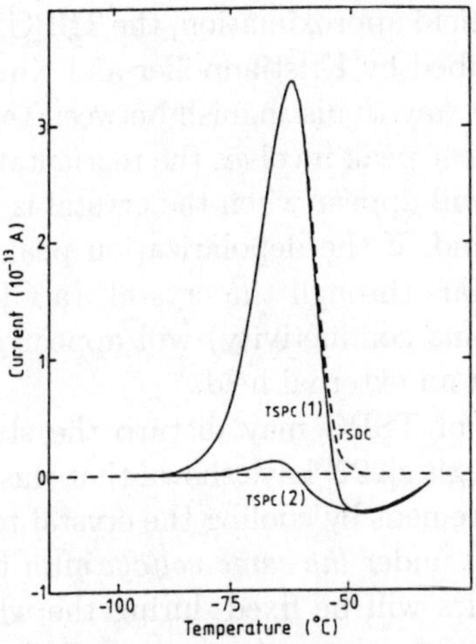

Figure 8.10 TSPC (1 and 2) and TSDC curves for Mg-Li(vacancy) dipoles in LiF:Mg, showing the TSPC current reversal on the high temperature side of the peak.

(After McKeever and Lilley[225].)

current density is given by

$$j(T) = \left[\tau^{-1}(T)\exp\left\{-\int_{T_0}^{T}\frac{dT}{\beta\tau(T)}\right\}\right]\int_{T_0}^{T}\frac{P_o(\theta)}{\beta\tau(\theta)}\exp\left\{\int_{T_0}^{\theta}\frac{d\phi}{\beta\tau(\phi)}\right\}d\theta - \frac{P_o(T)}{\tau(T)} \tag{8.67}$$

where P_o is given by $P_o(T) = N\mu^2 E_p\alpha/(kT)$ and $\tau(T) = \tau_o \exp\{E/kT\}$. Computed expressions of Eq. 8.67 showed a typical current reversal due to the temperature dependence of the Langevin function. The reversal was indeed observed in MOS structures[224] and later in alkali halides[225]. A comparison of experimental TSPC and TSDC curves in LiF:Mg is shown in Fig. 8.10.. The TSPC peak has the expected form of Eq. 8.67 (shown as TSPC(1)), including the current reversal on the high temperature side of the peak; the reversal is not seen with the TSDC peak (described by Eq. 8.29). Vanderschueren and Gasiot[226] further showed that if the applied field is maintained during cooling after performing a TSPC measurement, and the sample is reheated still with the field applied, a small current peak will be observed, followed again by a current reversal. An example of such an experimental peak is shown in Fig. 8.10 as TSPC(2). This is also a consequence of the temperature dependence of the Langevin function (Eq. 8.27).

A theoretical analysis of TSPC and TSD due to dipoles in doped alkali halides crystals (NaCl-type) has been presented by Berg and Poźniak[227]. They found that in the weak field approximation, the TSPC has the same shape as the TSD current. As described by Kristianpoller and Kirsh[228], the measurement of TSPC may provide a way to distinguish between various types of ITC peaks. If a depolarization current peak involves the reorientation of dipoles (DO-ITC), a parallel TSPC peak will appear when the crystal is heated under an external field. On the other hand, if the depolarization peak is due to the motion of vacancies and interstitials through the crystal, (a CD signal), an exponential rise of conductivity (ionic conductivity) will appear at the same temperature range during heating in an external field.

The occurrence of TSPC may disturb the study of radiation-induced TSC. McKeever and Hughes[222] have shown that this disturbance can be eliminated in a TSC measurement by cooling the crystal to a temperature T_0 under a voltage, and heating it under *the same voltage* after the irradiation at T_0. The orientation of the dipoles will be fixed during the whole cycle of cooling and heating, and TSPC will not appear. As shown by Kristianpoller and Kirsh[228], if an electrical field is applied, during cooling, to a crystal containing dipoles, the same field but in the *reverse direction* is applied to it during heating under voltage, the polarization alignment will change during the heating by 180 °, and the peak will appear twice as high. This may serve as another test to TSPC peaks; only a peak which involves dipole reorientation shows this effect.

In a later work, Vanderschueren et al.[229] developed a theoretical expression for TSPC in organic materials with single dipolar polarization. In their model, they studied the influence of thermal expansion on TSPC and TSDC by using the bistable model of Frölich, considering especially the case of materials characterized by expansion coefficients markedly different above and below the relaxation range. Good qualitative and quantitative agreement was obtained between theory and experiment in elastomeric materials, and they propose that thermal expansion is the main factor responsible for the appearance of current reversals in thermally stimulated polarization processes.

TSPC due to electronic charge carriers

Simmons and Taylor[212] found pronounced peaks in the current vs. temperature plots of thin film silicone nitride insulators during heating under a constant voltage. They developed the following equation to describe their experimental results

$$I(T) = \frac{e\lambda}{4}\frac{N}{t_r}\exp\left\{-\frac{kT^2}{\beta(E+kT)t_r}\right\} \tag{8.68}$$

with $t_r = \nu^{-1}\exp\{E/kT\}$ and where e is the electron charge, λ is the space charge region width, N is the density of donor traps, and β is the heating rate. Servini and Jonscher[230] performed similar measurements on evaporated SiO_2

films. Müller[209] showed that under certain conditions, the TSPC peak due to the motion of electrons and holes is identical in shape to the TSDC process, and can be described by Eq. 8.57. However, measurements with thin films of As_2Te_3 and As_2Se_3 showed some differences between the TSD and TSPC peaks[211]. Böhm et al.[231] used TSPC measurements to investigate the energy band diagrams of metal-insulator-metal systems. Taylor and Morgan[232] studied the local band distortion caused by defects near Schottky barriers, and showed that TSPC and TSC measurements can be used in this study, particularly when exact density values are not required. Finally, a review paper by Saffell et al.[233] compares the phenomena of TSDC and TSPC with other thermal analytical methods in the characterization of the amorphous phase of polymers.

8.4 Deep Level Transient Spectroscopy (DLTS)

Sah et al.[234] described a capacitance transients method which was used to obtain information about impurity levels in the depletion region of a Schottky barrier or a p-n junction. This was done by observing the capacitance transient associated with the return to thermal equilibrium of the occupancy of the level, following an initial non-equilibrium condition. One can measure the time constant of this transient as a function of temperature, and find the activation energy. Lang[235],[236] utilized the same concepts in a more sophisticated way in *Deep Level Transient Spectroscopy (DLTS)*. This method, which has been utilized by many investigators during the last twenty years, enables the evaluation of trapping parameters in semiconductors.

By the principle of detailed balance, the emission rate of minority carriers e_1 from a level is given by

$$e_1 = (\sigma_1 v_1 N_1 / g_1) \exp\{-\Delta E/kT\} \tag{8.69}$$

where σ_1 is the minority capture cross-section, v_1 is the mean thermal velocity of minority carriers, N_1 is the effective density of states in the minority carriers band, g_1 is the degeneracy of the trap level and ΔE is the energy depth of the level. A similar equation holds for e_2, the emission rate of majority carriers. Instead of the previously standard way of plotting $\ln e_1$, or $\ln e_2$ vs. 1000/T, and extracting ΔE out of the emerging straight line, Lang[235] has advocated the following alternative. He suggested the use of a dual gated signal averager (a double boxcar) which determines an emission rate window and enhances the signal-to-noise ratio. The capacitance transients are observed on a fast response capacitance bridge. A series of capacitance transients is shown schematically on the left hand side of Fig. 8.11. The emission rate is seen to be small for low temperature and faster at higher temperatures. The transient signals are fed into the double boxcar with gates at t_1 and t_2 as shown in Fig. 8.11. The

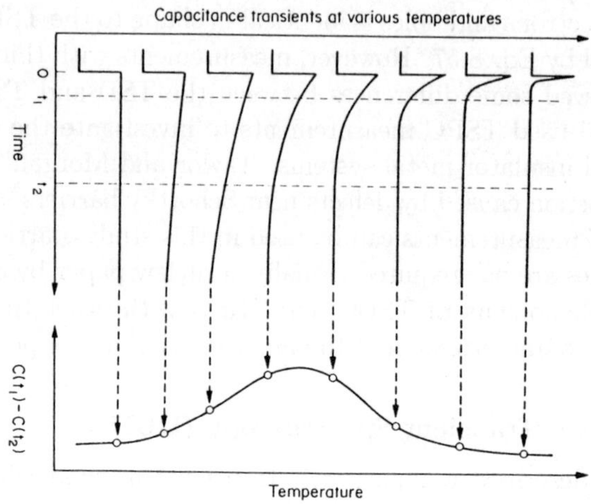

Figure 8.11 DLTS signal as a function of temperature (on the right hand side), resulting from using a double boxcar, to display the difference between the capacitances at times t_1 and t_2 as a function of temperature.
(After Lang[235].)

quantity $C(t_1) - C(t_2)$ is the signal which is applied from the output of the boxcar into a recording device. This quantity goes through a maximum when τ, the inverse of the transition rate constant is of the order of $t_2 - t_1$.

If we define the normalized DLTS signal

$$S(T) = [C(t_1) - C(t_2)] / \Delta C(0) \tag{8.70}$$

where $\Delta C(0)$ is the capacitance change due to the pulse at $t = 0$, we have for exponential transients

$$S(T) = \exp\{-t_1/\tau\} - \exp\{t_2/\tau\} \tag{8.71}$$

where $\tau = 1/e_1$. Differentiating $S(T)$ with respect to τ and setting the derivative to zero, one gets

$$\tau_{\max} = (t_1 - t_2) [\ln(t_1/t_2)]^{-1} . \tag{8.72}$$

At the maximum of the DLTS signal, one can measure the temperature and calculate $\tau_{\max}$ from Eq. 8.72 to get a point on the $\ln e_1$ vs. $1000/T$ plot. Other points are similarly found, and thus a plot yielding the value of ΔE is obtained. The method has been extended by Lang[235] to enable the calculation of the concentration and cross-section of the trap in question by methods originally

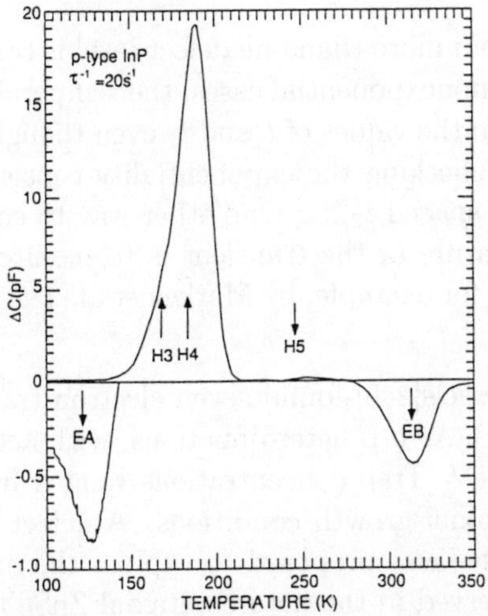

Figure 8.12 DLTS spectra for hole and electron traps in gamma-irradiated (174 Mrad(Si)) p-type MOCVD InP. Three hole traps (H3, H4 and H5) and two electron traps (EA and EB) are seen. The net acceptor concentration in the base region of the n^+p diode was 2.7×10^{22} m^{-3}.
(After McKeever et al.[237].)

developed for capacitance transient measurements at fixed temperature. A benefit of this method is that one can differentiate between trapped electrons and holes by the sign of the capacitance transient - i.e. by the sign of the DLTS signal. In pn-junction devices, both minority and majority carrier injection (and, therefore, both electron and hole trapping) can be achieved, whereas in Schottky barrier devices only majority carrier injection can be studied. Figure 8.12 illustrates both hole and electron DLTS signals from an n^+p InP diode following γ irradiation[237].

This method has been successfully used for GaAs samples by Lang[235] and for several other semiconductors by different researchers during the years. An important drawback of the method, which has been pointed out by Lang himself, is that one *assumes* an exponential decay. As discussed already with reference to phosphorescence decay, this is by no means necessarily the case. Therefore, the quantitative results may not be very accurate. Since in each decay, only two points are taken into account, one cannot check whether the decay is exponential or not. Thurber et al.[238] further discussed this issue of nonexponential transients in DLTS. They studied DLTS in different Si wafers. In addition to samples with known exponential decay, they also used samples in which incomplete collapse of the depletion region produces a nonexponential transient, samples in which high doping effects distort the transients, and samples where emission from more than one defect level is contributing to the transient. In each of these nonexponential cases, the temperature of the DLTS peak was found to depend on the values of $t_1$and t_2 even though $\tau_{\max}$ was unchanged. Thus, the method for checking the exponentiality consists of recording two or three curves at widely spaced t_2/t_1. One other way to confirm the exponential (or nonexponential) nature of the transient is to monitor the full capacitance transient, as was done, for example, by Markey et al.[239] for DLTS from n-type ZnSe Schottky diodes.

Besomi and Wessels[240] found seven electron traps in Au/ZnSe Schottky diodes and ZnSe/GaAs n-p heterojunctions, with activation energies ranging from 0.24 to 0.86 eV. Trap concentrations ranged from 10^{17} to 10^{20} m^{-3}, depending on the epitaxial growth conditions. A defect model, based on selenium vacancy complexes was proposed to explain the nature and behavior of the different traps observed in the heteroepitaxial ZnSe films. Endo et al.[241] studied deep levels in 10 MeV electron-irradiated GaP by DLTS. The DLTS curve depended strongly on the specific samples. A total of seven electron traps was detected in the various samples, with activation energies from 0.16 to 0.71 eV. In this material, the sum density of these traps, I_1-I_7, is much smaller than the sum density N_a of compensating defects obtained from C-V measurements at the same dose. It appears that the majority of the undetected traps or compensating centers lie deeper than 0.8 eV. These traps or centers may be related to very complex defects or damaged regions. Chretien et al.[242] reported the determination of the valence band offset between strained $Si_{1-x}Ge_x$ and unstrained Si layers by DLTS on Si/$Si_{1-x}Ge_x$/Si quantum well structures. In these samples a DLTS signal related to hole emission can only be obtained if one uses selective growth and good Schottky contacts with a low leakage current at low temperatures since the holes must be stored for a sufficiently long period of time ($>$ 1 ms) to observe the slow thermal emission. These DLTS results are favorably comparable to admittance spectroscopy results and photoluminescence measurements. Hacke et al.[243] investigated deep levels in Mg-doped n-type GaN films using DLTS. Deep levels at 0.26 and 0.62 eV below the conduction band were found in relatively low concentration of $\sim 2 \times 10^{19}$ m^{-3} in undoped GaN. The addition of small quantities of Mg increased the concentration of the 0.62 eV level, whereas the shallower level remained unchanged. Hacke et al.[243] concluded that these deep levels may detrimentally affect optical and electrical properties, increasing the resistivity of p-type GaN and decreasing the efficiency of the desired luminescent recombinations. Martin et al.[244] discussed DLTS signals from Mg-ion-implanted p-n junctions in undoped InP. Five deep levels located in the upper-half of the band gap were detected in the junctions. Three of these, at 0.425, 0.45 and 0.6 eV appeared to be due to rapid thermal annealing. The shallower ones at 0.285 and 0.31 eV were ascribed to the implantation damage. The DLTS measurements showed that the concentration of deep levels decreased after long annealing times, and that the concentration of deep levels due to the implantation increased with additional P or Si implantations. Ghosh and Kumar[245] reported on the use of DLTS to investigate the Si-related DX center in heavily doped ion-implanted GaAs. Two distinct peaks with activation energies of 0.27 and 0.315 eV were observed. The authors explain the peaks to result from splitting of the DX ground state through a Coulombic contribution to the inter-valley scattering in the L-band of GaAs. Okamoto et al.[246] suggested a wave-form analysis version of CVCT-DLTS (constant voltage-capacitance DLTS), using an iterative feedback system. The emission rate spectrum was analyzed by the feedback voltage transient wave form by using a nonlinear least-squares program. The method was tested on Au-doped Si p^+n diode. The method is expected to be effective in resolving the problem of closely spaced multiple deep levels, even if the sample has a high trap concentration and/or a non-abrupt junction profile.

Alejos et al.[247] present a new robust version of the DLTS technique. Here, a single temperature scan enables the accurate determination of the trap parameters for isolated DLTS peaks. These parameters, the activation energy, the amplitude and the pre-exponential factor, can be directly computed from the experimental data by means of a set of closed-form expressions. In particular, they assume a dependence of the pre-exponential factor on a power of the absolute temperature, in a similar way to that discussed with respect to TL. The technique allows a complete automation of the DLTS measuring system. The usefulness of the method was tested by analyzing the deep-level signal of palladium silicon diode. Benzohra et al.[248] have described a new method of measuring the energy levels and cross-sections, based on DLTS measurements and simplex linear programming analysis. Ingham et al.[249] presented a novel linear prediction based parameter estimation algorithm for analyzing DLTS signals.

8.5 Chemical Thermally Stimulated Phenomena

A number of methods used in chemistry are based on measuring chemical or physical changes in a sample while its temperature is varied in a pre-determined manner. They are termed methods of *thermal analysis* or *thermoanalytical* methods. These include *thermogravimetry* (TG), *derivative thermogravimetry* (DTG), *differential thermal analysis* (DTA), *differential scanning calorimetry* (DSC), *thermal desorption* as well as some other, less known methods. These phenomena can be considered as close relatives to thermoluminescence in the sense that a property of the sample is measured as a function of temperature. Also, in the relatively simple cases of these phenomena and of TL, there is a very close resemblance between the equations governing the processes. Thus, one can speak of first-order and general-order kinetics, about an activation energy, and about a pre-exponential factor. It is to be noted that both in TL and in the methods of thermal analysis, these terms are associated with the relatively simple models governing the process. The analogy does not go as far as dealing with the simultaneous differential rate equations in TL on the one hand, and the more complicated rate equations governing the more complex situations in the chemical processes on the other. In the present section, we will mention briefly the main thermoanalytical methods, and point out the similarities and some of the differences between them and TL.

According to the definition by the International Confederation for Thermal Analysis (ICTA), the term *thermal analysis* covers a group of techniques in which a physical property of a substance and/or its reaction products are measured as functions of temperature, while the substance is subjected to a controlled temperature program. In each thermoanalytical method, the outcome of the measurement is a curve, sometimes called a thermogram, which describes the sample in terms of one specific physical property, such as its heat content or its mass. Measuring through temperature ranges over which changes of this property take place, and evaluating the rates at which these changes occur, enable certain chemical analysis and provide information about microscopic processes occurring in the substance under investigation. Some of these methods are quite widely used in diverse fields of science and technology. Thousands of studies are carried out every year on the instrumentation, experimental procedures, theory and applications of these methods of thermal analysis. According to Liptay[250], in the decade between 1977 and 1987, well over 2000 publications appeared per year in the scientific literature on the different aspects of thermal analysis.

Concerning the history of thermal analysis, Mackenzie[251] described the origin of the subject as follows. Rudberg[252] seems to have been the first who observed inverse cooling-rate phenomena in 1829. Frankenheim[253] followed by determining cooling curves in 1837. Hannay[254] was the first to introduce mass-change curves in 1877. Later, Le Chatelier devised quasi-isothermal heating curves in 1883[255], and heating-rate curves in 1887[256].

Two international scientific journals are devoted entirely to thermal analysis. The *Journal of Thermal Analysis* (published by Akadémiai Kiadó, Budapest and Heyden, London since 1969) and *Thermochimica Acta* (published by Elsevier, Amsterdam since 1970). A large number of books, monographs and extended review papers deal with the methods of thermal analysis and their applications. These include the works of Garn[257], Einhorn[258], Menis[259], Daniels[260], Šesták et al.[261], Blažek[262], Kambe and Garn[263], Wendlandt[264],[265], Duval[266],Todor[267], Wendlandt and Collins[268], Šesták

[269], Wunderlich[270] and Haines[271].

8.5.1 Thermogravimetry (TG) and Differential Thermogravimetry (DTG)

In the method of *thermogravimetry* (TG), the mass of a sample is recorded as a function of time or temperature by measuring its weight while its environment is heated (or, in some cases, cooled) at a controlled rate. The measurement is performed with a thermobalance which enables the continuous weighing of the sample during the process. The plot of the sample mass M, or relative mass M/M_o, where M_o is the initial mass, vs. the time or the temperature of the sample is called the thermogravimetric (TG) curve. If the first derivative of M, with respect to time or temperature, is recorded, the technique is called *derivative thermogravimetry* (DTG). If the sample is held at a constant temperature and a decay curve is recorded, the term used is *isothermal mass-change determination* or *isothermal TG*. It is quite obvious that DTG is analogous to TL, isothermal DTG is similar to phosphorescence, and the isothermal mass-change is similar to the decay with temperature of the concentration of electrons or holes in traps or centers. This important point will be elaborated upon below.

The thermogravimetric techniques and their applications are described in the books mentioned at the beginning of this section. Some more books, dealing with TG only are by Duval[272], Vallet[273] and by Keattch and Dollimore[274]. Although isobaric and isothermal mass-change determinations, sometimes called *equilibrium techniques*, were extensively used in the past, interest at present concentrates mainly on the TG and DTG methods, which are sometimes termed *dynamic techniques*.

The history of the development of TG and DTG was presented in detail by Keattch and Dollimore[274]. The term *thermobalance* appeared first in a paper by the Japanese scientist Kôtaro Honda, published in 1915. In spite of the great impact of this work on the development of thermogravimetry, the wide scale use of TG and the production of commercial instruments for its measurement took off only in the mid-1950's.

TG measurements provide information on the thermal stability and the composition of the sample under investigation, and may be used for detecting many types of physical changes or chemical reactions. In some cases, a gain in weight may occur, resulting from adsorption and absorption of gases and vapors. Loss in weight, which is much more common during heating, and which has close analogy to TL, may be the result of desorption of gases and vapors, vaporization of liquids, sublimation of solids, dehydration, evolution of dissolved gases and a chemical decomposition of the sample. Modern TG equipment provide the possibility of applying complementary techniques such as differential thermal analysis (DTA) and evolved gas analysis (EGA).

When the mass of the sample is recorded vs. time or temperature, the resulting TG curves, $M = f(t)$ or $M = f(T)$ are stepwise curves as seen on

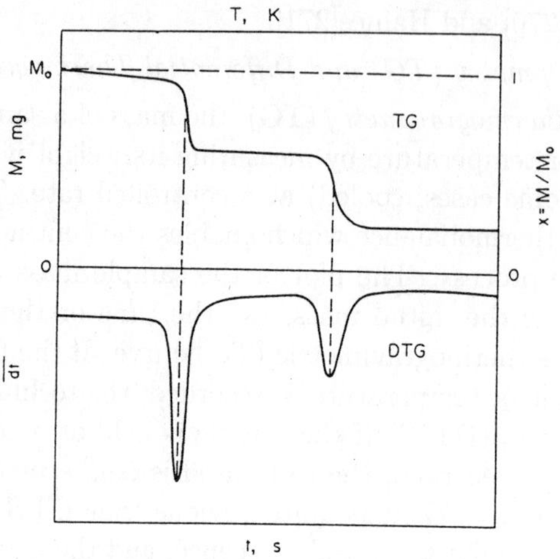

Figure 8.13 Schematic typical TG and DTG curves. (After Chen and Kirsh[1].)

the upper half of Fig. 8.13. Each step represents a process in which the loss (or, in more rare cases, the gain) of mass is involved. According to ICTA recommendations (see Lombardi[275]), the mass or weight should be plotted on the ordinate with mass decreasing downwards and t and T on the abscissa from left to right, as seen in Fig. 8.13. In a DTG curve, the loss or gain of the mass is expressed by a peak. This is seen as negative peaks on the lower part of Fig. 8.13. The analogy with TL, however, is very close. The TG curve is analogous to the occupancy of traps or centers, and the DTG curve to TL. The fact that the DTG curve is inverted as compared with TL is obvious since TL is associated with $-dn/dt$ where n is the relevant carrier concentration. The area under the DTG peak is equal to the total change in mass, again in analogy with the association of the area under a TL peak and the carrier concentration. Actually, the DTG curve contains the same information as the TG curve. However, in order to evaluate the temperature of maximum change, and for extracting information from the details of the shape of the curve, the DTG peak is more useful. Several authors mentioned that in a DTG measurement, very small changes in mass can be observed, and adjacent changes can be rather easily be separated (see, for example, Erdey et al.[276], Wilburn and Hesford[277], Paulik et al.[278], and Oswald and Dubler[279]). Székely et al.[280] used the second derivative of the TG curve (DDTG) and found that the method permits more accurate evaluation of the kinetic parameters, despite the fact that the second derivative is very sensitive to experimental errors.

As agreed by many investigators (see e.g. Coats and Redfern[281], Ozawa

[282], Manche and Carroll[283] and Agherghinei[284]), the kinetics of the thermal decomposition of many materials which have volatile products can be described by

$$\frac{d\alpha}{dt} = Z \exp\{-E/RT\}(1-\alpha)^b \tag{8.73}$$

where α is the fraction of the sample which has been decomposed ($0 \leq \alpha \leq 1$). α is defined as the weight loss at time t divided by the total weight. E is the activation energy, traditionally measured in Joules or kcal per mole (rather than in eV per particle, as in the phenomena discussed so far). b is the order of the reaction and R is the ideal-gas constant in kcal per mole per degree. It is clear that E/kT utilized in the previous chapters and E/RT discussed here are exactly the same; the latter is obtained by multiplying the numerator and the denominator of the former by Avogadro's number. The pre-exponential factor Z may be temperature dependent like T^m[284], in a similar way as in TL theory. Replacing $1-\alpha$ by x, Eq. 8.73 is changed to

$$-\frac{dx}{dt} = Zx^b \exp\{-E/RT\}, \tag{8.74}$$

in complete analogy with the general-order kinetics in TL, including the first-order kinetics for $b = 1$. In the particular case where Z is temperature independent, one gets for the TG signal, and first order kinetics

$$TG(T) = x(T) = \exp\left\{-\frac{Z}{\beta}\int_{T_0}^{T} \exp\{-E/R\theta\}\, d\theta\right\} \tag{8.75}$$

and for DTG

$$DTG(T) = \frac{dx}{dt} = -Z\exp\{-E/RT\}\exp\left\{-\frac{Z}{\beta}\int_{T_0}^{T} \exp\{-E/R\theta\}\, d\theta\right\}, \tag{8.76}$$

which is the same as Eq. 2.28 in TL theory, with the minus sign being the only difference. The analogy with Eq. 2.33 for the case of non-first-order kinetics follows immediately. The expressions for the maximum conditions for first-order and other-order kinetics are identical to those in TL, and so is the case for the different methods for evaluating the activation energy by the shape methods, different heating rates, curve fitting, etc.

It is to be noted that the analogy between TL and DTG is limited to the rather simple cases of the latter, and in fact, also to a simplified version of the former. A more general equation governing TG is

$$d\alpha/dt = D(T)f(\alpha) \tag{8.77}$$

where $D(T) = Z\exp\{-E/RT\}$ and where $f(\alpha)$ assumes different forms. Following Zsakó[285] and Šatava[286] we can mention some

$$f(\alpha) = \alpha^c(1-\alpha)^b, \tag{8.78}$$

$$f(\alpha) = (1-\alpha)^b\,[-\ln(1-\alpha)]^p\,, \tag{8.79}$$

as well as other, rather complicated expressions[287]. Similarly to the discussion concerning general-order kinetics in TL, Šatava[286] suggests that the various equations presented, including Eq. 8.74 which is exactly analogous to TL, are merely a phenomenological description of more complex situations. Many researchers still find it very useful to characterize their DTG results in terms of the parameters E, Z and b. Some publications dealing with the reaction mechanism of DTG are the works by Zuru[288], Dollimore et al.[289] and by Chen et al.[290].

8.5.2 Differential Thermal Analysis (DTA)

In the differential thermal analysis (DTA) technique, the temperature of the sample is compared to the temperature of a reference material when both are heated (or cooled) under similar conditions. The reference is a material with a heat capacity which does not undergo abrupt changes in the temperature range of the measurement, and matches that of the sample. The difference between the temperature of the sample (T_s) and that of the reference (T_r) is recorded as a function of time or of the reference temperature, while both are maintained in a furnace and heated. Usually, the sample and the reference are held in two cells or containers in close proximity, permitting moderately free exchange of heat between them. Thus, the temperature difference $\Delta T = T_s - T_r$ is nearly constant, unless abrupt enthalpic processes occur in the sample. Processes such as endothermic or exothermic chemical reactions, phase transitions and changes of the crystalline structure cause deviations of ΔT from the nearly horizontal line recorded otherwise. When the enthalpic process is completed, the temperature difference reduces to its baseline. Thus, most of the deviations of the DTA curve from the horizontal baseline form peaks, either in the upward or downward directions.

According to the recommendations of ICTA (see Lombardi[275]), the temperature difference ΔT should be plotted on the ordinate with endothermic reactions downwards, and t or T on the abscissa, increasing from left to right. A typical DTA curve is shown in Fig. 8.14. DTA curves are often utilized for the qualitative identification of organic and inorganic compounds, since the number, temperatures and shapes of the peaks give an indication to the composition of the sample. Many publications of the DTA technique involve the determination of the temperature over which the enthalpic process occurs, and the heat of the

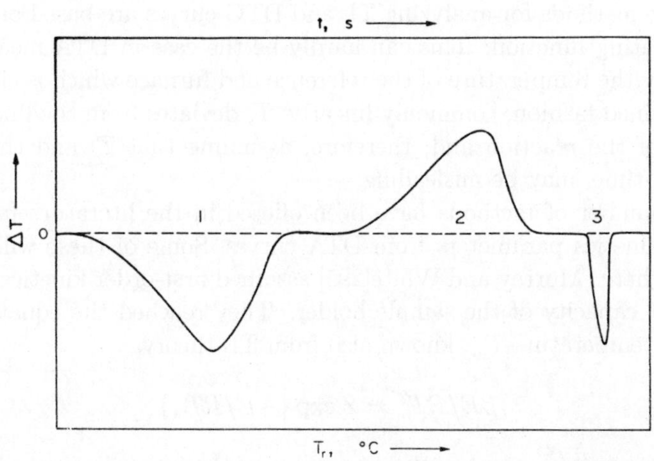

Figure 8.14 A typical DTA curve. The three peaks describe schematically: (1) an endothermic process; (2) an exothermic process; (3) a phase transition. (After Chen and Kirsh[1].)

reaction ΔH (the enthalpic change) which is nearly proportional to the area under the peak. Also, the kinetics of the reaction can be determined.

The instrumentation and experimental procedure of DTA measurements have been described in several books, some of which were mentioned at the beginning of this section. Some more, specific to DTA, are by Mackenzie[291],[292], Smykatz-Kloss[293], and by Pope and Judd[294].

Many reactions studied by DTA follow general-order kinetics[295]-[299]. This means that Eq. 8.73 governing TG, namely,

$$\frac{d\alpha}{dt} = Z \exp\{-E/RT\}(1-\alpha)^b \tag{8.80}$$

governs the process where α here is the fractional extent of the reaction[296] and the other magnitudes keep the same meaning as in the TG case. Here too, more complicated functions of α may replace $(1-\alpha)^b$ on the right hand side of Eq. 8.80. However, methods for evaluating the parameters of DTG or a TL peak cannot be applied to a DTA peak without a careful consideration regarding the following points. In TL and DTG, the recorded signal is considered to be proportional to the rate of reaction $d\alpha/dt$ (or $-dx/dt$ in the alternative expression). It is usually assumed that dH/dt, the heat generated (or absorbed) by the sample in a unit time, is proportional to $d\alpha/dt$. Following Borchardt and Daniels[300] and Gray[301], the DTA signal, ΔT, is approximately proportional to dH/dt only if the heat capacity of the sample holder is very small. In addition, some of the methods for analyzing TL and DTG curves are based on the use of a linear heating function. This can hardly be the case in DTA measurements. It is usually the temperature of the reference and furnace which is changed in a pre-determined fashion, commonly linearly. T_s deviates from the linear heating in course of the reaction and, therefore, assuming that T_s and thus ΔT are linear with time, may be misleading.

A number of methods have been offered in the literature for the evaluation of kinetics parameters from DTA curves. Some of these will be briefly mentioned here. Murray and White[295] assumed first-order kinetics and a negligible heat capacity of the sample holder. They reached the equation for the maximum temperature T_m, known also from TL theory,

$$\beta E/RT_m^2 = Z \exp\{-E/RT_m\} \tag{8.81}$$

and, by an approximation, to the expression of the intensity at the maximum

$$-\ln(x_m) = 1 - 2RT_m/E \tag{8.82}$$

where x_m is the fraction of the sample not reacted upon at T_m. Thus, E can be evaluated, and, by insertion into Eq. 8.81, Z can be found.

Kissinger[302] suggested the evaluation of the kinetics parameters of a first-order DTA peak by measuring the variation of the peak maximum, T_m, with the heating rate, β. As was shown in section 3.5, the same method was developed independently for TL and other thermally stimulated phenomena. As shown by Kissinger, by using Eq. 8.81, one gets

$$\frac{d\left[\ln(\beta/T_m^2)\right]}{d(1/T_m)} = -\frac{E}{R}. \tag{8.83}$$

Thus, plotting $\ln(\beta/T_m^2)$ vs. $1/T_m$ for different heating rates should yield a straight line with a slope of $-E/R$, from which E is readily found. In a later paper, Kissinger[303] extended the heating rates method to non-first-order kinetics, and showed that Eq. 8.83 still holds as a good approximation. He also suggested the computation of the order of kinetics, b, by a semi-empirical expression

$$b = 1.26S^{1/2} \tag{8.84}$$

where S, the shape index, is the absolute value of the ratio of the slopes of the tangents to the curve at the inflection points, where $d^2x/dt^2 = 0$ as shown in Fig. 8.15.

Due to its simplicity, Kissinger's heating rates method became one of the most used techniques for evaluating kinetic parameters from DTA. However, several authors claimed that this method leads to large errors in the values of E

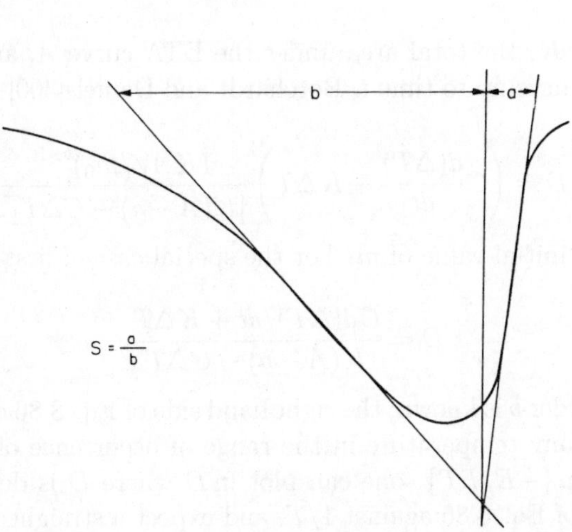

Figure 8.15 The shape index S in Kissinger's method. (After Kissinger[303].)

and b. Some of the critical accounts were by Reed et al.[304], Melling et al.[305], Sharp[306] and Balarin[307]. Most of these researchers blamed the assumption that the maximum rate of reaction occurs at the tip of the DTA peak. This has been explained above to be inaccurate for DTA peaks, as opposed to TL where it is normally known to be correct. Some more difficulties with the Kissinger method have to do with the fact that slight errors in T_m cause large errors in the evaluated E, and with a lag between the temperature of the sample and that of the thermocouple, which increases with increasing heating rate. These two points are valid in TL and DTG peaks as well. Finally, an important element specific to DTA, is that a control of the sample temperature during the reaction is, in fact, impossible. Therefore, it is difficult to obtain precisely the heating rate of the sample at T_m since the used β is the heating rate of the reference. This point is taken care of in the sister method of DSC.

A method which is considered to have broader validity is that by Borchardt and Daniels[300]. Let us take $x = 1 - \alpha$ in Eq. 8.80 as the concentration of the reactant. If V is the volume of the solution containing the reactant and m is the number of moles of the reactant at time t, then $m(t) = Vx(t)$. Eq. 8.80 will then take the form

$$-\frac{1}{V}\frac{dm}{dt} = D\left(\frac{m}{V}\right)^b \tag{8.85}$$

where $D = Z\exp\{-E/RT\}$. Taking into account the magnitudes relevant to DTA, namely, the heat capacity of the sample C, the heat transfer coefficient of the container K, the total area under the DTA curve A, and a, the partial area under the curve up to time t, Borchardt and Daniels[300] found, using Eq. 8.85,

$$D = \left(C\frac{d(\Delta T)}{dt} + K\Delta T \right) \frac{(KAV/m_0)^{b-1}}{[K(A-a) - C\Delta T]^b} \tag{8.86}$$

where m_0 is the initial value of m. For the special case of first-order DTA, this reduces to

$$D = \frac{Cd(\Delta T)/dt + K\Delta T}{K(A-a) - C\Delta T}. \tag{8.87}$$

If the reaction order b is known, the right hand side of Eq. 8.86 can be calculated numerically for any temperature in the range of occurrence of the DTA peak. Since $D = Z\exp\{-E/RT\}$, one can plot $\ln D$ where D is determined by the right hand side of Eq. 8.86 against $1/T$, and expect a straight line with a slope of $-E/R$. If b is not known, the procedure should be repeated for different values of b, and the best straight line would yield the kinetic order, and the slope would give the activation energy E.

For further reading on the specific possibilities and difficulties in DTA and the evaluation of parameters thereof, the reader is referred to the works by Kieffer and Bonivardi[308], Reading et al.[309], Málek[310], Mianowski and Radko[311], Málek et al.[312], Flynn[313], and Liu et al.[314]. Three important review papers by Dollimore[315], Dollimore and Reading[316], and by Gallagher[317] are also to be noted. The works by Peregudov et al.[318], McGhie[319], and by Zhong and Gallagher[320], dealing with simultaneous measurements of TG (or DTG) and DTA, should also be mentioned, along with a review paper by Dollimore[321].

8.5.3 Differential Scanning Calorimetry (DSC)

The DSC method is very closely related to DTA. Here, the sample and reference are maintained at the same temperature as both are heated (or cooled) at a pre-determined, usually uniform, rate. The heat flow required to sustain a zero temperature difference, $T_s - T_r = 0$ is recorded as a function of time or temperature. The measurement is performed with a *Differential Scanning Calorimeter*, in which two separate heaters supply heat to the sample and reference. Details can be found in the literature, e.g. a review paper by McNaughton and Mortimer[322]. In this apparatus, the environment of the sample and reference holders is inert, so that the electrical heaters are the only sources of heat. They serve to raise the temperature of the sample and reference while maintaining the same temperature in both, compensating for the losses, or gains, in the sample. In the apparatus, there are two separate control circuits. One is for the control of the average temperature, which regulates the average power applied to the heaters, according to a pre-set heating program. The other control unit corrects any temperature difference between the sample and reference, by adjusting the differential power increment applied to the two heaters. The temperatures, T_s and T_r, are measured continuously, usually by platinum-resistance thermometers, the readings of which are fed into the respective control circuit. A signal proportional to the difference in power applied to the two heaters, which is the DSC signal, is transferred to the recording device. The curve obtained is that of the heat flow dH/dt versus either time or average temperature. The heat flux dH/dt required to maintain a null temperature difference is equal to the difference between the change in enthalpy of the sample and reference, provided the heat losses of the two specimens are identical. Since the reference is a thermally inert material, changes in dH/dt indicate changes in the heat capacity of the sample.

The DSC curve is very similar to the DTA curve of the same material; however, the DSC peaks may be somewhat different in position and shape, and may exhibit an internal structure not apparent in the DTA curve. A major advantage of DSC over DTA is that, in the former, the heat flow is measured directly. If dH/dt is recorded vs. time under favorable conditions of rather similar thermal properties of sample and reference, and with a uniform heating of the reference, the area under a DSC peak is equal to the heat of reaction involved. Usually, the output signal is in arbitrary units and calibration is still needed in order to convert the area to joules or calories, however, the calibration coefficient is a constant of the instrument and the reference, and does not vary with temperature as in the DTA method.

DSC is preferable to DTA for obtaining kinetic data since it provides a plot of dH/dt rather than ΔT. It can usually be assumed that dH/dt is proportional to $-dx/dt$ where x is the fraction of the sample not yet reacted. Therefore, a DSC curve can be analyzed with the methods used for TL and other thermally stimulated processes. If the reaction under study obeys Eq. 8.80, the DSC peak has a shape similar to that of Eq. 8.76. If the assumptions of Borchardt and Daniels[300] hold, dH/dt is related to the other quantities by

$$\frac{dH}{dt} = (A-a)^b D/(AV/m_0)^{b-1} \tag{8.88}$$

where the symbols have the same meaning as in the previous sub-section. In the model by Gray[301], the relevant equation for DSC is

$$\frac{dH}{dt} = -\frac{dq}{dt} + (C_s - C_r)\frac{dT}{dt} - \frac{C_s}{K}\frac{d^2q}{dt^2} \tag{8.89}$$

where the recorded signal, dq/dt, is the rate at which thermal energy is generated in the sample, and dT/dt can be taken as the constant heating rate β. Since the third term in Eq. 8.89, which arises from thermal lag, can be negligibly small, then when $C_s \simeq C_r$ we really have $dq/dt \cong -dH/dt$. Manche and

Carroll[283] pointed out that in DSC, C_s/K can be reduced without a loss of sensitivity, whereas in DTA, the recorded signal itself is proportional to $1/K$, thus an increase of K would reduce the sensitivity. The DSC method is used for evaluating temperatures of transition, such as glass transition, crystallization and melting phenomena[323]. According to Manche and Carroll[283], the transition temperature can usually be taken as the point of first observable deflection from the base-line.

As pointed out above, the method of various heating rates as used in the analysis of DTA peaks is rather questionable, since the evaluation of the heating rate of the sample itself is problematic. This is remedied in the study of DSC as explained by Chataing and Vergnaud[324] and by Armand and Vergnaud[325]. They show that this technique helps in the separation of different peaks related to different processes. Further work on different heating rates was reported by Kaiserberger and Opfermann[326] who applied the method to the decomposition of hexogen and the vulcanization of synthetic rubber, and found the relevant activation energies and the order of kinetics. Graydon et al.[327] studied the crystallization of amorphous metals such as $Ni_{72}Co_2Mo_6B_{20}$ by the use of DSC with various heating rates. A general account on the use of DSC to obtain thermodynamic and kinetic data was given by Cassell and DiVito[328].

Finally, we briefly mention a number of works describing either simultaneous measurements of DSC and other thermoanalytical techniques, or comparing DSC with those techniques. Le Parlouër[329] describes an apparatus for TG-DSC simultaneous measurement and explains its advantages over TG and DSC run separately. Brown and Galwey [330] investigated isothermal DSC and isothermal DTG and showed how the combined measurement yields more reliable information of the studied sample. Hoffman and Pan[331] showed how to measure the heats of reaction by combining DSC and TG. Mentel and Anderson[332] used non-linear optimization for the evaluation of parameters of TG, DTA and DSC. Gallagher[333] compared the sensitivity of DSC and other thermoanalytical methods, and advocated the use of simultaneous TG/DTA/DSC measurements. He also described the concept of *Controlled Rate Thermal Analysis* (CRTA), previously suggested by Rouquerol[334]. Griffin and Laye[335] also discussed the advantages of combining DTA, DSC and TG measurements. Speyer[336] described a method for the deconvolution of superimposed DTA/DSC peaks by using the simplex algorithm.

8.5.4 Thermal Annealing; the Release of Stored Energy and Optical Absorption

A method closely related to DTA and DSC is that of a calorimetric measurement during the heating of a sample, following an irradiation which may be x-rays, or other forms of ionizing radiation. Such an irradiation may create different kinds of defects, or fill existing defects or impurities. The annealing of the defects upon heating involves the release of energy stored in the sample during irradiation. Quite often, the released energy can be detected as TL, as explained above. A more direct measurement of the thermal annealing process involves calorimetric methods which directly record the rate of release of stored energy as a function of temperature. Wehr et al.[337] measured the release of energy stored in solid argon by reactor irradiation at 4.6 K. They used a differential *heat flow calorimeter* which enables the measurement of heat flow into the sample, as a function of its rising temperature. A measurement of a reference unirradiated sample in parallel permits the evaluation of the energy released by annealing of the defects. Similar measurements on KCl and KBr crystals were performed by Cruz-Vidal et al.[338],[339]. Further work along the same lines was reported by Delgado and Alvarez-Rivas[340] who studied the relation between the stored energy and TL in γ- and electron-irradiated NaCl samples. They showed that the TL peak always occurred at a temperature higher than that of the main stored-energy peak. They concluded that different recombination processes were involved in the two phenomena. In a later work, Delgado and Alvarez-Rivas[341] extended this work to other alkali halide samples. Jiménez de Castro and Alvarez-Rivas[342] found a stored energy of about 10^{20} eVg^{-1} in NaCl and KCl samples, following plastic deformation or quenching. The peaks appearing up to 700 °C were the same as those seen after a heavy irradiation of the samples. They proposed that the radiation-induced color centers are only a small part of the total radiation-induced defects.

In all these measurements it was assumed that the recorded signal, which is the rate of released energy, was proportional to the rate of the defect annihilation. The general equation governing the process was given by Damask and Dienes[343] as

$$-dn/dt = F(n)K_o \exp\{-E/kT\} \tag{8.90}$$

where n is the concentration of defects, $F(n)$ is a continuous function of n, K_o is the pre-exponential factor, and E is the activation energy. Cruz-Vidal et al.[338],[339] suggested that it is very realistic to assume first- or second-order kinetics, namely, $F(n) \sim n$ or $F(n) \sim n^2$. Once such a behavior is established, by measuring the shape factor of the peak, μ_g, for example, analysis methods developed for TL and TG peaks can be utilized. Wehr et al.[337] used the first-order equation for the recombination of pairs of interstitials and vacancies, and the second-order equation for recombinations of interstitials and vacancies after free migration.

An alternative method of studying thermal annealing is based on the optical absorption of the defects. Itoh et al.[344] and Saidoh and Itoh[345] investigated the annealing process of irradiated crystals by measuring the optical density of the irradiated sample in a range of intrinsic transparency. During the heating of the sample, its optical absorption changes and thus, the defect concentration $n = n(T)$ can be evaluated. The differentiation of this curve with respect to temperature gives an annealing curve, analogous to a TL curve. Itoh et al.[344] used an equation similar to Eq. 8.90 to describe their results; however, they assumed cases of general-order kinetics, namely $F(n) \sim n^b$ where b is not necessarily 1 or 2.

Damask and Dienes[343] showed that if $F(n)$ is a "well behaved function" of n, a good approximation for the activation energy may be evaluated by the heating rates methods described in Section 3.5. Kelly[346] described several possible mechanisms of thermal annealing of irradiation-induced defects, and pointed out that the activation energy may depend on the defect concentration, due to the overlapping of strain fields surrounding the defects. This can be written as

$$E = E_o - \alpha n \tag{8.91}$$

where n is the number of remaining defects. This yields the annealing rate

$$\frac{dn}{dt} = An^b \exp\left\{-\frac{E_o}{kT}\right\} \exp\left\{\frac{\alpha n}{kT}\right\}. \tag{8.92}$$

The extra term has the effect of spreading the peak, and thus resembles a situation of an increase in the order of kinetics. A further discussion on the simultaneous measurements of TL and optical absorption is given in Ch. 9. Electric conductivity has also been utilized to monitor the recovery of radiation-induced defects. Such a measurement, on n-type germanium, was performed by Meese[347], who evaluated the conductivity at 4.2 K, following annealing at different higher temperatures, under a pulsed electric field.

8.5.5 Evolved Gas Analysis (EGA) and Thermal Desorption

Evolved gas analysis (EGA) and evolved gas detection (EGD) are two closely related techniques, sometimes used in conjunction with TG and DTA, in which the volatile products emitted by a sample during its heating are detected and analyzed. According to the ICTA recommendations (see Lombardi[275]), EGD is the technique in which *the evolution of gas* from a substance is detected as a function of temperature, while the substance is subjected to a controlled temperature program. Evolved gas analysis (EGA) is the technique in which *the nature and/or amount* of volatile products released by a substance are measured when the substance is subjected to a controlled temperature program. Several other terms were used for different versions of the EGD-EGA techniques. These are: Mass Spectrometric Analysis (Wilson and Hamaker[348]), Mass Spectrometric Thermal Analysis, MTA (Friedman et al.[349], Sakamoto et al.[350]), Thermal Volatilization Analysis, TVA (McNeil[351]) and Gas Effluent Analysis (Lodding[352]). The term *thermal evolution* is also in use as a comprehensive name for both EGA and EGD (Chiu and Palermo[353]). The same term is also used in cases in which the evolved gas has been injected into the sample prior to the heating. Different instruments have been developed to measure this phenomenon, and details of some of them can be found in the paper by Chiu and Palermo[353]. More modern EGA systems are often based on mass spectrometry, which also provides their composition, in addition to the rate of emission of the gases. An example can be seen in the work by Sakamoto et al.[350].

The kinetics of the EGA process and its parameters were studied in several investigations. The pioneers in this field, Rogers et al.[354], described conditions whereby the EGA peaks parallel those of DTG and, therefore, the same equations can be used and the same parameters - activation energies, pre-exponential factors and kinetic order - can be evaluated. More details can be found in the paper by Sakamoto et al.[350] and by Ozawa[355]. Several investigators used the EGA technique to investigate the evolution of gases which had been implanted in the sample before heating (see e.g. Donnelly and Ingram[356]). The implantation can be performed using ion beams at energies from a few eV up to several MeV. Inert gases, such as helium, neon and argon are usually employed. When the injected gas penetrates to a depth of only a few atomic layers, it is more appropriate to consider it as the thermal desorption of the gas from the surface, a subject which is discussed later in this subsection. If the penetration is deeper, then during heating it is released following its diffusion through the bulk of the sample or as a result of its pyrolysis. In some cases (see Donnelly and Ingram[356]), single step kinetics have been observed even when the gas was trapped several atomic layers below the surface, and simple first-order kinetics resulted. A probable explanation is that when a gas atom is released from its trapping site, it can diffuse to the surface as an interstitial, with an activation energy much smaller than that for detrapping. In other cases, the situation is more complex due to retrapping.

Donnelly and Armour[357] developed an expression for EGA, assuming that the activation energy for diffusion was the same as for retrapping. They assumed first that the gas atoms are initially located in one discrete plane. They found the rate at which the dissolved gas is evolved from the surface, while using a constant heating rate, to be

$$\rho = AT^{-3}\left[\exp\frac{Q}{RT}\right]^{1/2}\exp\left\{-BT^{-2}\exp\left\{\frac{Q}{RT}\right\}\right\}\exp\left\{-CT^{2}\exp\left\{-\frac{Q}{RT}\right\}\right\} \tag{8.93}$$

where Q is the activation energy for diffusion, and A, B and C are constants dependent on the physical parameters of the solid and the gas - such as the vibration frequency of the diffusive atoms, the inter-trap distance, and the boundary conditions. Qualitatively, this function describes a peak, with a maximum at ρ_m, which later decreases to zero when the trapped gas is exhausted. The shape of the peak is such that the fall-off part is longer than the low-temperature increase region, which resembles high-order TL peaks. In a real specimen, the diffusing atoms will not be initially located in one discrete plane, but rather, will have some distribution over several planes. This would certainly complicate the situation, and usually further broaden the observed peak.

Another technique, very similar to EGA is Emanation Thermal Analysis (ETA). Here, the release of radioactive gases from a sample is measured by an appropriate radioactivity detector. A review paper by Balek[358] describes the details of the method and its applications.

In the method of *thermal desorption*, sometimes termed *flash desorption* (see, e.g. Redhead[359]), the sample is rigorously cleaned in vacuum. (The sample is usually a metal in the form of a ribbon or a wire, but it can also be a crystal.) A known gas is introduced into the vacuum chamber, and allowed to adsorb on the sample. The gas is then desorbed by heating the sample in a controlled manner, and the pressure in the system is recorded vs. time. The resulting pressure-time (or -temperature) curve can be analyzed to yield information on various adsorption parameters. This includes the identification of various adsorbing phases and their populations, the activation energy of desorption of the different phases, and the order of the desorption kinetics when this parameter is relevant. The early measurements of thermal desorption were mainly concerned with evaporated tungsten films (see, for example, Hickmott and Ehrlich[360]). Later, desorption from silver (Wachs and Madix[361]), iron (Hill et al.[362]), silicon (Nguyen et al.[363]), platinum (Netzer[364]) and other metals was studied. A survey of the various methods that can be utilized for the determination of heats of adsorption appears in a review article by Gravelle[365]. The apparatus and experimental procedure used for thermal desorption measurements are similar to those used in EGA, but in thermal desorption, an ultrahigh vacuum is usually required.

The rate of desorption can be calculated, according to King[366] from the experimental curve by

$$-\frac{dn}{dt} = \frac{V}{AkT_g}\left[\frac{dP}{dt} + \left(\frac{S}{V}\right)P\right] \tag{8.94}$$

where n is the number of adsorbed molecules per unit area. P is the pressure increase above the background, V is the volume of the desorption chamber, S is the pumping speed of the system, A is the adsorbent area, and T_g is the temperature of the gas. Usually, the ratio of pumping speed to volume is large as compared to the rate of change of pressure, $P^{-1}(dP/dt) \ll S/V$, and therefore, the desorption rate is proportional to the pressure increase, P. The curve of pressure increase vs. time can therefore be treated as being proportional to dn/dt. For the analysis of thermal desorption experiments, it is usually assumed (see e.g. Redhead[359], King[366], Bornand[367], Vollmer and Träger[368], de

Jong and Niemantsverdriet[369], and García et al.[370]) that the rate of desorption can be written as

$$-dn/dt = \nu n^b \exp\{-E/RT\} \tag{8.95}$$

where b is the reaction order, ν is the pre-exponential factor, and E is the activation energy, in full analogy with the general-order TL and TG cases. The solutions of this equation, found before for the signal vs. temperature with a linear heating function for other thermally stimulated phenomena, are applicable here with no change.

A complex desorption curve, consisting of several peaks, is usually attributed to several desorption states, molecular or atomic, and to desorption from different crystal planes. The resolution between various peaks in a complex desorption curve is discussed by Holzapfel et al.[371]. The effect of readsorption is discussed by Edwards[372], and resembles retrapping in TL theory. The effect of a temperature gradient along the sample is described by Chiarena and Gillet[373].

In the same way as in EGA, mass spectrometers are being used which enable the determination of the individual desorption rate when several gases are desorbed simultaneously. The technique is called *thermal desorption mass spectrometry* (TMS). When the measurement is that of TMS, the spectrometer signal is

$$I(t) = -g\frac{dn}{dt}\left[T(t)\right]^{-1/2} \tag{8.96}$$

where g is a constant[374]. This is entirely analogous to a TL peak with a pre-exponential factor which depends on temperature like $T^{-1/2}$.

Falconer and Madix[375] investigated thermal desorption from nickel and found a first-order reaction, but with a slightly coverage-dependent activation energy

$$E = E_o - \gamma n(t). \tag{8.97}$$

Bauer et al.[376] studied thermal desorption of Cu, Ag and Au from clean surfaces of tungsten single crystals and found rather large changes of E and ν with coverage. King[366] analyzed theoretically the case where both E and ν are coverage-dependent, namely

$$-\frac{dn}{dt} = \nu(n)n^b\left[E(n)/RT\right] \tag{8.98}$$

and suggested a method for evaluating the parameters in this case. Other works on parameter determination are those of Zhou et al.[377], Seebauer[378], and Froitzheim et al.[379], and two review papers on thermal desorption kinetics, by Kreutzer[380] and by Kreutzer and Payne[381], are also mentioned.

8.6 References

1. R. Chen and Y. Kirsh, *Analysis of Thermally Stimulated Phenomena,* Pergamon Press, Oxford (1981).
2. R.C. Herman and R. Hofstadter, *Phys. Rev.* **57** (1940) 936.
3. P.R. Johnson, *J. Opt. Soc. Am.* **29** (1939) 387.
4. S.C. Agarwal and H. Fritzsche, *Phys. Rev. B* **10** (1974) 4351.
5. H.K. Henisch, *J. Electrostatics* **3** (1977) 233.
6. L.J. van-der Pauw, *Philips Res. Repts.* **13** (1958) 1.
7. P.M. Hemenger, *Rev. Sci. Inst.* **44** (1973) 698.
8. K. Kuriyama, S. Uematsu and H. Tahkahashi, *Appl. Phys. Lett.* **65** (1994) 746.
9. W.S. Lau, T.C. Chang, L.S. Tan, C.H. Goo and K.S. Goh, *Japan. J. Appl. Phys.* **11A** (1991) L1843.
10. C. Popescu and H.K. Henisch, *J. Phys. Chem. Solids* **37** (1976) 47.
11. C. Popescu and H.K. Henisch, *Phys. Rev. B* **14** (1976) 517.
12. H.K. Henisch and J. Zucker, *Rev. Sci. Inst.* **27** (1956) 261.
13. I.R. Weingarten and M.R. Rothberg, *J. Electrochem. Soc.* **108** (1961) 167.
14. C.A. Bryant and J. Gunn, *Rev. Sci. Inst.* **36** (1965) 1614.
15. D. Chatain, P. Gautier and C. Lacabanne, *Rev. Sci. Inst.* **41** (1970) 1610.
16. T. Ogawa, *J. Appl. Phys.* **32** (1961) 583.
17. P. Müller, *Phys. Stat. Sol. (a)* **78** (1983) 41.
18. A. Baba and K. Ikezaki, *J. Appl. Phys.* **57** (1985) 359.
19. B. Molnar and T.A. Kennedy, *J. Electrochem. Soc.* **125** (1978) 1318.
20. R.R. Haering and E.N. Adams, *Phys. Rev.* **117** (1960) 451.
21. P.N. Keating, *Proc. Phys. Soc.* **78** (1961) 1408.
22. K.H. Nicholas and J. Woods, *Brit. J. Appl. Phys.* **15** (1964) 783.
23. R.H. Bube, G.A. Dussel, C.T. Ho and L.D. Miller, *J. Appl. Phys.* **37** (1966) 21.
24. P. Devaux and M. Schott, *Phys. Stat. Sol. (a)* **20** (1967) 301.
25. T.A.T. Cowell and J. Woods, *Brit. J. Appl. Phys.* **18** (1967) 1045.
26. M.E. Haine and R.E. Carley-Read, *Brit. J. Appl. Phys. (J. Phys. D)* **1** (1968) 1257.
27. M.G. Buehler, *Sol. Stat. Elect.* **15** (1972) 69.
28. T.G. Garofano and S. Morelli, *Nuovo Cim.* **13B** (1973) 174.
29. T.W. Kang, C. Tong, L.Y. Leem and T.W. Kim, *J. Appl. Phys.* **69** (1991) 3115.
30. T. Datta, R. Noufi and S.K. Deb, *J. Appl. Phys.* **59** (1986) 1549.

31. P. Suryanarayana, H.N. Acharya and Y.F. Nicolau, *Semicond. Sci. Technol.* **7** (1992) 297.
32. T. Takizawa, J. Fuji and S. Nomura, *J. Phys.: Condens. Matter* **3** (1991) 6865.
33. S. Rabie and N. Rumin, *Appl. Phys. Lett.* **27** (1975) 29.
34. A.C. Lewandowski and S.W.S. McKeever, *Phys. Rev. B* **43** (1991) 8163.
35. A.C. Lewandowski, B.G. Markey and S.W.S. McKeever, *Phys. Rev. B* **49** (1994) 8029.
36. I.J. Saunders, *Brit. J. Appl. Phys.* **18** (1967) 1219.
37. R. Chen, *Brit. J. Appl. Phys. (J. Phys. D)* **2** (1969) 371.
38. R.K. Gartia, P.S. Mazumdar and Th. Subodh Chandra Singh, *J. Phys. D: Appl. Phys.* **22** (1989) 1728.
39. S. Dorenjit Singh and R.K. Gartia, *J. Phys. D: Appl. Phys.* **27** (1994) 861.
40. L.I. Filina and V.A. Margulis, *Semiconductors* **28** (1994) 820.
41. H. Wynands and M. Cocivera, *Solar Energy Materials and Solar Cells* **32** (1994) 187.
42. P. Kelly and P. Bräunlich, *Phys. Rev. B* **1** (1970) 1587.
43. R. Chen, *Chem. Phys. Lett.* **6** (1970) 125.
44. M. Lax, *Phys. Rev.* **119** (1960) 1502.
45. W. Siegel, H. Witte, G. Kühnel, H.A. Schneider and T. Flade, *Phys. Stat. Sol. (a)* **108** (1988) 689.
46. W. Tomaszewicz, *J. Phys.: Condens. Matter* **4** (1992) 3967.
47. R.H. Bube, *Electronic Properties of Crystalline Solids*, Academic Press, New-York (1974).
48. P. Kivits and H.J.L. Hagebeuk, *J. Lumin.* **15** (1977) 1.
49. D. Shaw and H.R. Hoghaddam, *Phys. Stat. Sol. (a)* **65** (1981) 167.
50. S.G. Elkomoss, M. Samimi, M. Hage-Ali and P. Siffert, *J. Appl. Phys.* **57** (1985) 5313.
51. V. Rinkevičius and G. Kavaliauskienė, *J. Phys. D: Appl. Phys.* **27** (1994) 1267.
52. V. Rinkevičius and G. Kavaliauskienė, *J. Phys. D: Appl. Phys.* **29** (1996) 1329.
53. R. Kiliulis, V. Rinkevičius, J. Storasta and J. Vaitkus, *Phys. Stat. Sol. (a)* **127** (1991) 415.
54. Z-Q. Fang and D.C. Look, *Appl. Phys. Lett.* **66** (1995) 3033.
55. D.C. Look, Z.Q. Fang, W. Kim, Ö. Aktas, A. Botchkaren, A. Salvador and H. Markoç, *Appl. Phys. Lett.* **68** (1996) 3775.
56. D. Bloom and S.W.S. McKeever, *J. Appl. Phys.* **77** (1995) 6511.
57. D. Bloom and S.W.S. McKeever, *J. Appl. Phys.* **77** (1995) 6521.

58. D.S. Misra, V.A. Singh and S.C. Agarwal, *Phys. Rev. B* **32** (1985) 4052.
59. N. Ibaraki and H. Fritsche, *J. Non-Crystalline Solids* **66** (1984) 231.
60. H. Fritsche and N. Ibaraki, *Phil. Mag. B* **52** (1985) 299.
61. S. Schrader and M. Kryszewski, *Phys. Stat. Sol. (a)* **91** (1985) 243.
62. J-H. Zhou and S.R. Elliott, *Phys. Rev. B* **46** (1992) 9792.
63. D.M. Fleetwood, R.A. Reber Jr. and P.S. Winokur, *Appl. Phys. Lett.* **60** (1992) 2008.
64. S.L. Miller, D.M. Fleetwood and P.J. McWhorter, *Phys. Rev. Lett.* **69** (1992) 820.
65. D.M. Fleetwood, S.L. Miller, R.A. Reber Jr. P.J. McWhorter, P.S. Winokur, M.R. Shaneyfeld and J.R. Schwank, *IEEE Trans. Nucl. Sci.* **39** (1992) 2192.
66. T. Mizutani, T. Nagota and M. Ieda, *J. Phys. D: Appl. Phys.* **17** (1984) 1883.
67. R.M. Faria, A. Jorge and O.N. Oliviera Jr., *J. Phys. D: Appl. Phys.* **23** (1990) 334.
68. A.G.J. Green, J.C. Bickerton, I.V.F. Viney, J.W. Brightwell and B. Ray, *Sol. St. Comm.* **62** (1987) 543.
69. M.F. Bulanyĭ, V.I. Klimenko and A.V. Likhoshva, *Semiconductors* **28** (1994) 457.
70. D.M. Pai, *J. Chem. Phys.* **52** (1970) 2285.
71. J.H. Slowik, *J. Appl. Phys.* **47** (1976) 2982.
72. I. Chen, *J. Appl. Phys.* **47** (1976) 2988.
73. B. Barland, M. Lemiti, S. Audisio and J.C. Bureau, *J. Electrochem. Soc.* **137** (1990) 2845.
74. F.S.W. Hwang and J.H. Fremlin, *Archaeometry* **12** (1970) 67.
75. J.P. Mitchell and D.G. DeNure, *IEEE Trans. Nucl. Sci.* **NS20** (1973) 67.
76. D.N. Bose and H.K. Henisch, *Sol. St. Elect.* **11** (1968) 273.
77. C. Bowlt, *J. Phys. D: Appl. Phys.* **17** (1976) 461.
78. L.A. Vermeulen and A. Halperin, *J. Phys. Chem. Solids* **48** (1981) 115.
79. P.R. Moran, E.B. Podgoršak, G.E. Fuller and G.D. Fullerton, *Med. Phys.* **1** (1974) 155.
80. G.D. Fullerton and P.R. Moran, *Med. Phys.* **1** (1974) 161.
81. A. Lörincz, *Phys. Stat. Sol. (a)* **61** (1980) K81.
82. L. Lassabatere, G. Seve and Mendolia, *Thin Solid Films* **6** (1970) 381.
83. A.F. Gomez and F.S. Sinencio, *Phys. Lett.* **34A** (1971) 130.
84. H.J. von Bardeleben, C. Schwab, C. Charager, J.C. Muller, P. Sifferet and R.S. Feigelson, *Phys. Stat. Sol.* **58** (1980) 143.
85. D.C. Gates, P. Krell and T.J. Magee, *Final Res. Rept. DNA 3515F*, Stanford Res. Inst. (1974).

86. M.W. Harper and B. Thomas, *Phys. Med. Biol.* **18** (1973) 409.
87. N.W. Ramsey and L.Y. Joesoef, *Biomedical Dosimetry,* IAEA Vienna (1975) 95.
88. M.M. Perlman and S. Unger, *J. Phys. D: Appl. Phys.* **5** (1972) 2115.
89. G. Lapke, W.F. Schmidt and J.G. Rabe, *J. Appl. Phys.* **53** (1982) 5763.
90. M. Grinter and C. Bowlt, *J. Phys. D: Appl. Phys.* **8** (1975) L159.
91. J. Conway, M.W. Harper and B. Thomas, *J. Phys. D: Appl. Phys.* **10** (1977) 1131.
92. R.A. Street and A.D. Yoffe, *Thin Solid Films* **11** (1972) 161.
93. C. Bowlt and D.J. Wagget, *Phys. Med. Biol.* **19** (1974) 534.
94. A. Scharmann, *Proc. III Int. Symp. Exoelect., Braunschweig, PTB Mitteilungen* **80** (1970) 319.
95. J. Kramer, *Z. Phys.* **125** (1949) 739.
96. H. Curie, *Comptes Rendus* **129** (1899) 739.
97. E. Rutherford, *Phil. Mag.* **43** (1900) 161.
98. J.W. Russel, *Proc. Roy. Soc.* **61** (1897) 424.
99. P. Villard, *Soc. Phys.* (1900) 37.
100. J. C. McLennan, *Phil. Mag.* **3** (1902) 195.
101. K. Becker, *CRC Critical Reviews in Sol. St. Sci.* **3** (1972) 39.
102. K. Becker, *Solid State Dosimetry,* CRC Press (1973).
103. J. Kramer, *Z. Phys.* **133** (1952) 629.
104. A. Bohun, *Czech. J. Phys.* **4** (1954) 91.
105. H. Nassenstein, *Z. Naturforsch.* **10a** (1955) 944.
106. P. Fintelman, *Proc. 4th Czech. Conf. on Electronics and Vacuum Physics,* Prague (1968) 395.
107. F.H. Attix, *Report of NRL Progress* (1970) 1.
108. A. Moreno, R.B. Gammage, R.B. Cheka, J.S. Nagpal and K. Becker, *ORNL-TM-3668, Oak Ridge National Laboratory* (1972).
109. A. Bohun, *Proc. III Int. Symp. Exoelect. Braunschweig, PTB Mitteilungen* **80** (1970) 321.
110. K. Becker, J.S. Cheka and R.B. Gammage, *Proc. III Int. Symp. Exoelect. Braunschweig, PTB Mitteillungen* **80** (1970) 335.
111. J.T. Randall and M.H.F. Wilkins, *Proc. Roy. Soc. (London)* **A184** (1945) 347.
112. A. Bohun, *Acta Phys. Austriaca* **10** (1957) 353.
113. G. Holzapfel, *Phys. Stat. Sol.* **33** (1969) 235.
114. W.T. Pimley and E.E. Francis, *J. Appl. Phys.* **32** (1961) 1729.

115. P. Kelly, *Phys. Rev. B* **5** (1972) 749.

116. A.G. Wintle, *J. Mater. Sci.* **9** (1974) 2059.

117. E.I. Tolpigo, K.B. Tolpigo and M.K. Scheinkman, *Izv. Akad. Nauk. SSSR, Ser. Fiz.* **30** (1966) 1901.

118. G. Holzapfel, *Proc. VI Int. Symp. Tech. Comm. Photon Detect.*, Siofok, Hungary (1974) 30.

119. M. Kawanishi, *Mem. Inst. Sci. Ind. Research, Osaka Univ.* **36** (1979) 51.

120. V.S. Kortov, T.S. Kortov, T.S. Bessonova, M.S. Akselrod and I.I. Milman, *Phys. Stat. Sol. (a)* **87** (1985) 629.

121. Y. Kutomi and A. Tomita, *Rad. Prot. Dosim.* **33** (1990) 347.

122. K. Biedrzycki, *Phys. Stat. Sol. (a)* **117** (1990) 313.

123. F. Golek, P.W. Klamut, B. Sujak, A.J. Zalenski, M. Ciszel and J. Olejniczak, *Mod. Phys. Letts. B* **4** (1990) 333.

124. Y. Watanabe, Y. Nakamura and H. Igarashi, *Japan. J. Appl. Phys.* **30** (1991) 143.

125. Y. Watanabe, Y. Nakamura, S. Hirayama and Y. Yamagachi, *J. Mater. Res.* **7** (1992) 1805.

126. Y. Momose, T. Yamamoto, M. Takeuchi and T. Sakurai, *J. Appl. Phys.* **73** (1993) 7482.

127. R. Huzimura and K. Nakahira, *Japan. J. Appl. Phys.* **33** (1994) 6336.

128. T. Sakurai and Y. Momose, *Phys. Lett. A* **179** (1993) 55.

129. T. Sakurai, *Phys. Stat. Sol (a)* **145** (1994) 145.

130. G. Rudlof and H. Glaefke, *Phys. Lett. A* **210** (1996) 433.

131. Z. Spurný and H. Käämbre, in *Techniques of Radiation Dosimetry"* K.M. Mahesh and D.R. Vij, editors (1985) 280.

132. P. Müller, *Phys. Stat. Sol. (a)* **23** (1974) 393.

133. J. van Turnhout, *Thermally Stimulated Discharge of Polymer Electrets,* Elsevier, Amsterdam (1975).

134. E.L. Kitts Jr., M. Ieka and J.H. Crawford Jr. *Phys. Rev. B* **8** (1973) 5840.

135. H. Frei and G. Groetzinger, *Phys. Z.* **37** (1936) 720.

136. B. Gross, *J. Chem. Phys.* **17** (1949) 866.

137. A.N. Gubkin and B.N. Matsonashvili, *Sov. Phys. Sol. St.* **4** (1962) 878.

138. P.V. Murphy and S. Costa-Ribeiro, *J. Appl. Phys.* **34** (1963) 2061.

139. M.C. Driver and G.T. Wright, *Proc. Phys. Soc.* **81** (1963) 141.

140. C. Bucci and R. Fieschi, *Phys. Rev. Lett.* **12** (1964) 16.

141. C. Bucci, R. Fieschi and G. Guidi, *Phys. Rev.* **148** (1966) 816.

142. R. Capelletti, R. Fieschi, A. Gainotti, C. Mora, L. Romanò and E. Zecchi, in *Defects in Insulating Crystals*, V.M. Tuchkevich and K.K. Shvarts, editors,

Springer Verlag, Berlin (1981) 675.

143. R. Capelletti, *Defects in Solids*, A.V. Chadwick and M. Terenzi, editors, Plenum Pub. Corp. (1986) 407.

144. V.F. Zolotaryov, D.G. Semak and D.V. Chepur, *Phys. Stat. Sol.* **21** (1967) 437.

145. G. Alzetta, G. Chella and R. Scarmozzino, *Lett. Nuovo Cim.* **3** (1972) 1.

146. D.E. Fields and P.R. Moran, *Phys. Rev. Lett.* **29** (1972) 721.

147. P.R. Moran and E.B. Podgoršak, *Phys. Lett.* **44A** (1973) 237.

148. M. Böhm and A. Scharmann, *Z. Phys. B* **22** (1975) 313.

149. J. van Turnhout, in *Electrets*, G.M. Sessler, editor, Springer-Verlag, Berlin (1978).

150. D.R. Figueroa, E. Laredo and M. Puma, *Sol. St. Comm.* **25** (1978) 509.

151. G.E. Matthews and J.H. Crawford Jr., *Phys. Rev. B* **15** (1977) 55.

152. I. Kunze and P. Müller, *Phys. Stat. Sol. (a)* **16** (1973) 55.

153. A. Kessler, *J. Physique* **34** (1973) C9-79.

154. G. Pfister and M.A. Abkovitz, *J. Appl. Phys.* **45** (1974) 1001.

155. T. Tanaka, S. Hiralayashi and K. Shilayama, *J. Appl. Phys.* **49** (1978) 784.

156. R. Shelley and G.R. Miller, *J. Sol. St. Chem.* **1** (1970) 218.

157. B. Gross, *J. Electrochem. Soc.* **115** (1968) 376.

158. B. Gross, *Appl. Optics Suppl.* **3** (1969) 176.

159. W. van Weperen, B.P.M. Lenting, E.J. Bijvank and H.W. den Hartog, *Phys. Rev. B* **16** (1977) 2953.

160. W. van Weperen and H.W. den Hartog, *Phys. Rev. B* **18** (1978) 2857.

161. E. Laredo, M. Puma, N. Suarez and D.R. Figueroa, *Phys. Rev. B* **23** (1981) 3009.

162. P. Dorenbos and H.W. den Hartog, *Phys. Rev. B* **31** (1985) 3932.

163. B. Jassemnejad and S.W.S. McKeever, *Phys. Rev. B* **36** (1987) 9769.

164. J. Vanderschueren, *Appl. Phys. Lett.* **25** (1974) 270.

165. J. Vanderschueren, *J. Polym. Sci.* **15** (1977) 873.

166. J. Vanderschueren, A. Linkens, B. Haas and E. Dellicour, *J. Macromol. Sci. Phys. B* **15** (1978) 449.

167. P. Fischler and P. Röhl, *J. Polym. Sci.* **14** (1976) 543.

168. E. Laredo, A. Bello, N. Suarez and M. Díaz, *Phys. Stat. Sol. (a)* **121** (1990) K121.

169. A.K. Jonscher, *Dielectric Relaxation in Solids,* Chelsea Dielectric Press, London (1983).

170. V. Halpern, *J. Phys. D: Appl. Phys.* **26** (1993) 307.

171. V. Halpern, *J. Phys. D: Appl. Phys.* **27** (1994) 2628.

172. G.F. Leal Ferreira, R.A. Moreno Alfaro and M.T. Figueiredo, *J. Phys. D: Appl. Phys.* **29** (1996) 3117.

173. H.A. Solunov and H.S. Ponevsky, *J. Polym. Sci., Polym. Phys.* **15** (1977) 969.

174. M. Zieliński, T. Swiderski and M. Kryszewski, *Polymer* **9** (1978) 883.

175. P. Hedvig, *Dielectric Spectroscopy of Polymers*, Adam Hilger, Bristol (1977).

176. A.K. Jonscher, *J. Electrostatics* **3** (1977) 53.

177. A.K. Jonscher, *J. Phys. D: Appl. Phys.* **24** (1991) 1633.

178. M.A. Careem and A.K. Jonscher, *J. Chem. Soc. Faraday Trans.* **2** (1986) 7.

179. K. Kato, M. Iwamoto and T. Hino, *Japan. J. Appl. Phys.* **26** (1987) 70.

180. D. Triantis, C. Anastadiadis and A. Markou, *J. Electrostatics* **22** (1989) 273.

181. D. Triantis, C. Anastadiadis and S. Orfanakos, *Phys. Stat. Sol. (a)* **114** (1989) 491.

182. R. Vila and M. Jiménez de Castro, *J. Phys. D: Appl. Phys.* **25** (1992) 1357.

183. J. Poźniak, *Acta Phys. Polon.* **86** (1994) 995.

184. R. Vila, A. Ibarra, M. Jiménez de Castro and F.W. Clinard, *Sol. St. Comm.* **79** (1991) 295.

185. L. Oliveira, P. Magna, N.J.H. Galo, E.C. Domenicucci and M. Siu Li, *Phys. Stat. Sol (b)* **171** (1992) 141.

186. L. Oliveira and M. Siu-Li, *Thin Solid Films,* **268** (1995) 30.

187. P. Pissis and D. Daoukaki-Diamanti, *J. Phys. Chem. Solids* **54** (1993) 701.

188. A.N. Papathanssiou, J. Grammatikakis and N.G. Borgis, *Phys. Rev. B* **48** (1993) 17715.

189. A. Algería, J.M. Banadriarán and J. Colmenero, *Phys. Stat. Sol. (b)* **125** (1984) 409.

190. C. Katyal, R.C. Verma and D. Kumar, *J. Mater. Sci. Lett.* **12** (1993) 847.

191. V.I. Mikla, I.P. Mikhalko and Y.Y. Nagy, *J. Phys.: Condens. Matter* **6** (1994) 8269.

192. F. Faubert, J.M. Gilli and P. Sixou, *Mol. Liq. Cryst.* **178** (1990) 133.

193. J.F. Mano, N.T. Correia and J.J. Moura-Ramos, *Polymer* **35** (1994) 3561.

194. A. Markiewicz, D.V. Balbachas and R.J. Fleming, *J. Therm. anal.* **37** (1991) 1137.

195. P. Pissis and L. Apekis, *J. Non-Crystalline Solids* **131** (1991) 95.

196. M. Aldana, E. Laredo, A. Bello and N. Suarez, *J. Polym. Sci. B* **32** (1994) 2197.

197. H.S. Faruque, *J. Mater. Sci.* **30** (1995) 5210.

198. I. Kunze and P. Müller, *Phys. Stat. Sol. (a)* **13** (1972) 197.

199. E.B. Podgoršak and P.R. Moran, *Phys. Rev. B* **8** (1973) 3405.

200. N. Kristianpoller and Y. Kirsh, *Phys. Stat. Sol. (a)* **21** (1974) 87.

201. K. Tanaka, Q. Zhuang, K. Tanaka, K. Kawana and R. Nakata, *J. Phys. Chem. Solids* **57** (1996) 307.

202. A.K. Agarwal and D.E. Day, *J. Am. Ceram. Soc.* **65** (1982) 111.

203. J. Meuldijk and H.W. den Hartog, *Phys. Rev. B* **28** (1983) 1036.

204. M. Ouwerkerk, F.J. Veldkamp and J. Schoonman, *Solid State Ionics* **22** (1987) 173.

205. K. Shindo, *IEEE Trans. Elect. Insulat.* **24** (1989) 481.

206. R.A. Creswell and M.M. Perlman, *J. Appl. Phys.* **41** (1970) 2365.

207. G.M. Sessler and J.W. West, *Phys. Rev. B* **10** (1974) 4488

208. P. Müller, *Phys. Stat. Sol. (a)* **23** (1974) 165.

209. P. Müller, *Phys. Stat. Sol. (a)* **23** (1974) 393.

210. I. Thurzo, M. Paliková and E. Mariani, *Czech. J. Phys. B* **25** (1974) 393.

211. P. Müller, *Amorphous Semiconductors '76*, Kósa Somogyi, editor, Akadémiai Kiadó, Budapest (1977).

212. J.G. Simmons and G.W. Taylor, *Phys. Rev. B* **5** (1972) 553.

213. J.G. Simmons and G.W. Taylor, *Phys. Rev. B* **6** (1972) 4804.

214. J.G. Simmons and G.S. Nadkarni, *Phys. Rev. B* **6** (1972) 4815.

215. H.A. Mar and J.G. Simmons, *Phys. Rev. B* **11** (1975) 775.

216. A.G. Zhdan, V.B. Sandomirskiĭ and A.D. Ozherov, *Sov. Phys. Semicond.* **2** (1968) 7.

217. A.G. Zhdan, V.B. Sandomirskiĭ and A.D. Ozherov, *Solid State Electronics* **11** (1968) 505.

218. A.G. Zhdan, V.B. Sandomirskiĭ and A.D. Ozherov, *Sov. Phys. Semicond.* **3** (1970) 1130.

219. A. Katzir and A. Halperin, *Sol. St. Electrostatics* **15** (1972) 573.

220. S.C. Agarwal, *Phys. Rev. B* **10** (1974) 4340.

221. G.M. Yang, *J. Phys. D: Appl. Phys.* **26** (1993) 690.

222. S.W.S. McKeever and D.M. Hughes, *J. Phys. D* **8** (1975) 1520.

223. S.W.S. McKeever and D.M. Hughes, *J. Phys. Chem. Solids* **39** (1978) 211.

224. J.C. Manifacier, J. Gasiot, P. Parot and J.P. Fillard, *J. Phys. C: Sol. St. Phys.* **11** (1978) 1011.

225. S.W.S. McKeever and E. Lilley, *J. Phys. C: Sol. St. Phys.* **14** (1981) 3547.

226. J. Vanderschueren and J. Gasiot, in *Thermally Stimulated Relaxation in Solids*, P. Bräunlich, editor, Springer-Verlag, Berlin, Heidelberg, New York (1979) 135.

227. G. Berg and J. Poźniak, *Phys. Stat. Sol. (a)* **38** (1976) 115.

228. N. Kristianpoller and Y. Kirsh, *J. Phys. C: Sol. St. Phys.* **12** (1979) 1079.
229. J. Vanderschueren, M. Ladang, J. Niezette and M. Corapci, *J. Appl. Phys.* **58** (1985) 4654.
230. A. Servini and A.K. Jonscher, *Thin Solid Films* **3** (1969) 341.
231. M. Böhm, O. Erb and A. Scharmann, *Phys. Stat. Sol. (a)* **41** (1977) 535.
232. P.D. Taylor and D.V. Morgan, *J. Phys. D: Appl. Phys.* **10** (1977) 73.
233. J.R. Saffell, A. Matthiesen, R. McIntyre and J.P. Ibar, *Thermochim. Acta* **192** (1991) 243.
234. C.T. Sah, W.W. Chan, H.S. Fu and J.W. Walker, *Appl. Phys. Lett.* **20** (1972) 193.
235. D.V. Lang, *J. Appl. Phys.* **45** (1974) 3023.
236. D.V. Lang, *Deep Centers in Semiconductors,* S.T. Pantelides, editor, Gordon and Breach, New-York (1986) 489.
237. S.W.S. McKeever, R.J. Walters, S.R. Messanger and G.P. Summers, *J. Appl. Phys.* **69** (1991) 1435.
238. W.R. Thurber, R.A. Forman and W.H. Phillips, *J. Appl. Phys.* **53** (1982) 7397.
239. B.G. Markey, S.W.S. McKeever and G. Cantwell, *Phys. Stat. Sol. (a)* **146** (1994) 735.
240. P. Besomi and B.W. Wessels, *J. Appl. Phys.* **53** (1982) 3076.
241. T. Endo, Y. Hirosaki, E. Uchida, H. Miyake and K. Sugiyama, *Japan. J. Appl. Phys.* **28** (1989) 1864.
242. O. Chretien, R. Apetz, L. Vescan, A. Souifi, H. Lüth, A. Schmalz and J.J. Koulman, *J. Appl. Phys.* **78** (1995) 5439.
243. P. Hacke, H. Nakayama, T. Detchprohm, K. Hiramatsu and N. Sawaki, *Appl. Phys. Lett.* **68** (1996) 1362.
244. J.M. Martin, J. García, I. Mártil, G. González-Díaz, E. Castán and S. Dueñas, *J. Appl. Phys.* **78** (1995) 5325.
245. S. Ghosh and V. Kumar, *Sol. St. Comm.* **98** (1996) 195.
246. Y. Okamoto, H. Yanekura, J. Morimoto and T. Miyakawa, *Rev. Sci. Inst.* **67** (1996) 809.
247. Ó. Alejos, C. De Francisco, P. Hernández, J. María Muñoz, J. Vincente and L. Enríquez, *Japan. J. Appl. Phys.* **35** (1996) 2196.
248. M. Benzohra, N. Medeghri-Rochedi and H. Hbib, *Japan. J. Appl. Phys.* **35** (1996) 2709.
249. E.A. Ingham, J.D. Scofield and M. Pachter, *J. Appl. Phys.* **80** (1996) 2805.
250. G. Liptay, *Thermochim. Acta* **150** (1989) 93.
251. R.C. Mackenzie, *Israel Journal Chem.* **22** (1982) 203.

252. F. Rudberg, *Ann. Phys. Chem.* **18** (1830) 240.

253. M.L. Frankenheim, *J. Prakt. Chem.* **16** (1839) 1.

254. J.B. Hannay, *J. Chem. Soc.* **32** (1877) 381.

255. H. Le Chatelier, *Compt. Rend.* **96** (1883) 1668.

256. H. Le Chatelier, *Bull. Soc. Fr. Minér. Cristallogr.* **10** (1887) 204.

257. P.D. Garn, *Thermoanalytical Methods of Investigation*, Academic Press, New-York (1965).

258. I. Einhorn, editor, *Thermal Analysis*, Polymer Conference Series, University of Utah (1970).

259. O. Menis, editor, *Status of Thermal Analysis*, National Bureau of Standards Special Publication, Washington D.C. **338** (1970).

260. T. Daniels, *Thermal Analysis*, Kogan Page, London (1973).

261. J. Šesták, V. Šatava and W.W. Wendlandt, Review in *Thermochim. Acta* **7** (1973) 333.

262. A. Blažek, *Thermal Analysis*, Van Norstrand-Reinhold, London (1974).

263. H. Kambe and P.D. Garn, editors, *Thermal Analysis: Comparative Studies on Materials,* Proc. of the U.S.-Japan Joint Seminar, Akron, Ohio; Kadansha, Wiley, Tokyo—New-York (1974).

264. W.W. Wendlandt, *Thermal Methods of Analysis,* 2nd edition, Wiley, New-York (1974).

265. W.W. Wendlandt, *Handbook of Commercial Scientific Instruments*, Vol. 2: *Thermoanalytical Techniques*, Marcel Dekker, New-York (1974).

266. C. Duval, *Thermal Methods in Analytical Chemistry*, Elsevier, Amsterdam (1976).

267. D.N. Todor, *Thermal Analysis of Minerals*, Abacus Press, Turnbridge Wells (1976).

268. W.W. Wendlandt and L.W. Collins, editors, *Benchmark Papers in Analytical Chemistry,* Vol. 2: *Thermal Analysis*, Dowden, Ross and Hutchinson, Stroudsbourg, Pennsylvania, USA (1976).

269. J. Šesták, Review in *Thermochim. Acta* **28** (1979) 197.

270. B. Wunderlich, *Thermal Analysis*, Academic Press (1990).

271. P.J. Haines, *Thermal Methods of Analysis: Principles, Applications and Problems*, Blakie Academic (1995).

272. C. Duval, *Inorganic Thermogravimetric Analysis,* 2nd edition, Elsevier, Amsterdam (1963).

273. P. Vallet, *Thermogravimetrie*, Gauthier-Villars, Paris (1972).

274. C.J. Keattch and D. Dollimore, *An Introduction to Thermogravimetry,* 2nd edition, Heyden, London (1975).

275. G. Lombardi, *For Better Thermal Analysis,* Inst. Mineral. e Petrogr., Città

Univ. 00100 Rome (1977; 2nd edition, 1980).

276. L. Erdey, F. Paulik and J. Paulik, *Nature* **175** (1954) 885.
277. F.W. Wilburn and J.R. Hesford, *J. Sci. Instr.* **40** (1963) 91.
278. F. Paulik, J. Paulik and L. Erdey, *Talanta* **13** (1966) 1405.
279. H.R. Oswald and E. Dubler, *Reviews on Analytical Chemistry,* Akadémiai Kiadó, Budapest (1976).
280. T. Székely, G. Várhegyi and F. Till, *J. Therm. Anal.* **5** (1973) 227.
281. A.W. Coats and J.P. Redfern, *J. Polym. Sci.* **3B** (1965) 917.
282. T. Ozawa, *Bull. Chem. Soc. Japan* **38** (1965) 1881.
283. E.P. Manche and B. Carroll, *Thermochim. Acta* **25** (1978) 77.
284. I. Agherghinei, *J. Therm. Anal.* **36** (1990) 473.
285. J. Zsakó, *J. Therm. Anal.* **2** (1970) 145.
286. V. Šatava, *J. Therm. Anal.* **5** (1973) 217.
287. D. Chen, X. Gao and D. Dollimore, *Thermochim. Acta* **215** (1993) 133.
288. A.A. Zuru, R. Whitehead and D.L. Griffiths, *Thermochim. Acta* **164** (1990) 285.
289. D. Dollimore, T.A. Evans, Y.F. Lee and F.W. Wilburn, *Thermochim Acta* **188** (1991) 77.
290. D. Chen, X. Gao and D. Dollimore, *Anal. Inst.* **20** (1992) 137.
291. R.C. Mackenzie, editor, *Differential Thermal Analysis,* Vol. 1: *Fundamental Aspects,* Academic Press, London (1970).
292. R.C. Mackenzie, editor, *Differential Thermal Analysis,* Vol. 2: *Applications,* Academic Press, London (1972).
293. W. Smykatz-Kloss, *Differential Thermal Analysis: Applications and Results in Mineralogy,* Springer Verlag, Berlin (1974).
294. M.I. Pope and M.D. Judd, *DTA: a Guide to the Technique and its Applications,* Heyden, London (1977).
295. P. Murray and J. White, *Trans. Brit. Ceram. Soc.* **54** (1955) 204.
296. M.W. Beck and M.E. Brown, *Thermochim. Acta* **164** (1990) 379.
297. C.C. Huang, T.S. Shien and A.L. Leu, *Thermochim. Acta* **188** (1991) 119.
298. C.C. Huang, and T.S. Wu, *Thermochim. Acta* **204** (1992) 239.
299. C.C. Ceidipor, R. Bucci and A.D. Magrí, *Thermochim. Acta* **199** (1992) 77.
300. H.J. Borchardt and F. Daniels, *J. Am. Chem. Soc.* **79** (1957) 41.
301. A.P. Gray, in *Analytical Calorimetry,* R.S. Porter and J.F. Johnson, editors, Plenum, New-York (1968).
302. H.E. Kissinger, *J. Nat. Bur. Standards* **57** (1956) 217.
303. H.E. Kissinger, *Anal. Chem.* **29** (1957) 1702.

304. R.L. Reed, L. Weber and B.S. Gottfried, *I. & EC Fundamentals* **4** (1965) 38.

305. R. Melling, F.W. Wilburn and R.M. McIntosh, *Anal. Chem.* **41** (1969) 1275.

306. J.H. Sharp, Ch. 28 in *Differential Thermal Analysis,* Vol. 2; R.C. Mackenzie, editor, Academic Press, London (1972).

307. M. Balarin, *Thermochim. Acta* **33** (1979) 341.

308. L.A. Kieffer and A.L. Bonivardi, *Thermochim. Acta* **171** (1990) 67.

309. M. Reading, D. Dollimore and R. Whitehead, *J. Therm. Anal.* **37** (1991) 2165.

310. J. Málek, *Thermochim. Acta* **200** (1992) 257.

311. A. Mianowski and T. Radko, *Thermochim. Acta* **204** (1992) 281.

312. J. Malék, J. Šesták, F. Rouquerol, J. Rouquerol, J.M. Criado and A. Ortega, *J. Therm. Anal.* **38** (1992) 71.

313. J.H. Flynn, *Thermochim. Acta* **203** (1992) 519.

314. J.S. Liu, X.C. Zeng, A.M. Tian and Y. Dang, *Thermochim. Acta* **231** (1994) 39.

315. D. Dollimore, *Anal. Chem.* **64** (1992) 147R.

316. D. Dollimore, and M. Reading, *Treatise on Analytical Chemistry,* Part 1, Vol. **13**, J.D. Winefordner, editor, Wiley & Sons (1993).

317. P.K. Gallagher, *Mater. Sci. Methods* **2A** (1992) 491.

318. A.N. Peregudov, T.V. Peregudova and L.G. Karpov, *Thermochim. Acta* **197** (1992) 21.

319. A.R. McGhie, *Thermochim. Acta* **234** (1994) 21.

320. Z. Zhong and P.K. Gallagher, *Thermochim. Acta* **186** (1991) 199.

321. D. Dollimore, *Anal. Chem.* **68** (1996) 63.

322. J.L. McNaughton and C.T. Mortimer, *IRS; Phys. Chem. Series 2* **10** (1975) 1.

323. D.D. Thornburg and R.I. Johnson, *J. Non-Crystalline Solids* **17** (1975) 2.

324. G. Chataing and J.M. Vergnaud, *Thermochim. Acta* **94** (1985) 379.

325. J.Y. Armand and J.M. Vergnaud, *Thermochim. Acta* **131** (1988) 15.

326. E. Kaiserberger and J. Opfermann, *Thermochim. Acta* **187** (1991) 151.

327. J.W. Graydon, S.J. Thorpe and D.W. Kirk, *Acta Metall. Mater.* **9** (1994) 3163.

328. B. Cassell and M.P. DiVito, *American Laboratory* **L** (1994) 1735.

329. P. Le Parlouër, *Thermochim. Acta* **121** (1987) 307.

330. M.E. Brown and A.K. Galwey, *Anal. Chem.* **61** (1989) 1136.

331. R. Hoffman and W.P. Pan, *Thermochim. Acta* **166** (1990) 251.

332. J. Mentel and H. Anderson, *Thermochim. Acta* **187** (1991) 121.

333. P.K. Gallagher, *Mikrochim. Acta* **II** (1991) 391.

334. J. Rouquerol, *Thermochim. Acta* **144** (1989) 209.

335. V.J. Griffin and P.G. Laye, *Spec. Publ. Roy. Soc. Chem.* **117** (1992) 17.

336. R.F. Speyer, *J. Mater. Sci.* **8** (1993) 675.

337. G. Wehr, K. Böning, J. Kalus, G. Vogl and G. Wenzel, *J. Phys. C: Sol. St. Phys.* **4** (1971) 324.

338. B.A. Cruz-Vidal, G. Wehr, K. Böning and G. Vogl, *J. Phys. Chem. Solids* **34** (1973) 1335.

339. B.A. Cruz-Vidal, K. Böning and G. Vogl, *J. Phys. Chem. Solids* **34** (1973) 1909.

340. L. Delgado and J.L. Alvarez-Rivas, *J. Phys. C: Solid St. Phys.* **12** (1979) 3159.

341. L. Delgado and J.L. Alvarez-Rivas, *J. Phys. C: Solid St. Phys.* **13** (1980) 1185.

342. M. Jiménez de Castro and J.L. Alvarez-Rivas, *J. Phys. C: Solid St. Phys.* **18** (1985) L1079.

343. A.C. Damask and G.J. Dienes, *Point Defects in Metals*, Gordon and Breach, New York (1964).

344. N. Itoh, B.S.H. Royce and R. Smoluchowski, *Phys. Rev.* **137** (1965) 1010.

345. A. Saidoh and N. Itoh, *J. Phys. Chem. Solids* **34** (1973) 1165.

346. B.T. Kelly, *Irradiation Damage to Solids*, Pergamon Press, Oxford (1966).

347. J.M. Meese, *Phys. Rev. B* **9** (1974) 4373.

348. D.E. Wilson and F.M. Hamaker, in *Thermal Analysis* Vol. 1, R.S. Schwenker and P.D. Garn, editors, Academic Press, New-York (1969).

349. H.L. Friedman, G.A. Griffith and H.W. Goldstein, in *Thermal Analysis* Vol. 1, R.S. Schwenker and P.D. Garn, editors, Academic Press, New-York (1969).

350. R. Sakamoto, T. Ozawa and M. Kanazashi, *Thermochim. Acta* **3** (1972) 291.

351. I.C. McNeill, in *Thermal Analysis* Vol. 1, R.S. Schwenker and P.D. Garn, editors, Academic Press, New-York (1969).

352. W. Lodding, editor, *Gas Effluent Analysis*, Marcel Dekker, New-York (1967).

353. J. Chiu and E.F. Palermo, *Anal. Chim. Acta* **81** (1976) 1.

354. R.N. Rogers, S.K. Yasuda and J. Zinn, *Anal. Chem.* **32** (1960) 672.

355. T. Ozawa, *J. Therm. Anal.* **5** (1973) 563.

356. S.E. Donnelly and D.C. Ingram, *Vacuum* **28** (1978) 69.

357. S.E. Donnelly and D.G. Armour, *Vacuum* **27** (1977) 21.

358. V. Balek, *Thermochim. Acta* **22** (1977) 1.

359. P.A. Redhead, *Vacuum* **12** (1962) 203.

360. T.W. Hickmott and G. Ehrlich, *J. Phys. Chem. Solids* **5** (1958) 47.

361. I.E. Wachs and R.J. Madix, *Appl. Surface Sci.* **1** (1978) 303.

362. R.V. Hill, E.K. Stefanakos and R.F. Tinker, *J. Appl. Phys.* **43** (1972) 1976.

363. T.T.A. Nguyen, *Surface Sci.* **43** (1974) 577.

364. F.P. Netzer, *Surface Sci.* **61** (1976) 343.

365. P.C. Gravelle, *J. Therm. Anal.* **14** (1978) 53.

366. D.A. King, *Surface Sci.* **47** (1975) 384.

367. E. Bornand, *Surface Sci.* **152/153** (1985) 314.

368. M. Vollmer and F. Träger, *Surface Sci.* **187** (1987) 445.

369. A.M. de Jong and J.W. Niemantsverdriet, *Surface Sci.* **223** (1990) 355.

370. V.J. García, J.M. Briceño-Valero and L. Martinez, *Surface Sci.* **339** (1995) 189.

371. G. Holzapfel, M. Krystek and L. Wolber, *2e Colloque Int. Phys. et Chim. Surface* **30A** (1975) 107.

372. D. Edwards Jr., *Surface Sci.* **49** (1975) 393.

373. J.C. Chiarena and E. Gillet, *J. Appl. Phys.* **48** (1977) 1712.

374. B. McCarroll, *J. Appl. Phys.* **40** (1969) 1.

375. J.L. Falconer and R,J. Madix, *Surface Sci.* **48** (1975) 393.

376. E. Bauer, F. Bonczek, H. Poppa and T. Todd, *Surface Sci.* **53** (1975) 87.

377. S. Zhou, G. Schulz-Ekhoff and D. Popovic, *J. Therm. Anal.* **41** (1994) 767.

378. E.G. Seebauer, *Surface Sci.* **316** (1994) 391.

379. F. Froitzheim, P. Schnek and G. Welder, *J. Vac. Sci. Technol.* **A11** (1993) 345.

380. H.J. Kreutzer, *Langmuir* **8** (1992) 774.

381. H.J. Kreutzer and S.H. Payne, in *Dynamics of Gas-Surface Interfaces*, C.T. Rettner and M.N.R. Ashfold, editors, The Royal Society of Chemistry (1991).

Chapter 9
SIMULTANEOUS MEASUREMENTS OF TL AND OTHER TSPS

9.1 Simultaneous TL and TSC Measurements

9.1.1 General Considerations

The basic equipment needed for measuring thermoluminescence and thermally stimulated conductivity consists of a vacuum cryostat in which the sample is held and provision is made for its heating and cooling, as described in detail by various authors (see e.g. Chen and Kirsh[1] and McKeever[2]; see also Ch. 11). In the TL measurements, a photomultiplier tube is used to measure the emitted light, whereas in TSC the sample has to have at least two electrical contacts, as described in Ch. 8, so that the conductivity of the sample can be monitored. Several investigators found it useful to have a set-up which permits the measurement of both TL and TSC. Since this kind of simultaneous measurement provides additional information, it has been performed by several groups on a wide variety of samples.

An important point to be remembered is that, at least with the prevalent two contact measurements, one measures the *current* through the sample. The translation of such results to the real conductivity is by no means simple due to the geometry of the sample and even more so, due to contact effects, mentioned in Ch. 8. The latter effects may, in some pathological instances, prevent the evaluation of the conductivity to any desirable accuracy. In this section, however, we assume that the conductivity $\sigma(T)$ can be evaluated to a reasonable accuracy from the TSC measurements. Moreover, we will usually assume that for the particular material under study, the mobility μ is known either as a constant or, which is more common, as a temperature dependent function $\mu(T)$; this we assume, can be determined separately, e.g. by Hall effect or by time-of-flight measurements.

In Ch. 8 we saw that some early investigators assumed that $I = n_c/\tau$ where I is the TL intensity, n_c is the concentration of free carriers (electrons in the conduction band or holes in the valence band), and τ is a constant free carrier lifetime. Under this assumption, one should expect the curves of $n_c(T)$

and $I(T)$ to look alike and, therefore, that the methods of analysis developed for TL peaks can be applied to TSC peaks. As already noted, however, this assumption is incorrect in most cases. It is quite obvious that if the assumption that τ is constant is correct, a simultaneous measurement would hardly provide any additional information. (Except, perhaps, that the recombination process occurs via transport of free charge carriers via a delocalized band.) In fact, had the TL measurements yielded TL results in absolute units (which is usually not the case), simultaneous measurements could give useful information about $\mu(T)$ when τ is constant.

As will be elaborated here, assuming $I = n_c/\tau$ with constant τ is an oversimplification, and interesting information can be extracted from the simultaneous measurement of TL and conductivity even in the common case where the TL results are on an arbitrary units scale. Another point that will be made is that one can usually expect, on theoretical grounds, a shift between a TL peak and its conductivity counterpart. Such a shift, which very frequently occurs experimentally (see examples below), is mainly related to the fact that the above-mentioned lifetime is not constant along the temperature range of the measurement. Since we prefer to talk in terms of the recombination probability A_{mn} rather than the lifetime τ we will start the discussion assuming that both the recombination probability A_m and the mobility μ of free carriers are temperature independent. The effect of their possible temperature dependence will be introduced later.

Bräunlich and Kelly[3] solved the relevant set of simultaneous differential equations using the assumption of quasiequilibrium, and came to the limited conclusion that from simultaneous measurements of TL and TSC, one can only decide: (1) whether there is charge carrier transport associated with the thermal release of trapped charge carriers and their subsequent recombination, and (2) whether there is a large concentration of the thermally disconnected traps. However, we will show how quantitative values of the recombination probability can be evaluated from such simultaneous measurements, and also how the cross-section for recombination can be found. Furthermore, a method will be described for evaluating the density of disconnected traps. In a later review article, Bräunlich et al.[4], referring to previous work by Fillard et al.[5], expressed more expansive views on the possibility of extracting valuable information from TL-TSC simultaneous measurements and mentioned how the temperature dependence of the cross-section for radiative transitions of free electrons in recombination centers can be evaluated.

We will also show how additional inferences concerning the validity of the quasiequilibrium assumption can also be made from such simultaneous measurements. Opanowicz and Przybyszewski[6] presented the results of numerical calculations of TL and TSC without making the quasiequilibrium (QE) assumption. They showed basically good agreement with calculations made with the QE assumption, thus indicating that, at least with the sets of parameters chosen, the assumption is valid. They also show that an almost first-order shape for both TL and TSC can be obtained for strong retrapping.

9.1.2 The Method for Evaluating the Recombination Probability

Let us reconsider Eq. 2.20, written in a very slightly different manner, where A_{mn} is now denoted, for the sake of brevity, by A, thus

$$I = -dm/dt = Amn_c \tag{9.1}$$

which, as discussed in Ch. 2 and elsewhere, is one of the three simultaneous differential equations, governing the behavior of a single TL peak. One of the functions in these equations is $n_c(t)$, the temperature dependent concentration of free electrons. As discussed above, $n_c(t)$ is determined, at least in principle, as part of the solution of the simultaneous equations; however, at present we will consider its role from a different point of view. As mentioned already, $I(t)$ and $n_c(t)$ given as functions of time can also be considered as functions of temperature $I(T)$ and $n_c(T)$ when the heating function $T = T(t)$ is known as a given monotonically increasing function.

Equation 9.1 merely states that the rate of recombination is proportional to both the concentrations of free electrons n_c and of holes in centers m, and that the TL intensity is equal (when measured in suitable units) to this rate. The factor A appears here as a proportionality constant with dimensions of m^3s^{-1}, and can be interpreted to be the product of the cross-section for recombination and the thermal velocity of the electrons. Of course, the same will hold true if the mobile carriers are holes, first trapped in trapping states during the excitation stage, then thermally excited into the valence band, where their concentration, $m_v(t)$, is analogous to $n_c(t)$; they then finally recombine with electrons trapped in recombination centers, the concentration of which is analogous to $m(t)$. For the sake of simplicity, we will continue the discussion in terms of mobile electrons recombining with holes in recombination centers. In fact, Eq. 9.1 has much more general significance[7] than merely describing the situation (or actually, part of the situation - being one equation out of three) where only one trapping state and one kind of recombination center are involved. As long as only transitions into one recombination center are *measured* in the course of heating in a certain temperature range, the equation holds even if there are simultaneous transitions into other centers as well. Such transitions may change the concentration of electrons in the conduction band, but Eq. 9.1 holds true for the *net* concentration n_c. Moreover, no information regarding the trap or traps contributing to n_c is needed as long as $n_c = n_c(t)$ can be evaluated separately, e.g. by conductivity.

A further examination of Eq. 9.1 shows immediately that the TL peak should normally appear at a lower temperature than its TSC counterpart. Assuming the recombination probability A to be temperature independent, one has $I \propto mn_c$ where $n_c(T)$ is a peak shaped function (exactly representing the TSC peak if the mobility μ is also temperature independent), and $m(T)$ is a decreasing function (since, during the heating stage, recombinations reduce the concentration of charges in centers). Unless $m(T)$ decreases extremely quickly, the multiplication of a function describing a peak by a function which is decreasing should give a peak shaped curve (i.e. $I(T)$) with a maximum at a lower temperature than that of the original peak (i.e. $n_c(T)$). This will be shown below to be the case more rigorously, and under more general conditions.

A simple way for evaluating A is by rewriting 9.1 as

$$A = I(T)/\left[m(T)n_c(T)\right] \tag{9.2}$$

which should hold true for any temperature T along the glow curve. By integrating the basic equation $I = -dm/dt$ we have

$$m - m_\infty = \int_t^\infty I(\tau)d\tau. \tag{9.3}$$

If one deals with only one peak, one has $m_\infty = 0$ and therefore, $m(t) = \int_t^\infty I(\tau)d\tau$ or, when the "usual " linear heating function is used,

$$m(T) = (1/\beta)\int_T^\infty I(\theta)d\theta \tag{9.4}$$

which can be evaluated from the area under the curve starting from a certain temperature T. This value can be inserted in Eq. 9.2, along with the measured $I(T)$ and $n_c(T)$ for the arbitrarily chosen T, and the value of A can be found. The reader is referred to Fig. 3.15 for the meaning of the integral in Eq. 9.4. T here corresponds to T_i in Fig 3.15 whereas the integration, as shown in the figure, goes to T_f - an arbitrarily chosen temperature in which the glow intensity is low enough - rather than to infinity. It is to be noted that one does not have to measure $I(T)$ in absolute units, since any proportionality constant appearing in $I(T)$ will be included in $m(T)$ as well, and should cancel out in the quotient in Eq. 9.2. Thus, the "arbitrary units" in which the light intensity is usually given are sufficient for this application. The obvious disadvantage of this method is that one has to assume that the peak is the only one in the glow curve or, at least, the one appearing at the highest temperature out of a series related to several trapping states and a single recombination center. In order to overcome this difficulty, Chen[7] developed the following method.

Assuming that $n_c(t)$ (or $n_c(T)$ with a given monotonically increasing heating function) is known, say, by conductivity measurements, Eq. 9.1 can be solved to yield

$$m = m_0 \exp\left\{-A\int_0^t n_c(\tau)d\tau\right\} \tag{9.5}$$

where m_0 is the initial concentration of holes in centers. On inserting Eq. 9.5 into Eq. 9.1 we get

$$I(t) = Am_0 n_c(t) \exp\left\{-A\int_0^t n_c(\tau)d\tau\right\}. \tag{9.6}$$

Equations 9.5 and 9.6 are valid for any heating function $T = T(t)$ including, of course, the linear function $T = T_0 + \beta t$ as well as the case of phosphorescence in which T is a constant. Writing Eq. 9.6 for two arbitrary points t_1 and t_2, and dividing one by the other, we get

$$I(t_2)/I(t_1) = [n_c(t_2)/n_c(t_1)] \exp\left\{-A\int_{t_1}^{t_2} n_c(t)dt\right\}. \tag{9.7}$$

For the linear heating function, this can be written as

$$I(T_2)/I(T_1) = [n_c(T_2)/n_c(T_1)] \exp\left\{(-A/\beta)\int_{T_1}^{T_2} n_c(T)dT\right\}. \tag{9.8}$$

All the quantities in Eq. 9.7 (or Eq. 9.8), except for A, are measurable and, thus, A can be evaluated from the equation.

For the determination of $n_c(T)$, we have to measure the conductivity curve $\sigma(T)$; this can be evaluated to a good accuracy in those cases where contact effects are negligible, from a measurement of the current through the sample, or by a four probe measurement as mentioned in section 8.1. The conductivity $\sigma(T)$ is given by $\sigma(T) = e\mu(T)n_c(T)$ where e is the electronic charge and μ the mobility of the charge carriers. Even in cases where μ is not constant, $n_c(T)$ is easily evaluated from $\sigma(T)$ provided $\mu = \mu(T)$ is known independently. The obvious advantage of using Eq. 9.7 rather than Eq. 9.2 for evaluating A is that the integration is performed here (on $n_c(T)$ rather than on $I(T)$, which does not add any additional difficulty) in a limited range, from T_1 to T_2 rather than to infinity. The complication due to the possible occurrence of peaks at higher temperatures is thus removed.

The discussion from this point on will alternately include the use of time and temperature functions. It is to be noted again, however, that as long as one

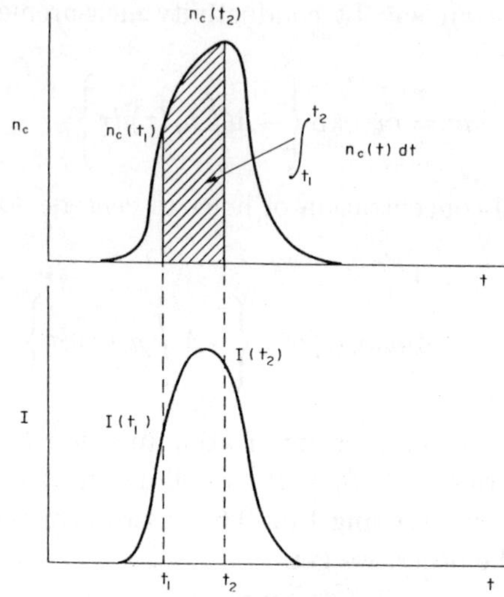

Figure 9.1 Quantities measured to find the recombination probability (schematic). (After Chen[7].)

has a known monotonically increasing function $T = T(t)$, the transition from one to the other is trivial and is immediate for the prevalent constant heating rate case.

Figure 9.1 shows schematically a peak of $n_c = n_c(t)$, the corresponding $I(t_1)$, $I(t_2)$, $n_c(t_1)$, $n_c(t_2)$ and $\int_{t_1}^{t_2} n_c(\tau)d\tau$, which are required for evaluating the recombination probability A. Similarly to the more primitive method which involves integration to infinity, only relative values of $I(t)$ are needed here as well, whereas absolute values of $n_c(t)$ are necessary (for the integral). A slight simplification can be achieved by choosing t_1 and t_2 so that $I(t_1) = I(t_2)$, or such that $n_c(t_1) = n_c(t_2)$.

9.1.3 *Examples: Insulating and Semiconducting Diamonds*

Chen[7] and Chen and Kirsh[1] reported the evaluation of the recombination probability of a TL peak at $\sim$ 350 K in a type IIa insulating diamond and for a TL peak at $\sim$ 250 K in a type IIb, semiconducting diamond. These are to be given here as an example of the method just described. Figure 9.2 shows the results of the measurement of the TL peak in the insulating diamond (curve a), and its $n_c(T)$ counterpart, evaluated from the TSC peak (curve b). The shift here is, indeed, as predicted, namely, the TL peak precedes the $n_c(T)$ peak. The values $n_c(T_1)$, $n_c(T_2)$, $I(T_1)$ and $I(T_2)$ are shown, as well as the shaded area

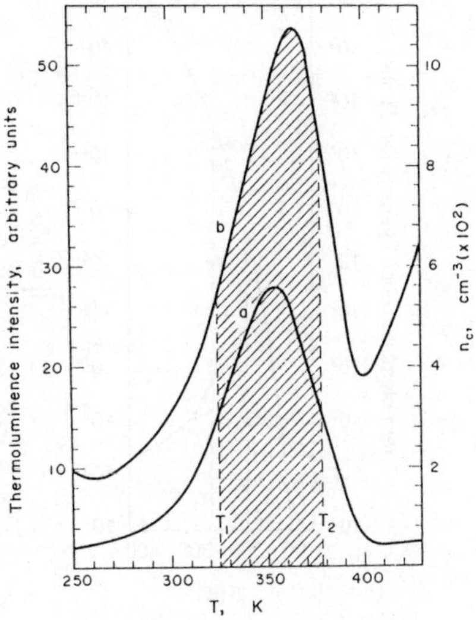

Figure 9.2 TL peak (a) and the accompanying $n_c(T)$ peak (b) as measured in a type IIa, insulating diamond.

(After Chen and Kirsh[1].)

which represents $\int_{T_1}^{T_2} n_c(T)dT$. Taking into account the constant heating rate which was 0.3 Ks^{-1}, and assuming the mobility to be $\sim$ 0.2 m^2/Vs, as known for diamonds at room temperature, one gets $A \cong 10^{-11}$ m^3s^{-1}. Uncertainties, mainly due to contact effects and to the geometry of the sample (see section 8.1), as well as the possible effect of the temperature dependent mobility, cause inaccuracies which are estimated to be up to a factor of 3. The cross-section for recombination S_m is found, using $A = vS_m$ where v, the thermal velocity, is given by $v = \sqrt{3kT/m^*}$ where m^* is the effective mass of the free carriers. A value of $\sim 10^{-16}$ m^2 was found for the cross-section, to the same accuracy of a factor of about 3.

The same method was used for semiconducting (type IIb) diamonds. These diamonds are extrinsic semiconductors, and their conductivity changes by more than ten orders of magnitude between 100 K and 400 K, as reflected by the dependence of the free carrier concentration shown in curve (b) of Fig. 9.3. This conductivity is known to be due to holes released from a level 0.35 eV above the valence band, whereas the band gap of diamonds (most of which are insulators) is 5.5 eV. The conductivity is not a TSC in the conventional sense since it shows up even without any prior excitation and, in fact, no peak occurs, but rather, the conductivity increases monotonically and very quickly. Following UV

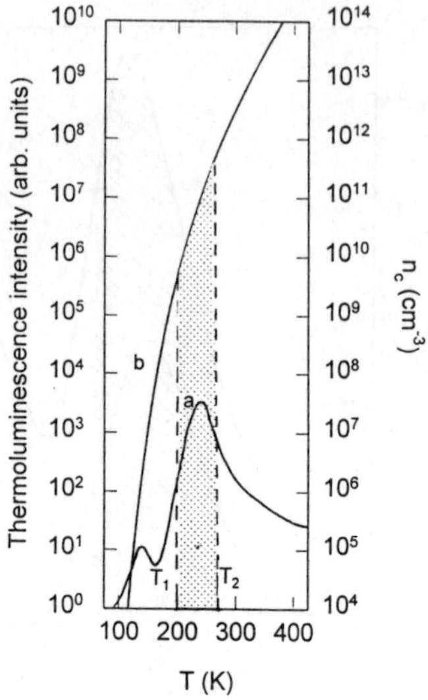

Figure 9.3 TL peak (a) and the $n_c(T)$ curve (b) as measured in a type IIb, semiconducting diamond.

(Redrawn from Chen and Kirsh[1].)

irradiation, a TL peak appears at ~250 K[8]; this peak has the same activation energy of 0.35 eV as the conductivity. It has, therefore, been concluded that the UV excitation fills a recombination center (from the emission spectrum it is determined to be at about the middle of the forbidden gap), with electrons; during heating, free holes recombine into the centers. As opposed to the usual concept of a TL peak which is related to the exhaustion of carriers in traps, this peak is due to the exhaustion of carriers (electrons in the present case) in centers. This possibility has been mentioned a number of times in the preceding chapters. Figure 9.3 depicts the results of a simultaneous measurement of the TL peak (curve a) and $n_c(T)$ (curve b) as calculated from the conductivity. T_1 and T_2 have been chosen arbitrarily; $I(T_1)$, $I(T_2)$, $n_c(T_1)$ and $n_c(T_2)$ are seen, as well as $\int_{T_1}^{T_2} n_c(T)dT$ which is the shaded area. The recombination probability A is evaluated to be $\sim 5 \times 10^{-19}$ m^{3}s^{-1} (again accurate to a factor of ~3 due to the same reasons as mentioned above) and the cross-section for recombination is $\sim 5 \times 10^{-24}$ m^2.

The difference of over seven orders of magnitude between the recombination probabilities of the insulating and semiconducting diamonds may appear somewhat strange for two kinds of diamond which are quite similar, apart from their conductivity properties. However, this may have been expected in the first place, since, for approximately the same dose of excitation, one gets roughly the same TL intensity whereas the conductivity is about eight orders of magnitude higher in the semiconducting diamond, in the relevant temperature range. Moreover, one can argue that both kinds of centers exist in both the IIa and IIb samples, which also have some other similar optical properties distinguishing them from the variety of type I diamonds. The center having the low cross-section, does not contribute to TL in the insulator due to the small number of available free carriers. On the other hand, the high cross-section center does not contribute to the TL in the semiconducting diamond since, due to the very large concentration of free carriers even at low temperature, the recombination may occur during and immediately after excitation, contributing to the fast decaying phosphorescence only.

9.1.4 The Shift between a TL Peak and the Accompanying TSC Peak

We noted above that a TL peak is expected to occur at a lower temperature than its TSC counterpart. The experimental evidence on this point will be discussed in a section to follow, and, in fact, the shift between the peaks may depend on the temperature dependencies of the recombination probability A and the carrier mobility μ. For the time being, we will assume that the former is temperature independent and in order to by-pass the possible temperature dependence of μ, we will limit the discussion here, as we have done so far, to the relation between the TL intensity $I(T)$ and the associated peak of $n_c(T)$. We are going to consider now in more detail the conditions under which the shift occurs, as given by Chen[7].

Consider again Eq. 9.6 which, for a given peak shaped $n_c(T)$ and a "reasonable" heating rate, should describe a peak shaped curve $I(t)$ (or $I(T)$). Setting the derivative of this expression to zero (or, for mathematical simplicity, one can set the derivative of the logarithm of the expression to zero), one gets

$$(dn_c/dt)_{\max} = A\left[n_c(t_{\max})\right]^2 \tag{9.9}$$

where the subscript *max* indicates the values taken at the maximum of the *thermoluminescence* peak. The right hand side of Eq. 9.9 is necessarily positive, and therefore, the left hand side should also be positive. This means that when the TL peak reaches the maximum point, the curve of $n_c(t)$ is still increasing, or, in other words, that $n_c(t)$ maximizes later (i.e. at a higher temperature) than $I(t)$. Note that Fields and Moran[9] reached the same conclusion regarding the order of occurrence of a TL and TSC peak when a single recombination center is involved, but by a slightly different way, and Böhm et al.[10] arrived at the same realization in yet another way.

An interesting extension to this proof has been given by Chen and Kirsh[1], namely, that even if several kinds of recombination centers are involved (in addition to the possibility mentioned above of several trapping states), the occurrence of the TL peak at a lower temperature than the conductivity one, should prevail. Let us denote the concentration of holes in centers 1, 2, ..., k by $m_1, m_2, ..., m_k$ and the respective recombination probabilities by A_1, A_2, ..., A_k. We now obtain

$$I_i(t) = -dm_i/dt = A_i m_i n_c \text{ for } i = 1, 2, ..., k \tag{9.10}$$

while the *same* n_c appears in all the equations. Each equation can now be solved independently, and the total measured emission should be the sum over i of the partial intensities $I_i(t)$. Since the detector may have different sensitivities for the various values of i (corresponding to different wavelengths), a weighted sum is taken, namely, each I_i is multiplied by an appropriate (positive) constant α_i and the total measured intensity is then

$$I(t) = \sum_{i=1}^{k} \alpha_i I_i(t) = \sum_{i=1}^{k} \alpha_i A_i m_{i0} n_c(t) \exp\left\{-A_i \int_0^t n_c(\tau)d\tau\right\}. \tag{9.11}$$

By setting the derivative of Eq. 9.11 to zero, one gets

$$\alpha(dn_c/dt)_{\max} = \gamma\left[n_c(t_{\max})\right]^2 \tag{9.12}$$

where

$$\alpha = \sum_{i=1}^{k} \alpha_i A_i m_{i0} \exp\left\{-A_i \int_0^{t_{\max}} n_c(\tau)d\tau\right\} \tag{9.13}$$

and

$$\gamma = \sum_{i=1}^{k} \alpha_i A_i^2 m_{i0} \exp\left\{-A_i \int_0^{t_{\max}} n_c(\tau)d\tau\right\}, \tag{9.14}$$

where m_{i0} is the initial filling of center i. Being the sums of positive factors, α and γ are always positive. The condition is now similar to that of Eq. 9.9 and bears similar implications, namely, that even under these general conditions, the TL peak should occur at a lower temperature than the corresponding peak for $n_c(T)$.

In a relatively small number of TL-TSC simultaneous measurements, the opposite order of appearance occurs, namely, the TSC peak preceeds the TL peak. This may be just sheer coincidence, i.e., that the TL and TSC peaks originate from different sources. For example, one is related to the release of trapped electrons and the other of trapped holes and the two peaks just happen to occur in the same temperature range. Another possibility is that the TSC peak is distorted due to contact or space charge effects as discussed by Henisch[11] and others. A third possibility considered by Chen and Fleming[12], takes into account the possible influence of the temperature dependence of the recombination probability A and the mobility μ. As noted, the recombination probability A is the product of the cross-section for recombination and the thermal velocity. According to Bemski[13], Lax[14] and Keating[15], the former varies like $T^{-\kappa}$ with $0 \leq \kappa \leq 4$; the thermal velocity is, obviously proportional to $T^{1/2}$. We can, therefore, write

$$A = BT^a; \ -7/2 \leq a \leq 1/2 \tag{9.15}$$

where B is a constant. Equation 9.9 will be written in the general case of $A = A(T)$ (and, through the given heating function $T = T(t)$, we also have $A = A(t)$), as

$$(dn_c/dt)_{\max} = A_{\max} n^2_{c\max} - n_{c\max} (d \ln A/dt)_{\max} \tag{9.16}$$

where the subscript max designates, again, values corresponding to the TL maximum. $A_{\max}$ in this context is the value of the recombination probability at the TL maximum. Equation 9.16 obviously reduces to Eq. 9.9 for a temperature-independent recombination probability. For any monotonically increasing heating function $T = T(t)$, Eq. 9.16 can also be written as

$$(dn_c/dT)_{\max} = A_{\max} n^2_{c\max}/\beta_{\max} - n_{c\max} (d \ln A/dT)_{\max} \tag{9.17}$$

where $\beta_{\max} = (dT/dt)_{\max}$ is the instantaneous heating rate at the TL maximum point. In order to determine whether the TL peak precedes the peak of $n_c(T)$ or not, it is necessary to examine the right-hand side of Eq. 9.17. If this expression is negative, then dn_c/dT is negative when the TL reaches its maximum value, and therefore, the $n_c(T)$ peak must occur at a lower temperature than that for TL. When $A = BT^a$ (where B is a constant), the right-hand side of Eq. 9.17 will be negative if

$$A_{\max} T_{\max} N_{c\max} < a\beta_{\max}. \tag{9.18}$$

A necessary condition for this inequality to hold is $a > 0$. Chen and Fleming[12] gave a hypothetical, though possible, example in which the recombination cross-section is temperature independent, and therefore, $a = \frac{1}{2}$. Choosing the cross-section to be $S_m = 10^{-22}$ m^2, $T_{\max} = 300$ K and $\beta_{\max} = 0.5$ Ks^{-1}, any $n_{c\max} < 7 \times 10^{13}$ m^{-3} would yield $(dn_c/dt)_{\max} < 0$, implying the occurrence of the maximum of the $n_c(T)$ peak at a temperature lower than that of the corresponding TL peak.

In the discussion so far, we have correlated TL and $n_c(T)$ peaks. The latter can be evaluated from conductivity measurements provided that the dependence of the mobility on temperature, $\mu = \mu(T)$ is separately known, and $n_c(T) = \sigma(T)/[e\mu(T)]$ is calculated. If one wishes to compare peak temperatures of TL and TSC directly, the possible dependence of the mobility on temperature should be included explicitly. According to Lax[14], the mobility of free carriers is given by

$$\mu = \mu_o T^b \tag{9.19}$$

where μ_o is a constant and b assumes values such as $-3/2$. Such a dependence of $\mu \propto T^{-3/2}$ was found, for example, by Benedict and Shockley[16] for electrons in germanium using contactless microwave measurements. Combining Eq. 9.1 and $\sigma = e\mu n_c$, we get

$$I(t) = -dm/dt = (A/\mu e)\sigma m = \bar{A}\sigma m \tag{9.20}$$

where $\bar{A} = A/\mu e$. Using Eqs. 9.15 and 9.19 we find

$$\bar{A} = A/\mu e = (B/\mu_o e)T^c \tag{9.21}$$

where $c = a - b$, and where $B/\mu_o e$ is a constant. One now has $-2 \le c \le 2$ for the previously mentioned temperature dependence of the recombination probability, with $b = -3/2$ being quite common. Because of the close similarity between Eqs. 9.1 and 9.20, the conclusion drawn from the former can be carried over to the latter. Thus, the equation analogous to Eq. 9.16 is

$$(d\sigma/dT)_{\max} = \bar{A}_{\max}\sigma^2_{\max}/\beta_{\max} - \sigma_{\max}(d\ln\bar{A}/dT)_{\max} \tag{9.22}$$

where $\bar{A}_{\max}$ is the value of $\bar{A}$ at the TL maximum and $\sigma_{\max}$ the conductivity at the same maximum. The order of occurrence of the TL and TSC peaks is governed by the sign of the right-hand side of Eq. 9.22. If the expression is negative, and assuming that $\bar{A}$ is given by Eq. 9.21, the condition for the TL peak to occur at a higher temperature than the $\sigma(T)$ counterpart is analogous to Eq. 9.18, namely

$$\bar{A}_{\max}T_{\max}\sigma_{\max} < c\beta_{\max}. \tag{9.23}$$

This is a somewhat less stringent condition than inequality 9.18 since the range of possible positive values of c is broader (the necessary condition now being $c > 0$).

To summarize this discussion, we should note that, usually, a TL peak and its corresponding $n_c(T)$ peak do not occur at exactly the same temperature. The normal case is that the TL peak appears at a lower temperature, which is substantiated by theoretical considerations and by experimental results (see below). The situation is quite similar if one compares the $I(T)$ and $\sigma(T)$ curves although the inclusion of the temperature dependent mobility shifts $\sigma(T)$ to a slightly lower temperature than that of $n_c(T)$.

A different explanation for the inversion of the order of appearance of TL and TSC peaks has been given by Oczkowski[17] for the case of semiconductors. Having in mind mainly materials like Zn-annealed ZnSe crystals, Oczkowski used a model with a deep acceptor (DA), deep donor (D), shallow donor (d) and a complex defect which is a metastable deep donor-acceptor (DA) pair. With the DA model, Oczkowski[17] showed that for such semiconductors and under the mentioned circumstances, the inversion of the TL and TSC peaks temperatures came out as a direct result of the trapping and compensation of excess minority carriers. It should be mentioned, however, that Opanowicz[18] has later challenged the Oczkowski model, and suggested that the TL and TSC in conductive ZnSe should be explained on the basis of the kinetic model with a deep electron trap, developed for a semi-insulating material. Work by Lee et al.[19] on the simultaneous measurements of TL and TSC peaks in ZnSe, showed that most of the peaks occur in the "correct" order, namely, the TL peak precedes the TSC peak. This was also found to be the case in ZnSe by McKeever et al.[20].

9.1.5 Information about Models from Simultaneous TL-TSC Measurements

McKeever et al.[20] also discuss how a simultaneous measurement of TL and TSC can be used to determine how close the system is to quasiequilibrium. We saw in Ch.2 that the closeness of the system to quasiequilibrium can be gauged from a consideration of the parameter $q(T)$, which was defined in Eq. 2.98 according to

$$-\frac{dn_{c(T)}}{dt} = q\left(T\right)\frac{n_c\left(T\right)}{\tau\left(T\right)}. \tag{9.24}$$

With a linear heating scheme this rearranges to become

$$q\left(T\right) = \left(\frac{\beta}{K}\right)\frac{d\sigma\left(T\right)}{dT}\frac{1}{I\left(T\right)} \tag{9.25}$$

where K is a constant encompassing the applied electric field, the sample area, the electronic charge, and a (constant) electron mobility μ. Equation 9.25 uses the fact that $n_c\left(T\right)/\tau\left(T\right) = dm\left(T\right)/dt = I\left(T\right)$ (assuming a luminescence efficiency of 1 - i.e. no non-radiative recombination pathways), and $n_c\left(T\right) \propto \sigma\left(T\right)$ (the TSC signal). Thus, Eq. 9.25 suggests that q may be evaluated from a simultaneous measurement of TL and TSC. Alternatively, one can rewrite Eq. 9.24 entirely in terms of TSC alone, namely

$$q\left(T\right) = \left(\beta\tau\left(T\right)\right)\frac{d\sigma\left(T\right)}{dT}\frac{1}{\sigma\left(T\right)} \tag{9.26}$$

where $\tau\left(T\right)$ is the free carrier lifetime. Thus, $q(T)$ can be determined from a simultaneous measurement of TL and TSC (for a luminescence efficiency of 1), or from a measurement of TSC alone (if the function $\tau\left(T\right)$ is known)[20],[21]. By measuring TL and TSC in semi-insulating ZnSe McKeever et al.[20] compared $q(T)$ as evaluated using Eqs. 9.25 and 9.26. They noted that strong TSC was observed at temperatures greater than ~ 210 K, but no TL was observed over this region, suggesting that non-radiative processes were occurring and that the luminescence efficiency $\neq 1$ in this system. They therefore preferred Eq. 9.26 over Eq. 9.25, using the $\tau(T)$ function obtained from photoconductivity measurements and assuming that μ is a constant in the temperature interval of interest. The TL and TSC data used by McKeever et al.[20] are shown in Fig. 9.4(a), while the $q\,(T)$-functions evaluated using Eq. 9.25 (curve q_9 in the figure) and Eq. 9.26 (curve q_{10} in the figure) are shown in Fig. 9.4(b). The zero value in the figure ($q(T) = 0$) indicates true quasi-equilibrium. However, no vertical scale is shown in Fig. 9.4(b) since $\tau(T)$ was not known in an absolute sense (only the shape of the function was known) and thus the extent of the departure from quasiequilibrium cannot be quantified. It is not possible to evaluate $q(T)$ at all temperatures since at very low T, or at high T, where no TSC or TL signals are observed, one only measures system noise. Furthermore, the exercise emphasizes that $q\,(T)$ can only be quantified if the system parameters and the recombination lifetime $\tau\,(T)$ can be evaluated separately.

Some further investigations on the correlation between TL and TSC have been conducted by Gasiot and Fillard[22] who measured the two simultaneously and evaluated

$$R(T) = \sigma(T)/I(T). \tag{9.27}$$

Since $I(T) = Amn_c$ and $\sigma(T) = e\mu n_c$, obviously

$$R(T) = (e\mu/A)(1/m) \tag{9.28}$$

and therefore, assuming $e\mu/A$ to be temperature independent,

$$I = -dm/dt = -(e\mu/A)\frac{d}{dt}\left(\frac{1}{R}\right). \tag{9.29}$$

Gasiot and Fillard compared the measured $I(T)$ and the calculated $d(1/R)/dt$ expecting them to be proportional to one another. Surprisingly, this did not happen in their samples. Their conclusion was that something should be wrong in the assumption $I \propto -dm/dt$ in these cases, however, the validity of the assumption that $e\mu/A$ is temperature independent should have been considered. Similar results of simultaneous measurements were given by Fillard and Gasiot[23], Jiménez-López[24] and Fillard et al.[25]. The main reason suggested by these authors for the discrepancy is that during the occurrence of the TL and TSC peaks, opposite sign carriers may channel into the recombination center involved. This possible element has not been included in the previous conditions while assuming that m decreased only by recombination with free carriers, which contributes to TL. Fillard et al.[25] also gave other possible reasons that may make the TL intensity $I(T)$ different from $-dm/dt$. These include changes

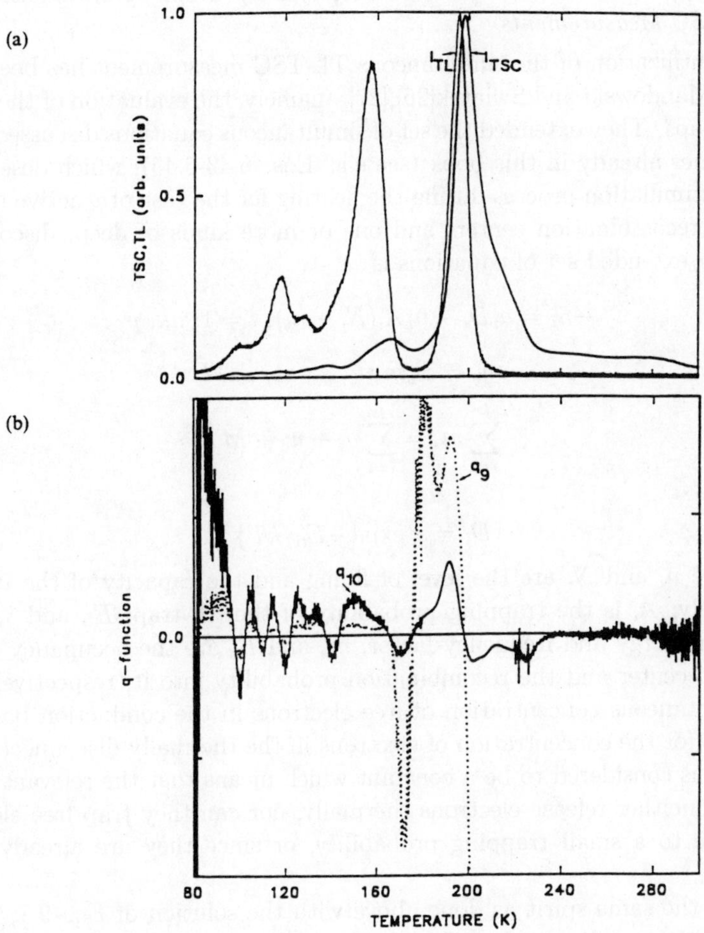

Figure 9.4 (a) TSC and TL curves from semi-insulating ZnSe. The peaks are believed to be due to electron detrapping. (b) Calculated curve shapes for $q(T)$ using two different methods of evaluation, as described in the text.
(After McKeever et al.[20].)

in the radiative yield with temperature (in ZnS), hole emission to the valence band (in ZnSe) and thermally stimulated ionic creation of recombination centers (in AgBr).

9.1.6 *The Determination of the Density of Deep Traps by Simultaneous TL-TSC Measurements*

Another utilization of the simultaneous TL-TSC measurement has been developed by Mandowski and Świątek[26],[27] - namely, the evaluation of the density of deep traps. They extended the set of simultaneous equations discussed a number of times already in this book (see e.g. Eqs. 6.43-6.45), which describe the thermal stimulation process during the heating for the case of p active trapping states, k recombination centers and one or more kinds of deep, disconnected trap. The extended set of equations is

$$-\dot{n}_i = n_i D_i - n_c A_i (N_i - n_i),\ i = 1,\ ....,\ p \tag{9.30}$$

$$-\dot{m}_s = B_s m_s n_c,\ s = 1,\ ....,\ k \tag{9.31}$$

$$\sum_{s=1}^{k} m_s = \sum_{i=1}^{p} n_i + n_c + M \tag{9.32}$$

where

$$D_i = s_i \exp\{-E_{ti}/kT\} \tag{9.33}$$

and where n_i and N_i are the level of filling and the capacity of the ith trap, respectively; A_i is the trapping probability of the ith trap; E_{ti} and s_i are its activation energy and frequency factor, m_s and B_s are the occupancy of holes in the sth center and the recombination probability into it, respectively, n_c is the instantaneous concentration of free electrons in the conduction band, and M stands for the concentration of electrons in the thermally disconnected deep traps. M is considered to be a constant which means that the relevant trap or traps can neither release electrons thermally, nor can they trap free electrons, either due to a small trapping probability, or since they are already full to capacity.

In the same spirit as done above with the solution of Eq. 9.1, i.e. addressing $n_c(t)$ as a known function of time (or temperature), which yielded Eq. 9.5, Mandowski and Świątek[26], [27] solved each of the equations in expression 9.30 for $i = 1,\ ....,\ p$ and found

$$n_i(t) = N_i \exp\left\{-\int_0^t [D_i(\tau) + n_c(\tau)A_i]\, d\tau\right\} \times$$

$$\left[\eta_{i0} + \int_0^t n_c(\tau) A_i \exp\left\{\int_0^\tau (D_i(\tau') + n_c(\tau')A_i)\, d\tau'\right\} d\tau\right], \tag{9.34}$$

where $\eta_{i0} = n_{i0}/N_i$, the relative initial filling of the ith trap and which, of course, may assume values between 0 and 1. Equation 9.5 is simply extended to be k solutions of Eqs. 9.31 as follows

$$m_s = m_{s0} \exp\left\{-B_s \int_0^t n_c(\tau) d\tau\right\}, \; s = 1, ..., k \tag{9.35}$$

where m_{s0} is the initial occupancy of the sth center. Mandowski and Świątek[27] defined the function

$$C(t) = \int_0^t n_c(\tau) d\tau \tag{9.36}$$

and its limit for very large values of t,

$$s = \lim_{t \to \infty} C(t). \tag{9.37}$$

Combining Eqs. 9.32, 9.34 and 9.34, they found the integral equation

$$\sum_{s=1}^{k} m_{s0} \exp\{-B_s C\} = \sum_{i=1}^{p} N_i \exp\{-A_i C\} \exp\left\{-\int_0^t D_i d\tau\right\} \times$$

$$\left[\eta_{i0} + \int_0^t A_i \dot{C} \exp\{A_i C\} \exp\left\{\int_0^\tau D_i d\tau'\right\} d\tau\right] + \dot{C} + M. \tag{9.38}$$

The solutions of this equation are expected to be the same as those of the set of equations 9.30-9.33. However, similarly to the remark made with respect to the work by Kantorovich et al.[28],[29] (discussed in Ch. 6), one should note that merely being able to write one equation, instead of $k + p + 1$ simultaneous equations, does not really help to solve them or even to draw some useful conclusions from them. Mandowski and Świątek[27] first turned to the relatively simple case of $p = 1$ and $k = 1$, and showed that for $t \to \infty$, one should take

$$\lim_{t \to \infty} n_c(t) = \lim_{t \to \infty} \dot{C}(t) = 0 \tag{9.39}$$

and therefore, Eq. 9.38 reduces to

$$m_0 \exp(-BS) = M. \tag{9.40}$$

The initial condition in this case is $m_0 = n_0 + M$, and with this, Eq. 9.40 can be written as

$$S = \frac{1}{B} \ln\left(1 + \frac{n_0}{M}\right), \tag{9.41}$$

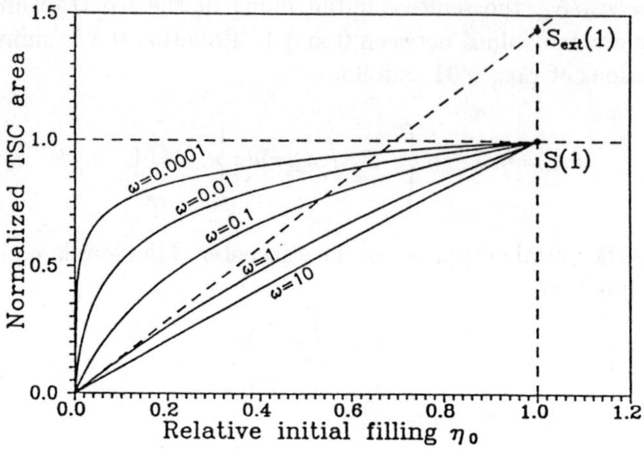

Figure 9.5 Plot of the normalized TSC area S vs. the initial relative filling of traps η_0 for different relative concentrations of traps ω. ω=(a) 10^{-4}; (b) 0.01; (c) 0.1; (d) 1.0; and (e) 10.0. $S(1)$ is the TSC area for η_0=1 and $S_{ext}(1)$ is the extrapolated value from the linear range.
(After Mandowski and Świątek[27].)

which gives the dependence of the area under the TSC curve on the initial concentration of trapped carriers. A slightly different way of writing this expression is

$$S(\eta_0) = \frac{1}{B} \ln \left(1 + \frac{\eta_0}{\omega}\right) \tag{9.42}$$

where $\eta_0 = n_0/N$ and $\omega = M/N$. Figure 9.5 shows some typical curves of $S(\eta_0)$.

Mandowski and Świątek then extended the discussion to the more interesting case of one kind of recombination center and a series of trapping states. Here they got instead of Eq. 9.41,

$$s = \frac{1}{B} \ln \left(1 + \sum_{i=1}^{p} n_{i0}/M\right). \tag{9.43}$$

Denoting now $\sum_{i-1}^{p} n_{i0}/\sum_{i=1}^{p} N_i$ by η and $M/\sum_{i=1}^{p} N_i$ by ω, they found the same equation as Eq. 9.42, with η replacing η_0, thus

$$S(\eta) = \frac{1}{B} \ln \left(1 + \frac{\eta}{\omega}\right). \tag{9.44}$$

From the simultaneous measurements of TL and TSC, one can determine S from the TSC curve and η from the TL curve[26]. The plot of $S(\eta)$ can, therefore, be evaluated experimentally and compared to Fig. 9.5. By curve fitting, the best of the curves in the figure can be chosen, and ω can be determined. Alternatively, one can extrapolate the linear range occurring at low values of η_0/ω so as to get $S_{ext}(1)$ as seen in Fig. 9.5. One gets now

$$\frac{S(1)}{S_{ext}(1)} = \omega \ln\left(1 + \frac{1}{\omega}\right), \tag{9.45}$$

and once $S(1)$ and $S_{ext}(1)$ are known experimentally, Eq. 9.45 can be solved numerically to yield ω. In addition, one can evaluate η_0/ω by measuring two areas S_1 and S_2 for the relative initial populations of carriers η_1 and η_2 respectively. In cases where the parameters $a = \eta_2/\eta_1$ and $\chi = S_1/S_2$ are known, the ratio $\eta_0/\omega (= n_0/M)$ can be found numerically from

$$\left(1 + a\frac{\eta_0}{\omega}\right)^{\chi} = \left(1 + \frac{\eta_0}{\omega}\right). \tag{9.46}$$

Thus, the solutions of Eqs. 9.45 and 9.46 together yield both $\eta_0 = n_0/N$ and $\omega = M/N$.

9.1.7 Experimental examples of some TL-TSC Measurements

Many authors reported on the results of simultaneous measurements of TL and TSC. In this subsection we shall mention some of these works. Figure 9.6 shows the results of such simultaneous data as measured by Coterón et al.[30] in Al_2O_3 x-irradiated at room temperature. Apart from a minor TL peak at 75 °C with no TSC counterpart, five TL peaks are seen to occur at 120, 145, 172, 235 and 197 °C. Their TSC associated peaks are seen at 122, 148, 180, 238 and 300 °C. Thus, all the peaks appear in the expected order of TL first and TSC at a higher temperature. A later work by Lapraz et al.[31] reports simultaneous measurements of α-Al_2O_3 x-irradiated at 77 K. Here, two pairs of TL-TSC peaks are seen at ~300 K and ~500 K, the latter being probably the same as the 235-238 °C peaks in the work by Coterón et al.[30]; both pairs also appear in the "correct" order of TL first, TSC next.

Another example of simultaneous measurements of the same sort is shown in Fig. 9.7 taken from Barland et al.[32]. The material used here was KI:Tl, x-irradiated at 77 K. Here, the first TL and TSC peaks coincide, at 123 K. A second TSC peak occurs at 140 K with no TL companion. The next two peaks are seen to be paired with relatively large shifts between TL and TSC; the temperatures being 205 K and 283 K in the former, and 240 K and 300 K in the latter.

Experimental results confirming that the TL peak is normally shifted to lower temperatures than the corresponding TSC peak were given by several researchers working on different materials. These included McKeever et

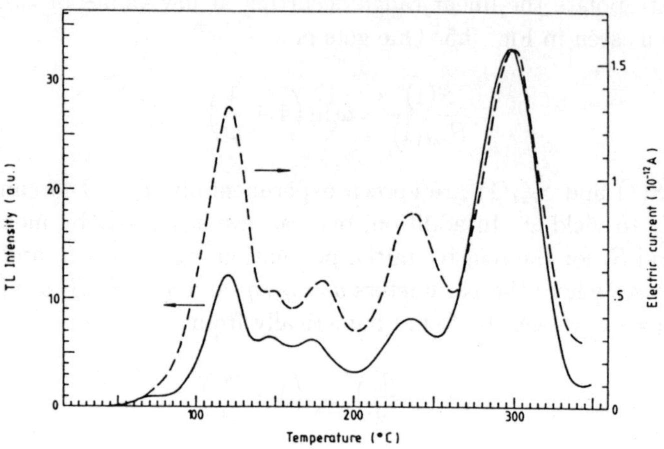

Figure 9.6 TL (solid line) and TSC (dashed line) curves in Al_2O_3, x-irradiated at room temperature.

(After Coterón et al.[30].)

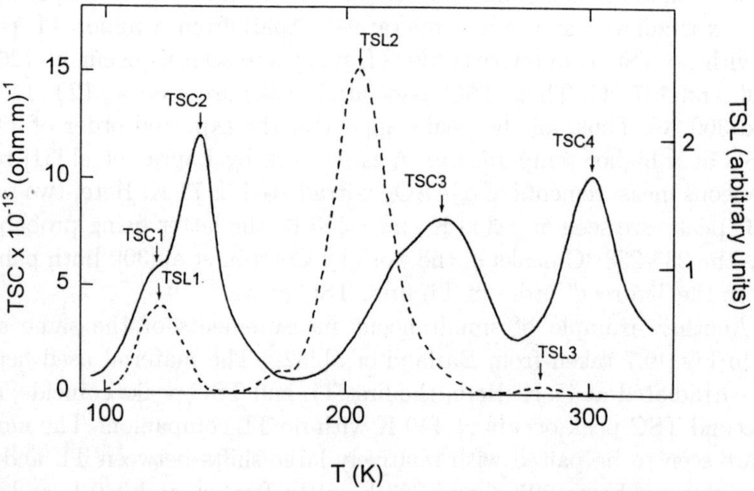

Figure 9.7 TL (dashed line) and TSC (solid line) in a KI crystal with 0.01 mole % of Tl, after irradiation at 77 K.

(After Barland et al.[32].)

al.[20], Hoogenstraaten[33], Broser and Broser-Warminsky[34], Bräunlich and Scharmann[35], Pipins and Grigas[36], Pearson and Cameron[37], Bose[38], and Sayer and Souder[39]. Böhm and Scharmann[40] reported a pair of TL-TSC peaks at ~140 K obtained from an x-rayed LiF crystal, where the TL peak preceded its TSC counterpart by ~3 K. The same authors, however, reported [41] two pairs of TL-TSC peaks in x-rayed KCl single crystals. The pair at ~85 K appears in the "correct" order, whereas the positions of the pair at ~140 K was inverted, namely, the TSC maximum preceded the TL maximum, probably due to one of the reasons mentioned above.

Hillhouse and Woods[42] studied TL and TSC, simultaneously measured in rutile (TiO_2) single crystals. They found two sets of 6 TL and TSC peaks in the temperature range of 100-330 K, and reported in each pair of TL-TSC peaks a shift between the maxima temperatures. This difference ranged from 1.9 to 7.1 K, and in all six pairs, the TL peak preceded its TSC companion.

Tuan et al.[43] studied TL and TSC peaks in x-irradiated KI crystals, and found a TL peak at 138 K and a TSC one at 144 K. They discuss this result in the framework of a previously developed model by Käämbre and Bohun[44] in which the Auger effect played an important role in these thermally stimulated processes. They also simulated a series of six TL-TSC pairs according to their Auger effect model, and found shifts of 2-6 K between the TL and TSC peaks, where, once again, TL occurred at lower temperatures.

A number of groups reported simultaneous TL-TSC measurements in different types of polyethylene. These include Nishitani et al.[45] who found four TSC and three TL peaks, in which three out of the four TSC peaks corresponded to the three TL maxima, and lagged behind them by several degrees. Ito and Nakakita[46] also found four peaks between 80 and 420 K, and a similar shift of the TL maxima was reported. Markiewicz and Fleming[47] and Fleming[48] discuss five TSC peaks, where only the first three are accompanied by TL. The shifts reported are in the "correct" direction, namely, TL first and TSC next.

Ikeya and Miki[49],[50] studied TL and TSC in LiH samples at low temperatures. A major TL peak was found at 57 K, which was accompanied by the disappearance of an ESR signal, but no TSC peak was found in this temperature range. A large TSC peak was found at ~187 K, accompanied by a TL peak at ~178 K.

Abdurazakov et al.[51] reported TL-TSC measurements in YAG:Nd^{3+} single crystals, and found 17 TL peaks and 14 TSC maxima, which appeared to correspond to 14 of the TL peaks. In three peaks, TSC preceded the TL counterpart by 3-6 K; in one case, the two appeared at the same temperature; and in 10 of the cases the TL peak appeared at lower temperature than the TSC peak, with shifts varying between 1 and 23 K.

It should be noted that important information can be revealed from simultaneous TL-TSC measurements even in cases where no correlation is found between the TL and TSC peaks. For example, Lapraz et al.[52] performed simultaneous measurements on calcium fluorapatite ($Ca_5(PO_4)_3F$:Mn^{2+}) single crystals, and found a TSC maximum at 70 K, accompanied by a TL peak at the same temperature. In addition, three TL peaks were found at 173, 263 and 298 K, with no corresponding TSC maxima. The authors concluded that these three TL peaks result from a localized recombination process.

Jiménez de Castro et al.[53] studied the TL-TSC curves in MgO x-irradiated at 80 K. They found six TL peaks associated with TSC maxima, and a major TSC peak at 312 K which was not accompanied by light emission. In the first three pairs, the TSC maxima, at 180, 230 and 252 K preceded their TL counterparts by 1-3 K. This is an inverse order, which can be explained by the above mentioned reasons. The TSC peaks at 287, 360 and 386 K, however, were preceded by the TL companions by 4-9 K, as is "normal". The authors explained that, since the emission spectrum of all six peaks was the same, all were associated with the same recombination center, and therefore, all six trapping states captured and released the same type of charge carriers, which in this case were holes. The seventh TSC peak appeared to be related to an electron trapping state. Two of these TL-TSC pairs at ~360 and ~380 K, were reported by Clement and Hodgson[54] in MgO, γ-irradiated at RT, and were explained to be associated with iron impurities.

Mariani et al.[55] reported simultaneous TL-TSC measurements in Al_2O_3, x-irradiated at 80 K. Four peaks were found in the TL glow curve, and only two in the TSC. At the low temperature range, TL occurred at 100 K, corresponding to a TSC peak at 105 K. The authors suggested that the next TL peaks, at 220, 255 and 275 K, corresponded to the same TSC maximum at 272 K. This would mean that for the last TL peak at 275 K, the TSC maximum preceded its counterpart. However, the situation is more complicated in the sense that this order changed to the expected one when the excitation dose was substantially decreased. Complementary work at lower temperatures was described by Böhm et al.[56] who reported simultaneous TL and TSC measurement in α-Al_2O_3 crystals, x-irradiated at 10 K. A main pair of TL-TSC peaks was found at ~60 K, where the TL maximum preceded its TSC companion by 1.5 K. Wong et al.[57] reported TL-TSC measurements in Al_2O_3:Ti^{4+} (Sapphire), UV irradiated at ~100 K. Two TL peaks were found, at 168 K and 256 K. In TSC, a minor peak was seen at ~170 K and a major one at ~267 K, which was explained as corresponding to the 256 K TL maximum with the shift being in the expected direction.

Castiglioni et al.[58] reported simultaneous measurements in synthetic crystalline quartz, x-irradiated at RT. TL peaks were found at 75 °C, 110 °C and 160 °C, accompanied by TSC peaks at 80 °C, 120 °C and 160 °C. For hydrogen-swept samples, additional TSC peaks were found at 180 °C and 275 °C which were suggested to be of ionic nature.

Schrimpf et al.[59] reported the results of simultaneous TL-TSC measurements in a rather unusual material, namely, solid Ar doped with gold or silver, along with O_2 molecules in the temperature range of 5-30 K, following synchrotron and x-ray irradiation. A peak at ~16 K was observed in the Au-doped material both in TL and TSC, with the former preceding the latter by ~1 K. The authors used a first-order kinetics model to extract binding energies, and values of the order of 50 meV were found.

9.2 Simultaneous Measurements of TL with Other Physical Phenomena

9.2.1 TL and TSEE Measurements

At first sight, one may think that since both in TL and TSEE the measured effect is related to the thermal release of electrons, the peaks showing up in the two effects should be very similar. As pointed out, however, in Ch. 8, there are reasons why this is not always the case:

1. In TSEE, obviously, only electrons can be released from the surface of the material whereas TL may just as well result from the thermal release of holes from traps into the valence band, and their subsequent recombination with electrons in recombination centers.

2. TL is, practically always, an effect occurring in the bulk of the sample whereas TSEE is a surface effect. It is therefore very interesting to follow those cases in which there is an obvious correlation between the two phenomena. Once this happens, one can conclude that the thermally released agents are electrons. Of course, even in this case, one does not expect an exact correlation between the TL and TSEE peaks. The shift expected between the TL and the $n_c(T)$ peak mentioned above, plays a role here also since the TSEE curve depends on $n_c(T)$. In addition, as discussed in Ch. 8, the function $\exp\{-\phi/kT\}$, associated with the work function ϕ, also causes a shift.

3. One cannot always conclude with certainty that the thermally released carriers are electrons. Auger processes may lead to the emission of electrons form the surface, as described in Ch. 8, but the trapped carriers may themselves be holes.

Figure 9.8 shows an example, taken from Bohun[60], of TL-TSEE curves measured in x-irradiated NaCl. The two peaks occurring both in TL and TSEE are at ~350 K and ~550 K. It is seen that the 350 K system has a single TL maximum and a peak with a shoulder in TSEE, whereas the 550 K system has a single TSEE maximum and a double TL peak. Yet, simple inspection of the two curves shows, with a high degree of certainty, that there is a strong correlation

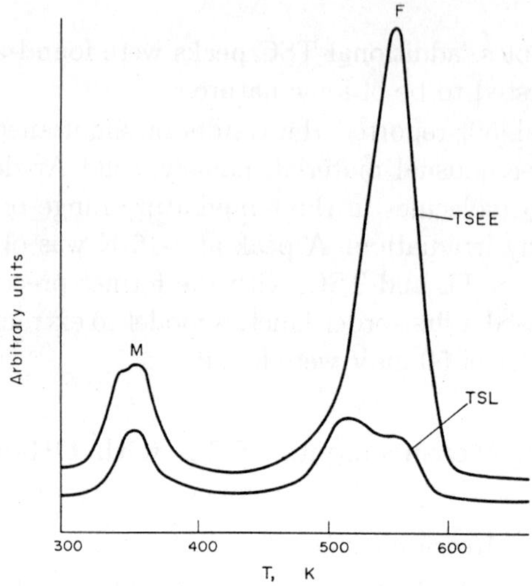

Figure 9.8 Results of a simultaneous measurement of TL and TSEE from x-irradiated NaCl.

(Redrawn from Bohun[60].)

between the two curves and, therefore, that both result (most likely) from the release trapped electrons into the conduction band.

Many other authors reported on simultaneous TL-TSEE measurements in different materials. Pipins and Grigas[36] reported such measurements in cathode-ray-excited NaCl-AgCl. The TL peaks in their measurements corresponded to the TSEE peaks, but were shifted by a few degrees to lower temperature. Thus, in this case, it seems that TSEE resembles $n_c(T)$, and the additional factor $\exp\{-\phi/kT\}$ (see Ch. 8) does not shift the TSEE to higher temperature by much. This indicates that the value of ϕ for this material is rather small. For an example in which the TSEE peaks precede their TL counterparts we can take the work by Kawamoto et al.[61] and Kutomi et al.[62]. These authors studied TL and TSEE in different kinds of $Li_2B_4O_7$:Cu glasses. An example is shown in Fig. 9.9. Here, four out of five TL peaks are accompanied by TSEE companions, two of the latter appearing at slightly lower temperatures than the former, and two at slightly higher temperatures. Here, again, it seems that $\exp\{-\phi/kT\}$ has a rather small influence on the TSEE maximum, and the order of appearance seems to be related to effects similar to those discussed in regard to the TL-TSC results. It also seems that the TL peak at ~160 °C, is related to the release of holes from a trapping state into the valence band, and their subsequent recombination with electrons into a center. Obviously, while the holes are in the valence band, they do not contribute to TSEE, as do the electrons released into the conduction band in the process

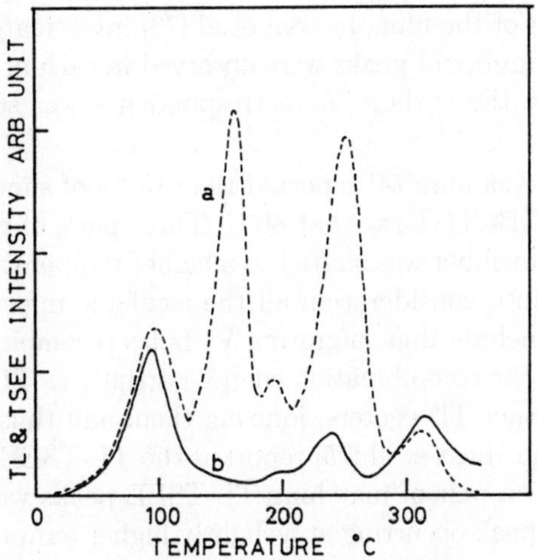

Figure 9.9 TL (curve a) and TSEE (curve b) in x-irradiated $Li_2B_4O_7$:Cu crystallized glass.
(After Kutomi et al.[62].)

leading to the appearance of the other peaks.

The same group of researchers reported on the results of such TL-TSEE measurements in different materials. Tomita et al.[63],[64] made such measurements in NaCl:Ca,Cu, which yielded a large number of peaks. In these results too, one can see TL peaks unaccompanied by TSEE, and which therefore seem to be related to the release of holes, as well as TL-TSEE pairs which definitely are associated with the thermal release of electrons. Also, some TSEE peaks are seen with no related TL and may be associated with surface effects. Kutomi et al.[65] reported on TL-TSEE simultaneous measurements in $Li_2B_4O_7$ single crystals. In this case, a number of peaks are measured in TL and TSEE, but in only one, at ~140 °C, does there seem to be correspondence between TL and TSEE. Further results of the same nature were reported by Fukuda et al.[66] in MgB_4O_7 samples, by Fukuda[67] in $Ca_3(PO_4)_2$:Ce samples, by Tomita et al.[68] in KCl:Ag single crystals, and by Tomita et al.[69] in $LiNbO_3$ single crystals.

Aoki et al.[70]-[72] studied simultaneous TL-TSEE in LiF and CaF_2 evaporated thin films. They reported a rather unusual effect of TL and TSEE observable on the first heating of the sample with no pre-excitation. The TSEE peaks in these samples were shifted by ~50 K above their TL counterparts. This seems excessive even when the two reasons for such a shift mentioned above are taken into consideration. The authors attribute these first heating peaks to the crystallization of the film evaporated in vacuum; that is, they seem to have close correlation with the growth of many microscopic bordered domains which appear on the surface of the film. Iacconi et al.[73] investigated TL and TSEE in ZrO_2 samples. A number of peaks were observed in each of the two phenomena but, as reported by the authors, no correspondence was seen between the TL and TSEE peaks.

Sorkin and Käämbre[74] reported the results of simultaneous TL-TSEE measurements in CsBr:Tl, x-rayed at 80 K. Three pairs of peaks were reported; in each, the TSEE member was shifted to a higher temperature compared to the TL peak. Taking into consideration all the available information in this sample, the authors conclude that migrating V_k holes recombine with Tl^+ trapped electrons, emitting the recombination energy partially as TL light, and partially transferring it to other Tl^o centers, ionizing them and thus initiating TSEE in the 120-170 K range. Lesz et al.[75] reported the TL-TSEE results of BeO thin film samples. Here, a pair of matching TL-TSEE peaks was reported at ~290 °C with the TSEE peak occurring at a slightly higher temperature. In addition, another TL peak appeared at ~200 °C, and a TSEE peak at ~400 °C. In later work, however, Lesz et al.[76] did not observe the ~300 °C peak of TL, and concluded that there is no relation between TSEE and TL in BeO thin film samples. Bichevin et al.[77] studied eximer-laser-induced TL-TSEE of alkali halides, as well as x-ray excited processes in the same materials. In KCl:Ag a main TL peak was seen at ~400 K and TSEE at ~430 K in the laser excited case. The authors associate the observed peaks with Ag^o centers.

Kortov et al.[78], [79] as well as Iacconi et al.[80] studied simultaneously TL and TSEE in Al_2O_3 samples. Kortov et al.[78] reported results from x-rayed Al_2O_3:Mg. Here, the main peaks are ascribed to the Mg impurity since they are not observed in the nominally pure corundum samples. The TSEE peak here is suggested to be related to thermally freed holes in an Auger process similar to that explained above. Akselrod and Kortov[79] performed similar measurements on α-Al_2O_3:C . The main TL-TSEE system here is at ~175 °C where the TL peak precedes its TSEE counterpart. Iacconi et al.[80] studied the TL and TSEE properties of α-Al_2O_3 samples following UV or x-ray excitations. They report that there is a reasonably good match between the TL and TSEE peaks, and that the TSEE maxima are at slightly higher temperatures. Here, there is a TL peak at ~87 K which is not associated with a TSEE maximum. Their conclusion is that this TL peak is associated with the thermal release of holes.

9.3 TL and the Concentration of Trapped Carriers as Reflected by Absorption and EPR

The signal measured in thermoluminescence is proportional to the first derivative of the concentration of carriers in certain trapping states - usually, but not necessarily always, the recombination center. In most of the discussion so far, we have not mentioned any direct measurement of this concentration itself, excluding the evaluation of this magnitude by integration of the TL curve up to very high temperature (in principle, to infinity). This may cause trouble if the glow curve includes more than one peak, since the evaluation of the relevant concentration by such integration is often incorrect if additional peaks are involved. In this section we discuss two methods for the direct evaluation of the concentration and its correlation with the measured TL. These are by optical absorption (OA) methods and by electron paramagnetic resonance (EPR; also known as electron spin resonance, ESR). A great importance of these correlative measurements is that one can often find the nature of the impurities or defects responsible for TL, and sometimes for TSC and TSEE as well, in a direct way, in those cases where the relation between the phenomena is clear. It should be noted, however, that these are not strictly simultaneous measurements in that one normally measures the absorption or EPR signal in one measurement, then measures the TL of a similarly treated sample in a separate measurement. Nevertheless, as long as the sample is in the same "state" before each measurement, the obtained data should be comparable.

We should also mention that there are two approaches to monitoring the OA or EPR as a function of sample temperature. For example, one can fix the wavelength at which the optical absorption is being measured and continually monitor the strength of this absorbance as the temperature of the sample is increased. When doing so, however, one has to correct for the changing width of the absorption peak due to phonon broadening. In EPR, the situation can be even worse in that the EPR signal may itself only be observable at low temperature (often less than room temperature) due to changes in the spin-lattice relaxation time. This may be true even though the defect itself is stable up to higher temperatures. As a result, an often-used procedure is to "pulse anneal" the sample to the desired high temperature, then to lower the sample temperature to that required for measurement of the OA or EPR signal. The pulse annealing/measurement cycle is then repeated many times, raising the temperature on each cycle to cover the temperature range of interest. In this way the signal strength (i.e. defect concentration) can be monitored as a function of the anneal temperature. The results obtained from such measurements should be compared with TL only with some caution, however, since the heating profile is entirely different in the two cases. In one case (TL) the temperature is increased from low temperature to high temperature in one continuous sweep. In the other case (absorption/EPR) the temperature is increased in a saw-tooth fashion. The result is that the TL signal obtained is not exactly the derivative of the concentration-versus-temperature profile obtained from the OA or EPR measurement. In particular, this derivative will reveal a peak at a significantly

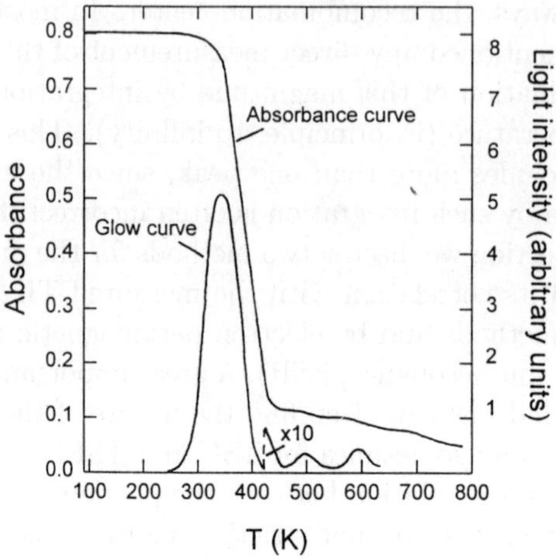

Figure 9.10 Bleaching curve and simultaneously measured glow peak in calcite. (After Medlin[81].)

lower temperature than the TL peak, even if the annealing of the center under observation is directly associated with the TL peak of interest.

9.3.1 TL and Optical Absorption Measurements

Figure 9.10 shows the results of simultaneous measurements of a glow curve and an absorbance curve of an x-ray excited calcite as measured in the integral range of 260 nm - 350 nm by Medlin[81]. The main maximum of TL at 360 K is clearly seen to be in the same temperature range as the fast change of absorbance. This indicates that for the derivative of the concentration with respect to temperature, one has $I \propto -dm/dT$ (or $-dn/dT$). For a constant heating rate, this is equivalent to $I \propto -dm/dt$, or $I \propto -dn/dt$, which is expected as discussed in Chapters 1 and 2. Medlin reports that the thermal bleaching of these color centers occurs in four stages, which is shown in the main variation of the absorbance curve at ~380 K and by the three wiggles at higher temperatures and, in parallel, the main TL peak and three smaller ones in the glow curve. Further, more detailed work by Medlin[81] on the calcite samples, based on polarized absorption spectra, indicated that the absorption bands are due to four kinds of trapped electron centers and one kind of trapped hole center. TL is produced when the recombination energy excites Mn^{2+} impurity ions. The results are consistent with an excitation mechanism involving nonradiative resonance transfer of energy between recombination sites and emission centers.

Braner and Halperin[82] reported the results of a similar investigation of x-ray colored KCl crystals. Figure 9.11 depicts some of their results where

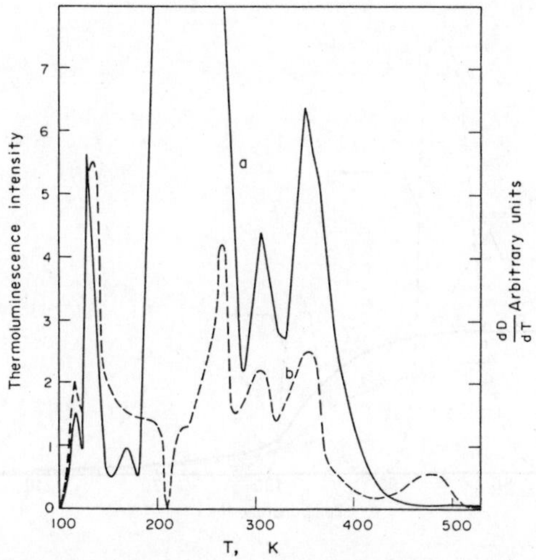

Figure 9.11 (a) Glow curve, and (b) temperature derivative of the thermal bleaching curve for the same crystal of x-irradiated KCl.
(Redrawn from Braner and Halperin[82].)

curve (a) is TL, and curve (b) is the negative first derivative of the absorption curve with respect to temperature. The agreement between the temperatures of occurrence of the peaks is very good. Somaiah et al.[83] described similar results of simultaneous measurements of TL and optical absorption in BaFCl crystals.

Diéguez and Cabrera[84] studied the TL and optical absorption of x-irradiated KH_2PO_4 (KDP) crystals. A strong dichroism is known to occur at low temperatures in the absorption spectra of the $\|c$ plates, and therefore, Diéguez and Cabrera[84] measured the thermal bleaching for the σ and π polarizations separately. This was done by recording the decay of the optical density maxima (at 550 nm and 510 nm respectively) on warming up the crystals at the rate used in the TL experiment, $\sim$5 Kmin^{-1}, following an excitation at 13 K. The results are shown in Fig. 9.12 where quite different behavior can be observed for the σ and π polarizations. In the π case, the absorption remains constant until 75 K, then it decreases down rather quickly to $\sim$125 K and bleaches out completely at about 230 K. The σ absorption has two steps of rapid decrease, at $\sim$40 K and $\sim$100 K; from 220 K upward, no optical absorption is detected. Two glow peaks, centered at $\sim$73 K and $\sim$123 K are seen. The 73 K peak seems to correlate with the σ polarization curve and the 123 K peak with both the σ and π polarization curves. It is to be noted that the latter temperature coincides with the Curie temperature of KDP. The dependence of the intensity of the glow peaks on the irradiation dose is similar to that found for the optical

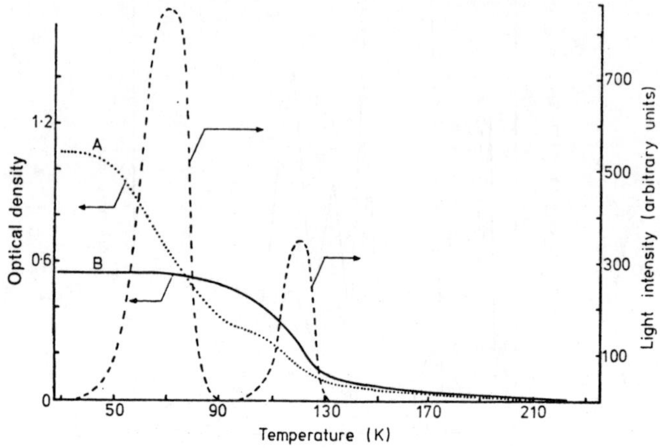

Figure 9.12 Thermal bleaching curves of a $\|$c plate. (A) σ polarization, $\lambda = 550$ nm. (B) π polarization, $\lambda = 510$ nm. The TL glow peaks are shown as dashed lines. The irradiation time for both experiments was about 360 min.
(After Diéguez and Cabrera[84].)

density. The authors[84] explain their results as being due to the existence of two centers, which they denote as the σ- and π-centers. They suggest that the σ-center does not contribute to the π absorption, whereas the π-center does make some contribution to the σ absorption. These authors also suggest that the σ- and π-centers are related to the A and B radicals previously observed by Jeffers et al.[85] in EPR. Apparently, the σ-center is associated with the B radical, $\dot{P}O_4H_2$ in KDP. The TL emission at 73 K seems to be associated with the reaction

$$\dot{P}O_4H_2 + e^- \rightarrow PO_4H_2^- + h\nu. \tag{9.47}$$

As for the π-center, Diéguez and Cabrera[84] suggest that since its light emission occurs at T_c, a close relationship seems to exist between the annihilation and the phase transition mechanism.

Delgado and Alvarez-Rivas[86] studied TL and the annealing of optical absorption in NaCl:Cu. They found a one-to-one correspondence between the TL peaks and the annealing steps of the F and Cu^- absorption bands, and with anion vacancies. In the latter case, short-lived V_k-centers are formed, which in turn recombine with Cu^- ions; thus, the crystal lattice is brought back to its pre-irradiation state.

Sagastibelza and Alvarez-Rivas[87] reported the results of similar TL - optical absorption measurements in x-irradiated LiF (TLD-100) samples. They studied the correlation between the TL curve and the negative derivative of the absorption curve, and found very close resemblance between the two curves. These authors[87] could identify this correlation between the corresponding curves, both in the F and M absorption bands.

López et al.[88] investigated TL and optical absorption of NaCl:Mn^{2+}. They found close correlation between the TL peaks at ~170 K and ~220 K, and the strong decrease in the absorption with temperature at the same temperatures. They associate the annealing of the absorption as well as the TL peaks with the annihilation of Mn^o B-centers. Further measurements on the same materials were performed by Delgado and Alvarez-Rivas[89]. They explained their results as being due to the recombination of thermally released interstitials with anion vacancies. The resulting short-lived V_k-centers recombine with electrons trapped at impurities, thus yielding the TL. Simultaneously, F-centers are annealed by luminescent recombination with interstitials.

Measurements of the same nature were reported by Prasad and Rao[90] in laser-excited KCl crystals, previously irradiated by x-rays. They found two TL peaks, at 143 °C and 255 °C, both associated with the destruction of the F-centers, previously formed by the x-ray irradiation.

Martín et al.[91] reported the results of a study of x-irradiation on tetrachlorides, K_2ZnCl_4, Rb_2ZnCl_4 and $[(CH_3)_4N]_2ZnCl_4$ (also termed $(TMA)_2ZnCl_4$). Both in Rb_2ZnCl_4 and $(TMA)_2ZnCl_4$, a major TL peak, at 200 K in the former and 150K in the latter, is associated with the very fast decrease of the optical absorption in the respective temperature dependent curves.

Lecoq et al.[92] studied the effects of different irradiations on $Bi_4Ge_3O_{12}$ (BGO) monocrystals. They used UV light, ^{60}Co γ rays as well as 18 MeV electrons for excitation, and studied simultaneously the variation of optical absorption with temperature, TL and TSC. In different samples, they found a TL peak at 100 °C, associated with a strong decrease of the optical absorption. Another decrease of the absorption was accompanied by a smaller TL peak at ~140 °C, as well as a major TSC peak at ~150 °C . The authors[92] suggest that the two TL peaks are related, one to a localized transition and the other to holes released into the valence band. The former seems to be associated with a Fe^{3+} ion substituting a lattice Ge^{4+}. The recombination center seems to be Bi^{5+} ions in close vicinity, which have captured holes during the excitation. The higher temperature peak was explained to be related to the same two impurities, however, in this case one considers Bi^{5+} and Fe^{3+} which are not in close proximity to each other in the first place. It appears that now, at a higher temperature (150 °C), the Bi^{5+} ions thermally release holes into the valence band which, in turn, recombine with electrons in Fe^{3+}, changing it into Fe^{2+} and emitting TL. This peak is associated with the mentioned TSC companion, observed while holes are in the valence band.

Bandyopadhyay and Summers[93] studied the optical absorption and TL in γ-irradiated NaCl:Cu. Here, the optical absorption decreases rather abruptly between 350 K and 430 K, and is entirely annealed above 433 K. Some difference was observed between unbleached samples and those optically bleached to remove F-centers before heating. In both, a TL peak occurring at ~450 K seemed to be uncorrelated with the optical absorption. In the bleached sample, another peak was seen at ~415 K which is in the range of strong decrease of the absorption. Additional, smaller peaks were seen at ~330 K in the unbleached sample and at ~530 K in the sample which was bleached with F band light. The authors also discuss the role of the Cu impurities.

Baba et al.[94] investigated the simultaneous results of TL and optical absorption of NaBr:S^{2-} crystals. They found that the glow peak, with an infrared emission of 815 nm, appears at 230 K and is accompanied by a significant change in the absorption. The authors ascribe the effect to the formation and decomposition of F_H-centers. These are F-centers associated with a neighboring sulfur ion.

Aleshin et al.[95] investigated TL and optical absorption in CaF_2:Na,Ce. They described two main TL maxima at 345 K and 370 K, as well as a weaker peak at 410 K. The emission maxima occurred at 350, 495 and 740 nm, and were ascribed to Ce^{3+}, Mn^{2+} and M_{Na}-centers, respectively. The most intense TL of the M_{Na}-centers was observed at the 340-370 K temperature range, in which temperature range all the color centers are still stable, showing only a slight uniform annealing. Since all the observed absorption bands belonged to electron color centers, such annealing was ascribed only to hole release. Thermoluminescence of the M_{Na}-centers at 410 K lies in the temperature range where annealing of the electron centers occurs.

Lakshmanan and Tiwari[96] reported on TL - optical absorption results in CaF_2:Tm crystals, irradiated at room temperature with γ rays. The TL peaks at 160 °C, 180 °C and 480 °C were found to be due to hole activation, giving rise to optical absorption near 310 nm. This was followed by recombination with Tm^{2+}, giving rise to optical absorption bands near 400 and 600 nm, and resulting in Tm^{3+} emission. The major TL peak at 280 °C, however, was not associated with any optical absorption band. The results indicate that this peak could arise from electron activation, recombination with holes and subsequent resonant energy transfer to Tm^{+3} ions.

Kishan-Kumar et al.[97] studied simultaneously TL and optical absorption in copper-doped alkali sulphate phosphors. In each of the three kinds of samples, three main peaks were observed following γ-excitation. The peaks occur at 353 K, 379 K and 421 K in Cs_2SO_4:Cu; 357 K, 393 K and 439 K in Rb_2SO_4:Cu; and at 362 K, 395 K and 468 K at K_2SO_4:Cu. The optical absorption shows that Cu^+ and Cu^{2+} act as electron traps, producing Cu^o and Cu^+

respectively. The TL process appears to result from the reactions

$$Cu^{o} + \text{hole} \rightarrow Cu^{+} + h\nu, \tag{9.48}$$

and

$$Cu^{+} + \text{hole} \rightarrow Cu^{2+} + h\nu. \tag{9.49}$$

Pokorný and Ibarra[98] studied TL and optical absorption in x-ray excited Al_2O_3:Cr,Ni. The dominant TL peak at 330 °C was shown to be associated with a strong decrease of the absorption bands at 6.1 eV, 4.4 eV and 2.7 eV as a function of temperature annealing. It was proposed that the high-intensity peak is due to oxygen vacancies, induced by the presence of Ni, and that the main recombination center is caused by Cr.

9.3.2 Simultaneous Measurements of TL and EPR (ESR)

As mentioned above, another absorption technique is EPR (ESR). Electron paramagnetic resonance is used for the study of materials which show paramagnetism because of the magnetic moment of unpaired electrons. EPR spectra are usually presented as plots of the absorption or dispersion of the energy of an oscillating magnetic field of fixed radio frequency versus the intensity of an applied static magnetic field. Among the wide variety of paramagnetic substances to which EPR spectroscopy has been applied, the one that is of interest to us here is that of impurity centers in solids, mainly single crystals. Of course, the same impurity centers may be associated with TL, either acting as the charge-carrier traps, or as recombination centers. Since EPR is normally capable of identifying the impurities in the crystal, in cases where the simultaneous TL-EPR measurements shows a direct relation between the two phenomena, the identification by EPR serves as a direct proof for the identity of the impurity involved in the TL process. Similarly to the optical absorption case, it is normal that the EPR signal shows a drop at a certain temperature range where the trapping state becomes unstable. The instability may very well be associated with either the thermal release of charge carriers from the paramagnetic impurity or, alternatively, the filling of paramagnetic impurities which serve as TL recombination centers. In both cases, it is expected that the TL peak will resemble the negative derivative of the EPR signal, again, with a close resemblance to optical absorption. Fig. 9.13 taken from Born et al.[99] shows the variation of the EPR signal with temperature (curve 1); its negative derivative with respect to temperature (curve 2) and the TL intensity (curve 3) in x-ray excited $CaWO_4$ crystals. It is seen that the peak of the negative of the first derivative occurs rather close to the TL peak, but is shifted by ~17 K to a lower temperature than its TL companion maximum. Part of the reason may be the use of different heating rates in this case where the measurement was not genuinely simultaneous.

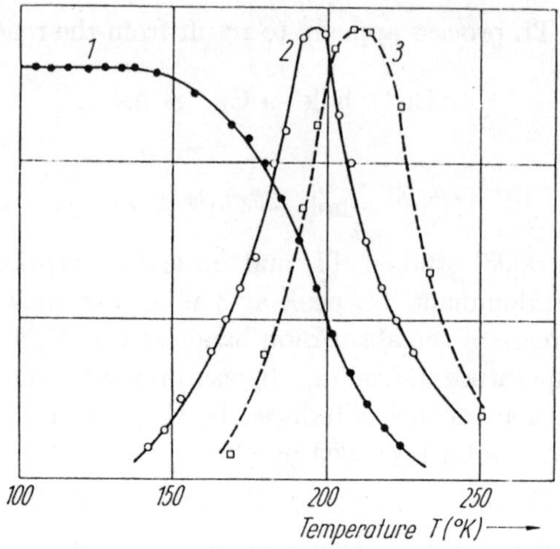

Figure 9.13 EPR and TL in $CaWO_4$. Curve 1: Intensity I of the EPR signals; curve 2: negative first derivative of curve 1, $-dI/dt$; curve 3: intensity of the emitted TL.

(After Born et al.[99].)

Another example, of simultaneous measurements of TL and EPR in CaF_2:Eu is shown in Fig. 9.14, taken from Tzalmona and Pershan[100]. Curve (a) shows two TL peaks, the solid line representing the peak for CdF_2:Eu (0.02%) and the dashed line for CdF_2:Eu (0.07%). Both curves peak at ~164 K. Curve (b) depicts the dependence of the EPR sample on the temperature. The range of strong decrease is seen at ~165 K, very close to the temperature of the glow peak.

Cooke and Alexander[101] studied the TL curve as well as the temperature dependence of EPR in L-alanine:Cr^{3+}; an example of their results is given in Fig. 9.15. Curve (a) is the measured TL peak at ~160 K. Curve (b) depicts the height of the EPR signals as a function of temperature; this represents the concentration of the cation in this material, x-irradiated at 77 K. Curve (c) shows the negative first derivative of curve (b); the similarity between (a) and (c) is obvious though the peak in (c) is shifted by about 20 K below that of (a). Cooke and Alexander attribute part of this shift to the different heating rates utilized in the measurements. Another indication to the common origin of TL and EPR was found in the way that if, following the x-irradiation, UV light is applied, then both signals are bleached quite substantially.

Some authors performed triple simultaneous measurements such as of TL, TSC and EPR. Katzir et al.[102] made such measurements in boron nitride. They observed both TL and TSC peaks and found that, as expected,

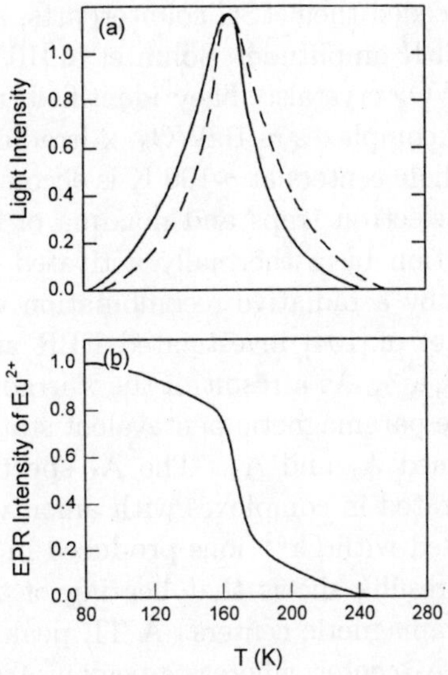

Figure 9.14 (a) TL and (b) EPR as functions of temperature in CdF_2:Eu. In (a), the solid line shows TL for CdF_2:Eu(0.02%) and the dashed line that for CdF_2:Eu(0.07%).

(After Tzalmona and Pershan[100].)

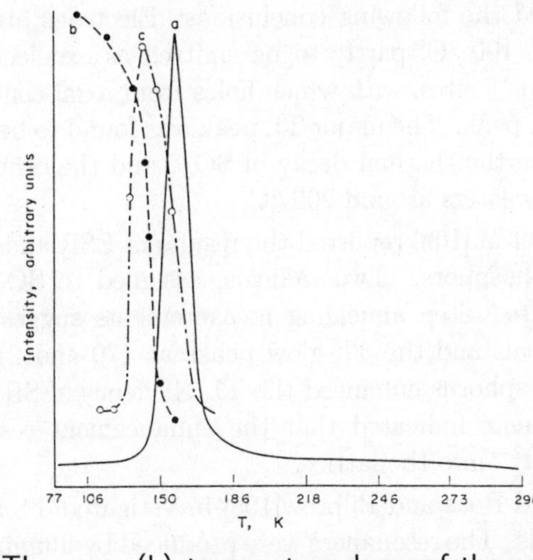

Figure 9.15 (a) TL curve; (b) temperature decay of the cation produced in x-irradiated L-alanine:Cr^{3+} and (c) the negative first derivative of the temperature decay.

(Redrawn from Cooke and Alexander[101].)

the two TL peaks preceded their TSC counterparts, and appeared in ranges of fast decrease in the EPR amplitude. Böhm et al.[103] ran TL, TSC and EPR measurements in $BaWO_4$ crystals. They identified, using EPR, electrons and holes trapped at WO_4 complexes in $BaWO_4$, x-irradiated at 80 K. The thermal decay of the intrinsic hole centers at ~100 K is accompanied by a simultaneous decrease of $(WO_4)^{3-}$ electron traps and maxima of both TL and TSC. They explained this connection by a thermally activated hopping of the $(WO_4)_2^{3-}$ hole centers, followed by a radiative recombination with electrons trapped at $(WO_4)^{3-}$. Alybakov et al.[104] investigated EPR and TL glow curves of x-irradiated KCl:CrO_4^-,Ca^{2+}. As a result of the x-irradiation, the Cr^{6+} impurity centers turned into the paramagnetic pentavalent state, and showed EPR spectra of two types, termed A_1 and A_2. The A_1 spectrum appeared to be due to the Cr^{5+} ions generated in complexes with anion vacancies, whereas the A_2 spectrum was associated with Cr^{5+} ions produced in complexes with Ca^{2+}. A comparison with TL results shows that heating of the sample results in the destruction of the paramagnetic centers. A TL peak at 436 K corresponds to the destruction of the A_1-center, whereas a peak at 486 K is correlated with the destruction of the A_2-center.

Huzimura and Atarashi[105] studied ESR, TL and TSEE in x-irradiated $CaSO_4$:Tm. They identified four types of electron center related to SO_3^- and three types of hole center associated with SO_4^-, and one type of axial center. By comparing the step-by-step annealing of these centers with the TL and TSEE curves, they reached the following conclusions: Electrons are thermally freed from SO_3^- at about 100 °C, partly to be emitted as exoelectrons and mostly to be trapped at Tm^{3+} sites, with which holes from axial centers recombine to produce a minor TL peak. The major TL peak was found to be composed of two parts, one related to the thermal decay of SO_4^-, and the other to a relaxation process of the SO_3^- centers around 200 °C.

Gundu Rao et al.[106] reported the results of ESR and TL in $BaSO_4$:Eu and $BaSO_4$:Eu,P phosphors. Two radicals, assigned to SO_2^-, were observed at room temperature. Step annealing measurements suggested a connection between these radicals and the TL glow peaks at 170 and 215 °C. Co-doping $BaSO_4$:Eu with phosphorus enhanced the TL efficiency. ESR and fluorescence emission measurements indicated that the enhancement results from an increased intake of Eu^{2+} into the lattice.

Boas[107] and Boas and Pilbrow[108] investigated TL and ESR in CaO and CaO:Mg crystals. The resonances were produced by illumination at $\lambda < 750$ nm and $T < 200$ K, and were associated with the TL peak at 180 K, with the emission spectrum centered at $\lambda_{max} = 690$ nm. The center involves three lattice sites along the <100> direction with an H^- ion at the anion-vacancy site next to the cation-vacancy site of a P^--center. It appears that this center arises as a result of the conditions during crystal-growth and is specifically associated with the Mg dopant. Orera et al.[109] studied EPR and TL in x-irradiated CaO at higher temperatures, and found two glow peaks, at 360 and 400 K. These were reported to be associated with the thermal destruction of two new donor-type paramagnetic defects of extrinsic character. The EPR spectrum of the defect which was associated with the first glow peak consisted of an intense central line and six weak transitions, corresponding to the hyperfine interaction with a ^{25}Mg nucleus. The second defect showed an interaction with an ^{27}Al nucleus. The thermal annealing of the EPR signals occurred between 350 and 400 K, the same range in which the two TL peaks were seen.

Tatumi et al.[110] investigated TL and ESR in aragonite, a trimorph of $CaCO_3$, together with calcite and vaterite. Natural aragonite shows a prominent TL peak at 613 K. Additional γ-irradiation enhanced this glow peak but shifted it down to 593 K, and induced two new peaks at 393 and 493 K. Pre-annealing at 653 K for 10 minutes and subsequent irradiation enhanced Pb^{3+} and CO_3^- hole centers observed in ESR. An isochronal thermal anneal sequence indicated that these ESR centers were related to the two glow peaks at 393 and 493 K. The experimental results indicated that the TL glow peaks in aragonite can be explained by second-order kinetics, and the concentrations of the Pb^{3+} and CO_3^- centers decreased with increasing anneal temperature up to 603 K, at which temperature the two TL glow peaks at 393 and 493K are exhausted. These glow peaks appeared to be related to a recombination between holes in Pb^{3+} ions and electrons thermally released from the CO_3^{3-} and CO_2^- electron traps. The difference in the dose dependence of the two peaks indicated that the different dependence of the filling of these two traps on the dose.

Jani and Halliburton[111] studied TL, ESR and optical absorption in $Bi_{12}SiO_{20}$ (BSO) and $Bi_{12}GeO_{20}$ (BGO). Excitation at 77 K with 350 nm converts Fe^{3+} to Fe^{2+} ions. The source of electrons in this process is unknown; the possibilities mentioned include transition-metal ion impurities, neutral oxygen vacancies, germanium or silicon vacancy complexes and antisite bismuth ions. Heating from 77 K to RT yielded a series of TL peaks as these photoinduced electron and hole pairs recombined. Comparison of ESR and TL in BGO showed that the recovery of the Fe^{3+} ESR and TL in BGO signal correlated with the 165 K peak. In BSO, the same three peaks as in BGO were found to occur at 145, 165 and 245 K, however, the peaks at 145 and 245 K were considerably larger in BSO. The ESR recovery step between 125 and 200 K correlated with the combined 145 and 165 K TL peaks, and the second Fe^{3+} recovery step closely paralleled the 245 K TL peak. Also, the photoinduced optical absorption decayed over the range from 125 to 200 K, which correlated well with the recovery of the Fe^{3+} ESR spectrum of the TL-related defects in BGO.

Porwal et al.[112] reported data from EPR and TL studies of γ-irradiated $K_2UO_2(SO_4)_2{\cdot}2H_2O$ and $Rb_2UO_2(SO_4)_2{\cdot}2H_2O$, excited by ^{60}Co γ rays. In the latter material, the temperature of the thermal destruction of SO_4^- (the α-center) and HO_2 (β-center), observed in the EPR measurements, coincided with the TL peak temperatures. In all these cases, the TL phenomenon was found to be associated with emission from UO_2^{2+} which acts as the recombination center. Whereas during γ-irradiation the process seems to be

$$(O_2U^{2+}\text{-}O_4S^-) \rightarrow (O_2U^+\text{-}O_4S^{2-}), \tag{9.50}$$

on heating, the photon producing process is

$$(O_2U^+\text{-}O_4S^{2-}) \rightarrow (O_2U^{2+}\text{-}O_4S^-) + h\nu. \tag{9.51}$$

It is to be noted that in these materials, which are uranium compounds, effects of self-excitation by the uranium are to be taken into account.

Andreev[113] investigated the annealing of radiation-generated paramagnetic centers and color centers in γ-irradiated $KTiOPO_4$ (KTP) single crystals by measuring TL, TSC, ESR and optical absorption. A complete annealing of all the radiation-induced paramagnetic centers and the color centers occurred at 430 K. A TL peak observed at $\sim$330 K was found to be due to annealing of radiation-produced hole centers of the PO_4^{2-} type, and a TSC peak at 360 K was determined to be due to annealing of radiation-induced paramagnetic electron Ti^{3+} centers.

Seshagiri et al.[114] discussed the results of TL in γ-irradiated, uranium-doped $K_2Ca_2(SO_4)_3$. Two glow peaks, at 400 K and 435 K, were observed. EPR measurements in these samples showed the formation of the radical ions SO_4^-, SO_3^-, SO_2^- and O_3^-. From the study of the thermal stability of these ions, it was found that the thermal destruction of the SO_2^- and SO_4^- radical ions were associated with the mentioned glow peaks at $\sim$400 K and $\sim$435 K, respectively. It was found that the uranate (UO_6^{6-}) ion serves as the luminescence center for the observed TL glow. As pointed out with regard to uranium compounds, self-excitation of TL and EPR may occur by the uranium impurity in this material.

9.4 References

1. R. Chen and Y. Kirsh, *Analysis of Thermally Stimulated Processes*, Pergamon Press (1981).
2. S.W.S. McKeever, *Thermoluminescence of Solids*, Cambridge University Press (1985).
3. P. Bräunlich and P. Kelly, *Phys. Rev. B* **1** (1970) 1596.
4. P. Bräunlich, P. Kelly and J.P. Fillard, in *Thermally Stimulated Relaxation in Solids*, P. Bräunlich, editor, Springer Verlag, Berlin, Heidelberg (1979).

5. J.P. Fillard, J. Gasiot and J.C. Manifacier, *Phys. Rev. B* **18** (1978) 4497.
6. A. Opanowicz and K. Przybyszewski, *Zestiy Naukowe Politechniki Lódzkeiej* Nr. 733, Fizyka **z.15** (1995) 53.
7. R. Chen, *J. Appl. Phys.* **42** (1971) 5899.
8. A. Halperin and R. Chen, *Phys. Rev.* **148** (1966) 839.
9. D.E. Fields and P.R. Moran, *Phys. Rev.* B **9** (1974) 1836.
10. M. Böhm, B. Cord, A. Hofstaetter and A. Scharmann, *J. Lumin.* **17** (1978) 291.
11. H.K. Henisch, *J. Electrostatics* **3** (1977) 323.
12. R. Chen and R.J. Fleming, *J. Appl. Phys.* **44** (1973) 1393.
13. G. Bemski, *Phys. Rev.* **111** (1958) 1515.
14. M. Lax, *Phys. Rev.* **119** (1960) 1502.
15. P.N. Keating, *Proc. Phys. Soc.* **78** (1961) 1408.
16. T.S. Benedict and W. Shockley, *Phys. Rev.* **89** (1957) 1152.
17. H.L. Oczkowski, *Phys. Stat. Sol (a)* **103** (1987) 499.
18. A. Opanowicz, *Phys. Stat. Sol. (a)* **108** (1988) K47.
19. C.H. Lee, G.N. Jeon, S.C. Yu and S.Y. Ko, *J. Phys. D: Appl. Phys.* **28** (1995) 1951.
20. S.W.S. McKeever, B.G. Markey and A.C. Lewandowski, *Nucl. Tracks Radiat. Meas.* **21** (1993) 57.
21. A.C. Lewandowski, B.G. Markey and S.W.S. McKeever, *Phys. Rev. B* **49** (1994) 8029.
22. J. Gasiot and J.P. Fillard, *J. Appl. Phys.* **48** (1977) 1371.
23. J.P. Fillard and J. Gasiot, *Phys. Stat. Sol. (a)* **32** (1975) K85.
24. M. Jiménez-López, *J. Electrostatics* **3** (1977) 133.
25. J.P. Fillard, J. Gasiot and M. de Murcia, *J. Electrostatics* **3** (1977) 49.
26. A. Mandowski and J. Świątek, *SPIE, Liquid and Solid State Crystals* **1845** (1992) 216.
27. A. Mandowski and J. Świątek, *J. Phys. D: Appl. Phys.* **25** (1992) 1829.
28. L.N. Kantorovich, G.M. Fogel and V.I. Gotlib, *J. Phys. D: Appl. Phys.* **23** (1990)1219.
29. L.N. Kantorovich, A.I. Livshicz and G.M. Fogel, *J. Phys.: Condens. Matter* **5** (1993) 7503.
30. M.J. Coterón, A. Ibarra and M. Jiménez de Castro, *Cryst. Latt. Def. and Amorph. Mat.* **17** (1987) 77.
31. D. Lapraz, S. Boutayeb, P. Iacconi, R. Bindi and P. Rostaing, *Phys. Stat. Sol. (a)* **136** (1993) 497.
32. M. Barland, E. Duval and A. Nouailhat, *J. Physique* **41** (1980) 75.

33. W. Hoogenstraaten, *Philips Res. Repts.* **13** (1958) 515.
34. I. Broser and R. Broser-Warminsky, *Brit. J. Appl. Phys. Suppl.* **4** (1955) 90.
35. P. Bräunlich and A. Scharmann, *Phys. Stat. Sol.* **18** (1966) 307.
36. P.A. Pipins and B.P. Grigas, *Opt. I Spekt.* **18** (1965) 43.
37. D.W. Pearson and J.R. Cameron, *AEC Report*, COO-1105-131 (1967).
38. D.N. Bose, *Phys. Stat. Sol.* **30** (1968) K57.
39. M. Sayer and A.D. Souder, *Can. J. Phys.* **47** (1969) 463.
40. M. Böhm and A. Scharmann, *Phys. Stat. Sol. (a)* **4** (1971) 99.
41. M. Böhm and A. Scharmann, *Phys. Stat. Sol (a)* **22** (1974) 143.
42. R.W.A. Hillhouse and J. Woods, *Phys. Stat. Sol. (a)* **6** (1981) 119.
43. B.T. Tuan, B. Velický and A. Bohun, *Z. Phys.* **251** ((1972) 289.
44. H. Käämbre and A. Bohun, *Czech. J. Phys. B* **14** (1964) 289.
45. T. Nishitani, K. Yoshino and Y. Inuishi, *Japan. J. Appl. Phys.* **14** (1975) 721.
46. D. Ito and T. Nakakita, *J. Appl. Phys.* **51** (1980) 349.
47. A. Markiewicz and R.J. Fleming, *J. Phys. D: Appl. Phys.* **21** (1988) 349.
48. R.J. Fleming, *IEEE Trans. Elect. Ins.* **24** (1989) 523.
49. M. Ikeya and T. Miki, *J. Physique,* **C6** (1980) 312.
50. T. Miki and M. Ikeya, *J. Phys. Soc. Japan,* **51** (1982) 2862.
51. A. Abdurazakov, V.A. Antonov and P.A. Arsen'ev, *Zhurnal Prid. Spekt.* **36** (1981) 26.
52. D. Lapraz, F. Gaume and M. Barland, *Phys. Stat. Sol (a)* **89** (1985) 249.
53. A. Jiménez de Castro, C. Carrascosa, A. Rascón, D.F. Mariani and J. L. Alvarez-Rivas, *J. Phys. C: Sol. St. Phys.* **18** (1985) 4863.
54. S. Clement and E.R. Hodgson, *Rad. Eff.* **97** (1986) 215.
55. D.F. Mariani and M. Jiménez de Castro, *J. Phys. Chem. Solids* **50** (1989) 125.
56. M. Böhm, P. Iacconi, K.D. Kromm and A. Scharmann, *Phys. Stat. Sol. (a)* **146** (1994) 757.
57. W.C. Wong, D.S. McClure, S.A. Basun and M.R. Kokta, *Phys. Rev. B* **51** (1995) 5693.
58. M. Castiglioni, M. Martini, G. Spinolo and A. Vedda, *Rad. Meas.* **23** (1994) 361.
59. A. Schrimpf, C. Boekstiegel, H.J. Stöckmann, T. Borenman, K. Ibbeken, J. Kraft and B. Herkert, *J. Phys.: Condens. Matter* **8** (1996) 3677.
60. A. Bohun, *III Int. Symp. Exoelect., Rraunschweig, PTB Mitteilungen* **80** (1970) 321.
61. T. Kawamoto, A. Tomita and M. Kawanishi, *Japan. J. Appl. Phys. Suppl.* **24** (1985) 242.

62. Y. Kutomi, A. Tomita and N. Takeuchi, *Rad. Prot. Dosim.* **17** (1986) 499.
63. A. Tomita, K. Inabe and N. Takeuchi, *Rad. Prot. Dosim.* **6** (1984) 359.
64. A. Tomita and N. Takeuchi, *Phys. Stat. Sol. (a)* **89** (1985) 609.
65. Y. Kutomi, A. Tomita and N. Takeuchi, *Rad. Eff. Express* **1** (1987) 63.
66. Y. Fukuda, A. Tomita and N. Takeuchi, *Phys. Stat. Sol. (a)* **114** (1989) K245.
67. Y. Fukuda, *Rad. Prot. Dosim.* **33** (1990) 151.
68. A. Tomita, Y. Fukuda, Y. Kutomi and N. Takeuchi, *Phys. Stat. Sol. (b)* **158** (1990) 383.
69. A. Tomita, Y. Fukuda and N. Takeuchi, *Phys. Stat. Sol. (a)* **134** (1992) 279.
70. M. Aoki, T. Nishikawa and S. Okabe, *Japan. J. Appl. Phys.* **21** (1982) L95.
71. M. Aoki, T. Nishikawa and S. Okabe, *Rad. Prot. Dosim.* **4** (1983) 216.
72. M. Aoki, T. Nishikawa and S. Okabe, *Phys. Stat. Sol (a)* **83** (1984) 85.
73. P. Iacconi, D. Lapraz, P. Keller and M. Petel, *Rad. Prot. Dosim.* **4** (1983) 219.
74. B. Sorkin and H. Käämbre, *Phys. Stat. Sol.* **82** (1984) K149.
75. J. Lesz, A. Scharmann and G. Holzapfel, *Japan. J. Appl. Phys. Suppl.* **24** (1985) 513.
76. J. Lesz, W. Kriegseis and A. Scharmann, *Rad. Prot. Dosim.* **17** (1986) 513.
77. V. Bichevin, G. Alseitov, M. Gapanov and H. Käämbre, *Phys. Stat. Sol. (a)* **114** (1989) K241.
78. V.S. Kortov, T.S. Bessanova, M.S. Akselrod and I.I. Milman, *Phys. Stat. Sol. (a)* **87** (1985) 629.
79. M.S. Akselrod and V.S. Kortov, *Rad. Prot. Dosim.* **39** (1991) 135.
80. P. Iacconi, F. Petel, D. Lapraz and R. Bindi, *Phys. Stat. Sol. (a)* **139** (1993) 489.
81. W.L. Medlin, *Phys. Rev.* **135** (1964) A1770.
82. A.A. Braner and A. Halperin, *Phys. Stat. Sol. (a)* **108** (1957) 932.
83. A.E. Somaiah, P. Veeresham, K.L.N. Prasad and V. Hari-Babu, *Phys. Stat. Sol. (a)* **56** (1979) 737.
84. E. Diéguez and J.M. Cabrera, *J. Phys. D: Appl. Phys.* **14** (1981) 91.
85. F. Jeffers, P.E. Wigen and J.A. Cowen, *Bull. Am. Phys. Soc.* **6** (1961) 118.
86. L. Delgado and J.L. Alvarez-Rivas, *Phys. Rev. B* **23** (1981) 6699.
87. F. Sagastibelza and J.L. Alvarez-Rivas, *J. Phys. C: Solid State Phys.* **14** (1981) 1873.
88. F.J. López, M. Aguilar and F. Agulló-López, *Phys. Rev. B* **23** (1981) 3041.
89. L. Delgado and J.L. Alvarez-Rivas, *J. Phys. C: Solid State Phys.* **15** (1982) 1591.

90. G. Prasad and K.V. Rao, *Phys. Stat. Sol. (a)* **94** (1986) 659.
91. A. Martín, F.J. López, E. Diéguez and F. Agulló-López, *Cryst. Latt. Def. and Amorph. Mat.* **17** (1987) 69.
92. P. Lecoq, P.J. Li and P. Rostaing, *Nucl. Inst. Meth. in Phys. Res.* **A300** (1991) 240.
93. P.K. Bandyopadhyay and G.P. Summers, *Sol. St. Comm.* **83** (1992) 227.
94. M. Baba, K. Shibata, Y. Kamei and T. Ikeda, *Japan. J. Appl. Phys.* **31** (1992) 1085.
95. A.I. Aleshin, V.E. Bozhevolnov, L.N. Ivanov and V.V. Karelin, *Opt. Spectrosc.* **73** (1992) 315.
96. A.R. Lakshmanan and S.S. Tiwari, *Rad. Prot. Dosim.* **47** (1993) 243.
97. V.S. Kishan-Kumar, B.S.V.S.R. Acharluyu and S.B.S. Sastry, *Phys. Stat. Sol. (b)* **175** (1993) 287.
98. P. Pokorný and A. Ibarra, *J. Phys.: Condens. Matter* **5** (1993) 7387.
99. G.K. Born, R.J. Grasser and A.O. Scharmann, *Phys. Stat. Sol.* **28** (1968) 583.
100. A. Tzalmona and P.S. Pershan, *J. Chem. Phys.* **55** (1971) 4804.
101. D.W. Cooke and C. Alexander Jr., *J. Chem. Phys.* **65** (1976) 3515.
102. A. Katzir, J.T. Süss, A. Zunger and A. Halperin, *Phys. Rev. B* **11** (1975) 2370.
103. M. Böhm, B. Cord, A. Hofstaetter, A. Scharmann and P. Parot, *J. Lumin.* **17** (1978) 291.
104. A.A. Alybakov, N. Toichiev and Sh. Akchalov, *Phys. Stat. Sol. (b)* **109** (1982) 295.
105. R. Huzimura and K. Atarashi, *Phys. Stat. Sol. (a)* **70** (1982) 649.
106. T.K. Gundu Rao, S.S. Shinde, B.C. Bhatt, J.K. Srivastava and K.S.V. Nambi, *J. Phys.: Condens. Matter* **7** (1995) 6569.
107. J.F. Boas, *Rad. Prot. Dosim.* **6** (1984) 58.
108. J.F. Boas and J.R. Pilbrow, *Phys. Rev. B* **32** (1985) 8258.
109. V.M. Orera, M.L. Sanjuàn and P.J. Alonso, *J. Phys. C: Solid State Phys.* **19** (1986) 4763.
110. S.H. Tatumi, T. Nagatamo, M. Matsuoka and S. Watanabe, *J. Phys. D: Appl. Phys.* **26** (1993) 1482.
111. M.G. Jani and L.E. Halliburton, *J. Appl. Phys.* **64** (1988) 2022.
112. N.K. Porwal, M.D. Sastry, S.V. Godbole and A.G. Page, *Molecular Physics* **72** (1991) 295.
113. B.V. Andreev, *Sov. Phys. Solid State* **34** (1992) 1304.
114. T.K. Seshagiri, V. Natarajan and M.D. Sastry, *Pramana* **44** (1995) 211.

Chapter 10
MISCELLANEOUS EFFECTS

In the present chapter we are going to include several different topics, mostly concerning thermoluminescence but some related to other thermally stimulated processes, which are important, but are not an integral part of the subjects discussed so far. We start with the important subjects of thermal fading and anomalous fading of TL which are of interest with respect to many thermoluminescent materials, and in particular, concerning those materials used for dosimetry and dating. We continue with other, more specific phenomena which are interesting due to the particular effect described, e.g. the extremely high values of E and s found for "peak 5" of LiF. Some miscellaneous TL effects in "exotic" materials such as food samples, extraterrestrial materials, and high temperature superconductors will also be discussed.

10.1 Thermal Fading

Investigators engaged in the study of either the physical properties of TL materials or in their potential and practical use as dosimeters or in TL dating, should be aware of the possibility of fading of the TL signal with time, following, or even during, the excitation. It is to be noted that if there is fading during the excitation of the sample, an apparent *dose-rate* effect can be expected, which was mentioned already in Ch. 4. It is obvious that if the same dose is imparted at a fast rate during a short period of time in one case or at a slow rate during a longer period of time in another case, then, if fading takes place during the excitation, the measured TL will be higher in the former situation. A related effect is thermal fading after the irradiation which, in principle, always occurs. Following a given excitation, the trapping states and recombination centers are filled (not necessarily to capacity) with carriers of opposite signs. Even at the temperature of excitation (say, room temperature) there is a finite probability that the carriers are thermally released from their trapping states into the delocalized bands, or any other excited state, from where they may recombine either radiatively or non-radiatively with carriers of the opposite sign in the recombination centers. If the transition is radiative, this is associated with a measurable phosphorescence which decays with time as the electrons and holes are depleted. In any case, the reduction in the number of trapped carriers results in a commensurate decrease in the intensity of the TL observed during subsequent heating, following a certain waiting period between irradiation and readout. In the case of first-order kinetics, it is easy to show that the decay curves of phosphorescence and the TL fading curves - namely, the intensity of the TL as a function of waiting time - look the same, provided that the TL intensity indeed directly reflects the remaining concentration of trapped carriers following the two stages of excitation and isothermal relaxation. Repeating Eq. 2.2, first-order kinetics yield

$$-\frac{dn}{dt} = ns\exp\left\{-\frac{E_t}{kT}\right\}, \tag{10.1}$$

the solution of which, at a constant temperature, is

$$n = n_0\exp\left\{-ts\exp\left\{-\frac{E_t}{kT}\right\}\right\}. \tag{10.2}$$

If, as suggested above, the measured TL following excitation to a concentration n_0, is proportional to the remaining concentration n after the waiting time t, the thermal decay of the TL intensity has the *lifetime* of $\tau = s^{-1}\exp\{E_t/kT\}$. If this time dependence function (Eq. 10.2) is re-inserted into Eq. 10.1, and if the phosphorescence intensity is proportional to the rate of change of the concentration of trapped carriers, the decay of phosphorescence as given by Eq. 2.4 is

$$I = I_0\exp\left\{-ts\exp\left\{-\frac{E_t}{kT}\right\}\right\}. \tag{10.3}$$

The point to stress is that in first-order kinetics, the thermal decay of TL and the phosphorescence decay must have the same functional shape which is also expressed by their having the same decay constant. This property is, however, unique to the first-order kinetics.

As an opposite example, let us consider second-order kinetics in which the basic equation is

$$-\frac{dn}{dt} = s'\exp\left\{-\frac{E_t}{kT}\right\}n^2 \tag{10.4}$$

which is solved to yield

$$n = \frac{n_0}{1 + n_0 s'\exp\{-E_t/kT\}t}. \tag{10.5}$$

This, of course, is an entirely different decay curve for which the concepts of *lifetime* or *half-life* are meaningless since the decay is not exponential. If the underlying assumption of second-order kinetics is justified, the thermal decay of TL should look like Eq. 10.5. If, on the other hand, Eq. 10.5 is inserted into

10.4, and under the assumption that the phosphorescence is associated with the rate of change of the concentration, we get

$$I(t) = \frac{I_0}{[n_0 s' \exp\{-E_t/kT\}t + 1]^2} \tag{10.6}$$

which has already been mentioned as Eq. 2.6. The two expressions 10.5 and 10.6 are not proportional to each other, and neither of them has a uniquely defined "half-life" since they are not exponential. The situation is further complicated in the general case governed by the three simultaneous differential equations 2.17, 2.19 and 2.20, as well as the relation expressed in Eq.2.22. Here, phosphorescence decay is related to $-dm/dt$ at a constant temperature as a function of the time t, whereas the remaining TL after a relaxation time t should depend on the minimum of n or m ($\min(n, m)$), where n and m are the remaining concentrations of trapped carriers of both signs in traps and centers respectively, and in some instances, on an even more complicated combination of n and m. The main general feature, however, is the same, namely, that there is thermal fading along with the phosphorescence decay, and excluding the first-order case, the two decay functions are different. In fact, the thermal decay is observed quite often when no phosphorescence is measured at all. The reason is either that the recombinations occurring during the decay are radiationless, or that the decay is relatively slow, and the accompanying phosphorescence is too weak to be measured. A further complication of the situation is in cases where several levels, sometimes a continuum of levels, are involved. Again, both thermal fading and phosphorescence are represented by decaying functions which depend on the same parameters, but in a different way.

It is obvious that the temperature dependence of $\exp\{-E_t/kT\}$ is crucial in all these expressions. Excluding the possibility of direct recombination, say via tunneling (which should include the element of the spatial proximity of the trapped electron and hole), we can distinguish between stable and unstable peaks in a qualitative way by the relation between the temperature of their occurrence and the temperature at which the sample is held prior to heating. Peaks appearing at a significantly higher temperature than the excitation temperature (or, rather, the storage temperature), are expected to be stable for long periods of time. Peaks appearing rather close to the storage temperature are susceptible to thermal fading. *Anomalous fading* discussed in the next section is anomalous in the sense that fading occurs even while the sample is held at a temperature much lower than that of the TL peak in question.

A number of examples of thermal fading of TL will be described and discussed here. In the 1966 book by Cameron et al.[1], the authors give a number of examples for the fading in several TLD materials. These include <5% in 12 weeks for a peak at 195 °C in LiF, ~10% in the first month for a peak at 200 °C in $Li_2B_4O_7$:Mn, ~10% in the first month of a peak at 260 °C in CaF_2:Mn (which appears to be anomalous fading). No detectable fading is found of a peak at the same temperature in natural CaF_2. On the other hand, a decay of 50-60% in the first 24 hours of a peak at 110 °C in $CaSO_4$:Mn is reported, which is just a normal decay at room-temperature. Understanding the effect of thermal fading is, of course, important with regard to the applications in dosimetry and dating, and can be used for the simultaneous estimation of the absorbed dose and the elapsed time following irradiation. This subject, which was studied by a number of groups and summarized by Horowitz and Yossian[2], will be discussed in some detail below. The main material used for TL dosimetry, namely LiF:Mg,Ti (TLD-100), shows this kind of *thermally induced fading* and different authors reported different rates of the fading. A summary of this problem was given by Horowitz[3]. The variety of different results includes a 7% decay in two weeks as reported by Becker[4], a 10% decay in one month reported by Mahesh et al.[5], decays of 5±3% after 84 days, 5% after 50 days, 5% after 20 days and 0-2% after 50 days reported by Horowitz[6], 5% decay found after one year reported by Oberhofer and Scharmann[7] and a 1% decay per year reported by DeWerd[8]. Horowitz[3] explains the large differences in decay rates in TLD-100 by suggesting that the different authors measured different proportions between the rapidly fading low temperature peaks numbers 2 and 3 in this material, and the more stable, high temperature peaks 4 and 5. Also, the occurrence of the anomalously high values of the E and s parameters found in peak 5, seem to play a role on the fading rate; this subject of anomalously high E and s values will be elaborated upon below. (Note, at this stage we refer only to a generic E, an "activation energy", rather than a specific E_t, a trap depth, since, as we will see, the E measured in TL analysis in these cases is not necessarily the trap depth.)

Another cause of thermal fading from LiF:Mg,Ti is not related to the release of charge from the traps at all, but rather to the transformation of the traps responsible for the TL signal. The main peak used in dosimetry in LiF:Mg,Ti is believed to be related to clusters of Mg impurities and Li vacancies, known as trimer defects. However, precipitation of the Mg from solid solution to form secondary phases occurs even at room temperature over a sufficiently long period. Thus, the signal can decrease in size as a function of time, despite the fact that the trapped charge is stable at the trimer defects. Such defect aggregation/precipitation processes are difficult to predict since they depend upon the precise defect equilibria which in turn is dependent upon the impurity concentration, annealing temperatures and times, cooling rates and storage temperatures. Under certain circumstances the main TL peak (so-called peak number 5) has even been observed to grow upon storage, rather than fade. The details of the complex behavior exhibited by this peak are reviewed by

McKeever et al.[9].

Kathuria and Sunta[10] presented experimental evidence suggesting that the isothermal decay of peak 5 in LiF:Ti,Mg is governed by non-first-order kinetics. This was based on the experimental result that the thermal decay of the TL peak was not exponential, which was evidenced by the lack of a straight line when plotting the results as $\ln I$ vs. time. Lilley and Taylor[11] gave another interpretation to these results, arguing that it is the contribution of peak 4 which explains the deviation from the exponential decay. A further debate between these two groups[12]-[14] did not resolve the issue unequivocally. Further work on the isothermal decay of peak 5 has been performed by Delgado et al.[15],[16]. Delgado and Barreiro[15] tried to resolve the question of the isothermal decay of peak 5 by measuring the isothermal decay at 165 °C. At this temperature, it appears that the isothermal decay curve is still not exponential; on the semi-log presentation the decay curve starts at higher values than the descending straight line occurring at longer times. However, if the same measurements are performed following an annealing at either 140 °C or, even better, at 165 °C, the decay curve becomes exponential, exhibiting a straight line on a semi-log scale. Therefore, Delgado and Barreiro support Lilley and Taylor[11],[13] in concluding that peak 5 is, indeed, of first order and that the deviation from the exponential decay at the beginning of the decay is related to the contribution of the lower temperature peaks.

In a later work, Delgado and Gómez Ros[16] studied the same decay of the TL peak after holding the sample for different periods of time at 165 °C while measuring phototransferred TL (PTTL). They consider this to be very convenient since the UV light, used for the phototransfer does not excite peak 4, thus, one expects only peak 5 to play a role. The effect of deviation from the exponential decay at short times is smaller in PTTL than in TL, but still exists. The authors[16], therefore, revised the previous conclusion of Delgado and Barreiro[15], suggesting that the short-time initial component is dominated by thermally activated impurity aggregation and precipitation processes modifying the structure of the traps associated with peak 5. Delgado et al.[17] reported similar measurements in the more sensitive LiF:Mg,Cu,P material. Here, the main peak is peak 4 which resembles to some extent peak 5 in LiF:Ti,Mg. These investigators performed isothermal decay experiments at temperatures between 170 and 190 °C, and found pure exponential decays, exhibited as straight lines on the semi-log scale. Furthermore, the slopes of these lines depended strongly on the temperature, as expected from the lifetime τ being dependent upon temperature with first-order kinetics, like $\tau = s^{-1} \exp\{E_t/kT\}$. As discussed in Ch. 3, a plot of $\ln \tau$ vs. $1/T$ should yield a straight line with a slope of E_t/k, from which E_t can be evaluated. Following this procedure, Delgado et al.[17] found values of an activation energy $E = 2.1 \pm 0.2$ eV from the slope, from which

$s \approx 10^{21}$ s^{-1} is evaluated. (Again , we specifically write E for the experimental parameter, since it is not clear that the activation energy evaluated is in fact the trap depth.) These anomalously high values agree with values found by the same authors using the peak shape methods, as well as parameters usually found for the main peak 5 in LiF:Ti,Mg, which appears to be the parallel of the one discussed here. This will be further discussed in section 10.3.

Lin et al.[18] reported the effects of various storage times on the effective kinetic parameters of the glow curves of CaF_2:Tm (TLD-300). In addition to the thermal decay of the TL intensity, they find a "drift" of the effective kinetic order which is different for various peaks. The most significant is the change from an order of ~1.5 to ~1.8 for peak 5 (at ~550 K). Lin et al.[18] suggest that the interpretation can be given either following Piters and Bos[19] who described the conversion from one kind of trap to another; or by the effect of competing non-radiative transitions as suggested by Kitis[20]; or by the assertion by Christodoulides[21] that general-order kinetics results from traps with the same activation energies, but a distribution of the frequency-factors. A change in this distribution of the frequency factors may thus alter the effective kinetic order.

Horowitz and Yossian[2] discussed in detail the possibility of simultaneous estimation of the absorbed dose and the elapsed time following irradiation, which may be pertinent to the appropriate medical treatment following accidental exposure. As early as 1969, Nakajima and Hashizame[22] used the fading ratios of the peak heights of two, well separated, low and high temperature peaks in single crystal CaF_2, to estimate elapsed times of up to 7 days following irradiation. Wang et al.[23] pointed out that an estimation of the elapsed time would require some knowledge of the temperature storage profile following the accidental exposure. Bos et al.[24] mentioned another complication, namely, that the fading properties may depend on the elapsed time between prior readout or annealing and the accidental exposure, the dose of which is to be evaluated. This occurs in LiF:Mg,Ti due to the thermal interaction between the low temperature peaks 2 and 3 and the dosimetry peaks 4 and 5, as already mentioned. Horowitz and Yossian[2] discuss further complications which include the possibilities that the fading properties depend on the dose for non-first-order peaks, and that the glow curve shape and time dependence may depend on the average temperature of the radiation field and the presence of neutrons. Furetta et al.[25] considered the possibility that the accidental exposure is superimposed on a significant constant background dose, in which case the estimation of the elapsed time from the shape of the glow curve is problematic due to the simultaneous decay of the TL signal and its growth due to the background dose.

Horowitz and Yossian[2] also elaborated upon the use of deconvolution of glow peaks, as discussed in Ch. 6, for the estimation of elapsed time following an irradiation, and summarized the previous work on this subject. Bacci et al.[26] seem to have been the first to use a crude form of deconvolution for this purpose, using $CaSO_4$:Dy ribbons (TLD-900). Wang et al.[23] used home-made $CaSO_4$:Dy samples and showed that the peak ratios, measured by peak height, are neither a function of the dose, nor a function of energy of the γ radiation field between 50 keV and 10 MeV. The observed variation of the peak ratio, however, with storage temperature was extremely strong. At room temperature (22 °C), the ratio P_2/P_1 varies strongly during the first 8 days, in agreement with Bacci et al.[26]. At 40 °C, the ratio P_2/P_1 increases monotonically for 30 days and the variations are significantly larger than at 22 °C. Lakshmanan et al.[27] used home-made $CaSO_4$:Dy,Na samples. The particular dopant concentration of 0.1 M% Dy and 0.5 M% Na gave well-separated peaks at 130 °C and 220 °C; the former faded rather quickly at room temperature whereas the latter was stable. The rate of decay of the low temperature peak was very sensitive to the temperature of storage/irradiation, and the authors concluded that the elapsed time can be accurately determined only if the storage temperature is known. Further work along the same lines in this material was given by Furetta et al.[25],[28].

Furetta et al.[25] also investigated the use of CaF_2:Tm (TLD-300) and CaF_2 (TLD-200) for the estimation of elapsed time following an exposure. The former material has a more well-defined glow structure, with peaks at 100 °C, 150 °C and 250 °C, which makes it a better candidate for this purpose than the latter materials with just two peak structures, at 140 °C and 200 °C. More work on the fading of TL in $CaSO_4$:Dy and CaF_2:Tm, using computerised glow curve deconvolution (CGCD) was performed by Ben-Shachar and Horowitz[29].

Moscovitch[30] investigated the applicability of LiF:Mg,Ti (TLD-100) to estimate the elapsed time using glow curve deconvolution. He made use of the fact that peaks 2 and 3 have half-lives of the order of several hours and a few months, respectively. As mentioned above concerning other materials, here too the results depended rather strongly on the storage temperature. The long-time fading of peaks 4 and 5 in LiF:Mg,Ti was described by Horowitz et al.[31]. They described different annealing/irradiation programs, and used CGCD for a better monitoring of the behaviors of the different peaks. It is to be noted that whereas peak 4, occurring below 500 K, decays in a year by only 3-4 %, peak 5, which is above 500 K decays by ~16 % in a year. This will be discussed below in connection with anomalous fading and the unusually high E and s values found for this peak. Delgado et al.[32] studied the fading of peaks 4 and 5 in TLD-100 at temperatures between 20 and 100 °C over periods of up to 30 days. The result of their study is that although peaks 4 and 5 separately behave sometimes strangely, namely that in some instances one grows with time on the expense of the other, the integral of peaks 4+5 is well behaved, namely it decreases monotonically by ~8 % per year at room temperature. Further work on the fading of peaks 4 and 5 following thermal neutron and α particle irradiation has been reported by Ben-Shachar and Horowitz[33].

Xie and Aitken[34] described the *mid-term* fading of TL in alkali feldspars. In order to explain the result that identical doses acquired during burial give smaller induced TL signals than those resulting from laboratory irradiation, they suggest the following. (i) The dating signal consists of an unstable component and a stable one. (ii) The lifetime of the latter is long compared with the age of the oldest samples of this kind. (iii) The lifetime of the unstable component is mid-term, i.e. long compared the laboratory time but short compared to the ages to be determined. Hornyak et al.[35] suggested that the decay behavior of these materials can be a result of a distribution, possibly Gaussian, of the activation energies of the traps. The shallower components of the distribution appear to be associated with the mid-term fading whereas the deeper part is responsible for the stable signal. Hornyak et al.[35] give numerical examples derived from this model which are very similar to the experimental results.

10.2 Anomalous Fading

The normal fading described so far, is associated with the thermal release of trapped carriers at ambient temperature. However, quite a number of researchers described a decay of the expected TL which occurs even after the sample is held at temperatures well below that of the TL peak. In other words, the decay was much faster than expected for the given trapping parameters and the temperature at which the sample is held. This effect was termed *anomalous, athermal* or *abnormal* fading, and, of course, it is detrimental to applications in dosimetry and dating. The first report of anomalous fading was given in 1950 by Bull and Garlick[36] for UV-stimulated TL in diamond. Two peaks, at 400 and 520 K, observed while heating the sample at 2.5 $^{\circ}Cs^{-1}$ were found to yield lower light sums if stored at 90 K for 6 hours before heating than if heated immediately. Another well-known researcher in the early stages of TL who described the effect was Hoogenstraaten[37] who reported the decay of light sums at low temperatures in ZnS samples doped with Cu, Co and Cl. He explained the effect of the decay as being due to a quantum mechanical tunneling process of electrons from traps to empty activation centers. Schulman et al.[38] reported a similar effect in CaF_2:Mn TL materials. Garlick and Robinson[39] and Blair et al.[40] found anomalous fading in lunar samples and, again, ascribed it to tunneling. Similar effects were found by Kieffer et al.[41] in organic glasses.

In the results mentioned so far, no temperature dependence of the anomalous fading was reported. Wintle[42],[43] measured the effect in various minerals at different temperatures and discussed the implications for dating archaeological samples. In particular, Wintle considered[43] the phenomenon in feldspar samples in which the anomalous fading was temperature dependent. Wintle suggested three alternative possibilities to explain anomalous fading.

(a) Defect diffusion which allows a non-radiative escape of trapped electrons when the diffusing entity encounters a trapping site. Wintle states, however, that it is difficult to explain the fading at low temperature (say, RT) as being due to the diffusion of defects, and this seems to suggest that diffusion is usually not the cause of anomalous fading. In any event, one would expect that the activation energy for defect diffusion would be similar to that for the thermal release of trapped electrons, and thus if one is not expected at low temperatures, neither is the other. Moharil et al.[44] suggested that anomalous fading in KBr may be explained in terms of the mobility of dislocations, but the same concerns apply here also.

(b) Direct transfer of electrons from a trap to an *adjacent* center, as suggested by Garlick and Robinson[39]. The temperature dependence reported by Wintle[43] of the anomalous fading in feldspars needs, however, a more complicated model.

(c) One should consider whether anomalous fading can be caused by the reduction in the number of effective recombination centers with time, due to a different mechanism which is not related to the active trapping state responsible for the TL process.

Hasan et al.[45] reported the results of anomalous fading of TL in meteorites. The TL seems to be associated with the low-density feldspar-rich fraction in the Dhajala meteorite, in those belonging to the shergottite class and probably also in the Kapaeta meteorite. It was explained that the data are consistent with a mechanism for anomalous fading which involves the overlap between the wave-functions for the excited and ground state electrons. The authors point out that, fortunately, most meteorites contain feldspar in the high-temperature form and studies of their natural TL should not be hampered by anomalous fading. Tyler and McKeever[46] challenged the conclusions of Hasan et al.[45]. They considered the localized transition model, previously suggested by Templer[47],[48] following an earlier discussion by Chen and Kirsh[50], and show that the TL decay for both the low-temperature and high-temperature forms is not consistent with tunneling, but rather, the qualitative features are well explained by the localized transition model. Templer[49] also claimed that the anomalous fading in zircon and pumice may be associated with the localized transition model rather than by tunneling. For more work on TL in meteorites, see later in this chapter.

Some additional studies of anomalous fading of TL and OSL in sanidine and other feldspars were reported by Spooner[51] and by Visocekas[52]. In fact, in the latter work, the anomalous fading of TL was accompanied by red and infra-red emission of light, attributed to tunneling radiative recombination. This effect of tunneling-related emission of light had previously been found by Visocekas et al.[53] in $CaSO_4$:Dy where a TL peak at ~250 °C anomalously decays with time at RT and below, down to LNT, practically independently of the temperature. A similar phenomenon was also reported by Visocekas[54] in labradorite. In both cases, the explanation offered is that the afterglow and anomalous fading result from a quantum mechanical tunneling effect.

Sanderson[55] also studied the anomalous fading of TL in feldspars and suggested that fading cannot be ascribed to simple proximity models based on randomly distributed traps and recombination centers. Visocekas et al.[56] reported the results of anomalous fading of TL, OSL and IRSL in different feldspar samples and attributed them to the tunneling effect. These authors suggest models, relating fading via tunneling to the disorder in the crystals. Zink and Visocekas[57] concentrated on sanidine feldspars and showed that the blue emission cannot be used for dating of the samples due to anomalous fading. On the other hand, the infra-red TL appears to be much more stable, and it can be used for the dating of lavas. A curious observation described by Zink and Visocekas[57], as well as by Zink et al.[58], however, is that the tunneling afterglow emission is in the IR, not the blue, whereas it is the blue TL emission which fades anomalously, not the TL emitting in the IR. No clear explanation of this effect has been offered. Fragoulis and Stoebe[59] ascribed at least part of the apparent anomalous fading observed in quartz to the strong anomalous fading taking place in feldspar since feldspar inclusions were observed in natural quartz grains taken from a variety of locations.

Clark and Templer[60], while referring to materials used in archaeological dating, suggested that although a large proportion of TL samples exhibit anomalous fading, it is possible to evaluate ages for such samples. They hypothesized that each mineral phase which suffers from anomalous fading consists of separable stable and unstable components. They showed that it is possible either to correct the age determination after standard TL measurements have been made, or to remove the unstable TL prior to performing measurements by holding the samples at elevated temperature. Stoneham and Winter[61] and Bowman[62] utilized the same technique to improve the reliability of authenticity testing. Göksu et al.[63] studied obsidians from various origins for the purpose of using them in high level radiation fields and at high temperatures. In the high dose studies they observed that above the saturation levels, which varied between samples from different sources, the TL intensity decreased with increasing dose. This appears to be related to the fact that close to the saturation dose, the decay of the high temperature peak is dose dependent. The authors suggest that at high doses, new centers are created which are less stable.

Visocekas[64] studied the tunneling afterglow following α and β irradiations in CaF_2:Mn and alumina TLDs, and explained the TL fading in these materials, along with the accompanying long-term afterglow, to be associated with the radiative effects. Also, Kitis et al.[65] performed different measurements in natural fluoroapatite, using different grain sizes, annealing temperatures, and irradiation temperatures as well as different pre-doses, and concluded that tunneling is the probable mechanism responsible for the anomalous fading in apatite.

Vlasov and Karezin[66] investigated the anomalous fading of $BaSO_4$:Eu. They suggested that when the Eu content exceeds 1.5 M%, clusters of 8-9 atoms are formed. A wavelike dependence of the fading on storage time at 300 K was interpreted as being related to electron retrapping at the Eu cluster. Many other descriptions and examples of anomalous fading can be found in the literature, and a summary is given by McKeever[67] of the major pre-1985 literature.

10.2.1 Models for Anomalous Fading

It is a characteristic of anomalous fading that the initial fading rate is quite rapid, followed by a slower decay at longer times. The conclusions reached by Visocekas and colleagues[68] from a study of the fading from calcite is that the intensity I of the afterglow observed during anomalous fading follows a hyperbolic law, namely $I \propto t^{-1}$, where t is time. A temperature independent hyperbolic law is indicative of a tunneling mechanism in accordance with Mikhailov's model[69] for tunneling. Here, an electron in a trap of depth E_t is separated by a distance r from a positive charge, at which recombination would produce the emission of a photon. The rate of tunneling of the electron through the potential barrier separating the two centers was given in Ch. 2 (Eq.2.117) as

$$P(r) = P_o \exp\{-\alpha r\} \tag{10.7}$$

where P_o is a frequency factor and α was given by

$$\alpha = \frac{2\sqrt{(2m^* E_1)}}{\hbar}. \tag{10.8}$$

This is obtained by approximating the electron trap to a square potential well, as shown in Fig.2.17, with an effective electron mass m^* and a barrier height E_1 at distance r. If the decrease in the number of trapped electrons n follows a first-order reaction, the luminescence (afterglow) intensity is given by

$$I = -\frac{dn}{dt} = \int_0^\infty n_0^2 \exp\{-P(r)\,t\}\, 4\pi r^2 dr \tag{10.9}$$

where n_0 is the initial trapped electron concentration and the total luminescence emitted during the afterglow is found by summing Eq.2.121 over all r. For large

$r \approx r_c$ (see Ch. 2), this becomes

$$I_{tot} \simeq -K/t \tag{10.10}$$

where K is approximately constant and thus the observed t^{-1} law is obtained[70].

Eq.10.10 holds if $t \gg t_i$ where t_i is the irradiation time and t is measured from the end of the irradiation at $t = 0$. If an extended irradiation time is taken into account, then

$$I = \frac{K}{t_i} \int_{-t_i}^{0} \frac{dt'}{(t - t')} = \frac{K}{t_i} \ln\left(1 + t/t_i\right). \tag{10.11}$$

Beyond $t = 10t_i$ the difference between Eq.10.10 and Eq.10.11 is negligible.

Alternatively, instead of monitoring the phosphorescence (afterglow) emitted during charge leakage from the trap, one could monitor the remnant TL - i.e. the TL remaining after the charge has escaped. If we define the parameter R to be the ratio of the TL at time t to the initial TL at time t_o, then we may write

$$R = \frac{TL\,(t)}{TL\,(t_o)} = -\frac{\ln\left(t/t_m\right)}{\ln\left(t_m/t_o\right)} \tag{10.12}$$

where t_m is the maximum time for which the tunneling mechanism holds. It is to be noted that since tunneling is more likely for the deeper traps (Fig.2.17, and Eqs.10.7 and 10.8) then, as pointed out by Templer[70], the loss of the TL signal is expected to be more rapid from the *high temperature side* of the glow curve than from the low temperature side. This is the opposite that one would expect from thermal fading or, as we will see below, from fading due to a localized transition model, and thereby provides an experimental tool by which fading due to tunneling can be recognized.

Visocekas et al.[71] also considered the possibility that the electron had to be thermally excited to an intermediate state (energy E_2 in Fig. 2.17) before tunneling could take place. The tunneling probability will now include a Boltzmann term and the probability of this "thermally assisted tunneling" will increase as the temperature increases. Such appears to be the case in calcite[71]. If n_2 is the population of the excited state and n_1 is the population of the ground state in Fig.2.17, then

$$\frac{n_2}{n_1} = \exp\left\{-\frac{E_2}{kT}\right\} \tag{10.13}$$

where the total charge density is $n_0 = n_1 + n_2$. The total tunneling probability is now

$$P\,(r) = \frac{P_o(\exp\{-\alpha_1 r\} + \exp\{-\alpha_2 r\})}{1 + \exp\{-E_2/kT\}} \tag{10.14}$$

where $\alpha_1 = 2\sqrt{(2m^* E_1)}/\hbar$ and $\alpha_2 = 2\sqrt{(2m^*(E_1 - E_2))}/\hbar$, and a temperature dependence is expected.

As noted above, and discussed in Ch. 2, Templer suggested an alternative to tunneling to describe anomalous fading[47],[48]. This was the localized transition model in which an alternative pathway to transport via the conduction band is available via an intermediate, localized state. The mathematics describing this case have already been described in Ch. 2. Here we note that the phosphorescence (afterglow) intensity expected from this case is obtained from Eq.2.126 and is

$$I = n_0 \left[\frac{\nu s}{\nu + s}\right] \exp\left\{-\frac{E_e}{kT}\right\} \exp\left\{-\left[\frac{\nu s}{\nu + s}\right] t \exp\left\{-\frac{E_e}{kT}\right\}\right\} \tag{10.15}$$

where all the terms have been previously defined in Ch. 2. Templer[70] describes the shape of the phosphorescence decay expected from the localized transition model if the traps are distributed in energy. For a general distribution of traps the afterglow intensity is given by

$$I = \int_0^\infty n(E_e) \left[\frac{\nu(E_e) s}{\nu(E_e) + s}\right] \exp\left\{-\frac{E_e}{kT}\right\} \exp\left\{-\left[\frac{\nu s}{\nu + s}\right] t \exp\left\{-\frac{E_e}{kT}\right\}\right\} dE_e. \tag{10.16}$$

For a uniform distribution $n(E_e)\, dE_e = n_0 dE_e$ we get

$$I = \frac{n_0 kT}{t}\left[\exp\left\{-\frac{t}{t_{\max}}\right\} - \exp\left\{-\frac{t}{t_{\min}}\right\}\right] \tag{10.17}$$

where $t_{\max}$ and $t_{\min}$ are the trapped charge lifetimes corresponding to the upper and lower limits of the energy distribution. Again, the $1/t$ component dominates except at short times when the exponential part causes a slower than $1/t$ decay. If the irradiation time t_i is to be accounted for we have

$$I = n_0 kT \int_{-t_i}^{0} \left[\frac{\exp\{-(t - t')/t_{\max}\} - \exp\{-(t - t')/t_{\min}\}}{t - t'}\right] dt'. \tag{10.18}$$

where t' is a dummy variable for time. Figure 10.1 shows an example of an attempt to fit Eqs. 10.10, 10.11 and 10.17 to the phosphorescence afterglow from irradiated oligoclase feldspar, after a post-irradiation anneal at 175 °C for 2 minutes to remove the thermally unstable component. Tyler and McKeever[46] argue that the localized transition model fits the data better than the tunneling model, and that the misfit between the data and the former model most likely arises from the assumption of a uniform trap distribution.

If the remnant TL is monitored, rather than the afterglow, one observes that the fading occurs from the *low temperature side* of the glow curve for the

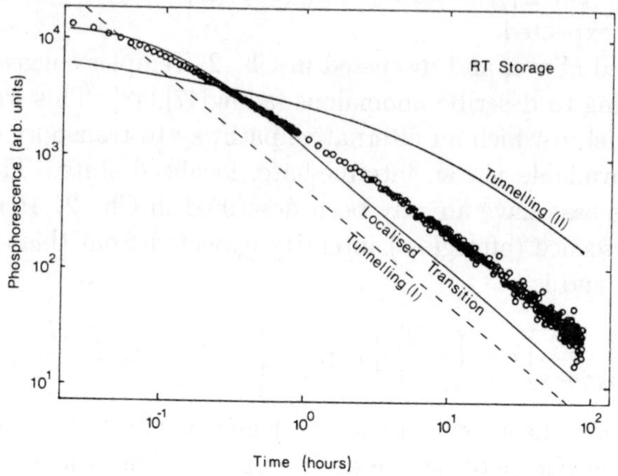

Figure 10.1 Phosphorescence decay from feldspar showing the best fits to the data using the localized transition model and tunneling models - with, II, or without, I, compensation for the irradiation time.
(After Tyler and McKeever [46].)

localized transition model, in contrast to that expected for tunneling[70]. In this case the remnant TL parameter R is described by

$$R = \frac{Ei\,[-t/t_{\min}] - Ei\,[-t/t_{\max}]}{Ei\,[-t_0/t_{\min}] - Ei\,[-t_0/t_{\max}]} \tag{10.19}$$

where $Ei\,[x]$ is the exponential integral function.

10.2.2 Anomalous Fading as Thermal Fading in Disguise

Chen and Hag-Yahya[72] presented the possibility that under some circumstances, "normal" thermal fading of TL may look like anomalous fading. This will be shown here in some detail. The very closely related subject of the possibility of explaining abnormally high E and s values will be discussed in the following section.

To illustrate this point, let us consider a first-order peak and use the shape method given by Eq. 3.60 as an example for all the shape methods,

$$E = 2.29kT_m^2/\omega \tag{10.20}$$

where $\omega = T_2 - T_1$ and where T_1 and T_2 are the half intensity temperatures, and T_m is the temperature of the maximum of the peak. It is evident that if by any artificial means, the peak looks narrower than otherwise expected, the *apparent* activation energy evaluated by Eq. 10.20 will be greater than the real

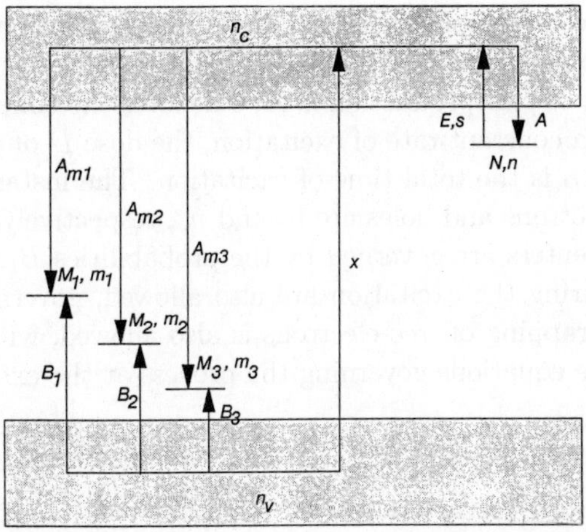

Figure 10.2 Energy levels model of one trapping state and three kinds of recombination centers. The transitions shown are those taking place during the excitation. The meaning of the different symbols is given in the text. (After Chen and Hag-Yahya[72].)

one. Similar results are expected with other shape methods for evaluating the activation energy. Curve fitting methods utilize basically the same features of the TL peak, and, therefore, similar results are expected. Obviously, once we get for a certain value of T_m a narrow peak with a high apparent value of E_t, denoted here by E_{app}, the accompanying frequency factor s_{app} will be orders of magnitude higher than the real one. This can be seen from the condition for the maximum of a first-order peak, 1.5, which can be re-written as

$$s = \beta E_t / \left(kT_m^2\right) \exp\left\{E_t/kT_m\right\}, \tag{10.21}$$

and since the energy term appears in the exponent, a large apparent value E_{app} (instead of the real value E_t) will result in an apparent frequency factor s_{app}, orders of magnitude higher than the actual value of s.

Let us consider now a model of one trapping state and three recombination centers as shown in Fig. 10.2. The main idea is that one of the centers, denoted by 2, is radiative whereas the two others act as non-radiative competitors and cause the peak to be anomalously narrow while preserving the characteristic first-order shape. We will write the appropriate simultaneous rate equations governing the process during excitation, relaxation and heating, solve them numerically and follow the shape of the visible glow-peak. In Fig. 10.2, one trapping state is shown with a given activation energy E_t and frequency factor s. The concentration of this trapping state is N and its instantaneous occupancy n. Three recombination centers with concentrations of M_1, M_2 and M_3 are assumed to participate in the process. During the excitation by external radiation, the rate of pair production is X; t denotes the time during the excitation. Assuming a constant rate of excitation, the dose D of excitation is thus $D = Xt_D$, where t_D is the total time of excitation. The instantaneous concentrations of free electrons and holes are n_c and n_v, respectively. The transitions of holes into the centers are governed by the probabilities B_1, B_2 and B_3, and recombinations during the excitation are also allowed, governed by A_{m1}, A_{m2} and A_{m3}. The retrapping of free electrons is also allowed, with the retrapping probability A. The equations governing the process at the excitation stage are

$$dm_1/dt = -A_{m1}m_1n_c + B_1(M_1 - m_1)n_v, \tag{10.22}$$

$$dm_2/dt = -A_{m2}m_2n_c + B_2(M_2 - m_2)n_v, \tag{10.23}$$

$$dm_3/dt = -A_{m3}m_3n_c + B_3(M_3 - m_3)n_v, \tag{10.24}$$

$$dn/dt = A(N - n)n_c, \tag{10.25}$$

$$dn_v/dt = X - B_1(M_1 - m_1)n_v - B_2(M_2 - m_2)n_v - B_3(M_3 - m_3)n_v, \tag{10.26}$$

$$dn_c/dt = dm_1/dt + dm_2/dt + dm_3/dt + dn_v/dt - dn/dt. \tag{10.27}$$

As usual, the numerical solution of these simultaneous equations can be performed by the use of a suitable Runge-Kutta algorithm.

Following the excitation, a relaxation period should be simulated using the same algorithm. At the end of the excitation, non-zero concentrations of electrons in the conduction band and holes in the valence band are expected. In order to follow the normal experimental procedure, we should consider the relaxation time at constant temperature, until the conduction and valence bands are practically empty. This can be done numerically by solving the set of equations 10.22-10.27 with the excitation intensity X set to zero.

The next stage is that of heating. The transitions involved are shown in Fig. 10.3. The trapped electrons are now thermally raised into the conduction band during heating and may recombine with the three centers M_1, M_2 and M_3. Also, retrapping of electrons with empty electron trapping states may take place. The equations governing the process are

$$-dm_1/dt = A_{m1}m_1n_c, \tag{10.28}$$

$$-dm_2/dt = A_{m2}m_2n_c, \tag{10.29}$$

$$-dm_3/dt = A_{m3}m_3n_c, \tag{10.30}$$

$$dn/dt = -sn\exp\{-E_t/kT\} + A(N - n)n_c, \tag{10.31}$$

$$dn_c/dt = dm_1/dt + dm_2/dt + dm_3/dt - dn/dt. \tag{10.32}$$

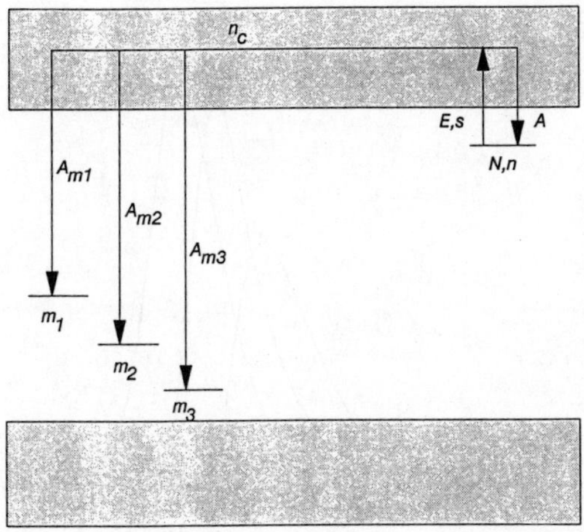

Figure 10.3 Transitions during the heating of a sample with three recombination centers, namely, the same model as in Fig. 10.1. The meaning of the symbols is also the same.

(After Chen and Hag-Yahya[72].)

Here too, the numerical solution should be performed with the final values of the previous stage serving as initial values in the present one. The intensity of emitted visible thermoluminescence is proportional to the rate of change of m_2, and assuming the proportionality factor to be unity (which merely defines a certain set of units for the emitted light), we can write the intensity of emission $I(T)$ as

$$I(T) = -dm_2/dt. \tag{10.33}$$

The effect of the two other centers, #1 and #3 is to make the visible TL peak (due to recombination at center #2) narrower than otherwise expected, thus resulting in higher *apparent* values of E_t and s than those inserted into the calculations as the physical parameters.

An example of the calculations, as given by Chen and Hag-Yahya[72], is given in Fig. 10.4. The parameters chosen for the trapping state are $E_t = 0.7$ eV and $s = 10^6$ s^{-1}. The decay time at room temperature (300 K) for this trap is $\tau = s^{-1}\exp\{E_t/kT\} = 5.75 \times 10^5$ s, which is just under a week. The other parameters chosen are the retrapping probability $A = 1 \times 10^{-22}$ m^{3}s^{-1}; the three recombination probabilities are $A_{m1} = 1.2 \times 10^{-18}$ m^{3}s^{-1}, $A_{m2} = 2.8 \times 10^{-19}$ m^{3}s^{-1}, $A_{m3} = 7.5 \times 10^{-20}$ m^{3}s^{-1}. Obviously, this is a case with nearly negligible retrapping. The trapping probabilities for holes into the respective recombination centers are $B_1 = 1 \times 10^{-21}$ m^{3}s^{-1}, $B_2 = 1.1 \times 10^{-21}$

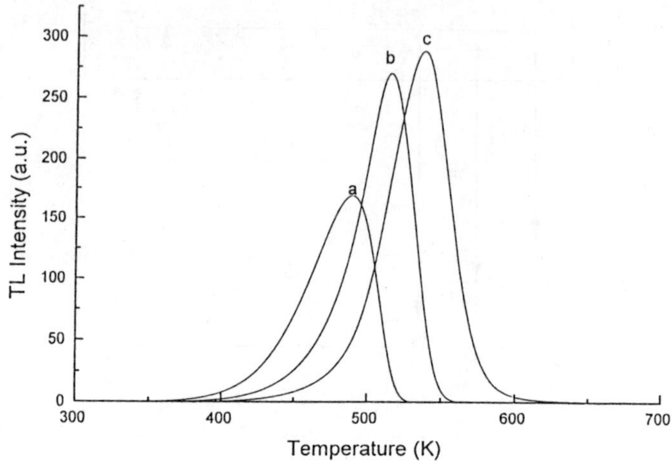

Figure 10.4 The rates of transitions into the three competing centers as a function of temperature. (a) $-dm_1/dt$; (b) $-dm_2/dt$; and (c) $-dm_3/dt$, as calculated by numerical solution of the rate equations. Curve (b) represents the measurable glow peak. The values of the parameters chosen are given in the text.
(After Chen and Hag-Yahya[72].)

m^3s^{-1}, $B_3 = 1.2 \times 10^{-21}$ m^3s^{-1}. The total concentration of traps is $N = 1.2 \times 10^{21}$ m^{-3} and the concentrations of the three centers are chosen to be $M_1 = M_2 = M_3 = 4 \times 10^{20}\ m^{-3}$. The radiation intensity in the appropriate units is $X = 5 \times 10^{20}\ m^{-3}s^{-1}$ and the irradiation time is 200 s. The simultaneous equations governing the process during the excitation and heating are solved numerically in three stages as explained before: excitation, relaxation - in which the excitation intensity X is set to zero - and finally, the heating of the sample at a heating rate of 5 Ks^{-1} is followed, yielding the rates of the three transitions, the visible $-dm_2/dt$ and the two invisible ones, $-dm_1/dt$ and $-dm_3/dt$. These are shown in Fig. 10.4; of course, one would like to focus on curve (b) which is assumed to be the only measurable one.

Peak (b) looks exactly like a simple TL peak. The maximum temperature is found at 515 K. The symmetry factor defined in Ch. 3, $\mu_g = \delta/\omega$ where $\delta = T_2 - T_m$ and $\omega = T_2 - T_1$, and where T_m is the temperature at the maximum, and T_1, T_2 are the low and high temperatures at half intensity, respectively. The value found here is $\mu_g = 0.424$ which is typical of first-order kinetics. Using the equations for evaluating E_t from a first-order peak yields $E_{app} = 1.3$ eV. Using the maximum condition 10.21 results in an apparent frequency factor of $s_{app} = 1.6 \times 10^{12}\ s^{-1}$. Using these values, the apparent life-time at 300 K is found to be $\tau_{app} = s_{app}^{-1} \exp\{E_{app}/kT\}$ which yields 4.3×10^9 s or ~ 137 years, nearly four orders of magnitude higher than the real one.

In a similar calculation, the activation energy chosen was 0.8 eV with all the other parameters kept the same. In this case, the life-time expected at room temperature is $\tau = s^{-1} \exp\{E_t/kT\} = 2.75 \times 10^7$ s, which is about 45 weeks. When the calculation is carried out in a similar way to that discussed above, solving numerically the same sets of equations, the radiative peak was found at 584 K, and looks like a first-order peak with a shape factor of 0.425. The apparent activation energy and frequency factor found here are, respectively, $E_{app} = 1.5$ eV and $s_{app} = 2 \times 10^{12}$ s^{-1}, yielding $\tau_{app} = 7.9 \times 10^{12}$ s, or $\sim 2.5 \times 10^5$ years. The ratio between the apparent and real life-times here was $\tau_{app}/\tau \approx 3 \times 10^5$. Obviously, this is a demonstration of a situation in which calculation of the trapping parameters from the peak shape leads one ot expect no decay of the peak at any measurable length of time, but an observable decay of $\sim$ 4.5 % per two weeks would be found.

This description is similar to the argument used in Ch. 2 (section 2.5) when describing the need for emission spectra in the analysis of TL peaks. In that case we discussed how emission at two different wavelengths will lead to two different TL peaks, each of distorted shape (specifically, narrower than they should be) if the TL is monitored at the two different wavelengths (Fig. 2.21). However, monitoring the total emission, with equal efficiency at each wavelength, will yield the true peak shape. In the present case, the extra recombinations (two of them) are both non-radiative and cannot be detected. This leads to a narrowed TL peak and an erroneous estimate for the trap depth and frequency factor. Another relevant factor, also discussed in section 2.5, is thermal quenching which, if undetected, can also lead to anomalously narrowed peaks and false trap depth values.

In summary, this description of anomalous fading as being normal fading in disguise, should be considered as one of the possibilities for explaining the effect. In any specific case where anomalous fading is observed, more detailed evaluation of the experimental results should be made to see what is the physical reason for the occurrence of the effect. In some cases the anomalous fading is temperature dependent and in others, not. It is quite obvious that the present explanation refers to the former case in which the rate of fading is highly temperature dependent.

10.3 The Occurrence of Very High E and s Values

According to Mott and Gurney[73], the frequency factor of a trapping state is related to the Debye lattice vibration frequency, which can vary in the range of $10^7 - 10^{13}$ s^{-1}. Lower values have been reported in the literature, so the "normal" range of s can be considered to be in the range of $10^6 - 10^{13}$ s^{-1}. In some special cases, values of s much beyond this, e.g. 10^{21} s^{-1} or even more have been reported, accompanied by relatively high values of the activation energy E. These were considered to be "non-physical" values, and some explanations were given to their occurrence. The most well known case is that of peak 5 in LiF:Mg,Ti (TLD-100). The experimental values of E and s appearing in the literature vary a lot, depending on the method used for the evaluation of the parameters. In some cases, even using the same methods by different authors resulted in significantly different results. Zimmerman et al.[74], Blak and Watanabe[75], Johnson[76] and Yossian et al.[77], using isothermal decay methods found values of $E \approx 1.25$ eV and $s \approx 4 \times 10^{10} - 2 \times 10^{12}$ s^{-1}. On the other hand, Taylor and Lilley[78] reported an activation energy of 2.06 eV and a frequency factor of 2×10^{20} s^{-1} for the isothermal decay results in peak 5. As for the method of different heating rates, Gorbics et al.[79] reported $E = 2.4$ eV and $s = 1.7 \times 10^{24}$ s^{-1} whereas Pohlit[80] gave the exceedingly high values of $E = 3.62$ eV and $s = 10^{42}$ s^{-1}. Fairchild et al.[81] suggested that the kinetics of this peak as well as other peaks with unusually large s values might actually be complex and that the apparent first-order behavior is a special case (or approximation) of the more complex kinetics situation. For the specific case of peak 5, McKeever[82] offered the explanation that the TL is related to trimer dissociation. This work is based on previous studies by Taylor and Lilley[83],[84] and McKeever[86]. Taylor and Lilley discussed the clustering of $Mg^{2+}V^-$ pairs into dimers and trimers, and showed that the latter clustering followed third-order kinetics, and a trimer binding energy of 0.89 eV was determined. Taylor and Lilley[84] also suggested that peak 2 in TLD-100, which had been ascribed to $Mg^{2+}V^-$ pairs, is in fact associated with an impurity other than Mg^{2+}. They postulated that it is due to Ti^{4+} associated with O_2^-, OH^- and/or vacancies, which they term Ti*. Peak 5 appears to be a combination of $(Mg^{2+}V^-)$ pairs and the peak 2 trap, which is now labelled $Ti^*(Mg^{2+}V^-)_n$. McKeever[82] proposed that the activation energy of peak 5 is the sum of the trimer binding energy and a term related to the release of charge from the dissociated dipoles, and the high value of s is thus derived from the high E_{app} value. Bos et al.[24],[87] studied in detail the effects of changing the cooling and heating rates, and of the annealing conditions on the calculated values of E_{app} and s_{app}. These authors showed that the slower the heating or cooling rate, the smaller the E_{app} value. These authors clearly show that the E_{app} and s_{app} values obtained are dependent upon the defect structure (concentrations of dipoles, trimers, etc.) existing in LiF:Mg,Ti during the time of heating and, furthermore, that this defect structure is non-constant during peak 5 readout - i.e. there is a dynamic lattice disorder. Tale et al.[88]-[90] discuss the perturbation to the electronic structure caused by changing the ionic defect structure during TL readout. They suggest that changes to the coulombic field from the vacancies, dipoles, trimers, etc., as a result of the dissociation/precipitation reactions, lead to perturbations of the activation energy and an increased value of E_{app} (and, consequently, of s_{app}). The experimental evidence for dynamic lattice disorder in LiF:Mg,Ti during all phases of use of these TLD materials (pre-irradiation storage, irradiation, readout and post-irradiation storage) is extremely strong. Measurements of the effects of clustering of defects under a variety of conditions upon the TL properties of this material have clearly established that ionic trap modification processes occur and have significant effects upon the TL behavior of this material.[9] In view of this, and in view of the evident perturbation on the real E_t-values it is surprising, perhaps, that the Randall-Wilkins model is found to describe the TL peak shape in such a satisfactory manner. All attempts at fitting the glow curve in LiF:Mg,Ti have concluded that Randall-Wilkins analysis fits the experimentally obtained glow peaks perfectly. However, this simple model is unable to account, for example, for the variation of E_{app} and s_{app} with heating or cooling rate[94]. Bos and Piters[91] modified the conventional Randall-Wilkins model and allowed for thermally stimulated defect interactions of the type discussed above (i.e. dynamic disorder) to take place during the heating cycle and examined the effects of changing the heating rate on the simulated glow curves. The results indicated that the E_{app} values obtained from fitting the simulated TL peaks with the Randall-Wilkins expressions for TL were governed not just by the trap depth, as assumed in the Randall-Wilkins model, but also by the activation energy for the thermally stimulated defect interaction (e.g. diffusion energy for the motion of ionic species). This is similar to the earlier suggestion of McKeever noted above[82], although the determined activation energy is not the simple sum of activation energies suggested by McKeever. Both the glow peak area and the determined value of E_{app} were observed to vary with heating rate in an ostensibly similar fashion to that observed in practice.[94]

Fairchild et al.[92] included in their analysis the possibility of simultaneous hole and electron release. If analyzed using the conventional expression for TL (vis. equation (2.6)) this will lead to an activation energy of $E_{app} = E_{te} + E_{th}$ and a frequency factor of $s_{app} = s_e s_h$, where the e and h subscripts refer to the hole and electron traps, respectively. It is difficult, however, to see how this mechanism fits in to what is known about the TL process in LiF:Mg,Ti. Moreover, in their equation, $s_e s_h$ is multiplied by a constant (which should have the dimensions of time, and may be a small quantity). Townsend et al.[93] suggested that the electron trap is a Mg-defect/V-center complex in LiF. A result of this is that the normal TL equation has to be multiplied by a term $s \exp\{-E_t/kT\}$ and this in turn results in an expression for TL which contains terms $E_{app} = 2E_t$ and $s_{app} = \tau_{ex} s^2$ where τ_{ex} is the lifetime for the V_3-center to remain separate from the Mg site. Here too, The high value of s_{app} is explained only if τ_{ex} is not very small.

Finally, an international effort to compare results of the glow curve in LiF by numerical deconvolution (see also Ch. 6), showed practically unanimously the same very high activation energy and frequency factor values for peak 5[94]. de Vries et al.[95] (see also Horowitz and Yossian[2]) showed that following a high ^{60}Co γ-irradiation as well as after irradiation by ^{244}Cm α particles, the glow curve changes substantially. A peak occurring at 379 K has high E_{app} and s_{app} values of 1.73 eV and 10^{22} s^{-1}, respectively, after γ rays, and 2.26 eV and 7×10^{21} s^{-1}, respectively, after α particles. Also, a peak at 498 K, now called peak 6, had $E_{app} = 2.24$ eV and $s_{app} = 3.2 \times 10^{21}$ s^{-1} following the α irradiation and $E_{app} = 2.26$ eV and $s_{app} = 6.9 \times 10^{21}$ s^{-1} following the high γ irradiation. Gartia and Dorendrajit-Singh[96] reported even higher effective values of $E_{app} = 2.53$ eV and $s_{app} = 2.5 \times 10^{25}$ s^{-1}. They point out that the exact kinetics are complicated and that the peak is not simple first order with just unusual values of the parameters. Also, Bacci et al.[97] reported peaks at $\sim$467 K with $E_{app} = 1.64$ eV, $s_{app} = 1.1 \times 10^{17}$ s^{-1} and at $\sim$508 K with $E_{app} = 2.00$ eV, $s_{app} \approx 10^{19}$ s^{-1} for high-sensitivity LiF:Mg,Cu,P. High E_{app} and s_{app} values were also reported for this material by McKeever et al.[98] who noted the possibility that the capture cross-section for the trap may be thermally activated (i.e. a repulsive center) according to

$$\sigma(T) = \sigma_\infty \exp\left\{-\frac{\Delta E}{kT}\right\} \tag{10.34}$$

where σ_∞ is a temperature independent capture cross-section and ΔE is the activation energy for electron capture. This relationship can arise when a large lattice relaxation occur via multiphonon interaction after the defect has trapped an electron. In considering the nature of the defect complexes responsible for the TL mechanisms in LiF:Mg,Ti, the idea of a large lattice relaxation is not difficult to accept since the Mg-clusters are held together by coulombic forces which will undoubtedly be modified during electron capture. Thus, a lattice relaxation in the vicinity of the defect after capture should not be unexpected. A consequence of this is that by performing the usual analysis of the glow curves, using the methods outlined in Ch. 2, what one actually measures is not the trap depth E_t, but $E_{app} = E_t + \Delta E$, with an appropriately higher frequency factor s_{app}. McKeever et al.[98] analyzed the glow curves in LiF:Mg,Cu,P and got slightly better fits when a capture cross-section of the form of Eq. 10.34 was assumed. However, because of the difficulty in analyzing the glow curve from LiF:Mg,Cu,P no definitive statements were possible and evidence for a multiphonon process is currently not conclusive.

The work of Chen and Hag-Yahya [72] offers a possible explanation for the occurrence of very high activation energies and frequency factors as a result of competition with non-radiative centers, using a model which is very similar to that given in the previous section for accounting for the anomalous fading. The basic idea is that in all shape methods for evaluating the activation energy, including curve-fitting, there is an inverse relation between the activation energy and the width of the peak, as exemplified by Eq. 10.20. Thus, if non-radiative competitors to the luminescent center cause an *artificial* narrowing of a peak, the apparent activation energy will be larger than expected. In order for the peak to keep looking like a first-order one, two non-radiative competing centers were assumed[72] which cause narrowing of the peak on its both high and low temperature sides. Once an apparent high activation energy is found with the maximum temperature practically unchanged, the use of Eq. 10.21 results in an anomalously high value of the apparent frequency factor. The energy level model shown in Fig. 10.2 can be used here for the excitation period of the simulation. The set of equations written by Chen and Hag-Yahya[72] is, in fact, 10.22-10.27; these can be solved numerically for a given excitation intensity X. In the same way as explained above, a relaxation period following the excitation can be simulated by setting the irradiation rate X to zero and solving the equations for a further period of time until the free carriers concentrations n_c and n_v go down to negligible values. Following this, the heating stage can be simulated by solving equations 10.28-10.32 which govern the transitions shown in Fig. 10.3 for a given heating rate β. The magnitudes of the transition rates into the three centers are given by $-dm_1/dt$, $-dm_2/dt$ and $-dm_3/dt$ which can be recorded as a function of temperature, keeping in mind that only the middle one is assumed to be measurable.

Solving the sets of simultaneous equations numerically using a fifth-order Runge-Kutta method led to the results of the intensities $I_1(T)$, $I_2(T)$ and $I_3(T)$, shown in curves A, B and C of Fig. 10.5, respectively. [72] The parameters used were $E_t = 1.2$ eV, $s = 2.5 \times 10^{11}$ s^{-1}, $A = 10^{-22}$ m^{3}s^{-1}, $N = 1.2 \times 10^{21}$ m^{-3}, $A_{m1} = 1.2 \times 10^{-18}$ m^{3}s^{-1}, $A_{m2} = 2.8 \times 10^{-19}$ m^{3}s^{-1}, $A_{m3} = 7.5 \times 10^{-20}$ m^{3}s^{-1}, $B_1 = 1.0 \times 10^{-21}$ m^{3}s^{-1}, $B_2 = 1.1 \times 10^{-21}$ m^{3}s^{-1}, $B_3 = 1.2 \times 10^{-21}$m^{3}s^{-1}, $M_1 = M_2 = M_3 = 4 \times 10^{20}$ m^{-3}. For the excitation it was assumed that $n_0 = m_{10} = m_{20} = m_{30} = n_{c0} = n_{v0} = 0$, $X = 10^{21}$ m^{-3}s^{-1}. The total excitation time was $t_D = 35$ s, and $T_0 = 300$ K. For the heating phase, $\beta = 6$ Ks^{-1} and the initial and final temperatures were $T_i = 300$ K and $T_f = 600$ K, respectively. As mentioned already, it was assumed that only curve B is measurable, and in this sense, it represents the glow peak of interest - i.e. a peak with unusually high values of E and s. This peak looks like a "normal" first-order peak with the symmetry factor $\mu_g \approx 0.42$. However, by using Eq. 10.20 for evaluating the activation energy, one gets $E_{app} = 2.236$ eV. Substituting this into Eq. 10.21 yields $s_{app} = 9.3 \times 10^{21}$ s^{-1}. These values should be compared with the original ones inserted into Eq. 10.31 which were here $E_t = 1.2$ eV and $s = 2.5 \times 10^{11}$ s^{-1}. Figure 10.6

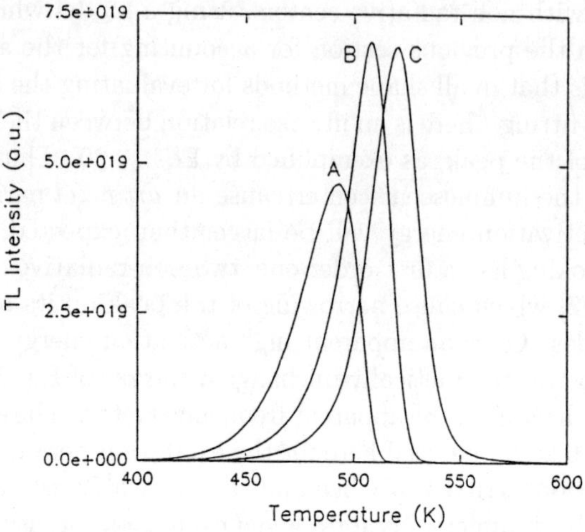

Figure 10.5 The rates of recombination into centers 1,2 and 3 calculated from the model with two competing non-radiative centers. The relevant parameters are given in the text. Curve B is repeated separately in Fig. 10.6.
(After Chen and Hag-Yahya[72].)

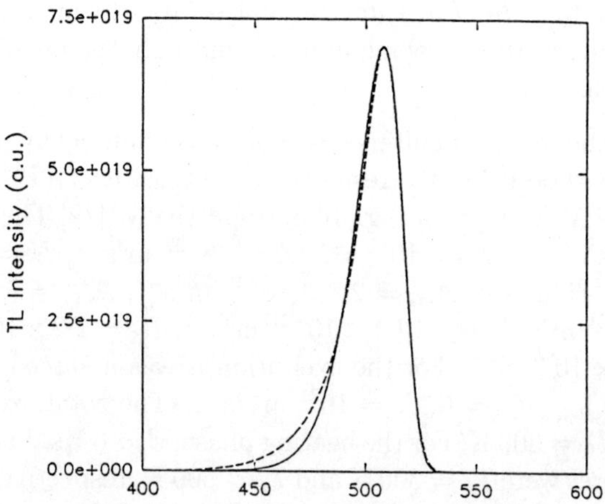

Figure 10.6 Curve B from Fig. 10.5 as simulated by the solution of the simultaneous differential equations is given by the dashed line. The best fitted first-order curve is shown as the solid line.
(After Chen and Hag-Yahya[72].)

shows a first-order best-fit to this peak, where the dashed line represents the peak calculated by the numerical solution of the equations, and the solid line is the best-fitted, first-order curve. Although at the low temperature range there is a slight discrepancy between the two, the agreement is reasonable. The effective parameters found by the best-fit method were $E_{app} = 2.115$ eV and $s_{app} = 5.6 \times 10^{20}$ s^{-1} which are in good agreement with those found by the half-width method (and, of course, with the experimentally found values in peak 5 of LiF:Mg,Ti). Thus, the simulation shows that such a competition model can explain the occurrence of the anomalously high activation energies and frequency factors which, according to this model, are merely apparent, effective values. For the specific case of LiF:Mg,Ti, however, it is not yet clear how this model can explain the changes in the evaluated parameters as a function of heating rate.

10.4 TL in Plastically Deformed Samples; Effects of High Pressure

Thermoluminescence and thermally stimulated conductivity are related to imperfections - impurities and defects - in crystals. It could be expected, therefore, that if the application of high pressure on a sample produces new defects it would influence the TL (and TSC) properties of the sample. This can be expressed in two main ways, one is that the high pressure produces TL and thus, if the sample is heated following its application, TL peaks will be observed. The other way is that a change is induced by the pressure in such a way that if the sample is irradiated following the deformation, the resulting TL will be different, usually higher, than if it is equally irradiated without the application of the pressure. These changes in the sample may or may not be reversible. For example, heating a sample during the TL readout causes a recovery of the sample in certain cases, and a further irradiation will result in a "regular" glow curve. Due to this sensitivity to pressure, TL has been used to investigate the mechanism of defect production in samples under pressure. Generally, TL induced by mechanically grinding, cutting or polishing of a TL material is known as tribothermoluminescence.

Gabrysh et al.[99] investigated the TL induced by high pressure in Al_2O_3 (sapphire and ruby) and found a similarity between the γ-induced and pressure-induced glow peaks. Aramu et al.[100] reported the results of TL in mechanically deformed ZnS and discussed the order of kinetics of the TL peaks as well as that of the decay of phosphorescence. Clark and Newman[101],[102] studied the effect in KBr and KCl. In KBr, which had undergone a uniaxial compression in a vice at 80 K, a series of TL peaks as well as a single TSC peak were observed during warming up to 290 K. Up to four new electron traps were produced by the dislocation motion in KBr. To some extent, these depended on the nature of the point defects (color centers) already present in the material. Recombination appeared to be into a hole center which may have been related to interstitial halogen. From similar measurements in KCl, it was concluded that the traps in this sample may be small anion-cation vacancy clusters produced by the jogged motion of dislocations during plastic deformation. KCl samples had previously been examined by Ueta et al.[103] who reported an increase of the response to UV light with the deformation. For a peak at 250 K they concluded that the increase in the TL intensity was due to an increase in the recombination probability. They also discussed the effect of plastic deformation after irradiation; it is not clear how the effect of pressure excitation of TL was eliminated from the combined effect. Similar work on γ-irradiated KCl was later described by Ausin and Alvarez-Rivas[104]. Also, Fieschi et al.[105] investigated the enhancement of x-ray-excited TL in KBr by the application of pressure, whereas the possibility of direct pressure inducement of TL was not mentioned.

Similar measurements have been performed on LiF. Petralia and Gnani [106] found an increase of peak 1 and a decrease of peak 5 as a result of deformation. Heating the sample during the measurement to 350 °C sufficed to anneal it so that its response to irradiation returned to normal. An effect of increase due to deformation was also observed in two peaks below room temperature. Niewiadomski[107] reported the results of the application of pressure on LiF samples before and after γ irradiation. He concluded that pressure produces different types of ion displacements in the crystal lattice. Some of the shifted ions create new electron traps while others impede the luminescence recombination process. On the other hand, pressure applied to the exposed phosphor seems to empty the filled traps. In a later paper, Kos and Mieke[108] studied TL in LiF:Ti deformed by uniaxial compression. The glow curve of these samples consists of only two peaks, at 120 °C and 170 °C, and the effect of pressure is basically to suppress the glow intensity. It was deduced that the TL decreases due to an enhancement of F-centers by anion vacancies created during the deformation, thus increasing non-radiative transitions.

Petit and Duval[109] tested the TL properties of plastically deformed ice. Again, the effect of the applied pressure was to reduce the TL signal, showing a relation between the number of traps and that of point or line defects by the deformation. Newton and Sibley[110] studied deformation-induced damage in MgO, and found that the blue TL is larger for deformed samples. Hook and Drickamer[111] studied the effect of pressure on variously doped ZnS samples. In ZnS:Cu,Cl, for example, two peaks, at $\sim$130 K and $\sim$240 K were found to increase with the applied pressure. These peaks were also found to shift with pressure, the 130 K peak to lower temperature and the 240 K peak to higher temperature. They associated the shift of the latter with an increase in the activation energy of the trap.

Joshi et al.[112] investigated deformation-induced TL in NaCl:Ca. Marked changes in the TL pattern were reported to be the result of thermal cycling. This, in turn, appeared to be due to the break-up of the impurity-vacancy dipole aggregates and the subsequent arrangement of the free dipoles. Mariani et al.[113] investigated TL and F-center thermal annealing in plastically deformed KBr samples, γ-irradiated at room temperature. The TL curve showed two characteristic peaks, not seen in the undeformed samples. The new peaks appeared to arise from recombination of F-centers with interstitial halogen atoms thermally released from deformation-induced traps. Baer and Drickamer[114] studied the TL glow curves of a number of x-irradiated alkali halides at pressures up to 3.5 GPa. The pressure-induced changes in the trap depth and in the TL from F-centers were found to be quite significant for NaCl, KCl and KBr. The effect of pressure on the TL of M-centers in the high-pressure phase of KCl and KBr was also shown. Discontinuities in the trap depths and the temperature at which the TL is at maximum intensity occur at the phase transition for both KCl and KBr.

Drickamer and his group also reported the effects of pressure on TL in other materials. In crystalline coronene ($C_{24}H_{12}$) they reported[115] that the general effect is to decrease the energy of charge carrier traps and to change their relative occupation. The energy of the shallowest occupied traps decreases, over a 40 kbar range, from 0.18 to 0.11 eV as a result of the difference in the perturbation of the charge-carrier-induced dipole interactions close to defects relative to those far away from them. Also, the occupation ratio of the deeper traps increases relative to that of the shallow traps with increasing pressure. Dreger et al.[116] studied the effect of pressure on the TL, photoluminescence and optical absorption of auramine O, both in the crystalline state and dispersed in poly(acrylic acid). In the crystalline samples, two discontinuous changes were observed, one was in photoluminescence and absorption spectra in the pressure region of 0-3 kbar; the other was mainly seen in the TL as a decrease in the trap energy for the charge carriers. Auramine O dispersed in the polymer exhibited only monotonic changes under pressure using any of these three methods. Finally, Baer and Drickamer[117] described the shift in the temperature of glow peaks of γ- and x-irradiated quartz as a function of pressure. Here the results indicated an electron trap which became shallower with increasing pressure.

10.5 Unusual Effects of Excitation of TL

In this section we describe two unique effects related to the excitation of TL. The first is the inability to excite a TL peak in a certain temperature range, below that of the peak, although it is found that the peak can be excited by

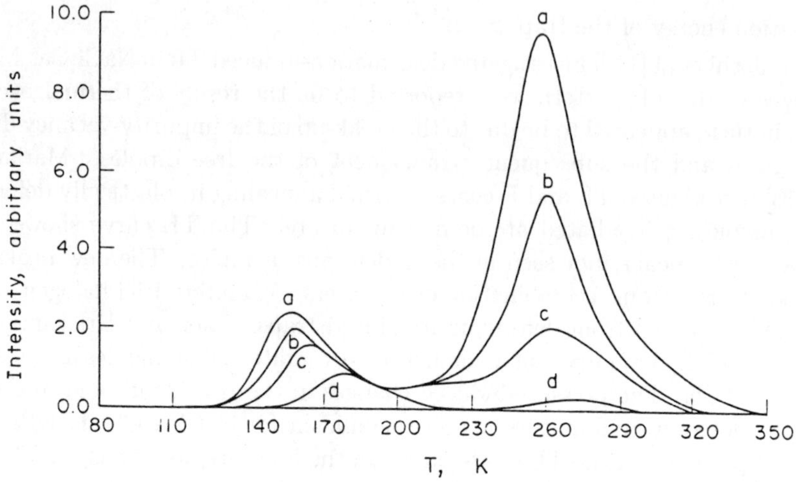

Figure 10.7 Excitation of TL at various temperatures in semiconducting (type IIb) diamond by 360 nm UV light at (a) 80 K; (b) 130 K; (c) 140 K; and (d) 150 K. (After Winer et al.[118].)

the same irradiation at yet lower temperatures. This effect was reported in $CaWO_4$ by Sayer and Souder[119] and in semiconducting diamonds by Winer et al.[118]. The other effect also occurs in semiconducting diamonds, and has to do with the pre-excitation of the sample at 77 K by light of 700 nm; this by itself does not excite any TL, but enhances substantially the TL recorded following a subsequent UV irradiation.

In most of the studies of thermoluminescence, as well as studies of other radiation-induced thermally stimulated processes - mainly TSC and TSEE - the peak in question is excitable at any temperature below the maximum of the peak. For excitation temperatures in the vicinity of the maximum, the observed intensity of the peak is a result of a combined effect of excitation on the one hand and thermal emptying, usually accompanied by phosphorescence, on the other. In this case, the intensity of the peak strongly depends on the time elapsed from the end of the irradiation and, actually, due to the thermal fading during excitation, it also depends on the length of time it has taken to excite the sample with the given dose. In other words, the result depends on the dose rate. The effect discussed here is certainly understandable in terms of the basic theory of thermally stimulated processes. However, Winer et al.[118] and Sayer and Souder[119] have found cases in which TL peaks were not excitable at temperatures substantially below that of the peak, although they could be excited at still lower temperatures.

As mentioned already in Ch. 4, the glow curve of UV-excited semicon-

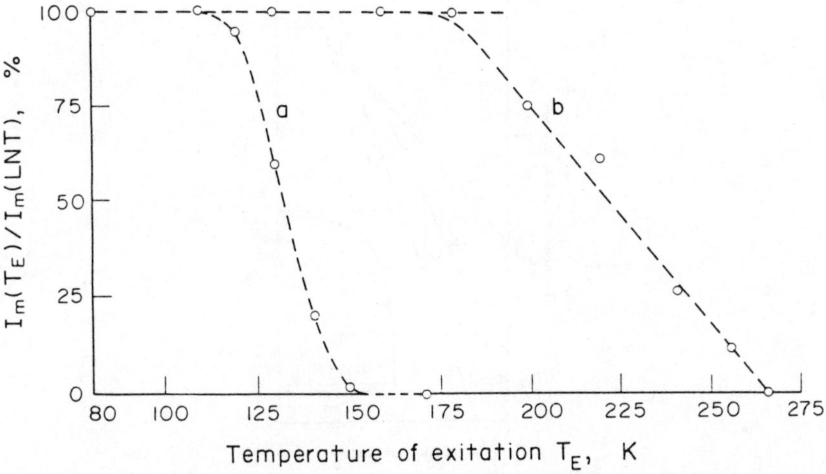

Figure 10.8 The dependence of the maximum intensity I_m of the 250 K blue peak in IIb diamond on the temperature of excitation for: (a) 360 nm; (b) 225 nm excitations.

(After Winer et al.[118].)

ducting diamonds (type IIb) consists of two peaks, at ~150 K and ~250 K; an example is shown in curve (a) in Fig. 10.7 which is the glow curve after irradiation with 360 nm at 77 K, taken from Winer et al.[118]. Curves (b), (c) and (d) are obtained after the same dose of irradiation at 130 K, 140 K and 150 K, respectively. As expected, the higher the temperature of excitation, the lower the low temperature peak; note that the maximum shifts to higher temperature. The surprising point is that the ~250 K peak decreases as well, and it can hardly be excited above 150 K. The maximum intensity as a function of excitation temperature is shown in curve (a) of Fig. 10.8. For the sake of comparison, similar measurements were performed when the excitation was by band-to-band (225 nm) UV light. These results are illustrated in curve (b) of Fig. 10.8 and the behavior is normal in the sense that the intensity decreases only when the temperature of excitation approaches that of the peak. The explanation of the effect[118] is connected with that of the strong superlinearity observed in these diamonds with the lower-than-band-gap photons, as described in Ch. 4. The model assumes a multistage transition of electrons from the valence to the conduction band for the 250 K peak. It appears that one of the intermediate levels hypothesized for the explanation of the superlinearity is the trapping state or recombination center which takes part in the occurrence of the 150 K peak. That is the reason why this level becomes unstable at 150 K and therefore, electrons cannot reach the conduction band and hence they are not

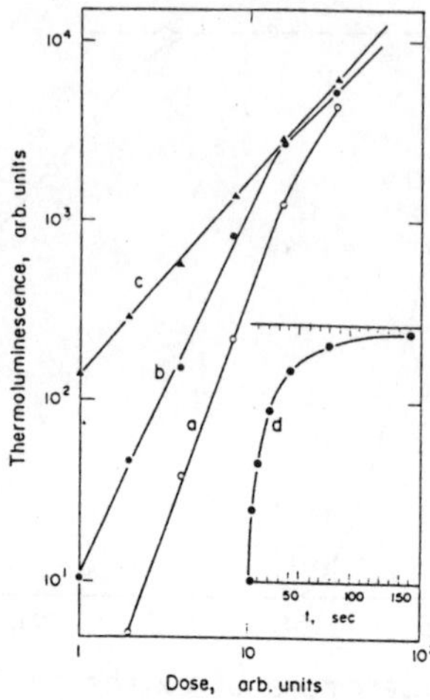

Figure 10.9 The effect of red (700 nm) pre-excitation, prior to the UV excitation (360 nm) of the 250 K peak in semiconducting diamond. (a) Excitation by 360 nm alone. (b) Same 360 nm doses, each preceded by a low 700 nm dose. (c) Same as (b), but with the 700 nm exposure 200 times higher. (d) The 250 K peak intensity as a function of the 700 nm pre-excitation dose.

(After Chen and Halperin[120].)

able to reach the electron recombination center. This picture is strengthened by the behavior under 225 nm excitation. In this case, the intermediate levels play no role since electrons can be directly raised from the valence to the conduction band and, therefore, this anomalous effect does not take place.

The results by Sayer and Souder[119] on $CaWO_4$ doped with Cu or Na, are surprisingly similar to those in semiconducting diamond. Two main peaks, at ~160 K and ~260 K are reported for excitations by UV light at 80 K. The authors show, however, that in both Cu- and Na-doped samples, the 260 K peak could not be excited above ~180 K. The comparison with semiconducting diamonds cannot be carried further, however, because the work was done with unfiltered UV light from a mercury lamp. Thus, no difference in the dose dependence could be observed with different wavelengths. Also, the results with x-rays cannot help in the study of this analogy since, in $CaWO_4$, x-rays induced the ~160 K peak but not the peak at ~260 K. The main feature, however, namely that with UV light the high temperature peak cannot be excited at temperatures above that of the low temperature peak, appears very clearly in the different $CaWO_4$ samples, and is the same as in semiconducting diamond.

Returning to semiconducting diamond, Chen and Halperin[120] found that the TL behavior changes if the sample is illuminated by longer wavelength light, e.g. 700 nm prior to the UV (300-400 nm) excitation. Although the 700 nm light does not excite TL by itself, the TL observed after the combination 700 nm + 360 nm, is much higher than after 360 nm alone, as shown in Fig. 10.9. Curve (a) is very similar to the dose dependence shown in Fig. 4.3. Apart from the fact that the points on curve (b) and (c) are much higher than in (a), it is to be noted that curve (c) is practically linear with the UV dose. The explanation given is that the 700 nm light can raise the electrons to the above-mentioned intermediate states associated with the multi-stage transition from the valence to the conduction band. This explains both the higher intensity for the same UV dose, and the linearity following a high 700 nm illumination.

10.6 Regeneration and Self-Dosing

A number of researchers reported the re-occurrence of TL in different materials following annealing of the sample to temperatures high enough to erase all the pre-existing TL. It appears as if, in these situations, just holding the sample at a certain temperature long enough is sufficient to re-excite certain TL peaks, without applying any external excitation (e.g. irradiation). The term *regeneration* has often been used to describe this effect; a term which should not be confused with the *regeneration method* discussed in Ch. 7 with respect to TL dating (a method which includes repeated irradiation of the sample; see, for example, Prószyńska-Bordas et al.[121]). The subject of regeneration should always be considered along with the possibility of self-dosing of samples which include radioactive materials such as uranium or thorium. This subject was briefly discussed with respect to the simultaneous TL-EPR measurements in uranium-containing samples in Ch. 9. Also, in extremely sensitive materials, cosmic rays, and possibly other background irradiation, may produce TL even in a well-sealed sample. The possibility of an inadvertent exposure to UV light, say, from a fluorescent lamp, which may excite TL, should also not be ruled out. However, in a number of works in the literature, a mysterious effect of regeneration is reported, which apparently is not related to the above-mentioned factors.

Templer[122] reported the *auto-regeneration* of TL in zircons, held at 77 K. Due to the high internal radioactivity - typically 10-1000 ppm of uranium and thorium - and a relatively high TL sensitivity, TL builds up in zircon when it is held at liquid nitrogen temperature. Templer showed that it was possible to detect auto-generated TL signals from single zircon grains stored for three weeks. Pashchenko et al.[123],[124], while trying to study the fading characteristics of TL in $KCl:Eu^{2+}$, found that self-irradiation by ^{40}K needs to be taken into consideration. These authors developed a method for calibration of $KCl:Eu^{2+}$ based on this self-irradiation. The calibration proved to be very useful for doses of the order of 0.5 mGy. Delgado et al.[125] investigated the self-dosing of LiF TLD-100 (LiF:Mg,Ti) and GR-200 (LiF:Mg,Cu,P) TL detectors. These TL materials had been exposed in a salt mine, where the background radiation was as low as 0.7 $nGyh^{-1}$. After six months, in the highly sensitive LiF GR-200, the intrinsic self-dose brought about a clear TL signal in excess of that caused by the background. The amount of self-dose was estimated to be $\sim$0.8 $nGyh^{-1}$. With LiF TLD-100, no appreciable TL signal could be detected after storage. One of the reasons for the difference seems to be a better stability of the GR-200 samples.

Practically all the reports of "genuine" non-radiation-induced (NRI) TL came from the group including Charalambous and colleagues. Petridou et al.[126] reported the appearance of a TL peak in LiF (TLD-100) at 130 °C followed by a steeply rising peak at higher temperatures. These were measured after a 50 kGy dose to the sample, followed by a first heating to 400 °C. A second heating to measure the non-radiation-induced peak was performed after an interval of time between 4 hours and 3 days in which the sample was held in the dark. The authors considered two possibilities for explaining the effect. One is that the heavy irradiation plus heating to 400 °C sensitized the sample so as to yield higher TL than expected otherwise. The source of irradiation could be cosmic rays or the background radiation field. This, however, was rejected by the result of another measurement in which the sample was exposed to a small cobalt source; the results did not agree with the hypothesis of an increased sensitivity. The other possibility considered was related to tribothermoluminescence, but this also was discarded due to the fact that the intensity of the NRI peak was nearly proportional to the time interval between successive heatings. The authors settled on a model in which carriers are released thermally from shallower traps and get trapped in deeper ones; however, this assertion is suspicious since the electrons in the shallow traps should have been depleted during the heating to 400 °C. Charalambous and Hasan[127] reported a similar regeneration effect in CaF_2:Dy. The sample had been irradiated with 30 kGy, heated to 300 °C, cooled to 77 K and held for 10 hours. In a subsequent heating, TL peaks were found at 40 °C and 83 °C. The model given was based on the formation of a peculiar composite defect composed of migrated defects and trapped electrons. Further results on the same material were given by Kitis et al.[128]. They mention briefly that the small signal observed can be considered to be the result of some kind of internal radiation source. Hasan et al.[129] reported the occurrence of regenerated TL in $CaSO_4$:Dy. The explanation given by them consisted of the postulation of "RTL triplets" or "composite traps". The three participants of the triplet are assumed to be a deep trap of electrons or holes, a trapped electron or hole and a slowly moving defect. The latter were believed to be created during the irradiation and to diffuse very slowly towards the surface. They assumed that such a defect can be captured by the field of the trapped carrier and produce a composite occupied trap which may have a smaller activation energy than the original trap. Thus, the trapped carriers may now be released at a relatively low temperature and so produce TL. Due to this effect, Kitis et al.[130],[131] recommended the use of pristine samples for measuring extremely low doses. Some further results on CaF_2:Dy were reported by Kitis and Charalambous[132]. In these measurements, an initial γ dose of 10 kGy was given at different temperatures from 77 K to 500 K. The results were considered in view of the same triplet model mentioned above.

10.7 TL in High Temperature Superconductors

A number of groups have studied the TL and TSEE properties of several high temperature superconductors, discovered in the mid-eighties. Fujiwara et al.[133] observed a characteristic x-ray-excited TL from the high-T_c superconducting Tl-Ba-Ca-Cu-O system. The TL glow curve consisted of 4 peaks, all occurring between 100 and 200 K, with activation energies between 0.1 eV and 0.3 eV, and emission spectra ranging from the UV to the visible. The range of temperatures of TL occurrence is to be compared with the onset of superconductivity at 120 K with zero resistance below 107 K. Fujiwara and Kobayashi[134] studied the TL of the Er-Ba-Cu-O and Nd-Ba-Cu-O systems. They ascribed their results to electronic transitions between weakly crystal-field split spin-orbit levels of trivalent Er and Nd ions. Also, TL peaks were found, again in the range of 100-200 K and with activation energies of 0.1-0.3 eV, in the Gd-Ba-Cu-O system when excited by x-rays. Furthermore, TL was also observed in the same temperature range in Ho-Ba-Cu-O and Bi-Sr-Ca-Cu-O.

Another group that dealt with TL of high-T_c superconductors is that of Cooke et al.[135]. They demonstrated that TL could be an important method for detecting insulating impurity phases in $YBa_2Cu_3O_{7-\delta}$. The same group[136] studied the TL properties of high purity Y_2O_3 which is used as a starting material for fabricating high-T_c superconductors. They irradiated the samples with ^{60}Co and x-rays, and found TL which was linearly dependent on the concentration of Tb impurities. Cooke et al.[137] reported the results of TL in different high temperature superconductors (HTS) following γ irradiation. The TL appears to be related to the granularity of the material and the consequent inclusion of insulating materials in grain boundaries. The high optical absorption of these opaque materials implies that TL emanates from within $\sim$1 μm of the surface. Therefore, TL can be used to assess the quality of HTS by detecting the presence of insulating surface impurities which, in general, harm the technological properties of the material. For example, the surface resistance R_s is determined by the HTS properties within the London penetration depth, which is the region probed by TL. The presence of insulating surface impurities contributes to the high-frequency losses as manifested by an enhanced R_s, and is directly correlated with TL. Jahan et al.[138] showed that TL is applicable to HTS films in detecting secondary insulating impurities such as Y_2O_3, $BaCO_3$, Y_2BaCuO_5 and $BaCuO_2$. They also report TL glow curves of the frequently used single crystal substrates such as Al_2O_3, MgO and $LaAlO_3$. The results show clearly the presence of insulating impurities in the HTS films by comparing the TL of film-on-substrate and substrate alone.

Katz et al.[139] studied the TL from ErBaCuO, YbBaCuO as well as (Bi,Pb)SrCaCuO, following x-irradiation at 20 K. A close similarity was found in the structure of the glow curves for all three compounds in the 20-300 K range. The strongest glow peaks appeared below the superconducting transition temperature. The origin of the glow peaks appears to be common in all these superconducting oxides. Most of the defects are associated with oxygen vacancies; in one glow peak at 80 K in ErBaCaO, the TL can be attributed to a specific kind of oxygen vacancy since it is demonstrated that the TL intensity is closely related to the oxygen stoichiometry of the material. Burrafato et al.[140] applied TL for the determination of precursor oxides in high T_c superconductors, perovskites of the (BiPb)SrCaCuO type.

Finally, two works on the related subject of TSEE should be mentioned. Yaskolko et al.[141] studied the thermally stimulated emission from $YBa_2Cu_3O_{7-\delta}$ and found a peak with a maximum T_m at 95 K, close to the critical temperature T_c of the material, and three other peaks at 148, 192 and 363 K. Chrzanowski[142] reported TSEE peaks in the same material at 330 K and 405 K.

10.8 TL of Extraterrestrial Materials

Thermoluminescence has been measured in observations of the Moon using telescopic equipment, as well as from lunar samples brought back by the Apollo astronauts. The emissions from the Moon observed visually for decades, mainly during lunar eclipses, were first referred to as luminescence by Link[143]. The suggestion was made by several authors, e.g. Ney et al.[144], Sydran[145] and Blair and Edgington[146] that the mechanism for this lunar emission could be TL and, therefore, that it should vary with the lunar phase. Chanin et al.[147] seem to have proven that this lunar emission is indeed due to TL. It appears that the energy stored by lunar dust and rocks during the polar night is released as luminescence during the increase in the temperature of the lunar surface after local dawn.

TL measurements have been performed on lunar samples in the laboratory following sample retrieval by US manned Apollo missions and the former

USSR unmanned Luna probes. Among the groups involved, Blair et al.[40] found rather strong anomalous fading in lunar material, which prevented the use of accumulated TL for dating the material. The TL intensity in these samples was also found to be too feeble to explain *transient lunar phenomena*, namely, the occurrence of an anomalous brightness of the lunar surface, as previously suggested by Blair and Edgington[146]. Inference has been made by Durrani et al.[148],[149] on the lunar temperature from TL measurements and in particular, on temperature differences between shaded and unshaded samples and between surface and deeper samples. In one case[149], special care was taken to keep the lunar sample refrigerated from the time of collection to the time of TL measurement. Several authors used the assumption that the TL from lunar material (primarily from the feldspar component) is in equilibrium in order to study TL as a function of depth. The assumption of equilibrium leads to the conclusion that the TL level is determined by the irradiation temperature and thus by studying TL as a function of depth beneath the lunar surface one can study the penetration of the diurnal heatwave, and the outflow of heat from the Moon's interior [150] -[152]. At thermal equilibrium we may set the rate of charge carrier build-up equal to the rate of thermal drainage with the result that the level of trap filling at equilibrium for first-order kinetics is

$$\frac{n}{N} = \frac{TL_{nat}}{TL_{sat}} = \frac{1}{[1 + (s/Rr)\exp\{E_t/kT\}]} \tag{10.35}$$

where r is the dose rate, R is the probability per unit dose of filling an empty trap, TL_{nat} is the natural TL level and TL_{sat} is the saturation TL level. The other terms have their usual meaning. There are many obvious limitations associated with using this simple equation (kinetic order, saturation TL limited by N, TL proportional to the level of trap filling, etc.) We have seen in the preceding chapters many examples of where these assumptions can breakdown. Nevertheless, with the additional assumption that r is invariant with depth one can use the TL-versus-depth profile to estimate the heat flow and subsurface temperature.

Chanin et al.[147] summarize the results of TL studies of lunar material. Among the conclusions reached is that low temperature storage of the TL was necessary to explain the high level of luminescence; samples taken in the shade exhibited a much larger luminescence than the sun-exposed material. Moreover, the behavior of the samples subject to temperature variations indicates that the maximum of luminescence should appear for a surface temperature of about 200 °C, which corresponds to an angle of 10° from the terminator (the dividing line between the illuminated and unilluminated part of the Moon's disk). This is consistent with the results and explains why the rise of luminescence can never be observed since the earliest data in the lunar cycle are observed around 10° from dawn.

Some work on thermally stimulated electron emission from lunar materials has been performed by Gammage and Becker[154]. It was found that the extreme radiation damage in the exterior region of regolith grains inhibits TSEE. Removal of the interfering outer layer by etching permits the observation of natural TSEE from deeper within. Artificial TSEE between 130 and 550 °C is reported to be a sensitive tool for studying surface oxidation and structural changes.

TL studies of meteorites have been extensive over the past few decades, particularly by Sears and colleagues[155], and have provided a rich vein of information about these extraterrestrial materials - including the rare Martian meteorites[156]. Examinations of the TL sensitivity of meteoritic materials have provided insight into the shock history of certain meteorite classes[157]-[159], the metamorphic history (and thereby the metamorphic classification)[160], and effects such as aqueous alteration[161]. Suggestions have also been made to use the shape of the glow curve obtained following irradiation in the laboratory as a paleothermometer[162].

Studies of the natural TL of meteorites are important because the natural TL signal contains information concerning the irradiation and thermal history of the material, albeit convoluted with the shock and metamorphic history. Most models of the TL from meteorites lead to the conclusion that the natural TL level is in thermal equilibrium. Estimates reveal that equilibrium is attained within $\sim 10^5$ years, whereas the cosmic ray exposure age of most meteorites is of the order of 10^6 years (thus leaving sufficient time for equilibrium to be reached)[155]. Using an expression such as Eq. 10.35 one can use the observed natural TL level to impose constraints upon the possible meteorite orbits[163]. However, this approach is only successful for those meteorites which are known to have fallen to Earth within the last year or two (i.e. meteorite "falls"). Once on the Earth, the natural TL level starts to re-equilibrate to a new level commensurate with the higher temperature and lower environmental dose rate experienced on the Earth's surface. Thus, the meteorite loses its in-space information and establishes a natural TL signal descriptive of its terrestrial location. One can then use the rate of decay of the natural TL from its high, in-space level (low T, high r) to its lower terrestrial level (high T, low r) to place bounds upon the length of time the meteorite has been on Earth. If the rate of decay from its in-space level to the new terrestrial can be established, and if one can place limits on the initial TL level, one can use the natural TL of older meteorites (i.e. "finds") to establish limits for the terrestrial age[164].

10.9 Thermoluminescence of Irradiated Food and Organic Materials

A somewhat unusual group of materials which has been studied using TL consists of various kinds of food. To mention briefly some early work, Chadwick and Oosterheert[165] reported TL from irradiated tomato seeds, Inoue et al.[166] studied the effect in wheat leaves, and Desai et al.[167] investigated the TL of spinach. Tatake[168] reviewed TL measurements in various bio-molecules such as nuclear acid bases and discussed the conclusions that one can draw from them regarding electronic transitions. In a later work, Jahan and Cooke[169] studied the TL of UV- and x-irradiated caffeine in the temperature range of 77-300 K. The x-induced glow curve produced peaks at 102, 128, 158 and 198 K, whereas the UV-excited curve is characterized by peaks at 120 and 208 K. These authors analyzed their TL results and found the trapping parameters, namely, the order of kinetics, the activation energy and the pre-exponential factor, and presented an energy-level diagram depicting the thermally stimulated process in irradiated caffeine.

Extensive work on the TL of different foodstuffs was carried out in the late eighties and into the nineties by a group which included Heide, Bögl, Helle and Schreiber. The main point in this work has to do with the use made of irradiation of food for the purpose of extending the shelf-life of the food by the reduction of pathogenic micro-organisms. The method is considered by many consumers with suspicion, however, one of the reasons being the lack of methods used by the food controlling agencies to detect irradiation treatment. Schreiber et al.[170] summarize in a review paper previous work on three methods in use, namely, TL, ESR and the detection of volatiles. In this review, Schreiber et al. describe a previous work by Sanderson et al.[171] which established the source of TL from spices and herbs. It appears that minerals contaminating the foodstuffs are often responsible for TL. The minerals can be separated from the organic components by density separation in a centrifuge in a high density solution. Sanderson et al. used the method of calibration employed in archaeological and geological dating. After the TL reading, the mineral sample is re-exposed to a known source, and a second TL curve is recorded. Schreiber et al.[172] investigated the TL of irradiated shellfish. Again, minerals were isolated from the guts of the animals. Normalized TL signals of non-irradiated samples were clearly separated. By calculation of the differences of TL intensities between non-irradiated and irradiated samples, an optimized integration range for glow curves was determined. Schreiber et al.[173],[174] reported the detection of low dose irradiation in fruits and vegetables by TL. Again, the isolated minerals contaminating the products were the source of TL.

Kispéter and Kiss[175] studied various food industry products by TL methods. These included milk protein concentrate powder, whole egg powder and vital wheat gluten powder. The authors explain their results by using the organic semiconductor band model, showing that the samples indeed behave as organic semiconductors. Khan and Ehlermann[176] describe the TL of quartz which is a common contaminant in food materials. In a later work, Khan and Delimcée investigated the detection of irradiation treatment of dates using the TL mineral contaminants. A comprehensive work by Pinnioja[178] summarizes the TL results of 300 lots of herbs, spices, berries, mushrooms and seafood. It appears that the calcite contaminant is the most reliable agent for detecting TL, even two years after irradiation. In the work by Calderón et al.[179]-[181], it was also evident that inorganic dust is the origin of TL observed in irradiated herbs and spices. The doses used in food irradiation vary considerably, depending upon the food type and upon the organism to be eliminated. Thus, doses as low as 100 Gy may be all that is required for the elimination of certain types of fruit flies from fresh fruits, while the preservation of frozen, packaged meats (as used, for example, in US and Russian space flights) requires doses as high as 40-45 kGy.

Concerning other organic materials, Barthe et al.[182] and Lapraz et al.[183] discuss the use of cotton in accident dosimetry. The former work explains the use of electron emission for this purpose, whereas the latter discusses TL, TSEE and photo-induced TL. Lapraz et al.[183] report a TL curve consisting of 4 peaks from a textile sample specially prepared to be a potential accident dosimeter for high risk personnel. Mauricio et al.[184] report the TL from dental restoration porcelains as potential dosimeters to estimate high doses in radiological accidents. Finally, Cooke[185] gives a review of charge trapping in biochemical systems. The results of TL studies of amino-acids, proteins, DNA, RNA and related compounds are reviewed. The overwhelming majority of these studies suggest that electronic de-excitation of the radiation-induced trapped charge occurs via the singlet and triplet manifolds of the molecule.

10.10 Field Effect Enhancement of TL and Electrically Excited TL

In the discussion of simultaneous measurements of TL and TSC in Ch. 9 it was implicitly assumed that the voltage applied to the sample did not influence the emitted TL. This appears to be the case when working with the low voltages normally utilized in these measurements. It seems, however, that this may not be the case when higher voltages are applied. Miyashita and Henisch[186] applied electric fields of 10^7 V/m to ZnS:Cu powder samples and found an enhancement of the 366 nm-excited TL by a factor of 100 or more. This enhancement is far beyond the additional light emission due to electroluminescence in the sample, stimulated by the applied field. The model given by Miyashita and Henisch consisted of the thermal release of electrons from traps and their subsequent acceleration by the field. The process terminates by impact ionization. This is consistent with the observed relationship between the enhancement factor and the applied voltage, which shows a zero effect over a wide initial voltage range followed by a sharp onset of enhancement at higher fields. The enhancement factor remains constant until breakdown occurs. Miyashita and Henisch[186] suggest the use of this effect, along with the accompanying TSC curve, as a tool for dosimetry. This idea is based upon the constancy of the enhancement factor over a broad range. It is to be noted that the work by Miyashita and Henisch followed previous investigations by Neumark[187] and Hahn[188] on *electro-thermoluminescence*, i.e. field-enhanced TL in a sample previously excited by ionizing radiation. A related effect of a shift of a TL peak with an applied dc voltage has also been described by Subrahmaniam and Rao[189]. They reported a shift to lower temperatures of TL peaks observed in x-irradiated KCl samples.

Levinson et al.[190] described results of TL excited solely by an applied electric field in semiconducting diamonds. This *electrically excited TL (EETL)* showed the same blue peak previously observed after UV excitation. The voltage dependence of EETL was found to be very similar to that of electroluminescence in the same crystals, namely, $I_m \propto \exp\left\{-b/V^{1/2}\right\}$. It was also found that the EETL was enhanced by illumination with light of wavelength ≥ 700 nm, with some similarity to the pre-excitation of TL, as described above. Levinson et al.[190] mentioned various possibilities for explaining EETL, namely:

1. Tunneling of electrons from the metal contact into the conduction band of the sample from where they may relax into luminescence centers.

2. Direct tunneling of electrons from the metal contact to the luminescence centers.

3. Tunneling followed by acceleration of the electrons in the field to energies high enough to impact-ionize valence band electrons into luminescence centers.

4. Impact ionization of the luminescence centers by free holes accelerated in the electric field.

The results by Levinson et al.[190] were not sufficient to enable a distinction to be made between the various processes and the authors concluded that, possibly, more than one of the above-mentioned processes may be taking place simultaneously.

10.11 Imaging by TL, OSL and TSEE

An early application of imaging TL emission was described by Fremlin and Srirath[191] who imaged the TL from a pottery sample using photographic film in order to illustrate the inhomogeneity of the emission. The use of TL as the basis of an image storage system was later suggested by Broadhead and Newman[192] and further discussed by Doi et al.[193]. Basically, the device consisted of a layer of thermoluminescent material on top of an electrically heated resistive layer, the heat from which induced the image appearance. The resistive layer was capable of generating a temperature of about 150 °C within a few minutes. The phosphor layer used by Broadhead and Newman was $CdSO_4$-Mn-Tl-NaF which has a TL peak at ~85 °C. The panel received x-ray exposures of the order of 1 R and good quality images were obtained. Doi et al.[193] improved the resolution by using a devitrified glass plate instead of the phosphor layer.

Color images of TL emission from a variety of natural quartz samples have been reported in a series of papers by Hashimoto and colleagues[194]-[197] using photographic film in a manner similar to that originally used by Fremlin and Srirath[191]. The TL Color Images (TLCI) produced by Hashimoto and colleagues were used to illustrate a variety of mineralogical features in natural quartzes, including alterations with annealing

Bailiff et al.[198] described a high resolution scanning system for measuring spatially resolved optically stimulated luminescence. As discussed in Ch. 5, the physics of the OSL method is very close to that of TL. In the setup of Bailiff et al. the output of a laser was directed through a focusing lens onto the surface of a sample. A computer controlled shutter in the beam path controlled the exposure of the sample. The luminescence was passed through selected optical filters to a photomultiplier tube (PMT) located above the sample. The sample was passed under the beam in a linear scanning, step-wise sequence and the output of the PMT was recorded. During the scanning, an intensity map was built up which reflected the spatial distribution of the luminescence from the sample. Such methods had been used previously in imaging systems for medical x-ray images, with a view to replace photographic plates[199],[200]. Known as photostimulable x-ray phosphors, the most useful materials appear to be BaFBr:Eu, $Ba_5SiO_4Br_6$:Eu, $Ba_3(PO_4)_2$:Eu and alkali halides RbI and KBr doped with Eu.[201]-[207]. These materials emit luminescence due to a 4f-5d transition in Eu^{2+} following stimulation with light (OSL). The basic mechanism is that the radiation causes conversion of Eu^{2+} to Eu^{3+} either directly or by trapping holes. Electrons are trapped by halogen ion vacancies (F^+-centers) to form F-centers. Illumination with visible light (usually in the green-red region of the spectrum) causes release of the trapped electrons and photoconversion of the Eu^{3+} ions back to Eu^{2+}, with the emission of luminescence. Photoconductivity is observed during the illumination confirming the excitation of charge into the delocalized bands during the illumination phase. Gasiot et al.[208] describe a patent for a real-time OSL imaging system based on XS:Y, where X = Ba, Ca, Sr or Mg, and Y = Eu or Sm.

Bräunlich[209] reported images obtained from a single crystalline dielectric material using thermally stimulated electron emission (TSEE). He made use of a micro-channel plate, a detector which permits the study of exoelectrons not only as a function of temperature but also as a function of position on the surface. Electrons emitted from various points on the specimen are collected by the micro-channel bundle and multiplied. The electron beams are accelerated in a DC electric field and made visible by the aid of a phosphor screen. Using electrostatic and magnetic lenses, a part of the crystal surface can be imaged onto the bundle and thus magnified.

10.12 TL by Laser Heating

One of the methods used for heating an irradiated sample is via localized, non-contact heating by a laser beam, as first suggested by Luckey[210] and later elaborated upon by Gasiot et al.[211] and Bräunlich et al.[212]. The heating rates attained in the devices described by these authors, as well as by others[213]-[215], were reported to be as high as 10^4 Ks^{-1}, or even more. Of course, such high heating rates are not suitable for TL kinetic analysis, but in dosimetry applications the high heating rates ensure a large TL signal, and thereby allow for the detection of very low doses. Another important feature of laser heating is that the lasers may be rapidly scanned across the TL layers for positional heating. In this way, the present method can be utilized for the readout of TL images in a similar fashion to that described in the previous section. A difficulty with the experimental work involving laser heating of TL materials is that the thermal response cannot be fully characterized because of limitations in the temperature measurement. Instead, a theoretical model was developed to study the laser heating of 0.1-0.5 mm thick LiF plates by Grupen and Kearfott[216]. This numerical analysis demonstrated the feasibility of multiple spot laser heating of TL materials, with luminescence temperatures reached for each spot within milliseconds for TL plates less than 0.5 mm thick. The combination of TL materials with rapid, positional laser heating could thus, at least in principle, enable the particularly interesting TL application of the rapid characterization of ionizing radiation fields. Abtahi et al.[217] and Kelly et al.[218] made studies of laser heated TL especially regarding the kinetics.

Grupen-Shemansky et al.[219] showed that it is possible to extract spatially resolved quantitative radiological information for TL detectors upon infrared laser stimulation. Acceptable spatial resolution for this TLD imaging procedure was attained by focusing the laser beam. These authors developed a two-dimensional, transient model to determine the axial and radial temperature profiles in a LiF layer heated with a focused CO_2 Gaussian laser beam. It was found that the focused laser heating is a strong function of the axial position despite the fact that LiF is a relatively weak absorber. Kearfott and Grupen-Shemansky[220],[221] described a laser-based system for the positional readout of TL plates to obtain dose profiles for a ^{60}Co radiotherapy beam. The experimentally obtained glow curves for LiF resulting from the high heating rates characteristic of laser heating, agreed well with results predicted using a first-order model. Excellent system linearity was obtained for sensitivity to ^{60}Co radiation. The useful dynamic range for this tissue-equivalent system could extend over 8 orders of magnitude. One of the advantages reported of the laser heating method of a given irradiated plate is that multiple spot heating can be performed on different portions of the sample so as to get a more reliable statistics for the results from a single sample detector.

Justus and colleagues[222],[223] described a laser heating system which requires only the use of a small IR-diode laser, rather than the cumbersome CO_2 lasers used in most other laser heating systems. The key to being able to use such small, and reliable, lasers for this application lies in the material which is fashioned from stable, transparent Vycor glass. Vycor glass is a porous glass host capable of storing nanocrystals of luminescent materials in nanometer size ($\sim$4 nm) pores or channels. The glass composition is $\sim$96% silica, $\sim$3% boron oxide and $<$ 1% aluminum oxide. The luminescent material is typically a II-VI compound. For TL applications ZnS nanocrystallites doped with Cu^+ were used[224]. For laser heating an IR absorbing dopant, Nd^{3+}, was added to the glass[222]. Absorption of the IR light from a 807 nm GaAs diode laser by the Nd^{3+} produces local heating throughout the bulk of the material and emission of TL from the luminescent activator results. The temperature of the bulk of the glass does not rise by more than a few degrees, but those luminescent traps located close to the absorbing Nd^{3+} centers are locally heated to produce the TL emission.

Setzkorn et al.[225] discussed a method of radiation dose mapping by scanning laser heated TL foils for the dose distribution analysis of β radiation fields. The laser scanning results are compared with results from measurements using very thin graphite-mixed MgB_4O_7:Dy TL detectors. The results show that detectors based on $CaSO_4$:Dy films are suitable as extremity dosemeters for monitoring in the β radiation field.

It is to be noted that in some of the above cases, part of the effect observed may be the result of infra-red stimulation of the luminescence - i.e. IR-stimulated OSL. For the dosimetric applications, however, the distinction between the two effects (TL versus OSL) may not be of great importance.

10.13 Miscellaneous Related Phenomena

We mention briefly here some non-luminescent, TL-related phenomena.

10.13.1 Dilatometry and other Thermomechanical Techniques

Dilatometry, or *thermodilatometry,* is a technique in which either the change in length or the change in volume of a sample is measured and recorded as a function of temperature. The terms *linear dilatometry* and *volume dilatometry* can be used to specify the dimension(s) measured. The plot of length or volume against temperature is called the dilatometric curve and its derivative, with respect to time or temperature, is called the derivative dilatometric curve. The broader term *thermomechanical analysis* (TMA) is used for the whole group of methods in which a mechanical property of the sample is measured as a function of its temperature, while the sample is under constant stress. The term *dynamic dilatometry* is used for techniques in which the dynamic mechanical properties are measured as a function of temperature. The definitions adopted by the International Confederation for Thermal Analysis (ICTA) for these techniques are as follows (see, for example, Lombardi[226] and Kambe[227]).

Thermodilatometry is a technique in which a dimension of a substance *under negligible load* is measured as a function of temperature while the substance is subjected to a controlled temperature program.

Thermomechanical Analysis (TMA) is a technique in which the deformation of a substance is measured under load as a function of temperature while the substance is subjected to a controlled temperature program. The mode of the stress applied, compression, tension, flexure or torsion should always be stated.

Dynamic thermomechanometry is a technique in which the *dynamic modulus* and/or the *damping* of a substance is measured under *oscillatory load* as a function of temperature while the substance is subjected to a controlled temperature program.

The first instruments for dilatometry appeared in the sixties (see e.g. Barrall et al.[228] and Wendlandt[229]). Dilatometry has been successfully applied to the study of polymers (see e.g. Gordon[230] and Miller[231]); often, measurements are performed simultaneously with other thermoanalytical methods (see Ch. 8).

Wendlandt[232] describes several methods to measure the sample dimensions as a function of its temperature. Most accurate are the measurements of the dependence of length on the temperature, which is generally approximately expressed by

$$\ell = \ell_0(1 + \alpha T) \tag{10.36}$$

where ℓ_0 is the length of the solid sample at some "low" temperature T_0 (usually 0 °C), ℓ is its length at temperature T, and α is the linear coefficient of thermal expansion. When ℓ is recorded vs. T in a temperature range where no phase transition occurs, the curve will have a slope of α. If the derivative of ℓ with respect to time is recorded versus T when the heating rate is constant (β Ks^{-1}), the recorded signal is proportional to α

$$d\ell/dt = \ell_0\alpha\,(dT/dt) = \alpha(\ell_0\beta) \tag{10.37}$$

and by calibration, the coefficient of thermal expansion can be read directly from the curve. Any phase transition would appear on this curve as a deviation from a straight line. A first-order transition would appear as a peak, the area under which is equal to the total change of ℓ. Several solid-to-solid and solid-to-liquid phase transitions have been detected and studied in this way (e.g. Paulik et al.[233] and Maeda and Kanetsuna[234]).

In some cases, the results of a dilatometry measurement can be described by an equation analogous to those of TL or TG. Kambe et al.[235] investigated the thermal shrinkage of aromatic polymer films and explained their results in terms of a two-state model. They assumed that in the as-received film, segments of the polymer were in a contracted (α) or an extended (β) state. During heating, a fraction of the segments in the β-state would shift to the α-state. It was found that the rate of contraction followed a first-order reaction, which was described by the equation

$$\frac{dx}{dt} = A\exp\{-\Delta E/RT\}\,(1-x) \tag{10.38}$$

where x is the fraction of segments shifted from the β-state to the α-state, and ΔE is the activation energy for the transition between the states. Equation 10.38 is similar to that of TG and DTG, and the authors were able to evaluate the activation energy for various samples, using a method developed for TG curves.

10.13.2 Thermally Stimulated Creep

Thermally stimulated creep is a technique of thermomechanical analysis in which the molecular movements in a substance are investigated by measuring the thermally stimulated response to a frozen stress or strain in the sample. The method was first proposed by Monpagens et al.[236] and its details are as follows:

1) The sample is subjected to a given mechanical stress at a given temperature for a period of time which allows some orientation of the mobile molecular units.

2) The temperature is lowered to a temperature T_0, at which the molecular motion is frozen; the outside stress is now removed.

3) The sample is heated at a constant rate so that the mobile units can disorient. The strain γ, its time derivative $\dot{\gamma}$, and the temperature of the sample are recorded as functions of time. The method was applied to several amorphous

polyolefines by Lacabanne et al.[237] and Monpagens et al.[238], and was found to be particularly suitable for the study of glass transitions. It was found that the relaxation process observed around the glass transition of each polyolefine can be described by a distribution of the retardation time τ, obeying the law

$$\tau(T) = \tau_c \exp\left\{\frac{\Delta H}{k}\left(\frac{1}{T} - \frac{1}{T_c}\right)\right\} \tag{10.39}$$

where

$$\tau(T) = |\gamma(T)/\dot{\gamma}(T)| \tag{10.40}$$

and τ_c and T_c are characteristic of the substance. It was found experimentally that $T_c \cong T_g + 27$ (in °C), where T_g is the glass transition temperature.

Equations 10.39 and 10.40 can be written as

$$-d\gamma/dt = \gamma\tau_o^{-1}\exp\{-\Delta H/kT\} \tag{10.41}$$

where $\tau_o = \tau_c \exp\{\Delta H/kT_c\}$. Since Eq.10.41 is entirely analogous to that of first-order TL and DTG, the methods developed for the analysis of TL or DTG, can be applied here as well.

Demont et al.[239] studied the retardation and relaxation behavior of poly(ether-block-amide)s PEBA copolymers using thermally stimulated creep, TSC and DSC. Two primary TSC and thermally stimulated creep peaks were found in the 150-300 K range, and an additional peak close to the DSC melting point of the poly(tetramethylene glycol) segments, was found in the thermally stimulated creep curve. Bernès et al.[240] utilized TSC and thermally stimulated creep for the investigation of amorphous phases of polymeric materials. The molecular mobility liberated at the glass transition temperature was analyzed. In linear polymers, the influence of chemical and physical structures was also studied.

10.13.3 Partial Thermoremanent Magnetization (PTRM)

The depolarization currents described in Ch. 8, are the result of a prior freezing of electrical polarization within the sample by cooling under an electric field. In a similar way, the magnetic moment can be frozen. When a magnetic specimen is cooled from some "high" temperature to room temperature in a magnetic field, it may acquire a *thermoremanent magnetization* (TRM) which is usually parallel to the applied field.

The thermal decay of TRM is usually studied in the following way. The sample in which the TRM has been induced is reheated to a predetermined temperature T, and allowed to cool to room temperature. The ratio of the remaining magnetic moment to the original moment is found, and plotted against T. This procedure is repeated until the sample is demagnetized. The curve obtained can be differentiated, to give the partial TRM (PTRM) curve. Such measurements

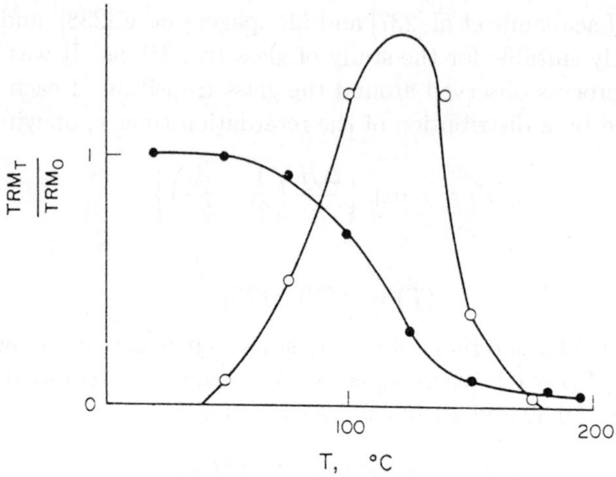

Figure 10.10 The TRM (closed circles) and PTRM (open circles) of $0.68FeTiO_3$:$0.32\alpha Fe_2O_3$. The applied magnetic field was 36 Oe. (After Westcott-Lewis[243].)

have been performed by Westcott-Lewis and colleagues[241]-[244] for synthetic and natural ilmenite-haematites and for iron-titanium oxides. Typical TRM and PTRM curves are shown in Fig. 10.10.

The theory of TRM and PTRM had previously been established by Néel[245] and by Stacey[246]. TRM is governed by the differential equation

$$-dM/dt = CM \exp\{-E/kT\} \tag{10.42}$$

where M is the magnetization of the sample after being heated to temperature T. E is the activation energy which is usually equal to the barrier height over which one has to move the walls of the magnetic domains in order to change the magnetization state of the sample. The pre-exponential factor C arises from the probability of a domain overcoming the barrier[246].

Equation 10.42 is analogous to the first-order TL and DTG equations. Therefore, as pointed out by Chen and Westcott-Lewis[247], the methods developed for analyzing other thermally stimulated processes can be applied here as well. It should be noted, as explained by Stacey[246], that C in Eq. 10.42 is proportional to T in the same way as is the s factor for certain TL peaks. The appropriate corrections developed for determining the activation energies of those TL peaks, as discussed in Ch. 3, can be used here in the same way.

10.14 References

1. J.R. Cameron, N. Suntharalingam and G.N. Kenney, *Thermoluminescent*

Dosimetry, The University of Wisconsin Press, Madison (1968).

2. Y.S. Horowitz and D. Yossian, *Rad. Prot. Dosim.* **60** (1995) 1.
3. Y.S. Horowitz, *Rad. Prot. Dosim.* **32** (1990) 147.
4. K. Becker, *Solid State Dosimetry*, CRC Press, Cleveland (1973).
5. K. Mahesh, P.S. Weng and C. Furetta, *Thermoluminescence in Solids and its Applications*, Ashford: Nucl. Tech. Pub. (1989).
6. Y.S. Horowitz, in *Thermoluminescence and Thermoluminescent Dosimetry, Vol. I,* Y.S. Horowitz, editor, CRC Press, Boca Raton (1984).
7. M. Oberhofer and A. Scharmann, editors, *Applied Thermoluminescence Dosimetry,* Adam Hilger, Bristol (1981).
8. L.A. DeWerd, in *Thermally Stimulated Relaxation in Solids*, P. Bräunlich, editors, Springer Verlag, Berlin (1979) 275.
9. S.W.S. McKeever, M. Moscovitch and P.D. Townsend, *Thermoluminescence Dosimetry Materials: Properties and Uses*, Nuclear Technology Publishing (1995).
10. S.P. Kathuria and C.M. Sunta, *J. Phys. D: Appl. Phys.* **12** (1979) 1573.
11. E. Lilley and G.C. Taylor, *J. Phys. D: Appl. Phys.* **14** (1981) L13.
12. S.P. Kathuria and C.M. Sunta, *J. Phys. D: Appl. Phys.* **15** (1982) 497.
13. E. Lilley and S.W.S. McKeever, *J. Phys. D: Appl. Phys.* **16** (1983) L39.
14. C.M. Sunta and S.P. Kathuria, *J. Phys. D: Appl. Phys.* **16** (1983) L163.
15. A. Delgado and J.C. Barreiro, *Cryst. Latt. Def. and Amorph. Mat.* **17** (1987) 205.
16. A. Delgado and J.M. Gómez Ros, *J. Phys. D: Appl. Phys.* **23** (1990) 571.
17. A. Delgado, J.M. Gómez Ros, C. Bacci, C. Furetta and B. Rispoli, *J. Phys. D: Appl. Phys.* **23** (1990) 1132.
18. S.W. Lin, P.S. Weng and P.C. Hsu, *Appl. Radiat. Isot.* **47** (1996) 83.
19. T.M. Piters and A.J.J. Bos, *J. Phys. D: Appl. Phys.* **26** (1993) 2255.
20. G. Kitis, M. Spiropulu, J. Papadopoulos and S. Charalambous, *Nucl. Inst. Meth.* **B73** (1993) 367.
21. C. Christodoulides, *Phys. Stat. Sol.* **A118** (1990) 333.
22. T. Nakajima and T. Hashizame, *Health Phys.* **16** (1969) 782.
23. T.K. Wang, P.C. Hsu and P.S. Weng, *Radiat. Prot. Dosim.* **18** (1987) 157.
24. A.J. Bos, R.N.M. Vijverberg, T.M. Piters and S.W.S. McKeever, *J. Phys. D: Appl. Phys.* **25** (1992) 1249.
25. C. Furetta, J.W.N. Tuyn, F. Louis, J. Azorin-Nieto, J. Guitérrez and C.M.H.. Driscoll, *Int. J. Radiat. Appl. Instrum.* **A39** (1988) 59.
26. C. Bacci, R. Bernabei, S. d'Angelo and C. Furetta, *Radiat. Eff.* **69** (1983) 127.
27. A.R. Lakshmanan, B.C. Bhatt and R.C. Bhatt, *Radiat. Prot. Dosim.* **27**

(1989) 15.

28. C. Furetta, J.W.N. Tuyn, F. Louis, J. Azorin and C.M.H. Driscoll, *Radiat. Prot. Dosim.* **17** (1986) 161.
29. B. Ben-Shachar and Y.S. Horowitz, *Radiat. Prot. Dosim.* **22** (1988) 87.
30. M. Moscovitch, *Radiat. Prot. Dosim.* **17** (1986) 165.
31. Y.S. Horowitz, B. Ben-Shachar and D. Yossian, *J. Phys. D: Appl. Phys.* **26** (1993) 1475.
32. A. Delgado, J.M. Gómez Ros, J.L. Muñiz, A.J. Bos and T.M. Piters, *Radiat. Prot. Dosim.* **47** (1993) 231.
33. B. Ben-Shachar and Y.S. Horowitz, *J. Phys. D: Appl. Phys.* **24** (1991) 1649.
34. J. Xie and M.J. Aitken, *Ancient TL* **9** (1991) 21.
35. W.F. Hornyak, A.D. Franklin and R. Chen, *Ancient TL* (1993) 21.
36. C. Bull and G.F.J. Garlick, *Proc. Phys. Soc. London* **63** (1950) 1283.
37. W. Hoogenstraaten, *Philips Res. Rep.* **13** (1958) 515.
38. J.H. Schulman, R.J. Ginther, S.G. Gorbics, A.E. Nash, E.J. West and F.H. Attix, *J. Appl. Rad. and Isotopes* **20** (1969) 523.
39. G.F.J. Garlick and I. Robinson, *The Moon*, S.K. Runcorn and H.C. Urey, ed., (1972) 324.
40. I.M. Blair, J.A. Edgington, R. Chen and R.A. Jahn, Proc. *III Lunar Science Conf.* **3** (1972) 2949.
41. F. Kieffer, C. Meyer and J. Rigaut, *Chem. Phys. Lett.* **11** (1971) 359.
42. A.G. Wintle, *Nature* **245** (1973) 143.
43. A.G. Wintle, *J. Lumin.* **15** (1977) 385.
44. S.V. Moharil, S.M. Patey and B.T. Deshmukh, *J. Lumin.* **42** (1989) 325.
45. F.A. Hasan, B.D. Keck, C. Hartmetz and D.W.G. Sears, *J. Lumin.* **34** (1986) 327.
46. S. Tyler and S.W.S. McKeever, *Nucl. Tracks Radiat. Meas.* **14** (1988) 149.
47. R.H. Templer, *Nucl. Tracks Radiat. Meas.* **10** (1985) 531.
48. R.H. Templer, *Radiat. Prot. Dosim.* **17** (1986) 139.
49. R.H. Templer, *Radiat. Prot. Dosim.* **17** (1986) 493.
50. R. Chen and Y. Kirsh, *Analysis of Thermally Stimulated Processes,* Pergamon Press, Oxford (1985).
51. N.A. Spooner, *Quat. Sci. Rev.* **11** (1992) 139.
52. R. Visocekas, *Nucl. Tracks Radiat. Meas.* **21** (1993) 175.
53. R. Visocekas, M. Ouchene and B. Gallois, *Nucl. Inst. and Methods* **214** (1983) 553.
54. R. Visocekas, *Nucl. Tracks Radiat. Meas.* **10** (1985) 521.

55. D.C. Sanderson, *Nucl. Tracks Radiat. Meas.* **14** (1988) 163.
56. R. Visocekas, N.A. Spooner, A. Zink and P. Blanc, *Radiat. Meas.* **23** (1994) 377.
57. A.J.C. Zink and R. Visocekas, *Radiat. Meas.* **27** (in press, 1997).
58. A. Zink, R. Visocekas and A.J.J. Bos, *Radiat. Meas.* **24** (1995) 513.
59. D. Fragoulis and T.G. Stoebe, *Radiat. Prot. Dosim.* **34** (1990) 65.
60. P.A. Clark and R.H. Templer, *Nucl. Tracks Radiat. Meas.* **14** (1988) 139.
61. D. Stoneham and M.B. Winter, *Nucl. Tracks Radiat. Meas.* **14** (1988) 127.
62. S.G.E. Bowman, *Nucl. Tracks Radiat. Meas.* **14** (1988) 131.
63. H.Y. Göksu, D.F. Regulla, A. Vogenauer and A. Wieser, *Nucl. Tracks Radiat. Meas.* **14** (1988) 143.
64. R. Visocekas, *Nucl. Tracks Radiat. Meas.* **14** (1988) 163.
65. G. Kitis, P. Bousbouras, C. Antypas and S. Charalambous, *Nucl. Tracks Radiat. Meas.* **18** (1991) 61.
66. V.K. Vlasov and V.V. Karezin, *Nucl. Tracks Radiat. Meas.* **21** (1993) 147.
67. S.W.S. McKeever, *Thermoluminescence of Solids*, University Press, Cambridge (1985).
68. R. Visocekas and A. Geoffroy, *Phys. Stat. Sol. (a)* **41** (1977) 499.
69. A.J. Mikhailov, *Dokl. Phys. Chem.* **197** (1971) 223.
70. R.H. Templer, *D. Phil Thesis,* University of Oxford (1986).
71. R. Visocekas, T. Ceva, C. Marti, F. Lefaucheux and M.C. Robert, *Phys. Stat. Sol. (a)* **35** (1976) 315.
72. R. Chen and A. Hag-Yahya, *Radiat. Meas.* **27** (in press, 1997).
73. N.F. Mott and R.W. Gurney, *Electronic Processes in Ionic Crystals*, Oxford University Press (1948).
74. D.W. Zimmerman, C.R. Rhyner and J.R. Cameron, *Health Phys.* **12** (1966) 525.
75. A.R. Blak and S. Watanabe, *Proc. 4th Int. Conf. Lumin. Dosim. Krakow* (1974) 169.
76. T.L. Johnson, *Proc. 4th Int. Conf. Lumin. Dosim. Krakow* (1974) 197.
77. D. Yossian, S. Mahajna, B. Ben-Shachar and Y.S. Horowitz, *Radiat. Prot. Dosim.* **47** (1993) 129.
78. G.C. Taylor and E. Lilley, *J. Phys. D: Appl. Phys.* **11** (1978) 567.
79. S.G. Gorbics, F.H. Attix and J.A. Pfaff, *Int. J. Appl. Radiat.* **18** (1967) 625.
80. W. Pohlit, *Biophysik* **5** (1969) 341.
81. R.G. Fairchild, P.L. Mattern, K. Lengweiler and P.W. Levy, *IEEE Trans. Nucl. Sci.* **NS-21** (1974) 366.
82. S.W.S. McKeever, *Radiat. Prot. Dosim.* **6** (1984) 49.

83. G.C. Taylor and E. Lilley, *J. Phys. D: Appl. Phys.* **15** (1982) 1243.
84. G.C. Taylor and E. Lilley, *J. Phys. D: Appl. Phys.* **15** (1982) 1253.
85. S.W.S. McKeever and E. Lilley, *J. Phys. Chem. Solids* **43** (1982) 885.
86. S.W.S. McKeever, *J. Appl. Phys.* **56** (1984) 2883.
87. A.J.J. Bos and T.M. Piters, *Nucl. Tracks Radiat. Meas.* **21** (1993) 163.
88. I. Tale, V. Tale and A. Nagornyi, *Radiat. Prot. Dosim.* **33** (1990) 71.
89. I. Tale, V. Tale and P. Kulis, *Radiat. Prot. Dosim.* **47** (1993) 107.
90. I. Tale, *Nucl. Tracks Radiat. Meas.* **21** (1993) 65.
91. A.J.J. Bos and T.M. Piters, *Radiat. Prot. Dosim.* **47** (1993) 41.
92. R.G. Fairchild, P.L. Mattern, K. Lengweiler and P.W. Levy, *J. Appl. Phys.* **49** (1978) 4512, 4523.
93. P.D. Townsend, G.C. Taylor and M.C. Wintersgill, *Radiat. Effects* **41** (1979) 11.
94. A.J.J. Bos, T.M. Piters, J.M. Gómez-Ros and A. Delgado, *GLOCANIN, An Intercomparison of Glow Curve Analysis Computer Programs*, IRI-CIEMAT-131-93-005 (1993).
95. W. de Vries, J.E. Hoogenboom, J.B. Dieldorf and A.J.J. Bos, *Proc. Symp. TL Dosim.* (1988) 31.
96. R.K. Gartia and S. Dorendrajit-Singh, *Phys. Stat. Sol. (a)* **135** (1993) K83.
97. C. Bacci, C. Damhlano, C. Furetta, B. Rispoli, J. Zhu, A. Delgado and J.M. Gómez-Ros, *CIEMAT Rep. No. 717* (1990).
98. J. McKeever, F.D. Walker and S.W.S. McKeever, *Nucl. Tracks Radiat. Meas.* **21** (1993) 179.
99. A.F. Gabrysh, J.M. Kennedy, H. Eyring and V.R. Johnson, *Phys. Rev.* **131** (1963) 1543.
100. F. Aramu, C. Muntoni and A. Serpi, *Rend. Semin. Fac. Univ. Cagliari* **38** (1968) 1.
101. C.D. Clark and D.H. Newman, *J. Phys. C: Solid St. Phys.* **2** (1969) 1726.
102. C.D. Clark and D.H. Newman, *J. Phys. C: Solid St. Phys.* **4** (1971) 1615.
103. M. Ueta, H. Sugimoto and I. Nagasawa, *J. Phys. Soc. Japan* **17** (1962) 1465.
104. V. Ausin and J.L. Alvarez-Rivas, *Phys. Rev.* **B6** (1972) 4828.
105. R. Fieschi, R. Oggioni and G. Spinolo, *Phys. Stat. Sol.* **3** (1967) 1207.
106. S. Petralia and G. Gnani, *Lett. Nuov. Cim.* **4** (1972) 483.
107. T. Niewiadomski, *Health Phys.* **31** (1976) 373.
108. H.J. Kos and S. Mieke, *Phys. Stat. Sol. (a)* **50** (1978) K165.
109. J.R. Petit and P. Duval, *Sol. St. Comm.* **19** (1976) 475.
110. R.D. Newton and W.A. Sibley, *Phys. Stat. Sol. (a)* **41** (1977) 569.

111. J.W. Hook III, and H.G. Drickamer, *J. Appl. Phys.* **49** (1978) 2503.

112. R.V. Joshi, K.D. Dhake and T.R. Joshi, *J. Lumin.* **31&32** (1984) 136.

113. D.F. Mariani, F. Brito, C. Oyarzun and J. Vignolo, *J. Phys. Chem. Solids* **48** (1987) 371.

114. B.J. Baer and H.G. Drickamer, *J. Phys. Chem.* **95** (1991) 9477.

115. Z.A. Dreger, J.M. Lang and H.G. Drickamer, *J. Lumin.* **50** (1991) 279.

116. Z.A. Dreger, J.M. Lang and H.G. Drickamer, *Chem. Phys. Lett.* **185** (1991) 184.

117. B.J. Baer and H.G. Drickamer, *Sol. St. Comm.* **83** (1992) 669.

118. S.A.A. Winer, N. Kristianpoller and R. Chen, *Luminescence of Crystals, Molecules and Solutions*, F. Williams, editor, Plenum (1973) 473.

119. M. Sayer and A.D. Souder, *Can. J. Phys.* **47** (1969) 463.

120. R. Chen and A. Halperin, *Proc. Int. Conf. Lumin.* Budapest (1966) 1414.

121. H. Prószyńska-Bordas, W. Stańska-Prószyńska and M. Prószyński, *Quater. Sci. Reviews* **7** (1988) 265.

122. R.H. Templer, *Nucl. Tracks* **10** (1985) 789.

123. L.P. Pashchenko, R. Pérez Sales, R. Aceves and M. Barboza Flores, *Appl. Phys. Lett.* **66** (1995) 3126.

124. L.P. Pashchenko, R. Pérez Sales, R. Aceves and M. Barboza Flores, *Appl. Phys. Lett.* **67** (1995) 3266.

125. A. Delgado, J.C. Saez Vergara, J.M. Gómez Ros and A. Romero, *Radiat. Prot. Dosim.* **58** (1995) 211.

126. C. Petridou, C. Christodoulides and S. Charalambous, *Nucl. Inst. Methods* **150** (1978) 247.

127. S. Charalambous and F. Hasan, *Phys. Lett.* **95A** (1983) 259.

128. G. Kitis, F. Hasan, M. Tsakiri and S. Charalambous, *Health Phys.* **49** (1985) 1193.

129. F. Hasan, G. Kitis and S. Charalambous, *J. Phys. C: Solid St. Phys.* **18** (1985) 1743.

130. G. Kitis, F. Hasan, M. Tsakiri and S. Charalambous, *Nucl. Tracks* **10** (1985) 571.

131. G. Kitis, F. Hasan, M. Tsakiri and S. Charalambous, *Nucl. Tracks* **10** (1985) 565.

132. G. Kitis and S. Charalambous, *Phys. Stat. Sol. (a)* **105** (1988) K175.

133. Y. Fujiwara, M. Tonouchi and T. Kobayashi, *Japan. J. Appl. Phys.* **27** (1988) L1706.

134. Y. Fujiwara and T. Kobayashi, *IEEE Trans. Magnetics* **25** (1989) 2563.

135. D.W. Cooke, M.S. Jahan, R.D. Brown, K.O. Ott, E.R. Gray, J.L. Smith, J.O.

Willis, B.L. Bennett, M.A. Maez, E.J. Peterson, W.L. Hults, J.Y. Coulter, A.M. Portis, H. Piel, N. Klein, G. Müller and M. Hein, *Appl. Phys. Lett.* **56** (1990) 2462.

136. M.S. Jahan, D.W. Cooke, B.L. Bennett, W.L. Hults, M.A. Maez, K.C. Ott and J.L. Smith, *Mat. Res. Soc. Symp. Proc.* **169** (1990) 417.
137. D.W. Cooke, M.S. Jahan, E.R Gray, J.L. Smith, B.L. Bennett, W.L. Hults and M.A. Maez, *J. Lumin.* **48&49** (1991) 819.
138. M.S. Jahan, D.W. Cooke, E.R. Gray, J.L. Smith, A. Mayer, W.L. Hults, B.L. Bennett and M.A. Maez, *J. Lumin.* **48&49** (1991) 823.
139. S. Katz, M. Roth, J. Golben and M. Vlasse, *J. Appl. Phys.* **75** (1994) 2712.
140. G. Burrafato, S.O. Troja, E. Turrisi, L. Arcidianoco, C. Caputa, A. Cro, G. Grasso, A.M. Gueli and M. Russo, *Nucl. Instr. Meth. Phys. Res. B* **116** (1996) 531.
141. V.Y. Yaskolko, L.N. Oster and K.M. Mukimov, *Phys. Stat. Sol. (a)* **123** (1991) K35.
142. J. Chrzanowski, *Mol. Phys. Repts.* **7** (1994) 209.
143. F. Link, *Comp. Roy. Acad. Sci.* **223** (1946) 976.
144. E.P. Ney, N.J. Woolf and R.J. Collins, *J. Geophys. Res.* **71** (1966) 1787.
145. M. Sydran, *J. Geophys. Res.* **73** (1968) 5195.
146. I.M. Blair and J.A. Edgington, *Nature* **217** (1968) 157.
147. M.L. Chanin, V. Lepine and J.E. Blamont, *The Moon and the Planets* **27** (1982) 143.
148. S.A. Durrani, W. Prachyabrued, W. Christodoulides, J.H. Fremlin, J.A. Edgington, R. Chen and R.A. Jahn, *Proc. III Lunar Sci. Conf.* **3** (1972) 2955.
149. S.A. Durrani, K.A.R. Khazal and A. Ali, *Proc. 7th Lunar Sci. Conf. (1976) 1157.*
150. G.B. Dalrymple and R.R. Doell, *Science* **167** (1970) 713.
151. R.M. Walker, D.W. Zimmerman and J. Zimmerman, *Proc. Conf. Lunar Geophys.* Houston (1971) 308.
152. H.P. Hoyt Jr., R.M. Walker, D.W. Zimmerman and J. Zimmerman, *Proc. III Lunar Sci. Conf.* **3** (1972) 2997.
153. C. Lalou, G. Valladas, U. Brito, A. Henri, T. Ceva and R. Visocekas, *Proc. III Lunar Sci. Conf.* **3** (1972) 2009.
154. R.B. Gammage and K. Becker, *Proc. II Lunar Sci. Conf.* **3** (1971) 2057.
155. D.W.G. Sears, in *Thermoluminescent Materials*, D.R. Vij, editor, Prentice Hall, Englewood Cliffs (1993) 99.
156. F.A. Hasan, M. Haq and D.W.G. Sears, *Geochim. Cosmochim. Acta* **50** (1986) 1031.
157. D.W. Sears, *Icarus* **44** (1980) 190.

158. D.W.G. Sears, M.H. Sparks and A.E. Rubin, *Geochim. Cosmochim. Acta* **48** (1984) 1189.

159. D.W.G. Sears, N. Bhaktiar, B.D. Keck and K.S. Weeks, *Geochim. Cosmochim. Acta* **48** (1984) 2265.

160. R.K. Guimon, G.E. Lofgren and D.W.G. Sears, *Lunar Planet. Sci. XVII* (1986) 297.

161. R. Hutchison, C.M.O. Alexander and D.J. Barber, *Geochim. Cosmochim. Acta* **51** (1987) 1875.

162. B.D. Keck, R.K. Guimon and D.W.G. Sears, *Earth Planet. Sci. Lett.* **77** (1986) 419.

163. P.H. Benoit, D.W.G. Sears and S.W.S. McKeever, *Icarus* **94** (1991) 311.

164. S.W.S. McKeever, *Earth Planet. Sci. Lett.* **58** (1982) 419.

165. K.H. Chadwick and W.R. Oosterheert, *Proc. Symp. Microdosim.* **EUR3747** (1967) 633.

166. Y. Inoue, T. Ichikawa and S. Kazuo, *Photochem. and Photobiol.* **23** (1976) 125.

167. T.S. Desai, P.V. Same and V.G. Tatake, *Photochem. and Photobiol.* **21** (1975) 345.

168. V.G. Tatake, *Proc. Natl. Symp. TL*, Madras (1975) 262.

169. M.S. Jahan and D.W. Cooke, *Photochem. Photobiol.* **41** (1985) 343.

170. G.A. Schreiber, N. Helle and K. Bögl, *Int. J. Radiat. Biol.* **63** (1993) 105.

171. D.C.W. Sanderson, C. Slater and K.J. Cairns, *Radiat. Phys. Chem.* **34** (1989) 915.

172. G.A. Schreiber, A. Hoffman, N. Helle and K.W. Bögl, *Radiat. Phys. Chem.* **43** (1994) 533.

173. G.A. Schreiber, B. Ziegelmann, M. Mayer, N. Helle and K.W. Bögl, *Proc. 3rd German Meeting Food Irrad.* (1993) 66.

174. G.A. Schreiber, B. Ziegelmann, G. Quitzsch, N. Helle and K.W. Bögl, *Food Structures* **12** (1993) 385.

175. J. Kispéter and L.I. Kiss, *Proc. 3rd German Meeting Food Irrad.* (1993) 71.

176. H.M. Khan and A.E. Ehlermann, *Proc. 3rd German Meeting Food Irrad.* (1993) 76.

177. H.M. Khan and H. Delimcée, *Radiat. Phys. Chem.* **46** (1995) 717.

178. S. Pinnioja, *Radiat. Phys. Chem.* **42** (1993) 397.

179. T. Calderón, H.M. Rendell, P. Beneitez, P.D. Townsend, A. Millán and R.A. Wood, *J. Food Sci.* **59** (1994) 1070.

180. P. Beneitez, V. Correcher, A. Millán and T. Calderón, *J. Radioanal. and Nucl. Chem.* **185** (1994) 401.

181. T. Calderón, V. Correcher, A. Millán, P. Beneitez, H.M. Rendell, M. Larsson,

P.D. Townsend and R.A. Wood, *J. Phys. D: Appl. Phys.* **28** (1995) 415.

182. J. Barthe, P. Iacconi, D. Lapraz, M. Petel, P. Keller and G. Portal, *Radiat. Prot. Dosim.* **6** (1984) 125.
183. D. Lapraz, P. Iacconi, M.F. Alessandri, J. Barthe and G. Portal, *Radiat. Prot. Dosim.* **33** (1990) 131.
184. C.L.P. Mauricio, L.A.R. Rosa and P.G. Cunha, *Radiat. Prot. Dosim.* **11** (1985) 185.
185. D. W. Cooke, *Radiat. Phys. Chem.* **36** (1990) 21.
186. K. Miyashita and H.K. Henisch, *Solid State Elect.* **9** (1966) 29.
187. G.F. Neumark, *Phys. Rev.* **103** (1956) 41.
188. D. Hahn, *J. Phys. Rad.* **17** (1956) 748.
189. A. Subrahmanyam and V.K. Rao, *Phys. Stat. Sol. (a)* **52** (1979) K147.
190. J. Levinson, A. Halperin and V. Bar, *J. Lumin.* **6** (1973) 1.
191. J.H. Fremlin and S. Srirath, *Archaeometry* **7** (1964) 58.
192. P. Broadhead and G.A. Newman, *J. Phys. E: Sci. Inst.* **5** (1972) 69.
193. A. Doi, T. Kanie and A. Naruse, *Japan. J. Appl. Phys.* **16** (1977) 2289.
194. T. Hashimoto, A. Koyanagi, K. Yokosaka and Y. Hayashi, *Geochem. J.* **20** (1986) 111.
195. T. Hashimoto, M. Fojima, N. Shirai and M. Ichino, *Radiat. Meas.* **21** (1993) 217.
196. T. Hashimoto, S. Sakaue, H. Akoi and M. Ichino, *Radiat. Meas.* **23** (1994) 293.
197. T. Hashimoto, S. Notoya, T. Ojima and M. Hoteida, *Radiat. Meas.* **24** (1995) 227.
198. I.K. Bailiff, R.J. Clark and S.M. Barnett, *Radiat. Meas.* **27** (in press, 1997).
199. M. Sonoda, M. Takano, J. Miyahara and H. Kato, *Radiology* **148** (1983) 833.
200. K. Takahashi and T. Nakamura, *US Patent*, 4,535,237 (1983).
201. K. Takahashi, K. Katsuhiro, J. Miyahara, Y. Kanemitsu, K. Amitani and S. Shionoya, *J. Lumin.* **31&32** (1984) 266.
202. H. von Seggern, T. Voigt, W. Knüpfer and G. Lange, *J. Appl. Phys.* **64** (1988) 1405.
203. M. Thoms, H. von Seggern and A. Winnacker, *Phys. Rev. B*, **44** (1991) 9240.
204. H. Nanto, K. Murayama, T. Usada, F. Endo, Y. Hirai, S. Taniguchi and N. Takeuchi, *J. Appl. Phys.* **74** (1993) 1445.
205. L.E. Trinkler, M.F. Trinkler and A.I. Popov, Phys. *Stat. Sol. (b)* **180** (1993) K31.
206. M. Thoms, H. von Seggern and A. Winnacker, *J. Appl. Phys.* **76** (1994) 1800.
207. Y. Iwabuchi, N. Mori, K. Takahashi, T. Matsuda and S. Shionoya, *Japan. J. Appl. Phys.* **33** (1994) 178.

208. J. Gasiot, P.F. Bräunlich and J.-P. Fillard, *US Patent* 4,517,463 (1985).

209. P. Bräunlich, *J. Appl. Phys.* **42** (1971) 495.

210. G.W. Luckey, *US Patent* 3,859,527 (1975).

211. J. Gasiot, P. Bräunlich and J.P. Fillard, *J. Appl. Phys.* **53** (1981) 5200.

212. P. Bräunlich, J. Gasiot, J.P. Fillard and M. Castagne, *Appl. Phys. Lett.* **39** (1981) 769.

213. Y. Yasuno, H. Tsutsui, O. Yamamoto and T. Yamashita, *Japan. J. Appl. Phys.* **21** (1982) 967.

214. Y. Yasuno, H. Tsutsui and O. Yamamoto, *Radiat. Prot. Dosim.* **6** (1984) 341.

215. K. Horikawa and S. Arakawa, *US Patent 4,527,061* (1985).

216. M.E. Grupen and K.J. Kearfott, *J. Appl. Phys.* **64** (1988) 1044.

217. A. Abtahi, P. Bräunlich, P. Kelly and J. Gasiot, *J. Appl. Phys.* **58** (1985) 1626.

218. P. Kelly, P. Bräunlich, A. Abtahi, S.C. Jones and M. DeMurcia, *Radiat. Prot. Dosim.* **4** (1984) 25.

219. M.E. Grupen-Shemansky, K.J. Kearfott and E.D. Hirleman, *J. Appl. Phys.* **66** (1989) 3407.

220. K.J. Kearfott and M.E. Grupen-Shemansky, *Health Phys.* **59** (1990) 421.

221. K.J. Kearfott and M.E. Grupen-Shemansky, *Med. Phys.* **17** (1990) 429.

222. B.L. Justus, A.L. Huston and T.L. Johnson, *Appl. Phys. Lett.* **69** (1996) 1.

223. A.L. Huston, B.L. Justus and T.L. Johnson, *Appl. Phys. Lett.* **68** (1996) 3377.

224. B.L. Justus, T.L. Johnson and A.L. Huston, *Nucl. Instrum. Meth. Phys. Res. B*, **95** (1995) 533.

225. R. Setzkorn, H. Prévost, J. Gasiot and P. Christensen, *Radiat. Prot. Dosim.* **66** (1996) 419.

226. G. Lombardi, *For Better Thermal Analysis*, Inst. Miner. e Petrograf., Rome (1980).

227. H. Kambe, *Thermochim. Acta* **26** (1978) 209.

228. E.M. Barrall, R.S. Porter and J.F. Johnson, *Anal. Chem.* **36** (1964) 2316.

229. W.W. Wendlandt, *Anal. Chim. Acta* **33** (1965) 98.

230. S.E. Gordon, *Thermal Analysis, Vol. 1,* R.S. Schwenker and P.D. Garn, editors, Academic Press, New York (1969) 667.

231. G.W. Miller, *Thermal Analysis, Vol. 1,* R.S. Schwenker and P.D. Garn, editors, Academic Press, New York (1969) 435.

232. W.W. Wendlandt, *Thermal Methods of Analysis*, Wiley, NY (1974).

233. F. Paulik, J. Paulik and J. Erdey, *Anal. Chim. Acta* **41** (1968) 170.

234. Y. Maeda and H. Kanetsuna, *Thermal Analysis: Comparative Studies on Materials,* H. Kambe and P.D. Garn, editors, Kadansha, Tokyo (1974).

235. H. Kambe, M. Kochi, T. Kato and M. Murakami, *Thermal Analysis: Comparative Studies on Materials,* H. Kambe and P.D. Garn, editors, Kadansha, Tokyo (1974).

236. J.C. Monpagens, D. Chatain and C. Lacabanne, *J. Electrostatics* **3** (1977) 87.

237. C. Lacabanne, D. Chatain, J.C. Monpagens, A. Hiltner and E. Baer, *Solid St. Comm.* **27** (1978) 1055.

238. J.C. Monpagens, C. Lacabanne, D. Chatain, A. Hiltner and E. Baer, *J. Macromol. Sci. Phys.* **B15** (1978) 503, 519.

239. P. Demont, D. Chatain, C. Lacabanne and M. Glotin, *Makromol. Chem. Macromol. Symp.* **25** (1989) 167.

240. A. Bernès, M. Martin, J.J. Martinez, J. Boye and C. Lacabanne, *J. Therm. Anal.* **38** (1992) 169.

241. M.F. Westcott and L.G. Parry, *J. Geophys. Res.* **73** (1968) 1269.

242. M.F. Westcott-Lewis, *Earth Planet. Sci. Lett.* **12** (1971) 124.

243. M.F. Westcott-Lewis and L.G. Parry, *Australian J. Phys.* **24** (1971) 735.

244. M.F. Westcott-Lewis, *Earth Planet. Sci. Lett.* **17** (1973) 408.

245. L. Néel, *Adv. Phys.* **4** (1955) 191.

246. F.D. Stacey, *Adv. Phys.* **12** (1963) 45.

247. R. Chen and M.F. Westcott-Lewis, *Australian J. Phys.* **26** (1973) 249.

Chapter 11
CONCLUDING REMARKS

Throughout this book, we have attempted to describe different experimental and theoretical approaches to the study of thermally stimulated phenomena. These approaches permit the utilization of these phenomena in, for example, the extraction of numerical values for certain physical parameters from the materials under study, or allow possible practical applications in a variety of arenas. We have mentioned some of the experimental difficulties inherent in the measurement of TL and its related phenomena - some of which can be measured simultaneously, or separately. In the present section we summarize, by way of conclusion, some of the main highlights of the preceeding chapters. We facilitate this by separating the discussion into the topics of *models*, *analysis*, *properties*, and *applications* - under the broad heading of *Prospects and Limitations.* Finally, since all the TL and related phenomena discussed in the book have to be recorded experimentally in some fashion, we conclude with a description of the *Instrumentation* used in the experimental measurement of TL glow curves.

11.1 Prospects and Limitations

11.1.1 Models

Progress in an understanding of TL and its related phenomena has hinged upon the establishment of acceptable models to describe the effect. The most ubiquitous model for TL utilizes the flat-band energy band diagram associated with semiconductors and insulators, featuring the lowermost empty conduction band and the uppermost filled valence band, with a distribution of localized states in the forbidden gap. In the simplest possible of all such models we have just one electron trap and one hole trap, and allow the thermal release of one type of charge carrier, in order to have it radiatively recombine with the trapped carrier of the opposite sign. However, we know of not one material in existence for which this simple picture is an accurate description of the TL process. Despite this obvious drawback, the energy band picture is an extraordinarily useful teaching aid in describing the essential themes of the TL process. Based on this uncomplicated model, the pivotal papers of Randall and Wilkins, and later, Garlick and Gibson became established as the cornerstones of the science of TL. Only later, as more and more complex behavior emerged (complex kinetic behavior, non-linear growth of TL with dose, sensitivity changes, athermal fading, and many more) did the necessity of introducing additional electron and hole traps become clear. With such additions, the concepts of deep, thermally disconnected traps, competition, and interactive kinetics could be explained both qualitatively and (at least) semi-quantitatively.

Perhaps, however, even in the course of this book, insufficient attention has been given to alternative models of TL production.[1],[2]. Certainly thermally stimulated luminescence due to the recombination of ionic species is possible in materials such as the alkali halides and the alkaline earth fluorides. It is worthwhile to linger here on this type of process in order to gain an appreciation of its possible relationship with the types of model (and subsequent analyses) discussed thus far. Consider irradiation of an ionic crystal; electrons trapped at anion vacancies (F, or F_A, or F_Z centers) and holes trapped at stable anion interstitial atoms (H, or H_A, H_Z centers) are created by excitonic processes. Such defect creation mechanisms have been described in the literature for alkali halide crystals (e.g. Itoh[3],[4]), oxides[5]-[7], and the alkaline earth halides[8]. The A and Z subscripts on the Frenkel pair defects noted above refer to F- and H-centers perturbed by the presence of monovalent or aliovalent ions, respectively. Unperturbed F- and H-centers are unlikely in impure crystals at room temperature. Heating of the material induces the restoration of the lattice according to

$$F_{A,Z} + H_{A,Z} \rightarrow \text{original lattice} + \text{photon}. \tag{11.1}$$

Here the thermally stimulated recombination of the ionic species takes place radiatively with the release of a photon, and hence TL is produced. The emission wavelengths are characteristic of the defects taking part in the process. Prompt emission, emitted during the irradiation itself, is characterized by a low energy triplet-state π and a high energy singlet-state σ, depending upon the orbital state of the excited electron. During TL emission, however, only the triplet state emission occurs. This is because the stable F-H pair is in an energy state which is higher than the low energy triplet state, but lower than the high energy singlet state[9]. Perturbation of the emission by impurities results in emission wavelengths shifted from the expected intrinsic triplet state emission. Many examples of such processes are described in the literature, especially for the alkali halides. The large body of work describing TL from the alkali halides has been reviewed by McKeever[1].

Often in interstitial-vacancy recombination processes the perturbed intrinsic emission does not occur in TL. Instead, emission characteristic of an impurity is all that is seen. The interpretation of this is that energy is transferred to the impurity. (Energy transfer is one way in which recombination at *different* recombination sites - either electron-hole recombination or interstitial-vacancy recombination - can nevertheless result in the *same* emission wavelengths.) CaF_2:Ce provides us with a useful example. In CaF_2:Ce the simple F-center is unstable at room temperature and does not form as a product of irradiation. However, when trivalent rare earth impurities are introduced impurity-vacancy pairs are formed following irradiation. These are of C_{3v} symmetry with the vacancy located in a <111> direction with respect to the impurity, with the whole complex having trapped 2 electrons. The complex can be considered as a divalent impurity with an F-center in a <111> nearest-neighbor position[10]. Jassemnejad and McKeever[11] investigated this in Ce-doped CaF_2 and showed that the perturbed intrinsic emission (at 280 nm) which results from this recombination is reabsorbed by the Ce^{3+}-F_{int} C_{4v} centers, and clusters of these centers, with the consequent emission of light at 320 nm and 340 nm, characteristic of the Ce^{3+} ions.

Clearly, processes such as those described here are entirely different from the usual electron-hole recombinations and thus the energy band model is totally inappropriate. Nevertheless, the main equations that can be written to describe the process remain the same, with just a different interpretation of the parameters. The probability for thermal release of the trapped interstitial atom from its interstitial site is of the form $p = s\exp\{-E/kT\}$ where s is an "attempt" frequency and E is an activation energy (enthalpy). The recombination rate dn_F/dt (where n_F is the concentration of F-centers) can be equated with the rate of release of the interstitials $dn_H/dt = -pn_H$ (where n_H is the concentration of interstitial atoms). Here we must bear in mind that geminate recombination only is allowed and first-order kinetics are expected. With the assumption that a photon is released during the recombination event, an expression for TL is obtained which is identical in form to the first-order Randall-Wilkins equation (Eq. 2.28).

Despite these considerations, it may be stated with certainty in many materials that the TL curves observed are due to electron-hole recombination processes. Such conclusions can be confidently drawn from the companion studies of TSC, TSEE, optical absorption, electron spin resonance, photoconductivity, optically stimulated conductivity and others. As we have noted in several places throughout the book, although it is possible to devise models and concoct parameter values to describe the behavior of the various TL peaks, the reverse problem, namely to take the experimental data and establish a definitive model, with evaluated parameters, is less straightforward owing to a certain degree of degeneracy. That is, more than one model (and/or parameter sets) may yield similar experimental TL curves. The validity of the analytical methods used in the analysis of the glow curves then has to be established (or, at least, supported). Here again the concomitant study of the above-mentioned techniques is essential to eliminate or confirm particular models and lend confidence to the subsequent analyses and conclusions.

11.1.2 Analysis

As pointed out in previous chapters (especially chapters 2 & 3), even in the relatively simple case of a single TL peak associated with one trapping state and one kind of recombination center, and for the case of electron-hole recombination within the confines of the energy band model, the number of parameters involved is eight. These include the activation energy (trap depth) E_t, the frequency factor s, the recombination and retrapping probabilities A_m and A_n respectively, the total concentration of trapping states N and the initial concentrations of electrons in traps n_0, of free electrons n_{c0}, and of holes in centers, m_0. The relative importance of the different parameters is such that the initial concentration of the free carriers, n_{c0}, is usually very unimportant for the results, thus reducing the effective number of relevant parameters to seven. Each of the other seven parameters is important in determining the shape of the peak for a given set of parameters. In the inverse problem of determining the parameters from the shape of the TL peak the practice of considering a smaller number of parameters (e.g. just four - E_t, s, b and a scaling parameter proportional to n_0) may be erroneous or, at least, its validity has to be carefully considered. The activation energy E_t plays a special role among the remaining seven parameters. Due to its appearance in the Arrhenius exponent, the shape and position of the TL peak depends very strongly on this parameter. Inversely, for the same reason, one can determine the value of E_t to a good accuracy without knowing the other parameters. This can be done in many cases by using the initial rise method which is rather insensitive to all the other parameters.

Ideally, an optimal method for evaluating the glow parameters should yield all the parameters simultaneously. So far, such a method does not exist and it may very well be that there will never be one. The situation is even further aggravated if there is an overlap of peaks which introduces more parameters that influence the results in the same temperature range. The solution, in principle, should be to find an aspect of the measured quantities which is strongly dependent on one of the parameters and as weakly as possible on the others, thus enabling the evaluation of this specific parameter. An example to this is the evaluation of the recombination probability A_m described in Ch. 9. Another idea that can be used is that once a parameter is satisfactorily evaluated, the total number of remaining parameters is reduced by one, and we should consider the problem of determining the reduced number of parameters. All in all, it is obvious that the task of determining *all* the relevant parameters is far from being straightforward. It is obvious that researchers who are just looking for the activation energy, frequency factor and sometimes the kinetic order, are usually oversimplifying a rather complicated situation.

On the other end of the spectrum of investigators are those who claim that due to the complexity of the processes involved, the results of these thermal techniques are of little value unless the pertinent parameters are independently measured by other experimental methods. In particular, two papers - by Bräunlich and Kelly[12] and by Kelly and Bräunlich[13] - suggested that thermally stimulated methods may provide only a means for checking previously known results. In these papers, Bräunlich and Kelly argued that, although a model of a solid with given parameters uniquely determines a TL curve, the curve by itself, gives little or no information about the model or parameters. To illustrate their point these authors established a model consisting one type of trap (concentration N), one type of recombination center, and one type of thermally disconnected trap (concentration M). They proceeded by choosing arbitrary sets of parameters describing the model and, *by making some simplifying approximations*, they numerically solved the set of simultaneous differential equations governing the TL process during heating, equivalent to Eqs. 2.17, 2.19, 2.20 and 2.22. They then plotted their results as $\ln I$ vs. $U = E_t/kT$ and found the results to be insensitive to the parameters $R = A_n/A_m$, N and M. Their conclusion was that, due to this insensitivity, one cannot extract these parameters from the experimental results, and, since the parameters characterize the model, one can tell practically nothing about the model and the parameters from TL measurements. In their work, they also calculated the $n_c(T)$ curve, associated with TSC, and again, argued that their results show that, even with simultaneous measurements of TL and TSC, one cannot evaluate any of the parameters except, perhaps, for the activation energy. This, they agree, can usually be evaluated but only by the initial rise method.

We maintain that Bräunlich and Kelly went much too far in their criticism. To begin with, the approximations made may have influenced the results, to some extent, and, nowadays, such work would have been done by numerical solutions of the equations without any approximations. In some cases, it has been shown that such approximations may have significant influence on the results. Some of the insensitivity noted by Bräunlich and Kelly may have resulted from the fact that they plotted $\ln I$ vs. U rather than I vs. T as usually measured [12]. Thus, changes that seem only minor on the $\ln I - U$ scale, may very well be more pronounced on a linear scale. Beyond that, had Bräunlich and Kelly only advocated caution regarding the use of oversimplified models and methods, we would go along with this advice, but we certainly do not agree with the general spirit of their arguments.

In a way, Bräunlich and Kelly[12],[13] construct a "game" with very limiting rules, that is - what can one get from the shape of a single TL peak or simultaneous TL and TSC peaks. Even within this framework, one can definitely evaluate the activation energy E_t and the recombination probability A_m (and the cross-section for recombination, thereof) as described in Ch. 9. The frequency factor s then follows. The process of extracting more information from thermally stimulated processes is, however, much more complicated and involves various elements described in this book. These include the excitation and emission spectra, phosphorescence decay, simultaneous measurements of two or more thermally stimulated phenomena, dose dependence measurements, bleaching, phototransfer and more measurements of the types mentioned already. The information that can be reached by the combined measurement of a TSP and other phenomena, such as Hall mobility, optical absorption, EPR, etc., some of which has been discussed in Ch. 9, is also of importance. In a later work, Bräunlich et al.[14] were less extreme in their pessimistic view. Here they only say that if one starts from a point of complete ignorance, the general complexity of the kinetics will usually prevent the extraction of quantitative trap spectroscopic data by solely employing presently available TSL and TSC techniques. Moreover, in a *Note added in Proof* they go further, saying that they are convinced that the future development of TSL/TSC methods as investigative tools of electronic properties of semiconductors and insulators will benefit greatly from skillful studies of correlated phenomena during thermally stimulated relaxations, such as those carried out by Fillard et al.[15] and Gasiot et al.[16].

Hagebeuk and Kivits[17] reported the construction of a mathematical *black box*, namely, a computer program, to extract the main five parameters from a simultaneous TL and TSC measurement. These are the activation energy, the frequency factor, the recombination probability, the ratio of the probabilities of retrapping and recombination, and the ratio of densities of disconnected and connected traps. Their method actually consists of obtaining the TL and TSC curves for a given set of parameters by numerically solving the three simultaneous differential equations. The parameters are determined by comparing the theoretical calculations with experimental data, using nonlinear regression analysis. Hagebeuk and Kivits[17] reported good results in simulated cases of synthetic TL and TSC curves, i.e., the original parameters were retrieved to a good accuracy by this method. This contradicts, at least in principle, the conclusion by Kelly and Bräunlich[13] that the parameters cannot be extracted by simultaneous TL and TSC peaks. On the other hand, Hagebeuk and Kivits do not report the application of their method to experimental results. As reported by Shenker and Chen[18], since the shapes of the peaks are rather insensitive to variations in the parameters, other than the activation energy E_t, small inaccuracies in the measurement may limit the method. This is so since, in the inverse problem of determining parameters from a given curve, there is a high sensitivity of the extracted parameters to small changes in the measured curves.

We maintain that in the present stage of development of TL and TSC, the appropriate attitude to be taken is somewhere between that of Bräunlich and Kelly[12] and that of Hagebeuk and Kivits[17]. We do not think that at present there exists an effective mathematical *black box* into which one can insert a glow curve and get out all the parameters involved. On the other hand, we do not share the pessimistic view of Bräunlich and Kelly that even a sophisticated experiment is highly unlikely to yield unique values of the parameters. The correct way to deal with an experimental curve, in our opinion, is to concentrate on a measurement or set of measurements which are highly sensitive to one parameter, and the least sensitive to all the others. The elements of this rather long procedure have been mentioned throughout this book, and have been summarized in Ch. 3. An important point that should be stressed is that when one considers the *information* that can be extracted from a glow curve, one is not limited to the evaluation of the parameters. Apart from the trap depths, the location of the recombination centers on the energy scale (as may sometimes be revealed by the emission spectrum), the recombination and retrapping probabilities and the concentrations and occupancies of imperfections, one is also interested in the identity of the various imperfections, the processes that take place during excitation and during the heating of the sample. Measurements of excitation spectrum, dose dependence, sensitization, etc., may all contribute significantly to the understanding of these processes. One should, therefore, always consider the information attainable from a thermogram in this broader context.

11.1.3 Properties

One of the most informative properties exhibited by a TL peak is its behavior as a function of dose. The linearity - or, specifically, the lack of linearity - of the dose response curve for a particular peak indicates much about the possible models to be used in the description of the TL process from the material under study. In Ch. 4 we discussed in depth the various models for describing the growth of TL as a function of dose, and these in turn have a significant influence on the models used to describe the glow curve shape. The most important aspect of these models is the extent of the competition that occurs between the defects both during irradiation and during heating to read the TL signal. It is essential to remember that TL does not reflect the defect states which exist within the material immediately after irradiation. This defect equilibrium is altered by heating, and since the act of measuring TL requires heating the sample, the defect equilibrium is altered by the readout of the TL itself. The net TL signal measured is, therefore, a result of convolving the initial irradiation-induced defect state with the thermally stimulated changes that occur during heating. Both processes (irradiation and heating) are required to produce the observed effect and often this simple, and rather obvious, consideration is overlooked when attempting to describe TL properties. An example of this problem is seen in the extensive discussions of the non-linear growth of TL from LiF available in the literature in which one body of opinion attempts to describe the observed properties purely in terms of what occurred during the radiation absorption stage, while ignoring the subsequent modifying effects of heating during the readout phase. (See papers by Horowtiz and colleagues[19],[20], and McKeever[21] for summary discussions.)

Other important properties to be included in a thorough study of TL include the emission spectrum. Emission spectra are not always easy to interpret, but in favorable circumstances one can at least identify the impurities, or other defects, involved in the TL process - an important step in itself. Characteristic emissions from rare earth or transition ions are easily identified. So too are intrinsic emissions from F- and V- and related centers in many materials. Estimates of the effects of self-absorption, and/or thermal quenching can also be extracted from a knowledge of the emission spectra.

By way of reinforcement of the sentiments of the previous section, we emphasize again here that a thorough analysis of TL properties, along with a reliable derivation of a model, requires the gathering of an extensive data set covering all aspects of the TL properties of the material.

11.1.4 Applications

Of course, the study of TL and related phenomena is not just an exercise in determining the most appropriate model to describe the effect. A proper (i.e. reliable) application of the technique in dating or dosimetry also requires an understanding of the TL production processes. For example, we need to understand how repeated heating and irradiation of a sample can effect the TL sensitivity (e.g. in the pre-dose effect, or in single aliquot techniques in TL dating). How does illumination of an irradiated specimen alter the TL signal? What exactly is the shape of the TL growth curve and what extrapolations should be used to determine equivalent doses from antiquity? These and other issues require a detailed examination of the TL process and should be mastered lest erroneous conclusions be made concerning the applicability of TL to different situations. For example, if certain correction procedures are to be accepted in, say, pre-dose dating, or if supralinearity correction algorithms are to confidently used in dosimetry, one needs to be assured that the correction procedure actually does what we think it does. Treatment of a sample as a "black-box" can only be stretched so far.

Our point is that TL has to be considered in a global sense. One cannot take one small aspect of the phenomenon and treat it in isolation from all the other TL properties exhibited by that sample. Because of the degeneracy problem inherent to the phenomenon the whole analysis must present a consistent picture if we are to make progress in truly understanding the theory of TL and related phenomena.

11.2 Instrumentation

11.2.1 Basic Principles

At various points in this book we have alluded to several experimental considerations that are necessary concerning the experimental measurement of TL and related phenomena (for example, emission spectra, temperature control and measurement, electrodes in TSC measurements, etc.). We conclude here with a broad description of the instrumentation used in the measurement of the phenomenon that has been the main focus of the present work - namely, thermoluminescence.

The simplicity of the necessary equipment for TL measurement is such that an accurate and reliable system can be set up for little expenditure. The basic requirements are an oven, a reliable temperature controller, a photodetector and a data storage medium. Additional design criteria are required if TL below room temperature is monitored, or if the spectrum of the TL emission is measured, or if simultaneous measurements of (say) TL & TSC, or TL & TSEE, are to be made.

11.2.2 Heating System

The heart of any TL instrument is the heating system. Three different types of heating method can be found in widespread use. The objectives are reproducible and controllable heating profiles. Ovens and sample chambers are most often filled with inert gas during heating to enhance thermal contact between the heater and sample, and to reduce spurious emissions of light from chemical effects.

Resistance Heating

Here one passes high current through a thin, low impedance load, such as a tantalum, nickel-chrome, or platinum alloy strip (the planchet). The sample (powder or solid chip) is placed in direct contact with the planchet, the temperature of which is monitored via thermocouples. Alternative designs involve the use of wire coils in contact with more massive metal blocks, or in direct contact with the sample itself. (An example of the latter is the US Navy's "bulb dosimeter" in which the sample is directly deposited onto the coil using adhesive and the whole encased in a glass bulb. Electrical leads protruding through the glass envelope enable electrical contact to be made.) The low thermal inertia of thin planchets or coils enable fine temperature control and rapid heating and cooling. The range of heating rates achievable with resistance heating techniques is from fractions of a degree per minute to several tens of degrees per second. An important consideration with these contact heating methods is the fact that the temperature monitored during the measurement is the temperature of the planchet, not the sample. Therefore, one has to ensure excellent thermal contact between the sample and the planchet. Methods to ensure this include bathing the sample in an inert gas atmosphere, the use of slow heating rates (≤ 1 K/s), and the use of a "glue" to ensure intimate contact between the sample and the planchet. Silicone grease is often used in the latter application. A detailed discussion of the potential problems associated with poor thermal contact has been given by Betts et al.[22].

Gas Heating

Rapid heating of solid samples, with precision temperature control, can be achieved via the use of hot gas (usually nitrogen) as a heat exchange medium. A jet of the hot gas is directed via nozzles, at a typical flow rate of a few liters per minute, onto the sample and extracted from the sample chamber at the same rate, thus preventing heating of the sample surroundings. Rapid and uniform heating of the sample can be achieved.

Optical Heating

Some early, commercial TL systems utilized the absorption of infra-red radiation from a tungsten lamp, filtered to remove unwanted visible and ultra-violet wavelengths. The absorption of the infra-red energy by the sample resulted in a controllable rise in temperature. Only small, thin samples of low thermal mass can be used effectively with such systems. A modern development of this concept is the use of infra-red (CO_2) cw lasers to heat either the sample, or a substrate to which the sample has been intimately bonded[23]-[25]. The active samples are thin (< 100 μm; 10-20 μg) and extremely high heating rates ($\geq 10^4$ K/s) can be achieved with short (10 - 1000 ms) laser pulses (at powers of a few watts). Such high heating rates result in a correspondingly high signal-to-noise ratio because of the combined effect of an enhancement of the TL output (see Chapter 2) and the discrimination against the noise counts from the detector. Potential limitations against the use of such high rates of heating stem from "thermal quenching" of the luminescence efficiency. As the heating rate increases, the temperature at which the TL is emitted also increases. Some phosphors lose luminescence efficiency with temperature increase and potential gains in output intensity are offset by quenching phenomena. Furthermore, for all materials the incandescence background also increases with temperature, sometimes to the point of intolerance.

11.2.3 Low Temperature Cryostats

Several low temperature cryostats are available commercially to enable the measurement of TL below room temperature. These include "cold finger" cryostats (in which the sample is cooled by being attached to a copper cold finger which extends outside the chamber and is immersed in a coolant - most usually liquid nitrogen). Other popular cryostat types include liquid helium or liquid nitrogen continuous flow systems, or closed-cycle helium refrigerators. Heating of the sample is normally achieved via resistance heating.

11.2.4 Temperature Control

The most popular type of temperature controller is a proportional-integral-derivative (PID) controller providing proportional feedback control to the system in order to ensure that the temperature of the planchet follows the desired heating (i.e. time-temperature) profile. Temperature sensors include a variety of thermocouple types, platinum resistors and diodes. A variety of continuous power or zero-crossing, pulsed power supplies may be used in conjunction with SCR power packs.

11.2.5 Light Detection System

The low intensity light emitted during a TL experiment is normally detected using photomultiplier tubes (PMTs). Either DC current or photon counting modes may be used, the former at high signal levels, and the latter at low signal levels. The key to high detectability is to match the energy response of the PMT photocathode to the wavelength of the emitted light. Most wide band gap materials emit in the visible, particularly blue, part of the spectrum and thus bialkali photocathodes are well matched to such signals. Longer wavelength emitters, however, require the use of extended response bialkali (e.g. S-20 or S-1), or GaAs photocathodes. High gain (10^6 - 10^7), low dark current (nA) PMTs are essential. Dark currents may be reduced by the use of Peltier cooled PMT housings to limit thermionic emission from the photocathode.

11.2.6 Emission Spectra

Real materials often contain several types of recombination site, each of which may be represented by its own characteristic emission wavelength. Additionally, materials containing rare earth or transition metal impurities often exhibit several emission wavelengths due to relaxation of the ions from a multiplicity of excited states to the ground state. Conventional measurements of TL glow curves use detectors which are sensitive over a broad range of wavelengths and make no effort to discriminate between the possible different emission colors. Detection through broad-band filters, centered in the visible, is the norm. Unfortunately, in doing this, much potential information is lost regarding the emission processes. In addition, one can easily arrive at false conclusions regarding the kinetics of the TL process and, furthermore, one may misinterpret observed changes in, for example, glow curve shape and intrinsic sample sensitivity following thermal annealing. (See Chapter 2 for some discussion of these effects.) Nevertheless, TL recorded through broad-band filters is still the most popular recording method and an additional dimension of information regarding the TL process is often lost.

Some early attempts at measurement of spectral data were made using narrow-band interference filters, but more recently greater detail has been obtained by measuring the TL intensity as a function of temperature and wavelength simultaneously. Such measurements lead to isometric plots of intensity-versus-temperature-versus-wavelength (so-called "3-D" plots). Various spectrometers have been described in the literature to achieve this, including simple rotating filter packs and rapid-scanning monochromators. However, most information is obtained by multiplexing the data and dispersing the signal over a defined wavelength range, and measuring the intensity over each wavelength interval separately, but simultaneously. Modern techniques include Fourier transform spectroscopy[26],[27] or, more popularly, position-sensitive photodetectors. In the latter, the light is dispersed over the wavelength range of interest (using a grating or prism spectrograph) and the dispersed light is detected over the active area of a device such as a CCD camera, diode array, or position-sensitive PMT. Such measurements are now quite common[28]-[31]. The utility of this detection mode has been reviewed by Townsend and colleagues[30],[32].

11.3 References

1. S.W.S. McKeever, *Thermoluminescence of Solids*, Cambridge University Press, Cambridge (1985).
2. S.W.S. McKeever, in *Spectroscopic Characterization of Minerals and Their Surfaces*, L.M. Coyne, S.W.S. McKeever and D.F. Blake, editors, American Chemical Society, Symposium Series 415 (1990) 166.
3. N. Itoh, *Radiat. Effects*, **64** (1982) 161.
4. N. Itoh, *Adv. Phys.* **31** (1982) 491.
5. A.E. Hughes and B. Henderson, in *Point Defects in Solids*, J.H. Crawford and L.M. Slifkin, editors, Plenum, New York (1972) 381.
6. J.H. Crawford, *J. Nucl. Mater.* **108/109** (1982) 644.
7. W.B. Fowler, *Semiconductors and Insulators* **5** (1982) 583.
8. W. Hayes and A.M. Stoneham, in *Crystals with the Fluorite Structure*, W. Hayes, editor, Clarendon, Oxford (1974) 185.
9. N. Itoh, A.M. Stoneham and C.K. Harker, *J. Phys. C: Sol. State Phys.* **10** (1977) 4197.
10. D.L. Staebler and S.E. Schnatterly, *Phys. Rev. B* **3** (1971) 516.

11. B. Jassemnejad and S.W.S. McKeever, *J. Phys. D: Appl. Phys.* **20** (1987) 323.
12. P. Bräunlich and P. Kelly, *Phys. Rev. B* **1** (1970) 1596.
13. P. Kelly and P. Bräunlich, *Phys. Rev. B* **1** (1970) 1587.
14. P. Bräunlich, P. Kelly and J.P. Fillard, in *Thermally Stimulated Relaxation in Solids,* P. Bräunlich, editor, Springer Verlag, Berlin, Heidelberg (1979).
15. J.P. Fillard, J. Gasiot and M. de Murcia, *J. Electrostatics* **3** (1977) 49.
16. J. Gasiot, M. de Murcia, M. Fillard and R. Chen, *J. Appl. Phys.* **48** (1977) 1371.
17. H.J.L. Hagebeuk and P. Kivits, *Physica* **83b** (1976) 289.
18. D. Shenker and R. Chen, *J. Comput. Phys.* **10** (1972) 272.
19. Y.S. Horowitz, *Radiat. Prot. Dosim.* **33** (1990) 81.
20. Y.S. Horowitz, S. Mahajna, M. Rosenkrantz and D. Yossian, *Radiat. Prot. Dosim.* **65** (1996) 7.
21. S.W.S. McKeever, *Radiat. Prot. Dosim.* **33** (1990) 83.
22. D.S. Betts, L. Couturier, A.H. Khayrat, B.J. Luff and P.D. Townsend, *J. Phys. D: Appl. Phys.* **26** (1993) 843, 849.
23. J. Gasiot, P. Bräunlich and J.P. Fillard, *J. Appl. Phys.* **53** (1982) 5200.
24. S.C. Jones, J.A. Sweet, P. Bräunlich, J.M. Hoffman and J.E. Hegland, *Radiat. Prot. Dosim.* **47** (1993) 525.
25. V.K. Mathur, J.H. Barkyoumb, R.J. Abundi, K. Chakrabarti, C.D. Bloomsburg, P.F. Bräunlich, J.E. Hegland, J.M. Hoffman, J.W. Hoelscher, A.E. Kelly, J.L. Thompson and W. Tetzlaff, *Radiat. Prot. Dosim.* **66** (1996) 79.
26. J.R. Prescott, P.J. Fox, R.A. Akber and H.E. Jensen, *Appl. Opt.* **27** (1988) 3496.
27. P. Haschberger, *Appl. Radiat. Isot.* **42** (1991) 797.
28. M. Kanemaki, K. Ninawa, I. Yamamoto, M. Nakagawa, T. Wada, Y. Yamashita and K. Endo, Nucl. Tracks Radiat. Meas. **18** (1991) 81.
29. B.J. Luff and P.D. Townsend, *Meas. Sci. Technol.* **4** (1993) 65.
30. P.D. Townsend, *Radiat. Meas.* **23** (1994) 341.
31. T.M. Piters, W.H. Meulemans and A.J.J. Bos, *Rev. Sci. Instrum.* **64** (1993) 109.
32. P.D. Townsend and Y. Kirsh, *Contemp. Phys.* **30** (1989) 337.

Subject Index

B

C

E

M

N

O

P

Q

Y

Z

Author Index

B

H

L

N

O

Q

R

T